Mathematical Analysis of Deterministic and Stochastic Problems in Complex Media Electromagnetics

The Princeton Series in Applied Mathematics publishes high quality advanced texts and monographs in all areas of applied mathematics. Books include those of a theoretical and general nature as well as those dealing with the mathematics of specific applications areas and real-world situations.

A list of titles in this series appears at the back of the book.

Mathematical Analysis of Deterministic and Stochastic Problems in Complex Media Electromagnetics

G. F. Roach
I. G. Stratis
A. N. Yannacopoulos

PRINCETON UNIVERSITY PRESS
PRINCETON AND OXFORD

Published by Princeton University Press
41 William Street, Princeton, New Jersey 08540

In the United Kingdom: Princeton University Press
6 Oxford Street, Woodstock, Oxfordshire, OX20 1TW

press.princeton.edu

Library of Congress Cataloging-in-Publication Data

Roach, G. F. (Gary Francis)
Mathematical analysis of deterministic and stochastic problems in complex media electromagnetics / Gary F. Roach, Ioannis Stratis, Athanasios N. Yannacopoulos.
p. cm. – (Princeton series in applied mathematics)
Includes bibliographical references and index.
ISBN 978-0-691-14217-3 (hardcover : alk. paper) 1. Electromagnetism–Mathematics. 2. Stochastic control in systems. 3. Mathematical analysis. I. Stratis, Ioannis G., 1955- II. Yannacopoulos, Athanasios N., 1968- III. Title.
QC760.4.M37R63 2012
537.01′519–dc23

2011038853

British Library Cataloging-in-Publication Data is available

This book has been composed in LaTeX

The publisher would like to acknowledge the authors of this volume for providing the camera-ready copy from which this book was printed.

Printed on acid-free paper. ∞

Printed in the United States of America

1 3 5 7 9 10 8 6 4 2

- To Isobel, for all her patience and understanding - G.F.R.

- To George, mainly for the little things, barely perceptible, that make us the father and son we are - I.G.S.

- To Electra, who magnetised me, for well posedness
  To Nikos and Jenny, for existence - A.N.Y.

Hence the difference between similar and equal things, which are yet not congruent (for instance, two symmetric helices), cannot be made intelligible by any concept, but only by the relation to the right and the left hands which immediately refers to intuition.

I. Kant
Prolegomena to Any Future Metaphysics (1783)

Contents

Preface xi

PART 1. MODELLING AND MATHEMATICAL PRELIMINARIES 1

Chapter 1. Complex Media 3

Chapter 2. The Maxwell Equations and Constitutive Relations 9

2.1 Introduction 9
2.2 Fundamentals 9
2.3 Constitutive relations 13
2.4 The Maxwell equations in complex media: A variety of problems 23

Chapter 3. Spaces and Operators 38

3.1 Introduction 38
3.2 Function spaces 38
3.3 Standard differential and trace operators 45
3.4 Function spaces for electromagnetics 48
3.5 Traces 51
3.6 Various decompositions 52
3.7 Compact embeddings 53
3.8 The operators of vector analysis revisited 54
3.9 The Maxwell operator 56

PART 2. TIME-HARMONIC DETERMINISTIC PROBLEMS 59

Chapter 4. Well Posedness 61

4.1 Introduction 61
4.2 Solvability of the interior problem 62
4.3 The eigenvalue problem 68
4.4 Low chirality behaviour 70
4.5 Comments on exterior domain problems 74
4.6 Towards numerics 77

Chapter 5. Scattering Problems: Beltrami Fields and Solvability 83

5.1 Introduction 83

5.2 Elliptic, circular and linear polarisation of waves 84
5.3 Beltrami fields - The Bohren decomposition 86
5.4 Scattering problems: Formulation 88
5.5 An introduction to BIEs 91
5.6 Properties of Beltrami fields 96
5.7 Solvability 99
5.8 Generalised Müller's BIEs 106
5.9 Low chirality approximations 108
5.10 Miscellanea 109

Chapter 6. Scattering Problems: A Variety of Topics 112
6.1 Introduction 112
6.2 Important concepts of scattering theory 113
6.3 Back to chiral media: Scattering relations and the far-field operator 118
6.4 Using dyadics 124
6.5 Herglotz wave functions 129
6.6 Domain derivative 136
6.7 Miscellanea 140

PART 3. TIME-DEPENDENT DETERMINISTIC PROBLEMS 149

Chapter 7. Well Posedness 151
7.1 Introduction 151
7.2 The Maxwell equations in the time domain 151
7.3 Functional framework and assumptions 152
7.4 Solvability 153
7.5 Other possible approaches to solvability 158
7.6 Miscellanea 162

Chapter 8. Controllability 163
8.1 Introduction 163
8.2 Formulation 163
8.3 Controllability of achiral media: The Hilbert Uniqueness method 165
8.4 The forward and backward problems 167
8.5 Controllability: Complex media 174
8.6 Miscellanea 176

Chapter 9. Homogenisation 180
9.1 Introduction 180
9.2 Formulation 181
9.3 A formal two-scale expansion 184
9.4 The optical response region 188
9.5 General bianisotropic media 199
9.6 Miscellanea 207

Chapter 10. Towards a Scattering Theory 212
10.1 Introduction 212
10.2 Formulation 213

10.3 Some basic strategies 214
10.4 On the construction of solutions 217
10.5 Wave operators and their construction 220
10.6 Complex media electromagnetics 225
10.7 Miscellanea 229

Chapter 11. Nonlinear Problems 231
11.1 Introduction 231
11.2 Formulation 231
11.3 Well posedness of the model 232
11.4 Miscellanea 241

PART 4. STOCHASTIC PROBLEMS 245

Chapter 12. Well Posedness 247
12.1 Introduction 247
12.2 Maxwell equations for random media 248
12.3 Functional setting 249
12.4 Well posedness 250
12.5 Other possible approaches to solvability 255
12.6 Miscellanea 261

Chapter 13. Controllability 263
13.1 Introduction 263
13.2 Formulation 263
13.3 Subtleties of stochastic controllability 264
13.4 Approximate controllability I: Random PDEs 266
13.5 Approximate controllability II: BSPDEs 269
13.6 Miscellanea 272

Chapter 14. Homogenisation 275
14.1 Introduction 275
14.2 Ergodic media 276
14.3 Formulation 279
14.4 A formal two-scale expansion 282
14.5 Homogenisation of the Maxwell system 284
14.6 Miscellanea 288

PART 5. APPENDICES 291

Appendix A. Some Facts from Functional Analysis 293
A.1 Duality 293
A.2 Strong, weak and weak-$*$ convergence 295
A.3 Calculus in Banach spaces 297
A.4 Basic elements of spectral theory 300
A.5 Compactness criteria 303
A.6 Compact operators 304

A.7 The Banach-Steinhaus theorem 308
A.8 Semigroups and the Cauchy problem 308
A.9 Some fixed point theorems 312
A.10 The Lax-Milgram lemma 313
A.11 Gronwall's inequality 314
A.12 Nonlinear operators 315

Appendix B. Some Facts from Stochastic Analysis 316
B.1 Probability in Hilbert spaces 316
B.2 Stochastic processes and random fields 318
B.3 Gaussian measures 319
B.4 The Q- and the cylindrical Wiener process 320
B.5 The Itō integral 321
B.6 Itō formula 324
B.7 Stochastic convolution 325
B.8 SDEs in Hilbert spaces 325
B.9 Martingale representation theorem 326

Appendix C. Some Facts from Elliptic Homogenisation Theory 327
C.1 Spaces of periodic functions 327
C.2 Compensated compactness 329
C.3 Homogenisation of elliptic equations 329
C.4 Random elliptic homogenisation theory 332

Appendix D. Some Facts from Dyadic Analysis (*by George Dassios*) 334

Appendix E. Notation and abbreviations 341

Bibliography 343

Index 377

Preface

Writing a book is an adventure. To begin with, it is a toy and an amusement; then it becomes a mistress, and then it becomes a master, and then a tyrant. The last phase is that just as you are about to be reconciled to your servitude, you kill the monster, and fling him out to the public.

W. Churchill

This book deals with the mathematical analysis of deterministic and stochastic problems arising in the theory of the electromagnetics in complex media.

The most primitive electrical and magnetic phenomena (the attraction of a husk of grain to rubbed amber[1] and of iron to lodestone) were undoubtedly observed before recorded history. However, these phenomena were first registered in antiquity by Thales of Miletus (ca. 585 B.C.). The first reported attempt at a "scientific" explanation of magnetism[2] was by Lucretius (first century B.C.). For a concise account of the history of electromagnetism, see, e.g., [83], [150], [348].

The fascinating story of complex media started in the last part of the nineteenth century and the early years of the twentieth. After Maxwell unified optics with electricity and magnetism, it became possible to establish the connection between optical activity and the electromagnetic parameters of materials (some historical comments can be found in Chapter 1). Until the 1960s, electromagnetic researchers focussed on vacuum, or metals, or dielectric media. Sporadic attention to general electromagnetic media emerged rather slowly and was limited to a theoretical level until the mid-1980s. Since then, though, the landscape has changed drastically, at first mainly because of the technological importance of chiral materials at microwave frequencies, more recently because of the vast progress being made in theoretical and experimental research in complex media electromagnetics. Intense research that has resulted in an impressively extensive bibliography on electromagnetic fields in complex (and in particular chiral) media appeared in

[1]The ancient Greek word for amber is ἤλεκτρον, pronounced "ílectron", meaning "shining".

[2]The Greek word for magnet is μαγνήτης, pronounced "magnítis". It originates from *Mágnites*, a tribe that inhabited — from the tenth century B.C. — the southeastern area (still called Magnesía) of central Greece. In the seventh century B.C. the tribe established two colonies (both bearing the name Magnesía) in Asia Minor; in the region of one of them natural magnetic minerals could be found and were called *magnesian stones*.

the applied physics and engineering communities. Results for complex media electromagnetics within a framework of rigorous mathematical analysis are much fewer, however, having started to appear only in the mid-1990s. Publications in this field have dealt mostly with time-harmonic deterministic problems, followed by deterministic problems in the time domain and finally problems related to the effects of randomness on the evolution and behaviour of electromagnetic fields. Although the research in these areas now constitutes a distinct, well-established branch of applied mathematics, no monograph has appeared treating this field as such. Our purpose in this book is to take a step toward filling this gap by describing the major developments in this field as presented in the work of many researchers, including the authors of this book. In addition, the book includes results (e.g., parts of Chapters 8, 9, 10 and 13, Chapter 11, and Chapter 14) that have not previously been published elsewhere.

The audience for this monograph is expected to comprise researchers and graduate students in applied mathematics, applied analysis, applied stochastic analysis, applied physics, electrical engineering, telecommunications, etc. Had this book been written thirty years ago, one could have said - lightheartedly - that it was addressed mainly to applied mathematicians of the French school. However, the concepts that were perhaps considered abstract applied analysis a few years ago have now entered the mainstream of applied mathematics (and rightly so). Techniques from functional analysis or the abstract theory of differential equations are now widely used in the study of applied problems, to gain qualitative information on the behaviour of systems and even to make progress in the numerical analysis of a variety of problems. This book should therefore be of interest to applied mathematicians desirous of using advanced concepts from mathematical analysis and the theory of partial differential equations (PDEs) in the study of physical and engineering problems, in particular problems in electromagnetics, and also to engineers and physicists who wish to obtain a deeper understanding of the physical models and their potentials and limitations. The latter turn out to be more important than the former in a number of cases! The concept of well posedness[3] proves to be extremely important in this respect: a physical model is often a formal statement of physical postulates expressed in mathematical language; however, unless one checks the grammar and the syntax of the statement and shows that it is correct, the statement is of little use to anyone. The well posedeness of a system (consisting of differential equations plus initial orboundary conditions) describing some model can thus be used as an internal consistency check of the modelling; if the system fails to be well posed, then one has to go back to the drawing board and even reconsider the physical assumptions employed in the model. This step is very important before one proceeds to the numerical treatment of the models. It is not coincidental that solvability methods, e.g., the Faedo-Galerkin

[3]A problem is *well posed* in the sense of Hadamard [177] if (a) a solution exists and is unique for the class of data of interest, and (b) the solution depends continuously on the data.

method, may turn into powerful numerical techniques for the numerical approximation of various problems. The other issues addressed in this book are also of dual interest, to applied mathematicians and engineers and physicists alike. Control theory is the mathematical study of how to manipulate the parameters affecting the behaviour of a system so as to produce a desired or optimal outcome. Homogenisation deals with material properties in heterogeneous media; they can frequently be described by their effective behaviour, i.e., there is a homogeneous medium, the effective medium, whose material properties are close to those of the real medium when measured on long space-time scales. A process of averaging, or homogenisation, takes place so that the complicated small-scale structure of the material is replaced by an asymptotically equivalent homogeneous structure. Homogenisation as a mathematical problem treats the asymptotic behaviour of solutions of PDEs with rapidly changing coefficients. A variety of computational aspects (Sections 4.6, 5.9, 6.4.2, 7.5.2, 12.5.2, 12.5.3) will probably be of interest to the book's readers.

As the title indicates, both deterministic and stochastic[4] problems are treated - an unusual combination for a single text.

The book is divided into five parts. Part 1 refers to preliminary concepts: Chapter 1 provides a general introduction and some historical comments on complex media, while in Chapter 2 we introduce the Maxwell equations and discuss constitutive relations; additionally, we describe the problems treated in this book. Chapter 3 deals with the function spaces and operators needed in mathematical electromagnetic theory. Part 2 deals with time-harmonic deterministic problems: in Chapter 4 we consider solvability issues for spatially dependent media, while in Chapters 5 and 6 we consider a variety of scattering problems in homogeneous media. Part 3 treats time-dependent deterministic problems: Chapter 7 provides a study of solvability of the Maxwell equations for the most general linear media in the time domain. In Chapter 8 we study controllability issues, and in Chapter 9 we deal with homogenisation problems. Chapter 10 is an introductory chapter on scattering theory in the time domain, while Chapter 11 is an introduction to the mathematical theory for nonlinear complex media. Part 4 deals with stochastic problems: Chapter 12 provides well-posedness results for stochastic media, Chapter 13 focuses on stochastic controllability issues and Chapter 14 deals with stochastic homogenisation. The trinity well posedness - controllability - homogenisation is the common underlying framework of these two parts. Since we aim at a rather broad and diverse audience, we have decided to include some material that should facilitate the access of mathematicians of a more applied persuasion or engineers and physicists to the material of the book. Hence, Part 5 contains five appendices that, for the convenience of the reader, include useful facts from functional analysis (Appendix A), stochastic analysis (Appendix B), elliptic homogenisation theory (Appendix

[4]A stochastic process is one referring to systems whose subsequent state is determined both by the process's predictable actions and by an unpredictable random element.

C) and dyadic analysis (Appendix D); Appendix E is a list of notations and abbreviations used throughout the book.

It would not be possible to end this preface without expressing our acknowledgements to a number of people. First, collectively:

- ▷ We are indebted to George Dassios, a good friend and great colleague, for writing Appendix D, containing a short introduction to dyadic analysis.
- ▷ Several friends and collaborators have been a constant source of inspiration. We wish to thank our co-authors of papers on topics related to the mathematical analysis of complex media electromagnetics: Christos E. Athanasiadis, Gerassimos Barbatis, George Costakis, Patrick Courilleau, Dimitri Frantzeskakis, Thierry Horsin, Andreas Ioannidis, Nikos Kavallaris, Gerhard Kristensson, Kostas Liaskos, Paul A. Martin, Roland Potthast and Thanassis Spyropoulos.
- ▷ It has been an intellectual pleasure and a mathematical experience to talk with Nicolas Charalambakis, Gerhard Kristensson, François Murat and George C. Papanicolaou.
- ▷ Special thanks to the press's two anonymous reviewers, whose suggestions and constructive criticism resulted in a substantially improved manuscript.
- ▷ We also thank the editorial staff of Princeton University Press, in particular Vickie Kearn, Stefani Wexler and Leslie Grundfest, for a most professional and friendly collaboration and for their patience and understanding. Further, we thank Marjorie Pannell for her superb copyediting work.
- ▷ Regarding the typesetting, the times when we referred to *More Math Into LaTeX*, by G. Grätzer (4th ed., Springer, New York, 2007) have indeed been uncountable!

Individual acknowledgements:

G.F.R. The influence of the pioneering work of Teruo Ikebe, Peter Lax and Ralph Phillips, and Calvin Wilcox on wave-scattering problems is gratefully recognized. In particular, a profound debt of gratitude is owed to Aldo Belleni-Morante (1938–2009), Rolf Leis and Calvin Wilcox (1924–2001) who have been such an inspiration over the years. I would like to express my gratitude to the many colleagues with whom I have had useful discussions; in particular, I thank Wilson Lamb and Christos E. Athanasiadis.

I.G.S. It is a great pleasure and a moral obligation to express my thanks to several persons, most of whom honour me with their friendship: first, to my oldest fellow traveller on many journeys in the mathematics of chiral media, Christos E. Athanasiadis. My introduction to the electromagnetics of complex media emerged originally from the work of, and later from the personal acquaintance and discussions with, Akhlesh Lakhtakia and Werner Weiglhofer (1962–2003). My understanding of applied mathematics has been enhanced by a long-standing collaboration with Paul A. Martin. Luis Castro, Martin Costabel, Thanassis Fokas, Vladislav Kravchenko, Gerhard Kristensson, David Natroshvili and Valeriy Serov have substantially contributed to my understanding of their areas of expertise. I have enjoyed and benefited from discussions with Habib Ammari and Andreas Kirsch. I owe exceptional thanks to Eleni Manolakaki for the time she spent talking to me about causality and determinism. It is always a pleasure to be in the same academic environment with Nick Alikakos, Christos E. Athanasiadis, Gerassimos Barbatis, Vassilis Dougalis, Grigoris Kalogeropoulos, Giorgos Sagias and Savva Avramovich Tersenov. Being thanked by a son in a book's preface is, I am sure, a great satisfaction for a parent. I am sorry that my father, George (1917–2007), is no longer with us, while my mother, Theodora, has not been granted by life a peaceful old age. I can only hope that in previous years I was successful in letting them feel how grateful I was to both of them for the way they were to me and Anastasia, the best sister one could wish for. As for precious *Μπέης*, there are no words to express my feelings. Finally, I could never imagine that my son would be so supportive, tolerant and understanding during the long hours I "betrayed" him for this book: thanks, George!

A.N.Y. I would like to thank my colleagues from AUEB, Professors Nikos Frangos, Epameinondas Panas, Anastasios Xepapadeas and Michael Zazanis and Stelios Psarakis, for useful and enjoyable discussions (mathematical or not), as well as for contributing to a fruitful and productive atmosphere within AUEB. I also thank Professors Dimitris Kravvaritis, Kyriakos Hizanidis and Claude Le Bris for stimulating discussions on nonlinear analysis, electromagnetics and stochastic homogenisation, respectively; as well as Stelios Xanthopoulos. I would also like to express my sincere acknowledgements to Professor Ioannis Karatzas, from whom I have learned a lot, both in stochastic analysis and in writing style. On a personal level, this work would not have been completed had it not been for Electra's constant encouragement and support both on the moral side and in quotidian matters, and for her advice on matters of style and presentation. The book absorbed a large part of our common time at home; luckily being a colleague - though not of the mathematical variety - she understands! Furthermore, I am indebted to my parents, Jenny and Nikos, for their constant support and to my sister Helen. Last but not least, I must thank Kalitsa. Finally, special thanks go to my good friend Lambros Boukas, for invaluable help on the information technology front!

Gary F. Roach (Fintry, Glasgow)
Ioannis G. Stratis (Halandri, Athens)
Athanasios N. Yannacopoulos (Pagrati, Athens)
October 2011

PART 1
Modelling and Mathematical Preliminaries

Chapter One

Complex Media

In recent years technology has replaced Hercules as far as the labours are concerned: the progress in theoretical studies, followed by impressive experimental work and achievements, is reaching the everyday lives of ordinary people and is rapidly changing our habits and lives.

A big part of this technological revolution, which emerged in the late twentieth century and is propagating with increasing speed and expanding front, is the result of complex media. Complex media are artificial materials exhibiting properties, based on their structure rather than their composition, superior to those in naturally existing materials. Nevertheless, there certainly do exist materials in nature displaying "exotic" properties.

A characteristic of a fast-growing research area, such as the one concerned with the study of complex media, is its interdisciplinary nature; scientists from a wide provenance spectrum, including electrical engineering, electromagnetics, solid state physics, microwave and antenna engineering, optoelectronics, classical optics, materials science, semiconductor engineering, and nanoscience, are engaged in this field. Of course, mathematics has its usual share, as well!

A discrimination between left and right has proved to be a fertile concept in the many branches of science that feed into electromagnetics: *handedness* is a term that is used extensively in the complex media[1] literature. There are actually three notions of handedness of interest in electromagnetics[2]:

Left-handedness: The term *left-handed* as a description of a certain class of metamaterials springs from the handedness of the vector triplet (E, H, K) (E being the electric field, H the magnetic field, and K the wave vector[3], respectively) of a linearly[4] polarised wave propagating in such media. This type of left-handedness refers to materials whose electric permittivity and magnetic permeability are both negative. The theoretical prediction of their existence was made by V. Veselago [420] in 1964.

---

[1]The very fashionable term *metamaterials* refers to a wide class of complex media. A thorough discussion on the use and meaning of this term can be found in [383].

[2]For more details see [384], on which the following discussion of the notions of handedness is based.

[3]Recall that for a three-dimensional travelling plane wave $\Psi(t,x) = A\,cos(K\cdot x - \varpi\, t + \phi)$, with position vector $x \in \mathbb{R}^3$, at time $t > 0$, of angular frequency ϖ, amplitude A, and "phase offset" ϕ, the vector K is the "wave vector" and its magnitude is the angular wave number $|K| = 2\pi/\lambda$, λ being the wavelength.

[4]See Section 5.2.

Handedness of a circularly[5] *polarised wave:* In the electrical engineering community, handedness is manifested in relation to polarisation, which refers to the direction and behaviour of the electric field vector, which in the case of circular (or elliptical) polarisation exhibits a form of helicity (or handedness). The wave propagates in a certain direction, and (for isotropic media) the electric field is transverse. In the transverse plane, the temporal oscillations of the field vector are described by an ellipse or a circle (in the case of linear polarisation, the ellipse shrinks to a straight line). Along its direction of propagation, the wave may rotate to the left or to the right. Of course, these notions are meaningless unless one of them is properly defined: according to the U.S. Federal Standard 1037C (http://www.its.bldrdoc.gov/fs-1037/), the polarisation is defined as right-handed if the temporal rotation is clockwise when viewed from the transmitter (in the propagation direction) and left-handed if the rotation is counterclockwise. By contrast, astronomers look towards the source (transmitter), and therefore in the direction opposite that in which the wave propagates; hence the terms "clockwise" and "counterclockwise" attribute meanings opposite to right- and left-handedness. Nevertheless, the handedness of a specific object remains invariant under orthogonal transformations.

Chirality and geometry: Handedness is a characteristic of material objects, such as corkscrews, doors, cookers, sinks, computer mice, keyboards, scissors, and a variety of construction tools. The mirror image of a right-handed object is the same as the original except that it is left-handed (the original image cannot be superimposed on its mirror image.) A nonhanded object remains the same within this mirror-image operation[6] since, after imaging, it can be brought into congruence with the original by simple translations and rotations. A handed object is called *chiral* (a term coined in 1888 by Kelvin[7], from the Greek word $\chi\varepsilon i\varrho$, meaning "hand"). Chiral media possess optical activity, or the ability to rotate the plane of polarisation of a beam of light passing through them. The relation between the chiral (micro)structure and the (macroscopic) optical rotation was discovered by Pasteur in the 1840s. The mirror-image operation is also called *parity transformation* (all spatial axes are reversed when parity is changed); it is a fundamental property of

---

[5]See Section 5.2.

[6]In a much more general setting, "mirror symmetry" is an example of a phenomenon known as *duality*, which occurs when two seemingly different physical systems are isomorphic in a nontrivial way. The nontriviality of this isomorphism makes quantum corrections necessary. In mathematics, an analogy is the Fourier transform: a local concept as the multiplication of two functions is equivalent to a convolution product, requiring integration over the whole space. Finding such dualities leads to solving complicated physical questions in terms of simple ones in the dual framework. A deep understanding of the inner mechanisms of duality symmetries is, in general, not yet feasible, with one exception: mirror symmetry. A mathematical framework to rigorise physical statements is already in an advanced stage of development. An excellent source elaborating aspects of this theory for physicists and mathematicians is [194].

[7]"I call any geometrical figure, or group of points, *chiral*, and say that it has chirality, if its image in a plane mirror, ideally realized, cannot be brought to coincide with itself" ([234], p. 619).

physics that parity symmetry is broken in subatomic interactions. On several different scales and levels of nature, parity is not balanced. From amino acids through bacteria, winding plants and right-handed human beings to spiral galaxies, one of the handednesses dominates the other. The handedness of an optically active substance is called *dextrorotary* (resp., *levorotary*) if polarised light is rotated clockwise (resp., anticlockwise) as the observer faces the substance, with the substance between the observer and the light source. The handedness is indicated by prefixing "*d*-" (resp., "*l*-") to the substance's name.[8]

In geometry, an object is chiral if it does not coincide with its image under rotations and translations: in three dimensions any object with a plane of symmetry or a centre of symmetry is not chiral, but there are objects that, although they have neither a plane of symmetry nor a centre of symmetry, they are nonchiral. In two dimensions, any bounded nonchiral figure has an axis of symmetry. Typical chiral ones in two dimensions are rhomboids and spirals, while in three dimensions they are irregular tetrahedra[9] and Möbius strips.

A right-handed object and its corresponding left-handed object would be considered identical by usual symmetry. So, in what sense do the three above ways of looking at handedness differ, as far as the left-right classification is concerned? Obviously, the circular-polarisation-based handedness property is fully symmetric. Although the conventions differ and the definitions of left- and right-handedness are not alike in different scientific fields, the handedness of the polarisation in dipole antennas is only a matter of phase shift. Only metamaterials (which according to certain definitions cannot exist naturally) can display material parameters that are both simultaneously negative. As *left-handed* materials they belong to a class of media that by no means can be considered to be identified with that of the *right-handed* ones. As far as structural chirality is concerned, if all DNA molecules[10] were to twist in the right-handed sense, there would be no chance of the opposite handedness surviving. This is the reason that justifies the use of the term dyssymmetry[11] for this specific type of partial asymmetry. This phenomenon was discovered in 1811 by Arago [16], experimenting with quartz crystals (an anisotropic material), and one year later by Biot [63],

---

[8]From the Latin words *dexter* meaning right and *laevus* meaning left.

[9]See *Chiral Polyhedra*, by E. W. Weisstein, in the framework of "The Wolfram Demonstrations Project" (http://demonstrations.wolfram.com/ChiralPolyhedra/).

[10]The DNA double helix is a spiral polymer of nucleic acids, held together by nucleotides that base pair together. In B-DNA, the most common double helical structure, the double helix is right-handed. Z-DNA is another of the many possible double helical structures of DNA. It is a left-handed double helical structure in which the double helix winds to the left in a zig-zag pattern. A-DNA is yet another of the possible double helical structures of DNA. A-DNA is thought to be one of three biologically active double helical structures, along with B- and Z-DNA. It is a right-handed double helix fairly similar to the B-DNA form, but with a shorter, more compact helical structure.

[11]From the Greek prefix $\delta\upsilon\sigma$- (dys-) meaning "difficult", "bad" or "ill" (it appears in many medical terms, e.g., dyspepsia, dysphagia, dyspnoea, etc.), and the word $\sigma\upsilon\mu\mu\varepsilon\tau\varrho\iota\alpha$, ("symmetry").

experimenting with turpentine vapour (an isotropic medium). Fresnel also examined optical activity in a chiral medium [154], as did in 1842 Cauchy [90]; this was the first mathematical study of chirality. The answer to the question of what is this strange property of media that makes them optically active was given by Pasteur in 1848 [345]: he noticed that two substances that were chemically identical in the classification scheme at the time but that had physical structures that were mirror images of each other exhibited different physical properties. Thus, Pasteur introduced geometry into chemistry and originated the branch of chemistry today called stereochemistry. Much more recently the studies of Prelog were extremely important; he shared[12] the 1975 Nobel Prize in Chemistry for his work in the field of natural compounds and stereochemistry. His lecture [351] at the Nobel Prize award ceremony regarding the rôle of chirality in chemistry, is very interesting.

Although they contain identical atoms in equal numbers, enantiomers[13] can, as mentioned above, have different properties. As Lakhtakia has written [272], "one enantiomer of the chiral compound thalidomide may be used to cure morning sickness, but its mirror image induces fetal malformation. Aspartame, a common artificial sweetener, is one of the four enantiomers of a dipeptive derivative. Of these four, one (i.e. aspartame) is sweet, another is bitter, while the remaining two are tasteless. Of the approximately 1850 natural, semisynthetic and synthetic drugs marketed these days, no less than 1045 can exist as two or more enantiomers; but only 570 were being marketed in the late 1980s as single enantiomers, 61 of which were totally synthetic. But since 1992, the U.S. Food and Drug Administration has insisted that only one enantiomer of a chiral drug be brought into market." Another example is mint flavored chewing gums containing chiral enantiomers; they create a different taste sensation to different people because the human taste sensors contain chiral molecules.

The great philosopher Kant was probably the first eminent scholar to point out the philosophical significance of mirror operations. The interested reader may refer to Section 13 of his 1783 "Prolegomena to Any Future Metaphysics" [222], where a most interesting discussion involving the notion of what is today called chirality is found.

Some of the history of the development of ideas about chirality may be found in the monographs [273], [268], [289] and the papers [272], [142], [213]. Also, the general audience oriented-books [158], [212] are very inspiring. See also [188].

The formalisation of the mathematical description of electrostatics took place around 1800, by giants such as J. L. Lagrange, P.-S. Laplace, S.-D. Poisson, G. Green and C. F. Gauss. However, there was no idea at the time of how electricity and magnetism were related. It was another giant, J. C.

---

[12]With Sir J. W. Cornforth.

[13]In chemistry, enantiomers (from the Greek words ϵνάντιος, meaning "opposite", and μέϱος, meaning "part" or "portion") are stereoisomers that are nonsuperimposable complete mirror images of each other.

Maxwell, who in the 1860s unified optics with electricity and magnetism in his monumental *A Treatise on Electricity and Magnetism*, first published in 1873 [306]. For a concise account of the history of electromagnetism, see, e.g., [83], [150], [348].

In the last part of the nineteenth century, after Maxwell's unification, it became possible to establish the connection between optical activity and the electromagnetic parameters of materials. In 1914, Lindman was the first to demonstrate the effect of a chiral medium on electromagnetic waves (his work in this field was about forty years ahead of that of other scientists); he devised a macroscopic model for the phenomenon of "optical" activity that used microwaves instead of light and wire spirals instead of chiral molecules. His related work was published in 1920 and 1922; for a very interesting account of Lindman's work, see [288].

At the macroscopic level, the Maxwell equations read

$$\begin{aligned} &\mathrm{curl} H = \partial_t D + J, && \text{Ampère's law,} \\ &\mathrm{curl} E = -\partial_t B, && \text{Faraday's law,} \\ &\mathrm{div} D = \rho, \ \mathrm{div} B = 0, && \text{Gauss's laws,} \end{aligned}$$

where E, H are the electric and the magnetic field, D, B are the electric and magnetic flux densities, J is the electric current density, and ρ is the density of the (externally impressed) electric charge.

This system contains eight equations (three from each of the first two "vector" laws and one from each of the "scalar" Gauss laws) but twelve unknowns (three components for each of the vector fields E, H, D, B). Constitutive relations, i.e., relations of the form

$$D = D(E, H)\,, \; B = B(E, H)\,,$$

must therefore be introduced. As is well known, constitutive relations are relations between physical quantities that are specific to a material or substance, and approximate the response of that material to external forces. Some constitutive equations are simply phenomenological; others are derived from first principles. This topic is discussed in Chapter 2.

Intensive research that has resulted in an impressively extensive bibliography on electromagnetic fields in complex (and in particular in chiral) media has appeared in the applied physics and engineering communities since the mid-1980s. By contrast, not so many rigorous mathematical contributions have appeared on the study of complex media. The large majority of these publications deal with time-harmonic electromagnetic fields in chiral media and appeared in the mid-1990s. The 1994 paper by Petri Ola [341] opened the way, followed initially by publications of the group at the Centre de Mathématiques Appliquées, École Polytechnique, Palaiseau, Paris, France, and the group at the Department of Mathematics of the National and Kapodistrian University of Athens, Greece. Of course, many other researchers gradually came onto the stage, so that the study of complex media in electromagnetics today forms an identifiable branch of applied mathematics. The rigorous mathematical analysis of time domain problems for complex media was the next step, and important progress in this field has been made. The vast majority of existing work deal with linear media, although recently advances have been made in nonlinear complex media. While most of the theory refers to deterministic complex media, its stochastic counterpart is not negligible.

Although there are books of different levels of mathematical rigour in the applied physics and engineering literature on the electromagnetics of complex media (e.g., [260], [266], [268], [271], [273], [289], [299], [378]), it seems that no books (apart from some parts of [91] and [145]) are devoted exclusively to the related mathematical theory. It is our intention to try to fill this gap by providing an introduction to the mathematical theory of complex media, linear and nonlinear, deterministic and stochastic. Of course, not all topics can be or are covered.

Chapter Two

The Maxwell Equations and Constitutive Relations

2.1 INTRODUCTION

The aim of this chapter is twofold: first, to introduce the constitutive relations which are commonly used in electromagnetic theory for the mathematical modelling of complex electromagnetic media. In the context of the present work these constitutive relations are to be understood as operators connecting the electric flux density and the magnetic flux density with the electric and the magnetic fields. These relations are considered as formal expressions of the physical laws that govern the electromagnetics of complex media. When they are introduced into the Maxwell equations, we obtain differential equations (PDEs) that govern the evolution of the electromagnetic fields; the treatment of these equations, in the rigorous mathematical sense, is the main object of this monograph. Through the treatment of these evolution equations we may model, understand and predict qualitative and quantitative phenomena related to complex media electromagnetics. The second goal of this chapter is to formulate and discuss the scope of the various problems related to the Maxwell equations that will be treated in this work. We introduce and formulate in terms of differential equations various problems of interest related to the Maxwell equations: time-harmonic problems, scattering problems, time-domain evolution problems, random and stochastic problems, controllability problems, homogenisation problems, etc. The mathematical analysis of these problems will be treated in detail in this book.

The structure of this chapter is as follows: in Section 2.2 we introduce the Maxwell equations, which are a set of PDEs that govern the evolution of electromagnetic fields in a general electromagnetic medium. In Section 2.3 we introduce a variety of constitutive relations that are used in the mathematical and physical modelling of complex electromagnetic media, while in Section 2.4 we introduce and discuss various problems related to the Maxwell equations in complex media that will be treated in the course of this book.

2.2 FUNDAMENTALS

Every electromagnetic phenomenon is specified by four vector quantities: the *electric field* E, the *magnetic field* H, the *electric flux density* D and the *magnetic flux density* B. These quantities are considered time-dependent vector fields on a domain $\mathcal{O}$ of $\mathbb{R}^3$, so they are vector-valued functions of the spatial variable $x \in \mathcal{O} \subset \mathbb{R}^3$ and the time variable $t \in \mathbb{R}$. The interdependence of these quantities is given by the celebrated *Maxwell system*, which at the macroscopic level is stated as

$$\begin{aligned} \mathrm{curl} H(t,x) &= \partial_t D(t,x) + J(t,x), \\ \mathrm{curl} E(t,x) &= -\partial_t B(t,x), \end{aligned} \tag{2.1}$$

where J is the electric current density. These equations are the so-called Ampère's law and Faraday's law, respectively. In addition to the above, we have the two laws of Gauss,

$$\begin{aligned} \mathrm{div} D(t,x) &= \rho(t,x), \\ \mathrm{div} B(t,x) &= 0, \end{aligned} \tag{2.2}$$

where ρ is the density of the externally impressed electric charge. For the time being, the differential operators curl and div are defined formally, in terms of their standard definitions used in vector calculus; we return to a more rigorous treatment of these operators in Chapter 3.

The Maxwell system must be supplemented with initial and boundary conditions. The initial conditions describe the initial state of the system at the time we assume that the observation starts (chosen to be without loss of generality $t = 0$) and are of the form

$$\begin{aligned} E(0,x) &= E_0(x)\,, \\ H(0,x) &= H_0(x)\,, \end{aligned} \tag{2.3}$$

for $x \in \mathcal{O}$. The boundary conditions, which describe the behaviour of the fields for all times t on the boundary of the considered domain $\partial\mathcal{O}$, can be of various types, depending on the physical situation we wish to model. In this work we make extensive but not exclusive use of the "perfect conductor" boundary condition

$$n(x) \times E(t,x) = 0, \text{ for } x \in \partial\mathcal{O} \text{ and } t \in I, \tag{2.4}$$

where I is a time interval. By $n(x)$ we denote the outward normal on $\partial\mathcal{O}$, which is assumed to be an appropriately smooth surface. In a number of instances we will make use of the "inhomogeneous" variant of the perfect conductor boundary condition, namely,

$$n(x) \times E(t,x) = f(t,x), \text{ for } x \in \partial\mathcal{O} \text{ and } t \in I, \tag{2.5}$$

REMARK 2.2.1 By a standard procedure[1], the general problem consisting of (2.1) and (2.5) can be reduced to the problem consisting of (2.1) and the perfect conductor boundary condition (2.4). This is possible by considering a suitable lifting f of f from $\partial\mathcal{O}$ to $\mathcal{O}$ and then appropriately modifying J in terms of f. In the same spirit, one can consider (2.3) with vanishing right-hand sides.

[1] Especially in the case of linear constitutive relations.

Concerning the interval $I \subset \mathbb{R}$, where the time variable takes values, we will restrict ourselves for most of this work to the case where $I \subset \mathbb{R}^+$. Furthermore, we consider that t is allowed only to increase, i.e., there is a definite direction in time. In doing so we accept the physical axiom of *causality*; stated simply, this means that only the past of the phenomenon is likely to affect the future; thus we neglect the behaviour of the fields for $t < 0$. The issue of causality is very important for the physical modelling of electromagnetic media, and we return to it in more detail in Section 2.3.

The Maxwell system (2.1) and the Gauss laws (2.2) form a differential system from which we wish to calculate the quadruplet (B, D, E, H), assuming that the vector J and the scalar ρ are known. The equation of continuity can then be considered a consistency condition between the known quantities. Thus, one has to calculate twelve scalar functions from a system of eight scalar equations. This means that the system is underdetermined. To overcome this deficiency, *constitutive relations* must be introduced:

$$D = D(E, H)\,, \quad B = B(E, H)\,. \tag{2.6}$$

These functional relations allow us to obtain a well-posed system of differential equations that, when combined with the initial and the boundary conditions, provides information on the evolution of the fields. The choice of constitutive relations is an extremely important step towards the modelling of complex media[2].

REMARK 2.2.2 Applying the divergence operator on equation (2.1) and using the first of (2.2), we obtain the *equation of continuity*

$$\partial_t \rho + \mathrm{div} J = 0. \tag{2.7}$$

Conversely, assume that (2.1) and (2.7) hold. Taking the divergence of (2.1) and using (2.7), we deduce that there exist $\phi(x), \psi(x)$ such that $\mathrm{div}D(t,x) = \rho(t,x) + \phi(x)$ and $\mathrm{div}B(t,x) = \psi(x)$. If $D(0,x) = D(E_0(x), H_0(x))$ and $B(0,x) = B(E_0(x), H_0(x))$ are such that $\mathrm{div}D(0,x) = \rho(0,x)$ and $\mathrm{div}B(0,x) = 0$, then Gauss's laws (2.2) hold for all t. This allows us to consider as "the Maxwell system" the set of equations (2.1) plus the constitutive relations (2.6), plus the equation of continuity (2.7), which in what follows will always be assumed to hold (although we will not explicitly state it). Furthermore, this argument points out that all four vector equations may *not* be independent.

REMARK 2.2.3 For conducting media, it is usual for the currents not to be entirely freely chosen but to obey a generalised Ohm's law, $J = J(E, H)$, instead. More precisely, the currents are expressed as the sum of a "constitutive" part and a "forced" part [145, p. 15], [370], $J = F(E, H) + J_{\mathrm{f}}$. Such a consideration does not essentially change the treatment of the problem, but it introduces some extra complications in notation and in the study of the energy of the medium. To avoid these complications, we will assume $F(E, H) = 0$.

[2]Some authors use the constitutive relations in the form $D = D(E, B)$, $H = H(E, B)$.

Often, especially when the main objective is the mathematical treatment of the Maxwell system, it is convenient to express the system in more compact form. This is done by using the so-called *six-vector notation.* Using the superscript *tr* to denote transposition, we define

- ▷ the *electromagnetic flux density* $\mathsf{d} := (D, B)^{tr}$,
- ▷ the *electromagnetic field* $u := (u_1, u_2)^{tr} := (E, H)^{tr}$,
- ▷ the *current* $j := (-J, 0)^{tr}$,
- ▷ the *initial state* $u_0 := (E_0, H_0)^{tr}$.

A linear operator acting on u is written as a 2×2 (block) matrix with linear operators as its entries. An important example, one that will be used very often in this work, is the *Maxwell operator*,

$$\mathsf{M} := \begin{pmatrix} 0 & \mathrm{curl} \\ -\mathrm{curl} & 0 \end{pmatrix}. \tag{2.8}$$

The constitutive relations are now modelled by an operator $\mathcal{L}$ and are understood as the functional equation

$$\mathsf{d} = \mathcal{L}u.$$

The properties of this operator reflect the physical properties of the medium in question. Various aspects of a "rigorous abstract modelling" approach to the properties of the operator $\mathcal{L}$ appear in [201], [202], [205], [206].

In view of the above, the Maxwell system can be written as an initial-value problem for an abstract evolution equation

$$\begin{aligned} (\mathcal{L}u)'(t) &= \mathsf{M}u(t) + j(t) \quad , \quad \text{for } t \geq 0\,, \\ u(0) &= u_0\,. \end{aligned} \tag{2.9}$$

The prime stands for the time derivative[3]. In fact, by using standard terminology, (2.9) is an inhomogeneous *neutral functional differential equation.*

A crucial first step towards the study of the solvability of (2.9) is the determination of its *state space*; this choice is again dictated by physical principles but does not have to be unique. One possibility would be to choose the state space of the problem as the space of spatially square integrable vector-valued functions of a real variable t. This choice is consistent with energy considerations for the Maxwell system (see Section 2.3.2). However, other choices are possible, depending on the properties of the electromagnetic fields we wish to address. Such functional framework alternatives are addressed in this work (see Chapter 3).

For thorough presentations of the theory of electromagnetism, we refer to the books [69], [145], [211], [245], [287], [328], [371], [396].

---

[3]The explicit dependence on the spatial variable x is omitted and it is considered that both u, j are functions mapping t into vector fields $u(t, x)$, $j(t, x)$ containing the explicit spatial dependence, which is suppressed. The operator M acts on the spatial part of the function u. Furthermore, the boundary conditions are included in the choice of function space that acts as the state space of the system and the definition of the operator M. In fact, equation (2.9) is considered an ordinary differential equation (ODE) in a properly selected function space setting. This is the approach adopted in this book.

2.3 CONSTITUTIVE RELATIONS

2.3.1 An "axiomatic" approach to constitutive relations

The constitutive relations for an electromagnetic medium reflect the physics that govern the phenomena and are expected to comply with the fundamental physical laws, which play the rôle of physical hypotheses, or postulates, concerning the properties of the material inside the domain $\mathcal{O}$.

We state the postulates that govern the evolution of the electromagnetic field in a complex medium. The approach followed, based on [176], [205], is system theoretic in the sense that we consider the electromagnetic field u as the *cause* and the electromagnetic flux density d as the *effect*. Compliance with these postulates dictates the form of the operator $\mathcal{L}$.

PLAUSIBLE PHYSICAL HYPOTHESES

- ▷ DETERMINISM For every cause, there exists exactly one effect.
- ▷ LINEARITY The effect is linearly related to its cause.
- ▷ CAUSALITY The effect cannot precede its cause[4].
- ▷ LOCALITY IN SPACE A cause at any particular spatial point produces an effect only at this point and not elsewhere.
- ▷ TIME-TRANSLATION INVARIANCE If the cause is advanced (or delayed) by some time interval, the same time shift occurs for the effect.

The above physical postulates have mathematical interpretations in terms of the properties of the operator $\mathcal{L}$, as follows:

- ▷ DETERMINISM $\mathcal{L}$ exists and is a single-valued nontrivial operator.
- ▷ LINEARITY $\mathcal{L}$ is a linear operator.
- ▷ CAUSALITY If $u(t,x) = 0$ for $t \leq \tau$, then $(\mathcal{L}u)(t,x) = 0$, for $t \leq \tau$.
- ▷ LOCALITY IN SPACE $\mathcal{L}$ is a local operator with respect to the spatial variables, i.e., $\mathcal{L}(u(\cdot,x))(\cdot,x) = \mathfrak{s}(\cdot,x)$, where $\mathfrak{s}$ is a local functional, allowing spatial derivatives of the electromagnetic fields, but not integrals with respect to the spatial variables. Locality with respect to temporal variables is not assumed; on the contrary, memory effects are allowed.
- ▷ TIME-TRANSLATION INVARIANCE For all $\varkappa \geq 0$, $\mathcal{L}$ commutes with the right $\varkappa$-shift operator $\tau_\varkappa$. Therefore, the time instant at which the observation starts does not play any significant rôle; the "present" can be chosen arbitrarily.

[4]The concept of causality is a very delicate concept that has generated extensive epistemological and philosophical discussion.

REMARK 2.3.1 We do not assume continuity since it follows by linearity and time-translation invariance. Note that continuity is not ascertained in the case where commutativity with the left shift operator is assumed instead of with the right shift operator (for details regarding both these comments, see [295]).

REMARK 2.3.2 The postulate of linearity will be abandoned in our study of nonlinear media (see Sections 2.3.5, 2.4.10 and Chapter 11). Furthermore, the postulate of time-translation invariance is abandoned by some authors, leading to more general constitutive relations (see, e.g., [70]). Most of the results in this book may be extended for such constitutive relations under additional technical assumptions.

The general form of the constitutive operator $\mathcal{L}$, consistent with the above physical postulates, is a continuous operator having the following convolution form ([226], [227]):

$$\mathsf{d}(t,x) = (\mathcal{L}u)(t,x) = \mathsf{A}_{\text{or}}(x)u(t,x) + \int_0^t \mathsf{G}_{\text{d}}(t-s,x)u(s,x)\,ds, \tag{2.10}$$

where

$$\mathsf{A}_{\text{or}}(x) := \begin{pmatrix} \varepsilon(x) & \xi(x) \\ \zeta(x) & \mu(x) \end{pmatrix}, \quad \mathsf{G}_{\text{d}}(t,x) := \begin{pmatrix} \varepsilon_{\text{d}}(t,x) & \xi_{\text{d}}(t,x) \\ \zeta_{\text{d}}(t,x) & \mu_{\text{d}}(t,x) \end{pmatrix}. \tag{2.11}$$

The entries of the above matrices are 3×3 matrices whose elements are essentially bounded functions[5] on $\mathcal{O}$. Note that each $\mathsf{A}_{\text{or}}(\cdot)$, $\mathsf{G}_{\text{d}}(t,\cdot)$ defines a multiplication operator in the state space. Equation (2.10) will often be abbreviated as

$$\mathsf{d} = \mathsf{A}_{\text{or}}u + \mathsf{G}_{\text{d}} \star u, \tag{2.12}$$

where $\star$ denotes temporal convolution. The local-in-time part A_{or} of the operator $\mathcal{L}$ models the instantaneous response of the medium, and A_{or} is accordingly called the *optical response* operator. The nonlocal-in-time part $\mathsf{G}_{\text{d}}\star$ of $\mathcal{L}$ models the dispersion phenomena. The kernel function G_{d} is called the *susceptibility kernel.*

REMARK 2.3.3 A necessary and sufficient condition for any linear operator to be a convolution operator is established, in an abstract distributional setting, by Zemanian (see Theorem 5.8-2 in [452]). This general result can be applied to justify the use of convolution operators of the form (2.10) to model the constitutive relations ([201], [205]).

Based on the above discussion, and in particular using the operator representation (2.10), we may provide a classification of the various complex electromagnetic media.

[5] See Chapter 3, Section 3.2.1.

DEFINITION 2.3.4 (MEDIA CLASSIFICATION) *A material is called*

- ISOTROPIC, *if* ε, μ, ε_d, μ_d *are scalar multiples of* $I_{3\times 3}$ *and* $\xi = \zeta = \xi_d = \zeta_d = 0$.
- ANISOTROPIC, *if the members of at least one of the pairs* $\varepsilon, \varepsilon_d$ *or* μ, μ_d *are not scalar multiples of* $I_{3\times 3}$ *and* $\xi = \zeta = \xi_d = \zeta_d = 0$.
- BIISOTROPIC, *if all the blocks of the matrices* A_{or}, G_{d} *are scalar multiples of* $I_{3\times 3}$.
- BIANISOTROPIC, *in all other cases.*

The following standing assumptions are used throughout this book.

ASSUMPTION 2.3.5 The optical response matrix A_{or} has essentially bounded entries and is *almost everywhere symmetric*, i.e., $\mathsf{A}_{\mathrm{or}}(x) = \mathsf{A}_{\mathrm{or}}(x)^{tr}$, for almost all $x \in \mathcal{O}$, and *almost everywhere uniformly coercive*, i.e., there exists a constant C such that $|\mathsf{y} \cdot \mathsf{A}_{\mathrm{or}}(x)\,\mathsf{y}| \geq C\,|\mathsf{y}|^2$ for almost all $x \in \mathcal{O}$ and all nonzero $\mathsf{y} \in \mathbb{R}^6$.

ASSUMPTION 2.3.6 The dispersion matrix $\mathsf{G}_{\mathrm{d}}(0,x)$ is *almost everywhere non-negative definite*, i.e., $\mathsf{G}_{\mathrm{d}}(0,x)\,\mathsf{y}\cdot\mathsf{y} \geq 0$, for almost all $x \in \mathcal{O}$ and all nonzero $\mathsf{y} \in \mathbb{R}^6$.

REMARK 2.3.7 Assumption 2.3.5 is related to the one adopted in [226], [227] and is consistent with physical arguments based on energy considerations (see also [145] for related arguments based on thermodynamics) and is convenient for the mathematical theory of complex media as developed in the present book; from the mathematical point of view, this assumption can be relaxed at the cost of (major) technical modifications. It is not hard to see that if Assumption 2.3.5 holds, then A_{or} is *almost everywhere positive definite* and *boundedly invertible*. Some comments on these energy considerations are presented in Section 2.3.2.

A large class of materials displays properties that may be modelled by a diagonal matrix A_{or}, i.e., $\xi(x) = \zeta(x) = 0$. To distinguish such materials we will employ the notation A_0 for the matrix A_{or} in such cases:

$$\mathsf{A}_0 = \mathsf{A}_0(x) = \begin{pmatrix} \varepsilon(x) & 0 \\ 0 & \mu(x) \end{pmatrix}. \tag{2.13}$$

Furthermore, it is common in the physics and engineering literature to scale the coefficients of the diagonal blocks of the matrix G_{d} in terms of ε and μ; this is done to emphasise the smallness of the dispersion terms as compared to the optical response terms. In such cases, instead of G_{d} we use the matrix

$$\mathsf{G}_0\,(t-s,x) = \begin{pmatrix} \varepsilon(x)\,\chi^{\mathrm{e}}(t-s) & \chi^{\mathrm{em}}(t-s) \\ \chi^{\mathrm{me}}(t-s) & \mu(x)\,\chi^{\mathrm{m}}(t-s) \end{pmatrix}. \tag{2.14}$$

This choice leads to constitutive relations of the general form

$$\mathsf{d}(t,x) = (\mathcal{L}u)(t,x) = \mathsf{A}_0 u(t,x) + \int_0^t \mathsf{G}_0\,(t-s,x)u(s,x)ds. \qquad (2.15)$$

It is important to state the physical meaning of the various terms appearing in equations (2.13) and (2.14). Here and in what follows, ε, μ are the *permittivity* and *permeability* of the medium, respectively. Depending on the setting they can either be constant (e.g., representing the permittivity and permeability of the vacuum, or the relative permittivity and permeability of a homogeneous medium) or functions of the position vector, in the case of spatially inhomogeneous media. Four susceptibility kernels appear in these equations: the dyadics[6] (3×3 matrices) $\chi^{\mathrm{e}}(t)$, $\chi^{\mathrm{m}}(t)$, $\chi^{\mathrm{em}}(t)$ and $\chi^{\mathrm{me}}(t)$ are respectively called the *dielectric susceptibility kernel*, the *magnetic susceptibility kernel* and the *magnetoelectric kernels.*

REMARK 2.3.8 Regarding notation, we will use (2.10) when dealing with general linear abstract complex media (e.g., in Chapters 9 or 14), whereas when treating special classes of complex media, e.g., chiral media, we will use (2.15), which is compatible with the widely used form adopted by the electromagnetic community.

2.3.2 Dissipative media

We now turn our attention to some general energy considerations for complex electromagnetic media. The discussion here is along similar lines to that in [226], [227]. The power of the electromagnetic field E, H in a domain $\mathcal{O}$ is

$$\int_{\mathcal{O}} (H \cdot \partial_t B + E \cdot \partial_t D)\; dx.$$

The integrand defines the power density of the material, and the total stored energy per volume is, therefore,

$$w(x) = \int_{-\infty}^t (H(s,x) \cdot \partial_s B(s,x) + E(s,x) \cdot \partial_s D(s,x))\; ds.$$

Recall that the *Poynting vector*, $S := E \times H$, models the power flux density. By using the identity $\mathrm{div}S = H \cdot \mathrm{curl}E - E \cdot \mathrm{curl}H$ and the Maxwell system, we obtain - in six-vector notation - the *Poynting Theorem*[7]

$$\mathrm{div}S + u \cdot \partial_t \mathsf{d} = u \cdot j\,.$$

The term $u \cdot \partial_t \mathsf{d}$ is the time rate of change of stored electric and magnetic energy density in the medium, whereas the term $u \cdot j$ is the power supplied by the current j. Note that the term $u \cdot \partial_t \mathsf{d}$ can be written as

$$u \cdot \partial_t \mathsf{d} = w_{\mathrm{opt}} + w_{\mathrm{disp}},$$

[6] A concise introduction to dyadic analysis is included in Appendix D.

[7] This theorem was independently discovered by J. H. Poynting, O. Heaviside and N. A. Umov.

where

$$w_{\rm opt} := u \cdot \mathsf{A}_{\rm or}\, \partial_t u, \quad w_{\rm disp} := u \cdot \partial_t(\mathsf{G}_{\rm d} \star u)$$

are the optical response part and the dispersive part, respectively.

Consider now a source-free medium, i.e., $j = 0$. The total stored energy per volume at a point x is

$$w(t) = w_{\rm opt}(t) + w_{\rm disp}(t).$$

In accordance with [156], [226] and [227], we introduce the following:

DEFINITION 2.3.9 (DISSIPATIVE (PASSIVE) MEDIA)

(*i*) *The medium is dissipative at a fixed point* $x \in \mathcal{O}$ *if and only if* $w(t) = w_{\rm opt}(t) + w_{\rm disp}(t) \geq 0$ *for every temporally localised sufficiently smooth six-vector* $u(t)$.

(*ii*) *The medium is dissipative in* $\mathcal{O}$ *if and only if it is dissipative for all* $x \in \mathcal{O}$.

REMARK 2.3.10 The above definition reflects that no net production of electromagnetic energy is possible in $\mathcal{O}$, and therefore the medium is *passive* ([156], [226], [227]). This is consistent with thermodynamic considerations (see [145]).

Whether a material is dissipative or not depends on the properties of $\mathsf{A}_{\rm or}$ and $\mathsf{G}_{\rm d}$. For instance, dissipativity is guaranteed in the optical response region if $\mathsf{A}_{\rm or}(x)$ is symmetric and positive semidefinite almost everywhere in $\mathcal{O}$ (see [156], [226], [227]). This may serve as a justification on physical grounds for Assumption 2.3.5. As for dissipation due to the dispersive part, a necessary condition, [227], is that $\mathsf{G}_{\rm d}(0, x)$ is a non-negative definite matrix; this serves as a justification on physical grounds for Assumption 2.3.6. Further, if $\mathsf{G}_{\rm d}(0, x) = 0$, it turns out ([227], [260]) that the function

$$\left(\partial_t \mathsf{G}_{\rm d}(t, x) + \partial_t \mathsf{G}_{\rm d}(-t, x)^{tr}\right) \mathsf{y} \cdot \mathsf{y}$$

is of *positive type*[8] with respect to t, for almost all $x \in \mathcal{O}$ and for all nonzero $\mathsf{y} \in \mathbb{R}^3 \times \mathbb{R}^3$. Let us note that the so-called *Lorentz materials* satisfy $\mathsf{G}_{\rm d}(0, x) = 0$, while the so-called *Debye materials* do not. Recall (see, e.g., [381]) that the susceptibility kernel for Debye materials is of the form $\alpha_{\rm D}\, \mathsf{h}(t) e^{-t/\tau}$, where τ is the relaxation time, while for Lorentz materials it is of the form $\alpha_{\rm L} \mathsf{h}(t) \sin(\nu_0\, t) e^{-\nu t/2}$, where ν is the damping amplitude, $\alpha_{\rm D}$, $\alpha_{\rm L}$, ν_0 are physical parameters, and h is the Heaviside step function.

[8] A complex-valued function $\vartheta \in C(\mathbb{R}; \mathbb{C})$ is of *positive type* if

$$\int_{\mathbb{R}} \left(\int_{\mathbb{R}} \vartheta(t-s)\, \overline{\theta(s)}\, ds \right) \theta(t)\, dt \geq 0, \;\; \forall\, \theta \in C(\mathbb{R}; \mathbb{C}).$$

Functions of positive type are characterised by a classical theorem due to Bochner, according to which they are exactly the functions of the form $\vartheta(t) = \int_{\mathbb{R}} e^{ikt} d\mu(k)$, where μ is a finite positive Borel measure.

REMARK 2.3.11 The integrated total stored energy in $\mathcal{O}$, i.e.,

$$\mathfrak{J}(t) := \int_0^t \int_{\mathcal{O}} w(t,x)\,dx\,dt = \int_0^t \int_{\mathcal{O}} u(t,x) \cdot \partial_t \mathsf{d}(t,x)\,dx\,dt,$$

is what usually measures the variation of electromagnetic energy for $t \geq 0$.

In view of the Poynting theorem and the Gauss divergence theorem we obtain[9]

$$\mathfrak{J}(t) = \int_0^t \int_{\mathcal{O}} u \cdot j - \int_0^t \int_{\partial\mathcal{O}} u_2 \cdot (n \times u_1)\ ,$$

a fact that shows the importance of the boundary conditions to the energy balance of parts of the medium.

2.3.3 Constitutive relations in the frequency domain

A very important special case is when the time dependence of the fields is harmonic. As is well known, this assumption leads to stationary problems. Assume that all time-dependent quantities are Fourier transformable,

$$\mathfrak{s}(t,x) = \frac{1}{2\pi} \int_{-\infty}^{+\infty} e^{-i\varpi t} \widetilde{\mathfrak{s}}(x,\varpi)\,d\varpi,$$

where $\mathfrak{s}$ is a proxy for the vector fields d, u, j and by ϖ we denote the angular frequency[10]. The integral is understood in the Bochner sense (see Appendix A, Section A.3.2). Using the properties of the convolution with respect to the Fourier transform, the Fourier transform of the constitutive relation becomes

$$\widetilde{\mathsf{d}} = \mathsf{A}_{\mathrm{or}} \widetilde{u} + \widetilde{\mathsf{G}_{\mathrm{d}}} \widetilde{u}\,, \tag{2.16}$$

where "$\widetilde{\mathfrak{s}}$" is used to denote the Fourier transform of $\mathfrak{s}$. Therefore, in the frequency domain, the constitutive relations are simplified by the transformation of the convolution operator to a multiplication operator.

We define

$$\widetilde{\mathsf{A}_{\mathrm{or}}} := \mathsf{A}_{\mathrm{or}} + \widetilde{\mathsf{G}_{\mathrm{d}}} = \begin{pmatrix} \varepsilon + \widetilde{\varepsilon_d} & \xi + \widetilde{\xi_d} \\ \zeta + \widetilde{\zeta_d} & \mu + \widetilde{\mu_d} \end{pmatrix} =: \begin{pmatrix} \varepsilon_{\mathfrak{F}} & \xi_{\mathfrak{F}} \\ \zeta_{\mathfrak{F}} & \mu_{\mathfrak{F}} \end{pmatrix}\,, \tag{2.17}$$

so the frequency domain constitutive relations are written as

$$\widetilde{\mathsf{d}} = \widetilde{\mathsf{A}_{\mathrm{or}}} \widetilde{u}\,, \tag{2.18}$$

where now in general $\widetilde{\mathsf{A}_{\mathrm{or}}} = \widetilde{\mathsf{A}_{\mathrm{or}}}(x;\varpi)$. We impose the following assumption on $\widetilde{\mathsf{A}_{\mathrm{or}}}$ in the frequency domain:

ASSUMPTION 2.3.12 Let c, C be positive constants. For any fixed frequency ϖ, the matrix $\widetilde{\mathsf{A}_{\mathrm{or}}} = \widetilde{\mathsf{A}_{\mathrm{or}}}(x;\varpi)$ satisfies the following properties[11]:

---

[9]Recall that $u = (u_1, u_2)^{tr} = (E, H)^{tr}$.

[10]We use the symbol ϖ instead of the usual symbol ω for the angular frequency because the latter will be used to denote an element of the sample space Ω in subsequent chapters referring to random media and stochastic problems.

[11]$|\cdot|$ denotes the modulus of a complex number and $||\cdot||$ denotes the Euclidean norm in $\mathbb{C}^6$.

(*i*) $\widetilde{\mathsf{A}_{\mathrm{or}}} \in L^\infty(\mathcal{O}, \mathbb{C}^{6\times 6})$.

(*ii*) $\bar{\mathsf{z}}^{tr} \cdot (\mathcal{I}m\, \widetilde{\mathsf{A}_{\mathrm{or}}})\, \mathsf{z} \geq c\,||\mathsf{z}||^2$, for all $\mathsf{z} \in \mathbb{C}^6$.

(*iii*) $|\mathsf{z}_1 \cdot \widetilde{\mathsf{A}_{\mathrm{or}}}\, \mathsf{z}_2| \leq C\,||\mathsf{z}_1||\,||\mathsf{z}_2||$, for all $\mathsf{z}_1, \mathsf{z}_2 \in \mathbb{C}^6$.

We now elaborate further on the classification of biisotropic media in the frequency domain. Let[12]

$$\xi_{\mathfrak{F}} = \kappa + i\chi\,, \;\; \zeta_{\mathfrak{F}} = \kappa - i\chi\,. \tag{2.19}$$

The *chirality parameter* χ measures the degree of handedness of the material; a change in the sign of χ corresponds to the consideration of the mirror image of the material. The other parameter κ describes the magnetoelectric effect; materials with $\kappa \neq 0$ are nonreciprocal.

The following terminology in the time-harmonic case is used. A medium is called:

- ▷ ISOTROPIC, if $\kappa = 0$ and $\chi = 0$, i.e., when $\xi_{\mathfrak{F}} = \zeta_{\mathfrak{F}} = 0$.
- ▷ NONRECIPROCAL NONCHIRAL, or TELLEGEN, if $\kappa \neq 0$ and $\chi = 0$, i.e., when $\xi_{\mathfrak{F}} = \zeta_{\mathfrak{F}}$.
- ▷ RECIPROCAL CHIRAL, or PASTEUR, if $\kappa = 0$ and $\chi \neq 0$, i.e., when $\xi_{\mathfrak{F}} = -\zeta_{\mathfrak{F}}$.
- ▷ NONRECIPROCAL CHIRAL or GENERAL BIISOTROPIC, if $\kappa \neq 0$ and $\chi \neq 0$, i.e., when $\xi_{\mathfrak{F}} \neq \zeta_{\mathfrak{F}}, -\zeta_{\mathfrak{F}}$.

Reciprocal chiral media will be studied mainly in Chapters 4, 5 and 6, and they will be simply referred to as *chiral* media there.

In the case of reciprocal chiral media the constitutive relations for time-harmonic fields were introduced by Chambers, and are usually written as

$$\widetilde{D} = \varepsilon_{\mathrm{T}}\widetilde{E} + \beta_{\mathrm{T}}\widetilde{H}, \quad \widetilde{B} = \mu_{\mathrm{T}}\widetilde{H} - \beta_{\mathrm{T}}\widetilde{E}, \tag{2.20}$$

(where $\varepsilon_{\mathrm{T}} := \varepsilon_{\mathfrak{F}}$, $\mu_{\mathrm{T}} := \mu_{\mathfrak{F}}$, $\beta_{\mathrm{T}} := \xi_{\mathfrak{F}} = -\zeta_{\mathfrak{F}}$). The chirality parameter β_{T} is an inverse speed ($\frac{m}{s}$).

Another set of constitutive relations for reciprocal chiral media is the *Drude-Born-Fedorov* (DBF) constitutive relations ([273], [268], [289]), introduced in 1959 by F. I. Fedorov as a modification of constitutive relations used in 1900 by P. K. L. Drude and in 1915 by M. Born. These read

$$\widetilde{D} = \varepsilon_{\mathrm{DBF}}(\widetilde{E} + \beta_{\mathrm{DBF}}\,\mathrm{curl}\widetilde{E}), \quad \widetilde{B} = \mu_{\mathrm{DBF}}(\widetilde{H} + \beta_{\mathrm{DBF}}\,\mathrm{curl}\widetilde{H}). \tag{2.21}$$

Here the medium is characterised by three (in general, complex) parameters, namely, the electric permittivity $\varepsilon_{\mathrm{DBF}}$, the magnetic permeability μ_{DBF}, and

[12] Care should be taken to avoid confusion regarding the use of the symbols κ, χ and $\xi_{\mathfrak{F}}$, $\zeta_{\mathfrak{F}}$ for the corresponding media parameters. In [289] the symbols κ, χ have the reverse meaning of the one we employ here; the authors of [289] adopted the notation employed in (2.19) in their publications some years after their book appeared. The current notation is more natural (since the term *chirality* comes from the word "hand" in Greek, and this word begins with the letter χ), and is nowadays used by many authors.

the chirality measure β_{DBF}, which carries the unit of length (m). For source-free regions the Tellegen constitutive parameters ε_{T}, μ_{T} and β_{T} are connected to the Drude-Born-Fedorov constitutive parameters $\varepsilon_{\mathrm{DBF}}$, μ_{DBF} and β_{DBF} via the relations

$$\begin{aligned}\varepsilon_{\mathrm{T}} &= \frac{\varepsilon_{\mathrm{DBF}}}{1-\varpi^2\varepsilon_{\mathrm{DBF}}\mu_{\mathrm{DBF}}\beta_{\mathrm{DBF}}^2}, \quad \mu_{\mathrm{T}} = \frac{\mu_{\mathrm{DBF}}}{1-\varpi^2\varepsilon_{\mathrm{DBF}}\mu_{\mathrm{DBF}}\beta_{\mathrm{DBF}}^2}, \\ \beta_{\mathrm{T}} &= i\varpi\varepsilon_{\mathrm{DBF}}\mu_{\mathrm{DBF}}\frac{\beta_{\mathrm{DBF}}}{1-\varpi^2\varepsilon_{\mathrm{DBF}}\mu_{\mathrm{DBF}}\beta_{\mathrm{DBF}}^2}.\end{aligned} \tag{2.22}$$

In order to consider the case where the electromagnetic field in a chiral region is irradiated by sources, both the electric and the magnetic sources in the Tellegen (resp. Drude-Born-Fedorov) description must be a combination of an electric and a magnetic source in the Drude-Born-Fedorov (resp. Tellegen) description; see [268].

In what follows we will drop the subscript "DBF" from the parameters of the DBF constitutive relations.

Remark 2.3.13 Apart from these constitutive relations, other sets have been introduced for time-harmonic fields, namely, the ones by Condon and those by Boys-Post. An interesting discussion concerning these relations can be found in the monograph by Lakhtakia [268]. The Tellegen constitutive relations are algebraically isomorphic to these other relations, and as such they can be considered representative of them. All these sets differ from the Drude-Born-Fedorov ones, (2.21), which contain differentials (the curl operator).

In this book, in the time-harmonic regime, we adopt the Drude-Born-Fedorov constitutive relations.

2.3.4 Constitutive relations for random media

In many applications (see, e.g., [209], [210], [322]), there is uncertainty concerning either the externally imposed sources or the nature of the medium under consideration. In such cases, it is useful to model the uncertain quantities as random variables, with a prescribed probability distribution. These random variables are now functions of the spatial variables and time, and can be considered random fields defined on a probability space $(\Omega, \mathscr{F}, P)$. The randomness is assumed to have as an effect that the repetition of different experiments on the medium will generate different outcomes ω, either of the medium parameters or the external sources. The set Ω contains the outcomes ω of all the possible experiments or all possible realisations of the medium, $\mathscr{F}$ is a σ-algebra on Ω and P is a probability measure on $\mathscr{F}$, quantifying the relative frequency of realisations of different outcomes in an (ideally infinite) repetition of experiments under identical experimental conditions. In turn, the randomness in the sources or the medium, is reflected in the resulting electromagnetic fields, which have to be modelled as random fields as well. It is of great interest to be able to predict the statistical properties of these fields, e.g., the distribution of realised values, statistical moments, etc., from knowledge of the relevant statistical properties of the data of the problem.

The constitutive relations for random media are of the form

$$\mathsf{d}_\omega = \mathsf{A}_{\mathrm{or},\omega} u_\omega + \mathsf{G}_\omega \star u_\omega, \tag{2.23}$$

where now the quantities d_ω, u_ω are considered to be random fields, depending on x and t, with the explicit dependence suppressed for simplicity. We include the explicit dependence on ω[13] to remind us that the values of these quantities depend on the particular realisation ω of the experiment performed (or the particular realisation of the random medium). Furthermore, $\mathsf{A}_{\mathrm{or},\omega}$, G_ω are (in general) random matrices whose elements consist of random fields, which model the random parameters of the medium. The medium may be spatially homogeneous or not, depending on the circumstances. Similarly, we consider the external source J a random field $J = J(t, x, \omega)$.

The description of random media in terms of random fields finds a number of interesting applications in the theory of composites, wave propagation or in scattering problems from rough surfaces (see, e.g., [49], [147], [149], [322]).

2.3.5 Nonlinear media

It is well known that linearity is simply an approximation for the properties of electromagnetic media since experimental evidence implies that the constitutive relations in many regimes of interest in physics and engineering can be nonlinear. A good example is the famous Kerr (cubic) nonlinearity in dielectric media which is widely used in nonlinear optics (see [315]) and has led to the prediction and subsequent experimental verification of soliton-type solutions, which find very important applications in telecommunications (optical fibres).

For a nonlinear complex medium, we must relax the second of the postulates introduced in Section 2.3.1, that of linearity. Therefore, the constitutive relation will be modelled as $\mathsf{d} = \mathcal{L}u$, where now $\mathcal{L}$ is allowed to be a *nonlinear operator*. Since we do not relax any other postulate apart from linearity, we assume that the nonlinear operator $\mathcal{L}$ has a representation in convolution form as in (2.10), with the important difference that now the entries of the matrices A_{or} and G_{d} depend on the components of the six-vector u. The physical implication of this is that now the "effect" (d) is no longer simply proportional to the "cause" (u).

We will assume this dependence to be on the magnitude of the electric and the magnetic field components of the six-vector u, so that $\mathfrak{s}(t, x, u) = \mathfrak{s}(t, x, |u_1|, |u_2|)$, where $\mathfrak{s}$ serves as a proxy for the components of the matrices A_{or} and G_{d}, and $|u_i|$ is the Euclidean norm of the three-vector u_i calculated pointwise in x and t, $(i = 1, 2)$. Furthermore, based on arguments related to the Taylor expansion of the general nonlinear constitutive relations, we

[13] The explicit dependence of a random field $\mathfrak{s}$ on the particular realisation of the experiment ω will be denoted either by $\mathfrak{s}_\omega(t)$ or by $\mathfrak{s}(t, x, \omega)$, which is a more common notation.

assume a decomposition of the form

$$\mathfrak{s}(t,x,u) = \mathfrak{s}_\ell(t,x) + \mathfrak{s}_{\mathrm{n}\ell}(t,x,u),$$

where $\mathfrak{s}_\ell$ (in general depending explicitly on t and x but *not* on the field u) is the linear approximation of $\mathfrak{s}$ and $\mathfrak{s}_{\mathrm{n}\ell}(t,x,u) = \mathfrak{s}_{\mathrm{n}\ell}(t,|u_1|,|u_2|)$ is the nonlinear part. The dependence of the nonlinear part $\mathfrak{s}_{\mathrm{n}\ell}$ on the field amplitudes $|u_i|$, $i = 1,2$, may be approximated by a polynomial. In the special case where this dependence is approximated by a quadratic polynomial, the above constitutive relations lead to generalisations of the Kerr nonlinearity. Other nonlinearities compatible with the above constitutive relations are the *saturable* nonlinearities ([243]), where $\mathfrak{s}_{\mathrm{n}\ell}$ is the ratio of polynomials of $|u_i|$, $i = 1,2$, chosen so that $\mathfrak{s}_{\mathrm{n}\ell} \to 0$ as $|u_i| \to \infty$.

In practice, the chirality effects and the nonlinear effects are considered to be weak, so that the dominant effect will be only one of the two. In line with this observation we will consider in this work *weakly dispersive - weakly nonlinear media*; this implies that since chirality is a weak effect, we may assume that the nonlinear chiral terms are negligible with respect to the other terms. Thus, we are led to simplifications of the general nonlinear constitutive relations. A nonlinear constitutive relation complying with the above assumptions is

$$\mathsf{d} = \mathcal{L}u = \mathsf{A}_0 u + \mathsf{G}_0 \star u + \mathsf{G}_{0,\mathrm{n}\ell} \star \mathsf{N}(u)u, \tag{2.24}$$

where A_0, G_0 are as in (2.13) and (2.14) and the nonlinear term is specified by

$$\mathsf{N}(u) := \begin{pmatrix} N_1|u_1|^q & 0 \\ 0 & N_2|u_2|^q \end{pmatrix}, \quad \mathsf{G}_{0,\mathrm{n}\ell}(t,x) := \begin{pmatrix} \chi^{\mathrm{e}}_{\mathrm{n}\ell}(t,x) & 0 \\ 0 & \chi^{\mathrm{m}}_{\mathrm{n}\ell}(t,x) \end{pmatrix}, \tag{2.25}$$

where $q \in \mathbb{N}$, $N_1, N_2 \in \mathbb{R}^{3\times 3}$ are matrices independent of the spatial and temporal variables, 0 stands for the zero 3×3 matrix, and $\mathsf{G}_{0,\mathrm{n}\ell}$ is the nonlinear convolution matrix kernel. Note that we do not break up the nonlinear part of the constitutive law explicitly into a local (nonlinear optical response) part and a nonlocal (dispersive) part[14]; this is obtained automatically if we assume singularities of $\mathsf{G}_{0,\mathrm{n}\ell}$ with respect to t.

REMARK 2.3.14 The "chiral constitutive relation" (2.24) is motivated by the Kerr-Debye model in dielectrics (described in [87]), which is often used in nonlinear optics and represents the *relaxation approximation* to the Kerr nonlinearity. For a dielectric, this model is stated in terms of the polarisation $P = \varepsilon_0 \vartheta E$, where

$$\partial_t \vartheta + \frac{1}{\tau}\vartheta = \frac{1}{\tau}\varepsilon_r |E|^2$$

and τ is a relaxation parameter that corresponds to the finite time response of the medium. Integrating this differential equation leads to a nonlinear convolutive constitutive relation $D = D(E)$ with an exponential kernel.

[14]As we have done for the linear part, to be consistent with the standard notation adopted in the vast literature on linear problems.

REMARK 2.3.15 Generalisations of the above nonlinear constitutive relations are possible, e.g., we may consider A_{or}, G_{d} instead of A_0, G_0 (with similar generalisations for the nonlinear terms).

2.4 THE MAXWELL EQUATIONS IN COMPLEX MEDIA: A VARIETY OF PROBLEMS

In this section we present various problems related to the Maxwell equations in chiral media that will be treated in this work.

2.4.1 Time-harmonic interior domain problems ↣ Ch. 4

The interior problem in the case of time harmonic fields of a given angular frequency ϖ is important in various applications. After taking the Fourier transform, the Maxwell equations become

$$\begin{aligned} \text{curl}\widetilde{H}(x) &= i\varpi\widetilde{D}(x) + \widetilde{J}(x), \\ \text{curl}\widetilde{E}(x) &= -i\varpi\widetilde{B}(x), \end{aligned} \tag{2.26}$$

the Gauss laws yield

$$\begin{aligned} \text{div}\widetilde{D}(x) &= \widetilde{\rho}(x), \\ \text{div}\widetilde{B}(x) &= 0, \end{aligned} \tag{2.27}$$

for $x \in \mathcal{O}$, and the boundary condition becomes

$$n(x) \times \widetilde{E}(x) = \widetilde{f}(x), \quad x \in \partial\mathcal{O}\,, \tag{2.28}$$

where "$\widetilde{\mathfrak{s}}$" denotes the Fourier transform of $\mathfrak{s}$. In the case of perfect conductor boundary condition, $\widetilde{f} = 0$.

As for the time-dependent case, using the constitutive relations we may obtain a system involving only the fields $\widetilde{E}$, $\widetilde{H}$. In terms of the six-vector notation and the Maxwell equations in the frequency domain become

$$i\varpi(\mathsf{A}_{\text{or}} + \widetilde{\mathsf{G}_{\text{d}}})\widetilde{u} = \mathsf{M}\widetilde{u} + \widetilde{j}\,, \tag{2.29}$$

where the boundary condition is included in the definition of the function space chosen as the state space of the system (see also Remark 2.2.1). This equation is a static equation.

In the frequency domain we may also employ the Drude-Born-Fedorov constitutive relations (2.21) in order to associate $\widetilde{D}$, $\widetilde{B}$ with $\widetilde{E}$, $\widetilde{H}$. Substituting these expressions in (2.26), we obtain[15]

$$\begin{aligned} \text{curl}\widetilde{H}(x) &= i\varpi\left(\varepsilon_{\mathfrak{F}}(\widetilde{E} + \beta\text{curl}\widetilde{E})\right) + \widetilde{J}(x), \\ \text{curl}\widetilde{E}(x) &= -i\varpi\left(\mu_{\mathfrak{F}}(\widetilde{H} + \beta\text{curl}\widetilde{H})\right), \end{aligned} \tag{2.30}$$

where we now omit the equations (2.27), which remain unaltered.

---

[15] After dropping the subscripts DBF.

We may rearrange equations (2.30), using standard algebraic manipulations, into the form

$$\begin{aligned}\operatorname{curl}\widetilde{E} &= \beta\gamma^2\widetilde{E} + i\varpi\mu_{\mathfrak{z}}\left(\frac{\gamma}{k}\right)^2\widetilde{H} + \widetilde{J},\\ \operatorname{curl}\widetilde{H} &= \beta\gamma^2\widetilde{H} - i\varpi\varepsilon_{\mathfrak{z}}\left(\frac{\gamma}{k}\right)^2\widetilde{E},\end{aligned} \tag{2.31}$$

in $\mathcal{O}$, where ϖ is the anqular frequency,

$$k^2 = \varpi^2\varepsilon_{\mathfrak{z}}\mu_{\mathfrak{z}} \quad , \quad \gamma^2 = k^2(1-\beta^2k^2)^{-1}, \tag{2.32}$$

and $\widetilde{J}$ is an external force. Note that k is just an abbreviation for $\varpi\sqrt{\varepsilon_{\mathfrak{z}}\mu_{\mathfrak{z}}}$ and *not* a wave number[16]. Additionally, these equations are complemented with the boundary condition (2.28).

REMARK 2.4.1 Note that, when written in six-vector notation, equation (2.31) coincides with (2.29). This motivates the relation (2.22) between the parameters of the medium in the DBF and the Telegen approximation, and in this way their equivalence.

Let us mention here that in the constant coefficients case (and assuming $\widetilde{J}=0$), one can easily eliminate either the electric or the magnetic field from (2.31) and obtain a *modified* vector Helmholtz equation for $\widetilde{U}$, namely,

$$\Delta\widetilde{U} + 2\beta\gamma^2\operatorname{curl}\widetilde{U} + \gamma^2\widetilde{U} = 0\,, \tag{2.33}$$

where $\widetilde{U}$ stands for either $\widetilde{H}$ or $\widetilde{E}$, respectively.
The appropriate boundary conditions are, respectively,

$$n\times(n\times\operatorname{curl}\widetilde{H}) = f_{\mathrm{H}} \text{ and } \quad n\cdot\widetilde{H} = g_{\mathrm{H}}, \text{ on } \partial\mathcal{O}, \tag{2.34}$$

and

$$n\times\widetilde{E} = f_{\mathrm{E}} \text{ and } \quad \operatorname{div}\widetilde{E} = g_{\mathrm{E}}, \text{ on } \partial\mathcal{O}\,. \tag{2.35}$$

REMARK 2.4.2 Another boundary condition often appearing in the study of electromagnetics in the frequency domain is the *impedance* or *Leontovich* condition,

$$\widetilde{E}\times n + i\,\varrho\,Z\,n\times(\widetilde{H}\times n) = q,$$

where ϱ is a constant of the order $\sigma^{-1/2}$ (σ being the conductivity), Z is the impedance and q is a given tangential field on $\partial\mathcal{O}$. We do not use this condition in the present work.

2.4.2 The resonating chiral cavity problem $\rightarrowtail$ Ch. 4

Suppose that in problem (2.29) we set $\widetilde{j}=0$, $\widetilde{f}=0$. Then the resulting problem is a homogeneous system that will have nontrivial solutions $\widetilde{u} = (\widetilde{E},\widetilde{H})^{tr}$ for specific values of the frequency ϖ (the other parameters of the problem are considered to be fixed). This is equivalent to an eigenvalue problem, that will reveal the resonant frequencies of $\mathcal{O}$, called the *cavity problem*.

---

[16]As we will later see, $\gamma_- = k\,(1-\beta k)^{-1}$ and $\gamma_+ = k\,(1+\beta k)^{-1}$ are the *two* wave numbers inherent in this problem.

2.4.3 Exterior domain problems $\rightarrowtail$ Ch. 4

An important class of problems is when the domain is the complement of a bounded domain $\mathcal{O}$, i.e., when we are working in $\mathcal{O}_e := \mathbb{R}^3 \setminus \mathcal{O}$. Such a set $\mathcal{O}_e$ is often called an *exterior domain.* The form of the Maxwell equations remains unchanged for such problems (i.e., (2.30) hold in $\mathcal{O}_e$). Now, in addition to the boundary condition imposed on $\partial\mathcal{O}$, we have to specify the behaviour of the fields at infinity. This "substitute" of a boundary condition is a so-called *radiation condition.* A typical choice of radiation conditions for time-harmonic fields in achiral electromagnetic media are the Silver-Müller radiation conditions, which read

$$\begin{aligned} &\lim_{|x|\to\infty} |x| \,(\sqrt{\mu_{\mathfrak{F},\mathrm{e}}}\widetilde{H} \times \widehat{x} - \sqrt{\varepsilon_{\mathfrak{F},\mathrm{e}}}\widetilde{E}) = 0, \\ &\lim_{|x|\to\infty} |x| \,(\sqrt{\varepsilon_{\mathfrak{F},\mathrm{e}}}\widetilde{E} \times \widehat{x} + \sqrt{\mu_{\mathfrak{F},\mathrm{e}}}\widetilde{H}) = 0\,, \end{aligned} \tag{2.36}$$

where $\varepsilon_{\mathfrak{F},\mathrm{e}}$, $\mu_{\mathfrak{F},\mathrm{e}}$ are the electric permittivity and magnetic permeability of the material (or the vacuum) in $\mathcal{O}_e$, respectively, and the convergence is assumed to be uniform over all directions $\widehat{x} = \frac{x}{|x|}$. It can be shown (see Section 5.7.4) that the Silver-Müller radiation conditions are adequate for the chiral case, too. On the boundary $\partial\mathcal{O}$ of $\mathcal{O}_e$ we shall impose, for instance, a boundary condition of the form (2.28).

Remark 2.4.3 In fact, it is well known that complementing the Maxwell equations in $\mathcal{O}_e$ with *either* of the Silver-Müller radiation conditions (2.36) is sufficient for obtaining uniqueness of the solution. Let us also note that the Cartesian components of any solution of the Maxwell equations satisfying the Silver-Müller radiation conditions (2.36) also satisfy the Sommerfeld radiation condition $\lim_{|x|\to\infty} (\widehat{x} \cdot \mathrm{grad} v(x) - ikv(x)) = 0$, uniformly over all directions $\widehat{x}$ (v is a solution of the scalar Helmholtz equation $\Delta v + k^2 v = 0$).

In the case that the coefficients are constant (and assuming $\widetilde{J} = 0$), one can again eliminate either the electric or the magnetic field and obtain (2.33) with boundary condition (2.34) when $\widetilde{U} = H$, or (2.35) when $\widetilde{U} = E$. The radiation conditions become

$$\lim_{|x|\to\infty} |x| \,(\mathrm{curl}\widetilde{U} \times n + n\, \mathrm{div}\widetilde{U} - ik\widetilde{U}) = 0\,.$$

2.4.4 Transmission problems $\rightarrowtail$ Chs. 5, 6

Consider now the case of a domain $\mathcal{O}$ filled with an electromagnetic medium (chiral or not). The medium is embedded in a different electromagnetic environment (again, chiral or not), i.e., $\mathcal{O}_e$ is considered to be filled with a different medium of infinite extent. Discontinuities of the tangential components of the electric and the magnetic field are allowed on the interface $\partial\mathcal{O}$ between $\mathcal{O}$ and $\mathcal{O}_e$. The problem of specifying the electromagnetic field in the whole of $\mathbb{R}^3$ is called the *transmission problem.*

Assume harmonic time dependence for the electromagnetic fields of angular frequency ϖ, that $\mathcal{O}$ is filled with a chiral medium with parameters $\varepsilon_{\mathfrak{z}}$, $\mu_{\mathfrak{z}}$, β and that $\mathcal{O}_e$ is filled with a medium (chiral or not) with parameters $\varepsilon_{\mathfrak{z},e}$, $\mu_{\mathfrak{z},e}$, β_e. Define

$$k^2 = \varpi^2 \varepsilon_{\mathfrak{z}} \mu_{\mathfrak{z}}, \qquad \gamma^2 = k^2(1-\beta^2 k^2)^{-1},$$
$$k_e^2 = \varpi^2 \varepsilon_{\mathfrak{z},e} \mu_{\mathfrak{z},e}, \qquad \gamma_e^2 = k_e^2(1-\beta_e^2 k_e^2)^{-1},$$

in the regions $\mathcal{O}$ and $\mathcal{O}_e$, respectively (see (2.32)).

The mathematical formulation of transmission problems consists in finding the electromagnetic field $(\widetilde{E}, \widetilde{H})$ in $\mathcal{O}$ and the electromagnetic field $(\widetilde{E}_e, \widetilde{H}_e)$ in $\mathcal{O}_e$ such that

$$\begin{aligned} \operatorname{curl} \widetilde{E} &= \beta\gamma^2 \widetilde{E} + i\varpi\mu_{\mathfrak{z}} \left(\tfrac{\gamma}{k}\right)^2 \widetilde{H}, \\ \operatorname{curl} \widetilde{H} &= \beta\gamma^2 \widetilde{H} - i\varpi\varepsilon_{\mathfrak{z}} \left(\tfrac{\gamma}{k}\right)^2 \widetilde{E}, \end{aligned} \qquad \text{in } \mathcal{O}, \tag{2.37}$$

and

$$\begin{aligned} \operatorname{curl} \widetilde{E}_e &= \beta_e\gamma_e^2 \widetilde{E}_e + i\varpi\mu_{\mathfrak{z},e} \left(\tfrac{\gamma_e}{k_e}\right)^2 \widetilde{H}_e, \\ \operatorname{curl} \widetilde{H}_e &= \beta_e\gamma_e^2 \widetilde{H}_e - i\varpi\varepsilon_{\mathfrak{z},e} \left(\tfrac{\gamma_e}{k_e}\right)^2 \widetilde{E}_e, \end{aligned} \qquad \text{in } \mathcal{O}_e. \tag{2.38}$$

The electromagnetic fields of the two different regions are related by the *transmission conditions* on the interface,

$$\begin{aligned} n \times \widetilde{E} - n \times \widetilde{E}_e &= \widetilde{f}_{\mathrm{E}}, \\ n \times \widetilde{H} - n \times \widetilde{H}_e &= \widetilde{f}_{\mathrm{H}}, \end{aligned} \qquad \text{on } \partial\mathcal{O}, \tag{2.39}$$

where n is the outward normal on $\partial\mathcal{O}$ and $\widetilde{f}_{\mathrm{E}}$, $\widetilde{f}_{\mathrm{H}}$ are given tangential fields. Further, one of the two Silver-Müller radiation conditions

$$\lim_{|x|\to\infty} |x| \left(\sqrt{\mu_{\mathfrak{z},e}}\, \widetilde{H}_e \times \widehat{x} - \sqrt{\varepsilon_{\mathfrak{z},e}}\, \widetilde{E}_e\right) = 0\,, \tag{2.40}$$

or

$$\lim_{|x|\to\infty} |x| \left(\sqrt{\varepsilon_{\mathfrak{z},e}}\, \widetilde{E}_e \times \widehat{x} + \sqrt{\mu_{\mathfrak{z},e}}\, \widetilde{H}_e\right) = 0\,, \tag{2.41}$$

is also assumed to hold, uniformly over all directions $\widehat{x}$.

REMARK 2.4.4 The above problem covers all four possible cases: the interior domain can be achiral or chiral, and similarly for its complement in three-dimensional space.

Clearly the corresponding electric and magnetic transmission problems can be also stated in terms of the modified Helmholtz equations.

2.4.5 Scattering Problems - Time-Harmonic Case ↣ Chs. 5, 6

A very important special case of exterior or transmission boundary value problems is that of *scattering* of electromagnetic waves. We consider the same setting (and adopt the same notation) as in Section 2.4.4. Assume now that the electromagnetic field, created by an *incident* field $(E^{\mathrm{inc}}, H^{\mathrm{inc}})$

which is applied in $\mathcal{O}_e$, interacts with the material in $\mathcal{O}$ (the *scatterer*). The incident field depends on the properties of the two different materials as well as on the geometry of $\mathcal{O}$, and generates a *total electromagnetic field* $(\widetilde{E}^t, \widetilde{H}^t)$ in $\mathcal{O}$ (and $\mathcal{O}_e$), part of which, called the *scattered field* $(\widetilde{E}^e, \widetilde{H}^e)$, is scattered in $\mathcal{O}_e$. In mathematical terms, the incident field $(E^{\text{inc}}, H^{\text{inc}})$ is a solution of the chiral Maxwell equations in the exterior unbounded domain $\mathcal{O}_e$ that does not satisfy the Silver-Müller radiation conditions. On the other hand, the *scattered* field $(\widetilde{E}^e, \widetilde{H}^e)$ satisfies the Silver-Müller radiation conditions.

The *total* electromagnetic field in $\mathcal{O}_e$,

$$\widetilde{E}^t = E^{\text{inc}} + \widetilde{E}^e, \quad \widetilde{H}^t = H^{\text{inc}} + \widetilde{H}^e,$$

satisfies the boundary condition under consideration (e.g., the perfect conductor boundary condition)[17] on the boundary $\partial\mathcal{O}$ of the domain $\mathcal{O}$ (the *scatterer*).

The perfect conductor scattering problem then reads: find the total exterior electromagnetic field $(\widetilde{E}^t, \widetilde{H}^t)$ so that

(*i*) $\widetilde{E}^t, \widetilde{H}^t$ satisfy the chiral Maxwell equations (2.38) in $\mathcal{O}_e$.

(*ii*) $\widetilde{E}$ satisfies the perfect conductor boundary condition $(n \times \widetilde{E}^t = 0)$ on $\partial\mathcal{O}$.

(*iii*) The scattered field $(\widetilde{E}^e, \widetilde{H}^e)$ satisfies either (2.40) or (2.41).

REMARK 2.4.5 Since the incident field $(E^{\text{inc}}, H^{\text{inc}})$ is a solution of the chiral Maxwell equations in $\mathcal{O}_e$, the above scattering problem can be restated in terms of the scattered field $(\widetilde{E}^e, \widetilde{H}^e)$. Indeed, $(\widetilde{E}^e, \widetilde{H}^e)$ must satisfy the chiral Maxwell equations in $\mathcal{O}_e$, one of the two above Silver-Müller radiation conditions, while the perfect conductor boundary condition becomes

$$\widetilde{E}^e \times n = -E^{\text{inc}} \times n\,. \tag{2.42}$$

Similarly, the scattering of a given incident electromagnetic wave $(E^{\text{inc}}, H^{\text{inc}})$ propagating in an environment $\mathcal{O}_e$ filled with a homogeneous chiral (or achiral) material of parameters $\varepsilon_{\mathfrak{F},e}, \mu_{\mathfrak{F},e}, \beta_e$ by a penetrable obstacle $\mathcal{O}$ filled with a different homogeneous chiral (or achiral) material of parameters $\varepsilon_{\mathfrak{F}}, \mu_{\mathfrak{F}}, \beta$ is described by the following transmission boundary value problem, which covers all four possible cases regarding the nature of the homogeneous materials which fill the obstacle and its surrounding space in terms of them being chiral or not: Find fields $\widetilde{E}, \widetilde{H}$ satisfying (2.37) in $\mathcal{O}$, and $\widetilde{E}^e, \widetilde{H}^e$ satisfying (2.38) in $\mathcal{O}_e$ the transmission conditions

$$\begin{aligned} \widetilde{E}^e \times n - \widetilde{E} \times n &= -E^{\text{inc}} \times n, \\ \widetilde{H}^e \times n - \widetilde{H} \times n &= -H^{\text{inc}} \times n, \end{aligned} \quad \text{on } \partial\mathcal{O}, \tag{2.43}$$

and one of the Silver-Müller radiation conditions (2.40) - (2.41).

[17] More general boundary conditions can be considered as long as they are *non-dissipative*; see [334].

2.4.6 Interior domain problems in the time domain $\rightarrowtail$ Ch. 7

An important class of problems comprises interior domain problems. In such problems we assume that $\mathcal{O}$ is a bounded domain, filled with a complex material. The aim is to describe the spatiotemporal evolution of the electromagnetic fields in the time domain. This amounts to solving the Maxwell equations in $\mathcal{O}$, supplemented with appropriate boundary conditions. A typical example of a boundary condition is the perfect conductor boundary condition (2.4).

So, the interior problem takes the form

$$\begin{aligned} \text{curl} H(t,x) &= \partial_t D(t,x) + J(t,x), \\ \text{curl} E(t,x) &= -\partial_t B(t,x), \end{aligned} \tag{2.44}$$

along with the Gauss laws

$$\begin{aligned} \text{div} D(t,x) &= \rho(t,x), \\ \text{div} B(t,x) &= 0, \end{aligned} \tag{2.45}$$

for $x \in \mathcal{O}$, where J is the electric current density and ρ is the density of the externally impressed electric charge. These equations are to be supplemented with the perfect conductor boundary condition

$$n \times E(t,x) = 0\,, \quad x \in \partial\mathcal{O}, \; t \in [0,T]\,,$$

where by n we denote the exterior normal to the boundary of $\mathcal{O}$. A more general boundary condition is the "nonhomogeneous" version of the perfect conductor boundary condition, i.e.,

$$n \times E(t,x) = f\,, \quad x \in \partial\mathcal{O}, \; t \in [0,T]\,,$$

where f is a given tangential field; this problem may be transformed into a problem with homogeneous boundary conditions with a modified source term (see Remark 2.2.1). The fields D, B are given in terms of the fields E, H by the constitutive relations (2.10) (or (2.15)). Inserting the constitutive relations in the Maxwell equations provides a system of PDEs determining the evolution of the fields E, H.

We now present this equation in the form of an abstract differential equation. Taking into account the comments in the footnote on page 12, the constitutive relations, e.g., (2.10) and the definition of the Maxwell operator M (see (2.8)) the Maxwell equations (2.44) assume the form

$$(\mathsf{A}_{\text{or}} u + \mathsf{G}_{\text{d}} \star u)' = \mathsf{M} u + j, \tag{2.46}$$

where the prime denotes differentiation with respect to time. The mathematical model for such problems is in the form of a Volterra-type integrodifferential boundary value problem. This equation can be further simplified if we assume that $\mathsf{G}_{\text{d}}(t,x)$ is weakly differentiable with respect to the temporal variable. Then we may differentiate the convolution integral, and by multiplying to the right by $\mathsf{A}_{\text{or}}^{-1}$ the equation can be written as

$$u' = \mathsf{M}_{\text{A}} u + \mathsf{G}_{\text{A}} \star u + \mathsf{J}_{\text{A}}, \tag{2.47}$$

where

$$\mathsf{G}_{\mathrm{A}} = -\mathsf{A}_{\mathrm{or}}^{-1}\mathsf{G}_{\mathrm{d}}', \ \mathsf{M}_{\mathrm{A}} = \mathsf{A}_{\mathrm{or}}^{-1}\mathsf{M}, \ \mathsf{J}_{\mathrm{A}} = \mathsf{A}_{\mathrm{or}}^{-1}\, j,$$

and we have assumed that $\mathsf{G}_{\mathrm{d}}(0, x) = 0$. Otherwise, there is an extra term[18] proportional to u, which may be incorporated in the definition of the operator M_{A}, leading to minor modifications to the mathematical treatment. The boundary conditions, as well as the divergence free character of the electromagnetic field, can be included in the definition of the operator M in appropriately selected function spaces.

The well posedness of such problems will be studied in Chapter 7.

2.4.7 Controllability and optimal control problems $\rightarrowtail$ Ch. 8

Consider a complex medium, confined within the domain $\mathcal{O} \subset \mathbb{R}^3$, considered to be sufficiently smooth. Assuming sufficient temporal smoothness of the convolution kernels G_{d}, we rewrite the Maxwell equations for the electromagnetic fields in the medium in the compact form (see Section 2.4.6, equation (2.47)):

$$u' = \mathsf{M}_{\mathrm{A}} u + \mathsf{G}_{\mathrm{A}} \star u + \mathsf{J}_{\mathrm{A}}. \tag{2.48}$$

We now assume that we have access to an *internal control* v, that acts on the system. This internal control is an externally determined function $v(\cdot)$, which may be chosen at will so as to impose on the system a specific type of behaviour. The terminology internal control is chosen to clarify that this process may act on the whole of the domain $\mathcal{O}$. Other choices for the control function, e.g., controls that are only allowed to act on $\partial\mathcal{O}$ (boundary controls), or internal controls that act on a subset $\mathcal{O}_c$ of the domain $\mathcal{O}$, can also be considered.

The effect that the control $v(\cdot)$ has on the system is modelled through an operator $\mathcal{B}$, which is usually called the *control to state operator*. In the general theory of control, it is usually assumed that the control function v "lives" in a space $\mathbb{V}$ (the *control space*), which is, in general, different from the space $\mathbb{H}$ in which the solution of the system "lives" (the *state space*). Therefore, the operator $\mathcal{B} : \mathbb{V} \to \mathbb{H}$ transfers the action of the control v from the control space to the state space. For simplicity of presentation, we identify the control space $\mathbb{V}$ with an appropriate subset of the state space $\mathbb{H}$. This choice is feasible within the context of internal control and allows us to set ideas involving as little technical complications as possible. The action of the control v on the state of the system is thus modelled by the evolution equation

$$u' = \mathsf{M}_{\mathrm{A}} u + \mathsf{J}_{\mathrm{A}} + \mathsf{G}_{\mathrm{A}} \star u + \mathcal{B}v. \tag{2.49}$$

The problem of controllability can now be stated as follows: *Given $T > 0$, an initial condition $u(0) = U_0$ and a final condition $u(T) = U_T$, can we find a control procedure $v^*(\cdot)$ such that the solution of the system (2.49) with $v(\cdot) = v^*(\cdot)$ satisfies $u(0) = U_0$ and $u(T) = U_T$?*

---

[18]Similar to the one that would arise if a linear Ohm's law were adopted for j.

2.4.8 Homogenisation problems in periodic media $\rightarrowtail$ Ch. 9

In practice, complex media may be spatially inhomogeneous, so that the constitutive relation is of the form

$$\mathsf{d} = \mathsf{A}_{\text{or}} u + \mathsf{G}_{\text{d}} * u,$$

with $\mathsf{A}_{\text{or}} = \mathsf{A}_{\text{or}}(x)$ and $\mathsf{G}_{\text{d}} = \mathsf{G}_{\text{d}}(t,x)$. In many applications of interest, e.g., crystals, periodic gratings, etc., the spatial dependence of the medium parameters can be periodic, i.e., there exists a vector ℓ (the period), such that $\mathfrak{s}(x+\ell) = \mathfrak{s}(x)$, where $\mathfrak{s}$ is a proxy for the functions A_{or} and G_{d}.

The spatial inhomogeneity of the medium may introduce interesting phenomena, e.g., quenching of waves ([437]). One interesting class of problems is when there is a periodic microstructure in the medium. Such situations may be modelled by assuming that the function $\mathfrak{s}$ has a spatial dependence of the form $\mathfrak{s}(\frac{x}{\epsilon})$, where ϵ is a small parameter. The small parameter ϵ is included to stress the fact that the coefficients of the medium present oscillations on a small spatial scale. This is a reasonable model for the medium microstructure.

For such media, the evolution of the fields is given by the Maxwell equations

$$(\mathsf{d}^\epsilon)' = \mathsf{M} u^\epsilon + j^\epsilon \tag{2.50}$$

with constitutive relations of the form

$$\mathsf{d}^\epsilon = \mathsf{A}^\epsilon_{\text{or}}(x) u^\epsilon + \mathsf{G}^\epsilon_{\text{d}}(x) \star u^\epsilon\,, \tag{2.51}$$

where $u^\epsilon = (E^\epsilon, H^\epsilon)^{tr}$, $\mathsf{d}^\epsilon = (D^\epsilon, B^\epsilon)^{tr}$ and $\mathsf{A}^\epsilon_{\text{or}}(x) = \mathsf{A}_{\text{or}}\left(\frac{x}{\epsilon}\right)$, $\mathsf{G}^\epsilon_{\text{d}}(\cdot,x) = \mathsf{G}_{\text{d}}\left(\cdot,\frac{x}{\epsilon}\right)$. The Maxwell equations are further supplemented with a suitable boundary condition, e.g., the perfect conductor boundary condition. In the above equations the notation $\mathfrak{s}^\epsilon = \mathfrak{s}^\epsilon(t,x)$, where $\mathfrak{s}$ is a proxy for d, u is used to show that for each choice of the parameter ϵ (which specifies the length scale of spatial periodicity of the medium), we obtain a particular field configuration.

An important question is what happens to this family of electromagnetic fields as $\epsilon \to 0$, that is, in the limit of very fast oscillations in space of the medium coefficients. Mathematically this amounts to considering the limit as $\epsilon \to 0$ of the sequence of functions $\{u^\epsilon\} = \{u^\epsilon(t,x)\}$, which are determined as the solutions of the set of differential equations (2.50) with the constitutive relations (2.51). The first issue to be studied is the existence of such a limit and in particular in which topology do we expect the sequence $\{u^\epsilon\}$ to have a limit. It turns out that this problem is not very well behaved in the sense that in most cases of interest, the limit exists in a very weak sense, namely, in the weak star topology[19]. The second issue to be studied is whether

[19] As a simple example, take the sequence of functions $\phi^\epsilon(x) = sin\left(\frac{x}{\epsilon}\right)$, which does not converge in the usual topology of $\mathbb{R}$ as $\epsilon \to 0$. However, the integral $\int_I \psi(x)\phi^\epsilon(x)dx$ converges as $\epsilon \to 0$, for any function ψ and any interval I. This convergence is identified as the weak star convergence for the particular example. For a thorough discussion of this topic, see, e.g., [97].

this limit can be thought of as satisfying a system of differential equations similar to the Maxwell equations. It can be shown that the limit satisfies the Maxwell equations with a constitutive relation similar to (2.51) but now with constant medium coefficients. This means that in the limit as $\epsilon \to 0$ we may replace the original spatially inhomogeneous medium by an "equivalent" spatially homogeneous medium, at least as far as the electromagnetic field properties are concerned. This approach is called homogenisation. It is very useful in a number of applications since it provides methods that allow us to compute the constant coefficients of the homogenised medium. Further, it is also useful for numerical treatment of these equations since it is extremely difficult and sometimes impossible to numerically approximate PDEs with changing coefficients on a fast scale.

The problem of homogenisation is meaningful also in the time-harmonic case, where we no longer monitor the temporal evolution of the fields, which is assumed to be time periodic with period $\frac{2\pi}{\varpi}$, but the spatial properties of the fields as influenced by the microstructure of the medium.

2.4.9 Some aspects of scattering theory $\rightarrowtail$ Ch. 10

Broadly speaking, scattering theory can be thought of as the study of the interaction of an propagating wave with an inhomogeneous medium. In developing a scattering theory, a comparison is made between the solutions of a perturbed system and the solutions of an unperturbed system. The asymptotic behaviour of the two systems is investigated as the time tends to $\pm\infty$ with the aim of determining whether or not the perturbed system appears to behave like the unperturbed system in the distant past or in the distant future. In this framework, under certain conditions, a relatively simple scattering theory involving achiral materials (unperturbed system) can be modified to accommodate problems involving chiral materials (perturbed system); this is the subject of Chapter 10.

2.4.10 Nonlinear problems $\rightarrowtail$ Ch. 11

The spatiotemporal evolution of the fields is given by the Maxwell equations where the constitutive relations $d = \mathcal{L}u$ are now nonlinear and of the form introduced in Section 2.3.5.

Using the notation introduced earlier (see Section 2.3.5 and in particular equation (2.24)), we may write the Maxwell system in a more compact form as

$$(\mathsf{A}_0 u + \mathsf{G}_0 \star u + \mathsf{G}_{0,n\ell} \star \mathsf{N}(u)u)' = \mathsf{M}u + j\,,$$

with initial condition $u(0) = u_0$ and appropriate boundary conditions (e.g., the perfect conductor boundary conditions).

This modification will turn the Maxwell system into a system of nonlinear partial differential equations. The mathematical theory of nonlinear PDEs is considerably more involved than that of linear systems, and the solutions of the former present new and exciting phenomena, e.g., blow-up phenomena, special classes of propagating wave solutions, etc. While some works on such systems, using formal asymptotics, predict the existence of interesting soliton-like solutions, to the best of our knowledge there are no other rigorous mathematical results in this direction.

2.4.11 Random media problems $\rightarrowtail$ Ch. 12

The evolution of the electromagnetic fields in random media will be modelled by the Maxwell equations complemented by the random constitutive relations introduced in Section 2.3.4. In these equations we will also allow the external sources terms J to be subject to some randomness, treating them as random fields. This leads to the Maxwell equations, suppressing as usual the x-dependence, in the form

$$(\mathsf{A}_{\mathrm{or},\omega} u + \mathsf{G}_{\omega} \star u)' = \mathsf{M} u + j_{\omega}\,, \tag{2.52}$$

where now $j_{\omega} = \mathrm{j}(t, x, \omega)$ is a random process defined on a probability space $(\Omega, \mathscr{F}, P)$. An important aspect of the problem is the choice of a convenient model for the randomness. There are two major classes of problems, presenting qualitatively different behaviour, depending on the choice of the random forcing term j_{ω},

- ▷ j_{ω} is of finite variation with respect to time,
- ▷ j_{ω} is of infinite variation with respect to time.

Both cases are equally important in modelling. As the first case is intuitively easier to understand, we focus on the interpretation of the second. Assume that we wish to model experimental uncertainty resulting from the accumulation of a large number of independent errors, identically distributed and each having finite statistical moments up to the second. Then, using arguments based on the central limit theorem, we expect Gaussian properties for the random fields modelling these errors. Consider the external source $\mathrm{j}(t, x, \omega)$ for instance: we assume that this may be decomposed into two contributions, one being the average behaviour of the random source and the other being the purely random fluctuations around this average behaviour. By average behaviour we understand some sort of mathematical expectation of the source term $\mathrm{j}(t, x, \omega)$; this will be expressed shortly in a more rigorous mathematical fashion as $\mathbb{E}_P[\mathrm{j}(t, x, \omega)]$, where $\mathbb{E}_P[\cdot]$ is the expectation operator over the probability measure P. The random fluctuation terms are expected to average to 0, i.e., they have vanishing expectation. This reflects the fact that we may observe realisations of the source terms exceeding the average behaviour by z and realisations receding the average behaviour by $-z$ with the same probability.

We now assume that the fluctuations part of $\mathrm{j}(t, x, \omega)$ can be modelled as a white noise type of process with respect to the temporal variation, which presents possible spatial correlations. Formally, if $j_{f\ell}(t, x, \omega)$ denotes the part of $\mathrm{j}(t, x, \omega)$ corresponding to the fluctuations, our modelling assumption for the randomness is that

$$\mathbb{E}_P[j_{\mathrm{f}\ell}(t,x,\omega)\, j_{\mathrm{f}\ell}(t',x',\omega)] = \varrho(x,x')\,\delta(t-t'),$$

where $\varrho(x,x')$ is a kernel function representing the spatial correlations of the fluctuations and δ is the Dirac delta function. A stochastic process that may satisfy, formally, these assumptions is the temporal derivative of a stochastic process, the Wiener process (or Brownian motion) (see, e.g., [224]), or rather its infinite-dimensional version, $W(t,\omega)$, called the Q-Wiener process (see, e.g., [120] or Appendix B). This is a Banach space - valued stochastic process, with almost surely continuous paths, and independent increments that are distributed by the normal distribution with mean 0 and variance $(t-s)Q$, where Q is a trace class operator that gives the structure of the spatial correlations of the fluctuations and $t-s$ is the time increment. This stochastic process is a very reasonable model as it is based on the assumption of normality of random deviations from the "mean" deterministic fields, a fact that can be supported by statistical arguments based on the law of large numbers. For an introduction to random variables on Banach spaces and the definition of the Q-Wiener process the reader may consult Appendix B, and in particular Section B.4.

Therefore, we consider the random source to be of the form

$$\mathsf{j}(t,x,\omega) = j(t,x) + \mathsf{Q}\frac{d}{dt}W(t,\omega), \tag{2.53}$$

where by $j(t,x)$ we denote the "average" source and the second term models the fluctuations around this average. Q is the spatial autocorrelation operator for the source term $\mathsf{j}(t,x,\omega)$ and is related to Q, the spatial autocorrelation operator for the Wiener process. However, we must stress that the expression of the random source in equation (2.53) is purely formal! It is well known that the temporal derivative of the Wiener process can be defined nowhere for almost all t and almost surely with respect to the probability measure P. The temporal derivative can be defined at best in terms of the theory of distributions.

The Maxwell equations with random constitutive relations and using the random field $\mathsf{j}(t,x,\omega)$ for the source term are random evolution equations the solutions of which will be random fields. If we assume that $\mathsf{j}(t,x,\omega)$ is of bounded variation with respect to t, then (2.52) can be treated as a partial integrodifferential equation with random coefficients, the solution of which is defined pointwise in ω, and which may be considered a random process itself. In this case one may extend the techniques employed for the study of deterministic evolution equations so as to treat (2.52). If we adopt the model (2.53) for the random source term, the nondifferentiability of the Wiener process calls for a special treatment of the random evolution equation. We first rewrite the random Maxwell equations in integral form as

$$\mathsf{d}(t,x,\omega) - \mathsf{d}(0,x) = \int_0^t \mathsf{M}u(s,x,\omega)ds + \int_0^t j(s,x)ds + \int_0^t \mathsf{Q}dW(s). \tag{2.54}$$

If necessary, the above can be supplemented with Gauss's laws, assuming the charge density to be a random field as well. The rigorous mathematical treatment of the term $\int_0^t \mathsf{Q}dW(s)$ has to be done in terms of the theory of the Itō integral. This is an integral designed to handle the problem of integrating a stochastic process over the paths of a Wiener process. Since the paths of the Wiener process have almost surely infinite variation, a theory of integration based on the Riemann-Stieltjes integral is not sufficient to deal with the rigorous modelling of the cumulative effects of the randomness on the electromagnetic fields. Instead, one has to exploit the fact that the Wiener process has finite quadratic variation and construct an integral based on the quadratic variation of the integrator. This theory, which is an infinite dimensional extension of Itō's theory of stochastic integration, finds important applications in a number of modelling situations. For a brief introduction to this theory the reader can consult Appendix B and in particular Section B.5. With the interpretation of the stochastic integral as an Itō integral, the random Maxwell equations become a set of stochastic partial differential equations (SPDEs).

We rewrite now the system in more compact form. Under the standing assumption, as in the deterministic case, that $\mathsf{A}_{\mathrm{or},\omega}$ is an invertible matrix almost surely in P we express (2.54) as the stochastic integrodifferential equation

$$u(t) + (\mathcal{K} \star u)(t) = \mathsf{A}_{\mathrm{or},\omega}^{-1} u(0) + \int_0^t \mathsf{M}_{\mathrm{A}} u(s)\, ds + \int_0^t \mathsf{J}_{\mathrm{A}}(s)\, ds + \int_0^t \mathsf{Q} dW(s),$$

where

$$\mathcal{K} = \mathsf{A}_{\mathrm{or},\omega}^{-1} \mathsf{G}_{\mathrm{d}}, \;\; \mathsf{M}_{\mathrm{A}} = \mathsf{A}_{\mathrm{or},\omega}^{-1} \mathsf{M}, \;\; \text{and } \mathsf{J}_{\mathrm{A}} = \mathsf{A}_{\mathrm{or},\omega}^{-1} j.$$

Assuming the differentiability with respect to time of the convolution kernel[20], this is equivalent to the integral equation

$$\begin{aligned} u(t) = u(0) + \int_0^t \mathsf{M}_{\mathrm{A}}\, u(s)\, ds + \int_0^t \left(\int_0^s \mathsf{G}_{\mathrm{A}}(s-r) u(r)\, dr \right) ds \\ + \int_0^t \mathsf{J}_{\mathrm{A}}(s)\, ds + \int_0^t \mathsf{Q}_{\mathrm{A}}(s,\omega)\, dW(s), \end{aligned} \tag{2.55}$$

where $\mathsf{G}_{\mathrm{A}} = -\mathsf{A}_{\mathrm{or},\omega}^{-1} \mathsf{G}'_{\omega}$ and $\mathsf{Q}_{\mathrm{A}} = \mathsf{A}_{\mathrm{or},\omega}^{-1} \mathsf{Q}$. In the above equations the integrals with respect to t are considered in the Riemann-Stieltjes sense, while the integrals with respect to the Wiener process are considered in the Itō sense.

In more general models, Q_{A} can be an operator-valued stochastic process that models the effect of spatial correlations of the fluctuating terms. Let us note that Q_{A} either may be independent of the electromagnetic field or may be a function of the electromagnetic field (linear or nonlinear). The first case, especially if G_{A} is not a random process, is called the *additive noise* case, whereas the second case is called the *multiplicative noise* case.

[20]So that $\mathcal{K} \star u = \int_0^t (\mathcal{K}' \star u)(s) ds$, where we have assumed without loss of generality that $\mathsf{G}_{\mathrm{d}}(0, x) = 0$.

The boundary condition is considered to be the perfect conductor boundary condition, as usual. In order to simplify the notation, we sometimes write the above integral equation in differential form as

$$du(t) = \big(\mathsf{M}_{\mathrm{A}} u(t) + (\mathsf{G}_{\mathrm{A}} \star u)(t) + \mathsf{J}_{\mathrm{A}}\big)\, dt + \mathsf{Q}_{\mathrm{A}}\, dW(t)\,. \tag{2.56}$$

The evolution equation for the electromagnetic fields is in the form of a stochastic integrodifferential equation of Volterra type. Similar issues as those treated in the deterministic case concerning well posedness arise also in the stochastic case, where of course different methods and techniques have to be employed. Furthermore, important issues concerning how the randomness in the sources and/or the medium affects the behaviour of observed fields arise, e.g., how are the spatial autocorrelation properties of the fields related to those of the sources, how is the regularity of the solutions affected by the regularity of the sources, etc. Such problems are treated in Chapter 12.

REMARK 2.4.6 (RANDOM VS STOCHASTIC)

- ▷ RANDOM The class of models for random media of the form (2.52) where the temporal part of the random fields involved (e.g., j_ω) is of bounded variation and there is possibly random spatial structure in the medium allows the evolution equations to be treated in the usual sense, i.e., in terms of the Riemann-Stieltjes or Bochner integration, as evolution equations with random coefficients. Such problems will be called *random problems* in what follows. In this case the solutions inherit the temporal regularity of the source terms, i.e., they are of bounded variation in time. Random problems find important applications in, e.g., the homogenisation of random media (see Section 2.4.13).

- ▷ STOCHASTIC The class of models where the randomness in the medium is introduced by a process whose temporal part displays unbounded variation, e.g., of the form (2.53), is possibly the most demanding case from the mathematical point of view as it requires the full arsenal of stochastic analysis for its treatment. Within this class of problems, we may either consider a well-behaved spatial structure of the noise terms (as modelled by the spatial autocorrelations operator Q) or a white noise structure in space as well, an assumption that leads to electromagnetic fields with rather rough spatial structure. We will use the terminology *stochastic problems* for this class. In this case the solutions inherit the pathological temporal behaviour of the forcing term, and in certain cases this has effects also on the spatial regularity.

2.4.12 Stochastic controllability ↣ Ch. 13

The controllability problems presented in Section 2.4.7 have natural counterparts for the stochastic Maxwell equations. However, the introduction of noise introduces complications related to the measurability of the control procedure with respect to the filtration generated by the Wiener process, thus calling for an approach that is different from the one employed in the deterministic case. In Chapter 13 we treat this problem offering two alternatives, one related to an approximate controllability approach based on the approximation of the final target state and one based on the theory of backward stochastic evolution equations (BSEEs).

2.4.13 Homogenisation problems in random media $\rightarrowtail$ Ch. 14

Often, complex electromagnetic media do not have a deterministic periodic structure but rather a statistical type of periodicity, in the sense that the statistical properties of the medium are invariant when one performs specific translations in space. Such cases may arise, for instance, when we consider media with a crystalline structure but with random imperfections, etc. If the medium exhibits some randomness, then the concept of periodicity is too restrictive and must be replaced by a more general condition that allows us to calculate limits of the form $\lim_{L\to\infty} \frac{1}{2L} \int_{2L} E(x,\omega)\, dx$. This concept is the concept of *ergodicity*.

Consider a linear random complex medium, modelled by constitutive relations of the form

$$\mathsf{d}_\omega = \mathsf{A}_{\mathrm{or},\omega} u_\omega + \mathsf{G}_\omega \star u_\omega .$$

For simplicity the randomness is assumed to be spatial only. In the above $u_\omega = u(\cdot,\omega)$ is a general notation for a random field $u : \Omega \to V$, where $(\Omega, \mathscr{F}, P)$ is a probability space and V is a properly selected function space. The probability space $(\Omega, \mathscr{F}, P)$ is a model for the spatial randomness of the medium; we will return to specific examples later on. The random nature of the medium is used to model imperfections of the medium due to its construction, experimental mis-specifications of the medium properties, etc.

We will consider the following framework. Let $(\Omega, \mathscr{F}, P)$ be a probability space and $\mathfrak{G}$ be a group on Ω such that the probability measure P is preserved under the action τ of this group, i.e., $P(\tau A) = P(A)$ for every $A \in \mathscr{F}$. The probability space $(\Omega, \mathscr{F}, P)$ is to be interpreted as follows: Each realisation ω corresponds to a particular configuration of the medium. In other words, each experiment we perform on a particular medium corresponds to a particular choice of $\omega \in \Omega$. However, it is not known beforehand and with certainty which medium is to be realised when the experiment is performed. The probability that a particular medium is realised is given by the probability measure P.

A particular case of interest is $\Omega = \mathbb{R}^3$ (i.e., each ω is identified with a point $x \in \mathbb{R}^3$), $\mathscr{F} = \mathscr{B}(\mathbb{R}^3)$ is the Borel σ-algebra, and P is any probability measure on $\mathscr{F}$. Then a typical choice is $\mathfrak{G} = (\mathbb{R}^3, +)$, the usual translation group with group action $\tau_y = x + y$. The preservation of the probability measure under the action of the group now reads

$$P(\tau_y A) = P(A), \quad \forall A \in \mathscr{F}, \ \forall y \in \mathbb{R}^3.$$

We assume certain conditions on the random coefficients. These are the conditions of *ergodicity* and *stationarity*. These properties guarantee that, in a statistical sense, parts of the material located at different positions will present the same properties, i.e., that the statistical properties of the medium are invariant under translations which are to be understood as the transformation τ_x. This fact, as we shall see shortly, allows us to look at average properties of the material at large scales and obtain nice expressions for these quantities. In fact, by the ergodic theorem (see, e.g., [67]) we may obtain that

$$F\left(\frac{x}{\epsilon},\omega\right) \rightharpoonup \mathbb{E}[F], \quad \text{as } \epsilon \to 0\,, \quad \text{a.s.} \quad \text{in } L^\infty(\mathbb{R}^3)\,,$$

where[21] $\mathbb{E}$ is the expectation over the measure P. The ergodic hypothesis implies that instead of looking at an ensemble average of media and averaging the properties of the medium on the ensemble average, we may consider a single realisation of the medium whose spatial dimensions are large and sample its properties by traversing this single realisation for large enough distances.

This approach is relevant for a variety of situations, e.g., for a medium exhibiting random imperfections whose numbers and centres are randomly distributed with the Poisson distribution (for details and other examples, see [344], [408]). It is important to note that deterministic periodic media fall easily within this general framework by taking Ω as the unit torus $\mathbb{T}$ in $\mathbb{R}^d$, the meaning of ω here being that we may consider any point in the unit torus (unit cell of the medium) as the origin of the coordinate system, chosen randomly according to the Lebesgue measure on the unit torus (see [344]).

The evolution of electromagnetic fields in this medium is given by the Maxwell equations in the usual form

$$(\mathsf{A}^\epsilon_{\text{or},\omega} u^\epsilon + \mathsf{G}^\epsilon_\omega \star u^\epsilon)' = \mathsf{M} u^\epsilon + \mathsf{J}_{\text{A}}\,, \tag{2.57}$$

in some domain $\mathcal{O} \subset \mathbb{R}^3$, complemented with the perfect conductor boundary condition

$$n \times u^\epsilon_1 = 0, \quad x \in \partial\mathcal{O},$$

where now all the involved quantities are considered random fields and the superscripts ϵ have the same meaning as in Section 2.4.8. We consider here the random terms to be of finite variation, so that (2.57) is understood as a differential equation with random coefficients rather than as a stochastic partial differential equation (see Remark 2.4.6).

The question to be addressed here is similar to the question addressed in Section 2.4.8 for the deterministic periodic case, i.e., (a) whether and in which sense do the limits of the sequence of random fields $\{u^\epsilon\}$, $\{\mathsf{d}^\epsilon\}$ exist and (b) do the limits of these fields u^* and d^* satisfy a differential equation similar in form to the Maxwell equation (2.57) and, if so, with what constitutive relations? We shall see in Chapter 14 that the answer to both questions is positive, and there exists a homogenised medium the coefficients of which may be determined using techniques from ergodic theory and elliptic homogenisation theory.

---

[21] The abbreviation a.s. stands for *almost surely*.

Chapter Three

Spaces and Operators

3.1 INTRODUCTION

Let $\mathcal{O}$ be an open set in $\mathbb{R}^N$ such that it is locally on one side of its boundary $\Gamma := \partial\mathcal{O}$, which is supposed to be bounded and Lipschitz. We shall be interested mainly in the case of $N = 3$, so in the following unless explicitly stated otherwise, we are considering this case. Further, without loss of generality, we suppose that Γ is connected (for otherwise, one could work separately at each connected component). Such a set $\mathcal{O}$ will be referred to as "regular" in what follows. Let n denote the outward unit normal vector to Γ. In addition, let $\mathcal{O}_e := \mathbb{R}^N \setminus \overline{\mathcal{O}}$. By $\mathbb{N}_0$ we denote the set $\mathbb{N} \cup \{0\}$.

Recall that if $\mathcal{T}$ is a topological space and $x, y \in \mathcal{T}$, a *path* from x to y is a continuous function $\mathfrak{f} : [0,1] \to \mathcal{T}$ such that $\mathfrak{f}(0) = x$ and $\mathfrak{f}(1) = y$. $\mathcal{T}$ is *pathwise connected* if there exists a path joining any two points in $\mathcal{T}$. Finally, $\mathcal{T}$ is *simply connected* if it is pathwise connected and every path between two points $x, y \in \mathcal{T}$ can be continuously transformed, staying within $\mathcal{T}$, into every other path while preserving the two end points. In three dimensions, let $\mathcal{T} = \mathcal{O} \subseteq \mathbb{R}^3$: a typical example of a simply connected set is any convex $\mathcal{O}$, while an example of a nonsimply connected set is the torus. In this work we will restrict ourselves to simply connected domains; however, many of the results may be generalised to nonsimply connected ones.

This chapter[1] is included for the convenience of the reader and consists mainly of definitions and various properties (without proofs) of spaces and operators used in this book; unless otherwise stated, for a full treatment of the contents of this chapter one can consult [91], [126]; see also [135], [324]. A large amount of related material can be also found in [79], [171].

3.2 FUNCTION SPACES

By $C^k(\mathcal{O})$, $k \in \mathbb{N}$, we denote the space of k-times continuously differentiable functions on $\mathcal{O}$, and by $C_0^k(\mathcal{O})$ the space of k-times continuously differentiable functions on $\mathcal{O}$ with compact support. By $C_b^k(\mathcal{O})$ we denote the

[1] We are fully aware that this chapter is somewhat "dry", but it is clear that otherwise a whole book should be written; we have no intention of undertaking such a burden, especially since the excellent books [91] and [126] are certainly much more than adequate. Nevertheless, we believe that it is convenient for the reader to have at her or his disposal the material of this chapter collected in one place.

space[2] of $C^k(\mathcal{O})$ functions whose derivatives of order up to and including k are bounded in $\mathcal{O}$, which is a Banach space with an appropriate norm (see [1]). By $C^{k,\alpha}(\mathcal{O})$, $\alpha \in (0,1)$ we denote the space of k-times differentiable functions with Hölder continuous derivatives in $\mathcal{O}$, by $C^{0,1}(\mathcal{O})$ the space of Lipschitz functions in $\mathcal{O}$, while by $C_0^\infty(\mathcal{O})$ (alternatively denoted by $\mathcal{D}(\mathcal{O})$) the space of $C^\infty(\mathcal{O})$ functions with compact support in $\mathcal{O}$, and by $C_0^{0,1}(\overline{\mathcal{O}})$ the space of restrictions to $\mathcal{O}$ of Lipschitz functions with compact support in $\mathbb{R}^3$. In addition, by $C_t(\partial\mathcal{O})$ we denote the space of continuous functions on the surface $\partial\mathcal{O}$. Finally, $\mathcal{D}'(\mathcal{O})$ is the set of distributions in $\mathcal{O}$.

3.2.1 Lebesgue spaces

In this section all definitions are given in the general case where $\mathcal{O} \subset \mathbb{R}^N$. By $L^p(\mathcal{O})$ we denote the usual space of p integrable functions $u : \mathcal{O} \to \mathbb{R}$; endowed with the norm

$$||u||_{L^p(\mathcal{O})} = \left(\int_{\mathcal{O}} |u(x)|^p \, dx \right)^{1/p},$$

it is a Banach space for $p \geq 1$. The case where $p = \infty$ corresponds to the case of essentially bounded functions and the norm is the essential supremum[3] of the function on $\mathcal{O}$. The particular case where $p = 2$ corresponds to a Hilbert space with inner product $(u,v) = \int_{\mathcal{O}} u(x)v(x)\,dx$. For complex-valued functions, the inner product is replaced by $(u,v) = \int_{\mathcal{O}} u(x)\overline{v(x)}\,dx$. For vector-valued functions $u : \mathcal{O} \to \mathbb{R}^3$, which is often the case of interest in electromagnetic theory, we will need the function spaces $(L^p(\mathcal{O}))^3$. Of particular interest is the case $p = 2$, which is again a Hilbert space, with inner product $(u,v) = \int_{\mathcal{O}} u(x) \cdot v(x)\,dx$, where "$\cdot$" denotes the standard dot product in $\mathbb{R}^3$. The extension to complex valued vector fields is straightforward. The meaning of the symbols $L^p(\mathcal{O}, \mathbb{R}^3)$ and $L^p(\mathcal{O}, \mathbb{C}^3)$ is obvious.

Let $p \in (1, \infty)$. The *conjugate exponent* p' of p is defined by $\frac{1}{p} + \frac{1}{p'} = 1$. In the case of $p = 1$, set $p' := \infty$. The space $L^{p'}(\mathcal{O})$ is the dual space of

---

[2]This space is larger than $C^k(\overline{\mathcal{O}})$ since its elements are not necessarily uniformly continuous in $\mathcal{O}$.

[3]The concept of *essential supremum* (respectively, *essential infimum*) is related to the notion of supremum (respectively, infimum); the former is more relevant in measure theory, where one often deals with statements that are not valid everywhere (i.e., for all elements in a set), but rather almost everywhere (i.e., except on a set of measure zero). To be precise, let $\mathscr{S}$ be a σ-algebra over a set X and $\mu : \mathscr{S} \to \mathbb{R} \cup \{\pm\infty\}$ be a measure; in the measure space $(X, \mathscr{S}, \mu)$ let $f : X \to \mathbb{R}$ be a not necessarily measurable function. A real number a is called an upper bound for f if $f(x) \leq a$ for all $x \in X$, that is, if $\{x \in X : f(x) > a\} = \emptyset$. In contrast, a is called an *essential upper bound* if the set $\{x \in X : f(x) > a\}$ is contained in a set of measure zero, that is to say, if $f(x) \leq a$ for almost all $x \in X$. Then, in the same way as the supremum of f is defined to be the smallest upper bound, the *essential supremum* is defined as the smallest essential upper bound. More formally, the essential supremum of f is defined by $\operatorname{ess\,sup} f = \inf\{a \in \mathbb{R} : \mu(\{x : f(x) > a\}) = 0\}$ if the set of essential upper bounds, $\{a \in \mathbb{R} : \mu(\{x : f(x) > a\}) = 0\}$, is nonempty, and $\operatorname{ess\,sup} f = +\infty$ otherwise.

$L^p(\mathcal{O})$, with duality pairing (see Section A.1 in Appendix A):

$$\langle v, w\rangle_{L^{p'}(\mathcal{O}),\, L^p(\mathcal{O})} = \int_{\mathcal{O}} v\, w\, dx, \quad \text{for } v \in L^{p'}(\mathcal{O}),\ w \in L^p(\mathcal{O}).$$

These spaces are related by the Lebesgue embedding theorem according to which if $\mathcal{O}$ has finite Lebesgue measure and $1 \leq q < p < \infty$, then

$$L^\infty(\mathcal{O}) \subset L^p(\mathcal{O}) \subset L^q(\mathcal{O}) \subset L^1(\mathcal{O}).$$

3.2.2 Sobolev spaces

In this section all definitions are given in the general case where $\mathcal{O} \subset \mathbb{R}^N$ is a nonempty open set. By $W^{m,p}(\mathcal{O})$, $m \in \mathbb{N}_0, p \in [1,\infty]$, we denote the standard Sobolev spaces defined by

$$W^{m,p}(\mathcal{O}) = \{u \in L^p(\mathcal{O}) \,:\, \partial^\alpha u \in L^p(\mathcal{O}) \text{ for all } |\alpha| \leq m\},$$

where the multi-index $\alpha = (\alpha_1, \ldots, \alpha_N) \in (\mathbb{N}_0)^N$, $|\alpha| = \sum_{i=1}^N \alpha_i$, and $\partial^\alpha u = \frac{\partial^{|\alpha|} u}{\partial x_1^{\alpha_1} \ldots \partial x_N^{\alpha_N}}$. These spaces contain $C_0^\infty(\mathcal{O})$, and in the case in which $\mathcal{O}$ is bounded they also contain $C^\infty(\overline{\mathcal{O}})$. When endowed with the norm

$$||u||_{W^{m,p}(\mathcal{O})} = \left(\sum_{|\alpha| \leq m} \int_{\mathcal{O}} |\partial^\alpha u|^p \, dx \right)^{1/p},$$

the Sobolev spaces become Banach spaces. For $p \in (1,\infty)$ the spaces $W^{m,p}(\mathcal{O})$ are reflexive, while for $p \in [1,\infty)$ they are separable. For $p = 1, \infty$, the corresponding spaces are not reflexive, while for $p = \infty$ the corresponding space is not separable. In the special case $p = 2$ these spaces become Hilbert spaces, with their norms generated by (obviously defined) inner products. In particular, we have

$$W^{0,p}(\mathcal{O}) = L^p(\mathcal{O}),$$
$$H^m(\mathcal{O}) := W^{m,2}(\mathcal{O}).$$

Of interest concerning the behaviour of functions satisfying homogeneous Dirichlet boundary conditions on a domain are the spaces

$$W_0^{m,p}(\mathcal{O}) = \text{the closure of } C_0^\infty(\mathcal{O}) \text{ in the } W^{m,p}(\mathcal{O}) \text{ norm.}$$

These functions satisfy the condition $u = 0$ on the boundary in the sense of traces. A usual notation for the special case $p = 2$ is $H_0^m(\mathcal{O})$. The extension of Sobolev spaces for vector fields in $\mathcal{O}$ is straightforward.

A useful result connecting the $L^p(\mathcal{O})$ norm of a function with those of its derivatives is the celebrated Friedrichs' inequality:

THEOREM 3.2.1 *Let $\mathcal{O}$ be a bounded subset of $\mathbb{R}^N$ with diameter d. Suppose that $u : \mathcal{O} \to \mathbb{R}$ lies in the Sobolev space $W_0^{m,p}(\mathcal{O})$. Then, for any multi-index α, we have*

$$||u||^p_{L^p(\mathcal{O})} \leq d^{m\,p} \sum_{|\alpha|=m} ||\partial^\alpha u||^p_{L^p(\mathcal{O})}.$$

The following embedding theorem is very useful in applications.

THEOREM 3.2.2 (SOBOLEV EMBEDDING THEOREM) *Assume that $\mathcal{O}$ is a domain (not necessarily bounded) with $C^{0,1}$ boundary and let $j, m \in \mathbb{N}_0$, $p \in [1,\infty)$.*

(i) *Let $m\,p < N$. Then $W^{j+m,p}(\mathcal{O}) \subset W^{j,q}(\mathcal{O})$ for $q \in [p, N\,p/(N-m\,p)]$. If $\mathcal{O}$ is bounded, then the above embedding holds additionally for $q \in [1,p)$.*

(ii) *Let $m\,p = N$. Then $W^{j+m,p}(\mathcal{O}) \subset W^{j,q}(\mathcal{O})$ for $q \in [p,\infty)$. If $\mathcal{O}$ is bounded, then the above embedding holds additionally for $q \in [1,p)$. If in particular $p = 1$, the embedding holds for $q = \infty$ as well.*

(iii) *Let $m\,p > N$. Assume that* one *of the following cases holds:*

   (a) *$N > (m-1)p$ and $\alpha \in (0, m-(N/p)]$.*

   (b) *$N = (m-1)p$ and $\alpha \in (0,1)$.*

   (c) *$N = m-1$, $p = 1$ and $\alpha \in (0,1]$.*

   Then $W^{j+m,p}(\mathcal{O}) \subset C^{j,\alpha}(\overline{\mathcal{O}})$.

To define the spaces $W^{\mathsf{s},p}(\mathcal{O})$ for $\mathsf{s} \in \mathbb{R}$ we first define, for $0 < \nu < 1$, the so-called *Hölder seminorms*

$$|u|_{\nu,p,\mathcal{O}} := \left(\int_{\mathcal{O}}\int_{\mathcal{O}} \frac{|u(x)-u(y)|^p}{|x-y|^{N+\nu p}}\,dx\,dy\right)^{1/p}, \ 1 \le p < \infty\,,$$

$$|u|_{\nu,\infty,\mathcal{O}} := \sup_{x,y\in\mathcal{O}} \frac{|u(x)-u(y)|}{|x-y|^{\nu}}\,.$$

For $\mathsf{s} = m + \nu$, $0 < \nu < 1$, we define

$$W^{\mathsf{s},p}(\mathcal{O}) = \{u \in W^{m,p}(\mathcal{O}) : |\partial^\alpha u|_{\nu,p,\mathcal{O}} < \infty, \text{ for } |\alpha| = m\}\,.$$

$W^{\mathsf{s},p}(\mathcal{O})$ becomes a Banach space when equipped with the norm

$$\|u\|_{W^{\mathsf{s},p}(\mathcal{O})} = \left(\|u\|^p_{W^{m,p}(\mathcal{O})} + \sum_{|\alpha|=m} |\partial^\alpha u|_{\nu,p,\mathcal{O}}\right)^{1/p}\,.$$

In particular, we have

$$H^{\mathsf{s}}(\mathcal{O}) := W^{\mathsf{s},2}(\mathcal{O})\,.$$

As above, for every $\mathsf{s} > 0$, we define

$$H_0^{\mathsf{s}}(\mathcal{O}) = \text{the closure of } C_0^\infty(\mathcal{O}) \text{ in the } H^{\mathsf{s}}(\mathcal{O}) \text{ norm,}$$

and

$$H^{-\mathsf{s}}(\mathcal{O}) = \text{the dual space of } H_0^s(\mathcal{O})\,.$$

To define H^{s} on Γ, assume first that Γ is of the special form

$$\Gamma_0 = \{(y, \mathfrak{y}(y)) : y \in \mathbb{R}^{N-1}\}\,,$$

where $\mathfrak{y} : \mathbb{R}^{N-1} \to \mathbb{R}$ is a $C^{k-1,1}$ function[4] ($k \in \mathbb{N}$). Define

$$u_{\mathfrak{y}}(y) := u(y, \mathfrak{y}(y)),\ y \in \mathbb{R}^{N-1}.$$

Then

$$H^{\mathsf{s}}(\Gamma_0) := \{u \in L^2(\Gamma_0) : u_{\mathfrak{y}} \in H^{\mathsf{s}}(\mathbb{R}^{N-1})\}, \text{ for } 0 \le |\mathsf{s}| \le k.$$

Equipped with the inner product $(u, v)_{H^{\mathsf{s}}(\Gamma_0)} := (u_{\mathfrak{y}}, v_{\mathfrak{y}})_{H^{\mathsf{s}}(\mathbb{R}^{N-1})}$, $H^{\mathsf{s}}(\Gamma_0)$ is a Hilbert space.

For $0 \le |\mathsf{s}| \le k$,

$$H^{-\mathsf{s}}(\Gamma_0) \text{ is the dual space of } H^{\mathsf{s}}(\Gamma_0).$$

Now consider the general case, in which Γ is assumed to be a $C^{k-1,1}$ boundary; then there exist finite families $\{\mathcal{W}_j\}$ and $\{\mathcal{O}_j\}$ such that each $\mathcal{W}_j$ is an open set in $\mathbb{R}^N$ and $\Gamma \subset \cup \mathcal{W}_j$, each $\mathcal{O}_j$ can be transformed by a rotation plus a translation to a set having a boundary of the form of Γ_0, and, for each j, $\mathcal{W}_j \cap \mathcal{O} = \mathcal{W}_j \cap \mathcal{O}_j$. Hence we can choose a *partition of unity*[5] $\{\psi_j\}$ associated with the cover $\{\mathcal{W}_j\}$. Let

$$\widetilde{\psi_j u} := \begin{cases} \psi_j u, & \text{in } \mathcal{W}_j \\ 0, & \text{in } \mathcal{O}_j \setminus \mathcal{W}_j. \end{cases}$$

Therefore we may define

$$u \in H^{\mathsf{s}}(\Gamma) \text{ if and only if } \widetilde{\psi_j u} \in H^{\mathsf{s}}(\partial \mathcal{O}_j).$$

Further, we define

$$\|u\|_{H^{\mathsf{s}}(\Gamma)} := \sum_j \|\widetilde{\psi_j u}\|_{H^{\mathsf{s}}(\partial \mathcal{O}_j)}.$$

This norm is independent of the choice of the partition of unity. Now, for $0 \le \mathsf{s} \le k$,

$$H^{-\mathsf{s}}(\Gamma) := \text{the dual space of } H^{\mathsf{s}}(\Gamma).$$

Finally, we define

$$H^{\mathsf{s}}_0(\Gamma) = \{\phi \in H^{\mathsf{s}}(\Gamma) : \langle 1, \phi \rangle_{H^{-\mathsf{s}}(\Gamma),\, H^{\mathsf{s}}(\Gamma)} = 0\},$$

where, $\langle \cdot, \cdot \rangle_{H^{-\mathsf{s}}(\Gamma),\, H^{\mathsf{s}}(\Gamma)}$ is the duality pairing between $H^{-\mathsf{s}}(\Gamma)$ and $H^{\mathsf{s}}(\Gamma)$.

The Sobolev embedding theorem may be generalised for fractional order Sobolev spaces.

THEOREM 3.2.3 *Let $\mathcal{O} \subset \mathbb{R}^N$ be a bounded domain with Lipschitz boundary, $\mathsf{s} > 0$, $1 < p < \infty$ and $\mathsf{r} := \mathsf{s} - N/p + N/q$. Under one of the following assumptions*

[4] The case $k = 1$ refers to Lipschitz boundaries.

[5] $\psi_j \in C_0^{k-1,1}(\mathcal{W}_j)$, such that $\sum_j \psi_j(x) = 1$ for all $x \in \Gamma$.

(i) $\mathsf{r} \geq 0$, $p < q$,

(ii) $\mathsf{r} > 0$ *and* $\mathsf{r} \notin \mathbb{N}$,

(iii) $\mathsf{r} \geq 0$ *and* $1 < p \leq 2$,

we have the embedding $W^{\mathsf{s},p}(\mathcal{O}) \subset W^{\mathsf{r},q}(\mathcal{O})$.

REMARK 3.2.4 This theorem holds for $\mathcal{O} = \mathbb{R}^N$ as well (see, e.g., [1]).

Regarding Sobolev spaces, one can consult, e.g., [1], [262], [406].

3.2.3 Lebesgue-Bochner spaces

In this section we consider functions from an interval $I := [0,T]$ to a vector space V, $u : I \to V$.

For $1 \leq p < \infty$, a Lebesgue-Bochner space $L^p(I,V)$ is the equivalence class of Bochner integrable[6] functions $u : I \to V$ such that $\int_0^T ||u||_V^p \, dt < \infty$. This is a Banach space when equipped with the norm

$$||u||_{L^p(I,V)} = \left(\int_0^T ||u||_V^p \, dt \right)^{1/p} .$$

If $p = \infty$, we must modify the norm to

$$||u||_{L^\infty(I,V)} = \text{ess sup}_{t\in I} ||u(t)||_V .$$

The following results concerning Lebesgue-Bochner spaces are useful.

THEOREM 3.2.5 $L^{p'}(I,V')) \subseteq (L^p(I,V))'$ *with equality if* V' *is separable. The duality pairing between the two spaces is defined as*

$$\langle v, u \rangle_{L^{p'}(I,V'),L^p(I,V)} = \int_0^T \langle v(t), u(t) \rangle_{V',V} \, dt .$$

The Young inequality, as well as certain interpolation inequalities, may be generalised in the context of Lebesgue-Bochner spaces.

THEOREM 3.2.6 *Let* $p_1, p_2, q_1, q_2 \in [1, +\infty]$, $\lambda \in [0,1]$, *and* $\frac{1}{p} = \frac{\lambda}{p_1} + \frac{1-\lambda}{p_2}$, $\frac{1}{q} = \frac{\lambda}{q_1} + \frac{1-\lambda}{q_2}$. *Then*

$$||v||_{L^p(I,L^q(\mathcal{O}))} \leq ||v||^{\lambda}_{L^{p_1}(I,L^{q_1}(\mathcal{O}))} ||v||^{1-\lambda}_{L^{p_2}(I,L^{q_2}(\mathcal{O}))} .$$

REMARK 3.2.7 The Lebesgue-Bochner spaces will be used in general in the treatment of time-dependent problems, where the space V may be either a Lebesgue space, a Sobolev space, or any of the spaces considered in Section 3.2.5 or 3.4.

Standard references for these spaces and their properties are, e.g., [262] and [373].

[6] See Section A.3.2 in Appendix A.

3.2.4 Sobolev-Bochner spaces

We now turn to the generalisation of Sobolev spaces in the context of Lebesgue-Bochner spaces.

The space $W^{1,p,q}(I, V_1, V_2)$, $V_1 \subset V_2$, is defined as the space

$$W^{1,p,q}(I, V_1, V_2) = \{u : u \in L^p(I, V_1), u' \in L^q(I, V_2)\},$$

where u' is the distributional derivative, which is the linear operator defined by $u'(\phi) = -\int_0^T u(t)\,\phi'(t)\,dt$ for all $\phi \in \mathcal{D}(I)$. This is a Banach space when equipped with the norm

$$||u||_{W^{1,p,q}(I,V_1,V_2)} = ||u||_{L^p(I,V_1)} + ||u'||_{L^q(I,V_2)}.$$

The choice $V_1 = V_2$ often appears in applications.

The following result is useful.

THEOREM 3.2.8 *Assume that $p, q \geq 1$ and that $V_1 \subset V_2$ continuously. Then*

(i) $W^{1,p,q}(I, V_1, V_2) \subset C(I, V_2)$ *continuously,*

(ii) $C^1(I, V_1) \subset W^{1,p,q}(I, V_1, V_2)$ *densely.*

(iii) Let $V \subset H \subset V'$ be a Gelfand triple[7]*. Then $W^{1,p,p'}(I, V, V') \subset C(I, H)$ continuously, where p' is the conjugate exponent of p, and the integration by parts formula*

$$(u(t_2), v(t_2)) - (u(t_1), v(t_1)) = \int_{t_1}^{t_2} \big(\langle u'(t), v(t)\rangle_{V',V} + \langle u(t), v'(t)\rangle_{V',V}\big)\,dt$$

holds for all $u, v \in W^{1,p,p'}(I, V, V')$ and all $t_1, t_2 \in I$.

REMARK 3.2.9 Sobolev-Bochner spaces will be used in the treatment of time-dependent linear or nonlinear problems. Similar remarks as in Remark 3.2.7 hold here as well.

For more details, one may consult [373].

3.2.5 Spaces of random fields

Stochastic problems involve random fields, which may be understood as stochastic processes indexed by a multidimensional parameter (typically $(t, x) \in \mathbb{R}^+ \times \mathbb{R}^3$). These random fields involve random variables with values on function spaces (typically spaces related to the function spaces described in Section 3.4 or 3.9.1) that have well-defined statistical moments.

Consider a probability space $(\Omega, \mathscr{F}, P)$ with a normal filtration $\{\mathscr{F}_t\}$, $t \geq 0$. The expectation of the random field u is defined by

$$\mathbb{E}[u] = \int_\Omega u\,dP,$$

[7] See Section A.1 in Appendix A.

where the integral is to be understood in the sense of the Bochner integral (see, e.g., [380]). In similar fashion we may define the statistical moments of the random fields, for instance, the pth moment is defined by

$$\mathbb{E}[||u||_{\mathbb{H}}^p] := \int_\Omega ||u||_{\mathbb{H}}^p \, dP\,,$$

where we consider the random field u as taking values in a function space $\mathbb{H}$.

Of particular importance in the study of stochastic evolution equations is the space of square integrable $\mathbb{H}$-valued $\mathscr{F}$-measurable random variables,

$$L^2(\Omega, \mathscr{F}, P; \mathbb{H}) = \{u : \Omega \to \mathbb{H} \, : \, u \in \text{m-}\mathscr{F} \text{ and } \mathbb{E}_P[||u||_{\mathbb{H}}^2] < \infty\},$$

where the notation $u \in \text{m-}\mathscr{F}$ means that u is measurable with respect to $\mathscr{F}$. This space may be understood as a special case of the Lebesgue-Bochner space, defined over the probability measure P. In cases where the σ-algebra $\mathscr{F}$ or the probability measure P are fixed, we may omit the explicit dependence on them in the notation of the space; in such cases we use the notation $L^2(\Omega, P; \mathbb{H})$ or $L^2(\Omega; \mathbb{H})$, respectively.

We will also employ the space of all continuous (in mean square) and square integrable predictable processes

$$\mathcal{C}([0,T], \Omega, \mathscr{P}_T, P; \mathbb{H}) = \{Y \in C([0,T]; L^2(\Omega, \mathbb{H}) : Y \in \text{m-}\mathscr{P}_T\}.$$

This space equipped with the norm

$$\|Y\|_{\mathcal{C}} = \sup_{t\in[0,T]} \left(\mathbb{E}\left[\|Y(t)\|_{\mathbb{H}}^2\right]\right)^{1/2}$$

is a Banach space. In cases where the probability space is fixed, we suppress the explicit dependence on Ω, $\mathscr{P}_T$, P and write $\mathcal{C}([0,T]; \mathbb{H})$.

Standard references for these spaces and their properties are, e.g., [120], [168], and [352].

3.3 STANDARD DIFFERENTIAL AND TRACE OPERATORS

As most of the function spaces used in mathematical electromagnetic theory are related to the action of the fundamental operators of vector analysis on vector-valued functions, let us recall the definitions of the basic linear[8] and trace operators that are essential in electromagnetism.

3.3.1 Differential operators

- ▷ $\text{grad} : \mathcal{D}'(\mathcal{O}) \longrightarrow (\mathcal{D}'(\mathcal{O}))^3\,,\ u \mapsto \text{grad}\, u := \left(\frac{\partial u}{\partial x_1}, \frac{\partial u}{\partial x_2}, \frac{\partial u}{\partial x_3}\right)^{tr}$,
- ▷ $\text{div} : (\mathcal{D}'(\mathcal{O}))^3 \longrightarrow \mathcal{D}'(\mathcal{O})\,,\ v \mapsto \text{div}\, v := \frac{\partial v_1}{\partial x_1} + \frac{\partial v_2}{\partial x_2} + \frac{\partial v_3}{\partial x_3}$,
- ▷ $\text{curl} : (\mathcal{D}'(\mathcal{O}))^3 \longrightarrow (\mathcal{D}'(\mathcal{O}))^3\,,\ w \mapsto \text{curl}\, w := \left(\frac{\partial w_3}{\partial x_2} - \frac{\partial w_2}{\partial x_3}, \frac{\partial w_1}{\partial x_3} - \frac{\partial w_3}{\partial x_1}, \frac{\partial w_2}{\partial x_1} - \frac{\partial w_1}{\partial x_2}\right)^{tr}$.

[8] The vector analysis symbols are often used, namely, ∇u, $\nabla \cdot v$ and $\nabla \times w$, instead of $\text{grad}\, u$, $\text{div}\, v$ and $\text{curl}\, w$, respectively. In this book we employ the latter.

These operators can be restricted to $C^k(\mathcal{O})$ and $C_0^\infty(\mathcal{O})$. The actions of these operators on various function spaces of interest can be summarised in the following diagrams:

$$\mathcal{D}'(\mathcal{O}) \overset{\text{grad}}{\longrightarrow} (\mathcal{D}'(\mathcal{O}))^3 \overset{\text{curl}}{\longrightarrow} (\mathcal{D}'(\mathcal{O}))^3 \overset{\text{div}}{\longrightarrow} \mathcal{D}'(\mathcal{O}),$$

$$C^{k+3}(\mathcal{O}) \overset{\text{grad}}{\longrightarrow} (C^{k+2}(\mathcal{O}))^3 \overset{\text{curl}}{\longrightarrow} (C^{k+1}(\mathcal{O}))^3 \overset{\text{div}}{\longrightarrow} C^k(\mathcal{O})\ , k \in \mathbb{N}\,,$$

$$C_0^\infty(\mathcal{O}) \overset{\text{grad}}{\longrightarrow} (C_0^\infty(\mathcal{O}))^3 \overset{\text{curl}}{\longrightarrow} (C_0^\infty(\mathcal{O}))^3 \overset{\text{div}}{\longrightarrow} C_0^\infty(\mathcal{O})\,.$$

It can be easily verified that $\operatorname{curl}\operatorname{grad} u = 0$ and $\operatorname{div}\operatorname{curl} w = 0$.

3.3.2 Trace operators

Trace theory connects the properties of functions defined in the interior of a domain $\mathcal{O}$ with their properties on its boundary Γ. Recall that for any vector field $v : \overline{\mathcal{O}} \to \mathbb{R}^3$, we have that

$$v|_\Gamma = (n \cdot v|_\Gamma)\, n + (n \times v|_\Gamma) \times n,$$

where n is the outward normal. This motivates the definition of the following operators, which are of fundamental importance in trace theory:

- ▷ The *normal trace operator* $\gamma_n : v \mapsto n \cdot v|_\Gamma$.
- ▷ The *tangential trace operator* $\gamma_\tau : v \mapsto n \times v|_\Gamma$.
- ▷ The *tangential components trace operator* $\pi_\Gamma : v \mapsto (n \times v|_\Gamma) \times n$.

The rigorous definition of these operators under weaker assumptions in various function spaces is the main objective of trace theory.

3.3.3 Integral identities

For sufficiently smooth vector fields $\mathfrak{u}, \mathfrak{v}, \mathfrak{w}$ and scalar ϕ, the following integral identities hold.

3.3.3.1 The divergence theorem

$$\int_{\mathcal{O}} \operatorname{div} \mathfrak{v}\, dx = \int_\Gamma \gamma_n(\mathfrak{v})\, ds\,.$$

3.3.3.2 Green's formulae

$$\int_{\mathcal{O}} \mathfrak{u} \cdot \operatorname{grad} \phi\, dx + \int_{\mathcal{O}} \operatorname{div} \mathfrak{u}\, \phi\, dx = \int_\Gamma \gamma_n(\mathfrak{u})\, \phi\, ds =: \langle \gamma_n(\mathfrak{u}), \phi \rangle\,,$$

$$\int_{\mathcal{O}} \operatorname{curl} \mathfrak{v} \cdot \mathfrak{w}\, dx - \int_{\mathcal{O}} \mathfrak{v} \cdot \operatorname{curl} \mathfrak{w}\, dx = \int_\Gamma \gamma_\tau(\mathfrak{v}) \cdot \pi_\Gamma(\mathfrak{w})\, ds =: \langle \gamma_\tau(\mathfrak{v}), \pi_\Gamma(\mathfrak{w}) \rangle\,.$$

Note that in the above equations we use the same notation for two different objects, namely, the surface integral of the product of two scalar quantities and the surface integral of the dot product of two vectors. The right-hand sides of the above will be used to define duality pairings between function spaces containing vector or scalar fields, less regular than what is required for the classical definition of the above integrals.

3.3.3.3 Jump relations

Let $\mathcal{O}$ be a bounded open set in $\mathbb{R}^3$ with an adequately smooth boundary Γ, and let $\Gamma^{\pm}$ be its exterior (i.e., from $\mathcal{O}_e$) and interior face (i.e., from $\mathcal{O}$), respectively. The (exterior) normal n to Γ is oriented from Γ^- to Γ^+. Let ϕ be a scalar function and $\mathfrak{u}$ a vector function, defined on $\mathbb{R}^3 \setminus \Gamma$ of sufficient regularity on each side, $\mathcal{O}$, $\mathcal{O}_e$ of Γ, having limits up to Γ. The jump of $\mathfrak{s}$ across Γ is denoted by

$$[\mathfrak{s}]_\Gamma := \mathfrak{s}|_{\Gamma^-} - \mathfrak{s}|_{\Gamma^+},$$

with $\mathfrak{s}|_{\Gamma^\pm}$ being the limit of $\mathfrak{s}$ as $x \to \Gamma^\pm$, respectively, where $\mathfrak{s}$ is either ϕ or $\mathfrak{u}$.

Then the following *jump relations*, easily derived from Green's formulae, hold (see [91]):

$$\int_{\mathbb{R}^3} \operatorname{grad}\phi \cdot \mathfrak{v}\, dx = \int_{\mathcal{O}} \operatorname{grad}\phi \cdot \mathfrak{v}\, dx + \int_{\mathcal{O}_e} \operatorname{grad}\phi \cdot \mathfrak{v}\, dx - \int_\Gamma [\phi]_\Gamma\, n \cdot \mathfrak{v}\, ds(x),$$

$$\int_{\mathbb{R}^3} \operatorname{div}\mathfrak{u}\, \psi\, dx = \int_{\mathcal{O}} \operatorname{div}\mathfrak{u}\, \psi\, dx + \int_{\mathcal{O}_e} \operatorname{div}\mathfrak{u}\, \psi\, dx - \int_\Gamma [n \cdot \mathfrak{u}]_\Gamma\, \psi\, ds(x),$$

$$\int_{\mathbb{R}^3} \operatorname{curl}\mathfrak{u} \cdot \mathfrak{v}\, dx = \int_{\mathcal{O}} \operatorname{curl}\mathfrak{u} \cdot \mathfrak{v}\, dx + \int_{\mathcal{O}_e} \operatorname{curl}\mathfrak{u} \cdot \mathfrak{v}\, dx - \int_\Gamma [n \times \mathfrak{u}]_\Gamma \cdot \mathfrak{v}\, ds(x),$$

for $\mathfrak{v} \in (C_0^\infty(\mathbb{R}^3))^3$ and $\psi \in C_0^\infty(\mathbb{R}^3)$.

The derivatives in the integrals on the left-hand sides of the above relations are to be understood in the distributional sense because of the discontinuities of ϕ or $\mathfrak{u}$ across Γ, while on the right-hand side they are to be understood in the classical sense.

3.3.4 The tangential differential operators

For a sufficiently smooth scalar function $\phi : \Gamma \to \mathbb{R}$, we consider an appropriate extension $\Phi : \overline{\mathcal{O}} \to \mathbb{R}$ such that $\Phi|_\Gamma = \phi$, and define the following:

$$\operatorname{Grad}\phi := \pi_\Gamma\,(\operatorname{grad}\Phi).$$

For a sufficiently smooth vector field $\mathfrak{v} : \Gamma \to \mathbb{R}^3$ we consider an appropriate extension $V : \overline{\mathcal{O}} \to \mathbb{R}^3$ such that $V|_\Gamma = \mathfrak{v}$ and define the following:

$$\operatorname{Curl}\mathfrak{v} = \gamma_n(\operatorname{curl}V).$$

Note that the operator Curl maps a vector field into a scalar function.

By Green's formulae it can be shown that the above two definitions are *independent* of the choice of the particular extension.

Using the above definition we can define, using duality, the following tangential operators:

$$\langle \overrightarrow{\mathrm{Curl}}\, \phi \,, \mathfrak{v} \,\rangle := -\,\langle\, \phi \,, \mathrm{Curl}\, \mathfrak{v} \,\rangle\,, \text{ for all regular vector fields}^9\ \mathfrak{v}.$$

$$\langle \mathrm{Div}\, \mathfrak{v} \,, \phi \,\rangle := -\,\langle \mathfrak{v} \,, \mathrm{Grad}\, \phi \,\rangle\,, \text{ for all regular scalar functions}^{10}\ \ \phi.$$

3.4 FUNCTION SPACES FOR ELECTROMAGNETICS

3.4.1 The space $H(\mathrm{div}, \mathcal{O})$

$$H(\mathrm{div}, \mathcal{O}) = \{u \in (L^2(\mathcal{O}))^3 \;:\; \mathrm{div}\, u \in L^2(\mathcal{O})\}.$$

When endowed with the (graph) norm

$$||u||_{H(\mathrm{div},\mathcal{O})} = \Big(||u||^2_{(L^2(\mathcal{O}))^3} + ||\mathrm{div}\, u||^2_{L^2(\mathcal{O})} \Big)^{1/2}$$

it is a Hilbert space. The associated inner product is defined by

$$(u, v)_{H(\mathrm{div},\mathcal{O})} = \int_{\mathcal{O}} u \cdot v \, dx + \int_{\mathcal{O}} (\mathrm{div}\, u)\, (\mathrm{div}\, v)\, dx.$$

If $\mathcal{O}$ is bounded, then

$$H(\mathrm{div}, \mathcal{O}) = \text{closure of } (C^\infty(\overline{\mathcal{O}}))^3 \text{ in the norm of } H(\mathrm{div}, \mathcal{O}).$$

Furthermore, $(C_0^{0,1}(\overline{\mathcal{O}}))^3$ and $(C_0^\infty(\overline{\mathcal{O}}))^3$ are dense in $H(\mathrm{div}, \mathcal{O})$, and there is a continuous extension from $H(\mathrm{div}, \mathcal{O})$ into $H(\mathrm{div}, \mathbb{R}^3)$.

3.4.2 The space $H_0(\mathrm{div}, \mathcal{O})$

$$H_0(\mathrm{div}, \mathcal{O}) = \text{closure of } (C_0^\infty(\overline{\mathcal{O}}))^3 \text{ in the norm of } H(\mathrm{div}\mathcal{O}).$$

If $\mathcal{O}$ is bounded, then

$$H_0(\mathrm{div}, \mathcal{O}) = \{u \in H(\mathrm{div}, \mathcal{O}) \;:\; u \cdot n|_\Gamma = 0\}.$$

3.4.3 The spaces $H(\mathrm{div}0, \mathcal{O})$ and $H_0(\mathrm{div}0, \mathcal{O})$

$$H(\mathrm{div}0, \mathcal{O}) = \big\{u \in (L^2(\mathcal{O}))^3 : \ \mathrm{div}\, u = 0\big\}.$$

$$H_0(\mathrm{div}0, \mathcal{O}) := \{u \in H(\mathrm{div}0, \mathcal{O}) \,:\, u \cdot n = 0 \text{ on } \Gamma\}.$$

---

[9] The symbol $\langle \cdot, \cdot \rangle$ on the left-hand side corresponds to the surface integral of the dot product of the involved vector fields, whereas the symbol $\langle \cdot, \cdot \rangle$ on the right-hand side corresponds to the surface integral of the (usual) product of the involved scalar fields.

[10] The symbol $\langle \cdot, \cdot \rangle$ on the left-hand side corresponds to the surface integral of the (usual) product of the involved scalar fields, whereas the symbol $\langle \cdot, \cdot \rangle$ on the right-hand side corresponds to the surface integral of the dot product of the involved vector fields.

3.4.4 The space $H(\mathrm{curl}, \mathcal{O})$

$$H(\mathrm{curl}, \mathcal{O}) = \{u \in (L^2(\mathcal{O}))^3 \ : \ \mathrm{curl}\, u \in (L^2(\mathcal{O}))^3\}.$$

When endowed with the (graph) norm

$$||u||_{H(\mathrm{curl},\mathcal{O})} = (||u||^2_{(L^2(\mathcal{O}))^3} + ||\mathrm{curl}\, u||^2_{(L^2(\mathcal{O}))^3})^{1/2}$$

it is a Hilbert space. The associated inner product is defined by

$$(u, v)_{H(\mathrm{curl},\mathcal{O})} = \int_{\mathcal{O}} u \cdot v \, dx + \int_{\mathcal{O}} \mathrm{curl}\, u \cdot \mathrm{curl}\, v \, dx.$$

An alternative characterisation of $H(\mathrm{curl}, \mathcal{O})$ is

$$H(\mathrm{curl}, \mathcal{O}) = \text{closure of } C^\infty(\overline{\mathcal{O}})) \text{ in the } H(\mathrm{curl}, \mathcal{O}) \text{ norm},$$

which is a density result.

Furthermore, $(C_0^{0,1}(\overline{\mathcal{O}}))^3$ and $(C_0^\infty(\overline{\mathcal{O}}))^3$ are dense in $H(\mathrm{curl}, \mathcal{O})$, and there is a continuous extension from $H(\mathrm{curl}, \mathcal{O})$ into $H(\mathrm{curl}, \mathbb{R}^3)$.

3.4.5 The space $H_0(\mathrm{curl}, \mathcal{O})$

$$H_0(\mathrm{curl}, \mathcal{O}) = \text{closure of } (C_0^\infty(\overline{\mathcal{O}}))^3 \text{ in the norm of } H(\mathrm{curl}, \mathcal{O}).$$

If $\mathcal{O}$ is bounded, then

$$H_0(\mathrm{curl}, \mathcal{O}) = \{u \in H(\mathrm{curl}, \mathcal{O}) \ : \ n \times u|_\Gamma = 0\}.$$

3.4.6 The spaces $H(\mathrm{curl}0, \mathcal{O})$ and $H_0(\mathrm{curl}0, \mathcal{O})$

$$H(\mathrm{curl}0, \mathcal{O}) = \big\{u \in (L^2(\mathcal{O}))^3 : \ \mathrm{curl}\, u = 0\big\}.$$

$$H_0(\mathrm{curl}0, \mathcal{O}) := \{u \in H(\mathrm{curl}0, \mathcal{O}) \ : \ u \times n = 0 \text{ on } \Gamma\}.$$

3.4.7 The spaces $H(\mathrm{curl}, \mathrm{div}, \mathcal{O})$, $H_0(\mathrm{curl}, \mathrm{div}, \mathcal{O})$ and $H(\mathrm{curl}, \mathrm{div}0, \mathcal{O})$

$$H(\mathrm{curl}, \mathrm{div}, \mathcal{O}) := H(\mathrm{curl}, \mathcal{O}) \cap H(\mathrm{div}, \mathcal{O})\}.$$

$$H_0(\mathrm{curl}, \mathrm{div}, \mathcal{O}) := H_0(\mathrm{curl}, \mathcal{O}) \cap H_0(\mathrm{div}, \mathcal{O}).$$

$$H(\mathrm{curl}, \mathrm{div}0, \mathcal{O}) := \{u \in H(\mathrm{curl}, \mathcal{O}) \ : \ \mathrm{div}\, u = 0\}.$$

When endowed with the norm

$$\Big(|| \, u \, ||^2_{(L^2(\mathcal{O}))^3} + || \, \mathrm{curl} u \, ||^2_{(L^2(\mathcal{O}))^3} + || \, \mathrm{div} u \, ||^2_{L^2(\mathcal{O})}\Big)^{1/2},$$

$H(\mathrm{curl}, \mathrm{div}, \mathcal{O})$ is a Hilbert space. Moreover[11], $H(\mathrm{curl}, \mathrm{div}, \mathcal{O}) \subset (H^1_{loc}(\mathcal{O}))^3$, and the following identification holds:

$$(H_0^1(\mathcal{O}))^3 = H_0(\mathrm{curl}, \mathrm{div}, \mathcal{O}).$$

[11] For the definition of $H^1_{loc}(\mathcal{O})$, see Section 3.4.12 below.

3.4.8 The spaces $H^1(\mathrm{div}, \mathcal{O}), H^1(\mathrm{curl}, \mathcal{O})$

$$H^1(\mathrm{div}, \mathcal{O}) = \{u \in (H^1(\mathcal{O}))^3 : \mathrm{div}\, u \in (H^1(\mathcal{O}))^3\}.$$

$$H^1(\mathrm{curl}, \mathcal{O}) = \{u \in (H^1(\mathcal{O}))^3 : \mathrm{curl}\, u \in (H^1(\mathcal{O}))^3\}.$$

3.4.9 Spaces of normal and tangential fields

$$C_t(\Gamma) := \{v \in (C(\Gamma))^3 \, : \, v \cdot n = 0 \text{ on } \Gamma\}\,.$$

$$L^2_t(\Gamma) := \{v \in (L^2(\Gamma))^3 \, : \, v \cdot n = 0 \text{ on } \Gamma\}\,.$$

$$L^2_n(\Gamma) := \{v \in (L^2(\Gamma))^3 \, : n \times (n \times v) = 0 \text{ on } \Gamma\}\,.$$

$$H^1_t(\mathcal{O}) := \{v \in H^1(\mathcal{O}) \, : \, v \cdot n = 0 \text{ on } \Gamma\}\,.$$

$$H^1_n(\mathcal{O}) := \{v \in H^1(\mathcal{O}) \, : \, n \times (n \times v) = 0 \text{ on } \Gamma\}\,.$$

$$H^{\mathsf{s}}_t(\Gamma) := \{v \in (H^{\mathsf{s}}(\Gamma))^3 \, : \, n \cdot v = 0 \text{ on } \Gamma\}, \ \mathsf{s} \in \mathbb{R}.$$

$H^{-\mathsf{s}}_t(\Gamma)$ is the dual space of $H^{\mathsf{s}}_t(\Gamma)$, with $L^2_t(\Gamma)$ as a pivot space[12].

3.4.10 The spaces $H^{-1/2}(\mathrm{div}, \Gamma)$ and $H^{-1/2}(\mathrm{curl}, \Gamma)$

Let $\mathcal{O}$ be a regular open set with $C^{1,1}$ boundary Γ. We define

$$H^{-1/2}(\mathrm{div}, \Gamma) = \{v \in (H^{-1/2}(\Gamma))^3 : n \cdot v = 0 \ , \ \mathrm{Div}\, v \in H^{-1/2}(\Gamma)\}.$$

$$H^{-1/2}(\mathrm{curl}, \Gamma) = \{v \in (H^{-1/2}(\Gamma))^3 : n \cdot v = 0 \ , \ \mathrm{Curl}\, v \in H^{-1/2}(\Gamma)\}.$$

These are both Hilbert spaces. These spaces are related in the sense that if $v \in H^{-1/2}(\mathrm{div}, \Gamma)$, then $n \times v \in H^{-1/2}(\mathrm{curl}, \Gamma)$, and if $v \in H^{-1/2}(\mathrm{curl}, \Gamma)$ then $n \times v \in H^{-1/2}(\mathrm{div}, \Gamma)$.

The space $H^{-1/2}(\mathrm{div}, \Gamma)$ is the dual space of $H^{-1/2}(\mathrm{curl}, \Gamma)$, with $L^2_t(\Gamma)$ the pivot space, and the following duality property holds:

$$\langle n \times u, v \rangle = - \langle u, n \times v \rangle \ , \ \forall u, v \in H^{-1/2}(\mathrm{curl}, \Gamma),$$

as verified by Green's formulae.

3.4.11 The spaces $H^{1/2}(\mathrm{div}, \Gamma)$ and $H^{1/2}(\mathrm{curl}, \Gamma)$

$$H^{1/2}(\mathrm{div}, \Gamma) = \{v \in (H^{1/2}(\Gamma))^3 : n \cdot v = 0 \ , \ \mathrm{Div}\, v \in H^{1/2}(\Gamma)\}.$$

$$H^{1/2}(\mathrm{curl}, \Gamma) = \{v \in (H^{1/2}(\Gamma))^3 : n \cdot v = 0 \ , \ \mathrm{Curl}\, v \in H^{1/2}(\Gamma)\}.$$

---

[12] See Section A.1 in Appendix A.

3.4.12 Spaces of locally square integrable functions

In the framework of exterior problems we will often make use of the spaces

$$L^2_{loc}(\mathcal{O}_e) = \{u : u \in L^2(Q \cap \mathcal{O}_e) \text{ for every cube } Q \subset \mathbb{R}^3\}.$$

$$L^2_{loc}(\text{curl}, \mathcal{O}_e) = L^2_{loc}(\mathcal{O}_e) \cap \{u : \text{curl} u \in L^2_{loc}(\mathcal{O}_e)\}.$$

$$L^2_{loc}(\text{curl curl}, \mathcal{O}_e) = L^2_{loc}(\text{curl}, \mathcal{O}_e) \cap \{u : \text{curl curl} u \in L^2_{loc}(\mathcal{O}_e)\}.$$

The meaning of, e.g., $H^1_{loc}(\mathcal{O}_e)$ is then clear.

We will further need the following spaces:

$$H^1_{\text{Div}}(\mathcal{O}) = \{u \in H^1(\mathcal{O}) : \text{ Div}(n \times u) \in H^{1/2}(\Gamma)\}.$$

$$H^1_{\text{Div},\, loc}(\mathcal{O}_e) = \{u \in H^1_{loc}(\mathcal{O}_e) : \text{ Div}(n \times u) \in H^{1/2}(\Gamma)\}.$$

$$H^1_{\text{Div},\, loc}(\Gamma) = \{u \in H^{1/2}_t(\Gamma) : \text{ Div} u \in H^{1/2}(\Gamma)\}.$$

3.5 TRACES

In what follows we collect some results concerning traces. For details, readers may consult [91], [126], [334].

In Section 3.3.2 the trace operators γ_n, γ_τ, π_Γ were defined formally, for sufficiently smooth functions (e.g., in $C^\infty(\overline{\mathcal{O}})$). These operators have continuous extensions in function spaces consisting of vector fields of less regularity. Next we summarise the properties of these continuous extensions.

3.5.1 Properties of the normal trace operator γ_n

PROPOSITION 3.5.1 *Assume that $\mathcal{O}$ is locally on one side of its boundary Γ, with Γ bounded and Lipschitz. Then*

(*i*) $\gamma_n : H(\text{div}, \mathcal{O}) \to H^{-1/2}(\Gamma)$ *is continuous and onto.*

(*ii*) $\ker \gamma_n = H_0(\text{div}, \mathcal{O})$.

3.5.2 Properties of the tangential trace operator γ_τ

PROPOSITION 3.5.2 *Assume that $\mathcal{O}$ is locally on one side of its boundary Γ, with Γ bounded and Lipschitz. Then*

(*i*) $\gamma_\tau : H(\text{curl}, \mathcal{O}) \to (H^{-1/2}(\Gamma))^3$ *is continuous but not necessarily onto.*

(*ii*) $\ker \gamma_\tau = H_0(\text{curl}, \mathcal{O})$.

PROPOSITION 3.5.3 *In addition to the assumptions of Proposition 3.5.2 assume further that*[13] *Γ is $C^{1,1}$. Then the following mappings are continuous and onto:*

[13] It is not known whether this regularity condition is optimal [91]. This comment applies wherever $C^{1,1}$ regularity is assumed in this section.

(*i*) $\gamma_\tau : H(\mathrm{curl}, \mathcal{O}) \to H^{-1/2}(\mathrm{div}, \Gamma)$.

(*ii*) $\gamma_\tau : H^1(\mathrm{curl}, \mathcal{O}) \to H^{1/2}(\mathrm{div}, \Gamma)$.

(*iii*) $\gamma_\tau : H(\mathrm{curl}, \mathrm{div}0, \mathcal{O}) \to H^{-1/2}(\mathrm{div}, \Gamma)$.

3.5.3 Properties of the tangential components trace operator π_Γ

PROPOSITION 3.5.4 *Assume that $\mathcal{O}$ is locally on one side of its boundary Γ, with Γ bounded and Lipschitz. Then*

(*i*) $\pi_\Gamma : H(\mathrm{curl}, \mathcal{O}) \to (H^{-1/2}(\Gamma))^3$ *is continuous.*

(*ii*) $\ker \pi_\Gamma = H_0(\mathrm{curl}, \mathcal{O})$.

PROPOSITION 3.5.5 *In addition to the assumptions of Proposition 3.5.4 assume further that Γ is $C^{1,1}$. Then the following mappings are continuous and onto:*

(*i*) $\pi_\Gamma : H(\mathrm{curl}, \mathcal{O}) \to H^{-1/2}(\mathrm{curl}, \Gamma)$.

(*ii*) $\pi_\Gamma : H^1(\mathrm{curl}, \mathcal{O}) \to H^{1/2}(\mathrm{curl}, \Gamma)$.

(*iii*) $\pi_\Gamma : H(\mathrm{curl}, \mathrm{div}0, \mathcal{O}) \to H^{-1/2}(\mathrm{curl}, \Gamma)$.

REMARK 3.5.6 ([91]) For the exterior domain $\mathcal{O}_e$, with $\Gamma \in C^{1,1}$, the trace mappings γ_τ, π_Γ defined on $H_{loc}(\mathrm{curl}, \overline{\mathcal{O}_e}) := \{v \in L^2_{loc}(\mathcal{O}_e, \mathbb{C}^3) : v\zeta \in H(\mathrm{curl}, \mathcal{O}_e),\ \forall \zeta \in C_0^\infty(\mathbb{R}^3)\}$ are also continuous and onto.

3.6 VARIOUS DECOMPOSITIONS

3.6.1 Orthogonal decompositions of $(L^2(\mathcal{O}))^3$

PROPOSITION 3.6.1 *Let $\mathcal{O}$ be a connected open set in $\mathbb{R}^3$, with a Lipschitz boundary Γ. Then we have the following orthogonal decompositions:*

(*i*) $(L^2(\mathcal{O}))^3 = \mathrm{grad}\, H^1(\mathcal{O}) \oplus H_0(\mathrm{div}0, \mathcal{O})$.

(*ii*) $(L^2(\mathcal{O}))^3 = \mathrm{grad}\, H_0^1(\mathcal{O}) \oplus H(\mathrm{div}0, \mathcal{O})$.

(*iii*) $(L^2(\mathcal{O}))^3 = \mathrm{grad}\, H_0^1(\mathcal{O}) \oplus \mathrm{curl}\,(H^1(\mathcal{O}))^3$.

(*iv*) $(L^2(\mathcal{O}))^3 = \mathrm{grad}\, H^1(\mathcal{O}) \oplus \mathrm{curl}\,(H_0^1(\mathcal{O}))^3$.

The above decompositions are customarily called *Helmholtz decompositions*. The typical form in which they are most often used is the following: consider, for example, the third case above; we can then uniquely decompose each element $u \in (L^2(\mathcal{O}))^3$ as $u = \mathrm{grad}\, p + \mathrm{curl}\, w$, where $p \in H_0^1(\mathcal{O})$ and $w \in (H^1(\mathcal{O}))^3$.

REMARK 3.6.2 These decompositions have generalisations for nonsimply connected domains, which are usually referred to as Hodge decompositions (see, e.g., [91]).

3.6.2 Decompositions in terms of normal and tangential spaces

For the definitions of the relevant spaces, see Section 3.4.9.

PROPOSITION 3.6.3 *The following orthogonal decomposition is true:*

$$(L^2(\Gamma))^3 = L_t^2(\Gamma) \oplus L_n^2(\Gamma)\,.$$

REMARK 3.6.4 If $\mathcal{O}$ is regular in the sense defined above, we cannot have a similar decomposition for $(H^{1/2}(\Gamma))^3$. But if $\mathcal{O}$ is $C^{1,\alpha}$ for $\alpha \in (\frac{1}{2}, 1)$, then

$$(H^{1/2}(\Gamma))^3 = H_t^{1/2}(\Gamma) \oplus H_n^{1/2}(\Gamma)\,,$$

while, by duality, an analogous decomposition is true for $(H^{-1/2}(\Gamma))^3$.

3.7 COMPACT EMBEDDINGS

Compact embeddings (denoted by $\overset{c}{\hookrightarrow}$) are certainly of sine qua non importance in establishing the solvability of boundary value problems. The embedding $X \overset{c}{\hookrightarrow} Y$ is compact if the injection $\mathfrak{J} : X \to Y$ is a compact operator. This allows turning, up to subsequences, a weak convergence in X into a strong convergence in Y (see Section A.2 in Appendix A).

We list here some compactness results, starting with the compactness version - bearing the name Rellich-Kondrachov theorem - of the famous *Sobolev embedding theorem.*

THEOREM 3.7.1 (RELLICH-KONDRACHOV THEOREM) *Let $\mathcal{O} \subset \mathbb{R}^N$ be a (not necessarily bounded) domain with Lipschitz boundary Γ, and let $\mathcal{O}_0 \subseteq \mathcal{O}$ be any bounded subdomain of $\mathcal{O}$. Let $m \in \mathbb{N}$, $j \in \mathbb{N}_0$ and $p \in [1, \infty)$.*

(i) If $m\,p \leq N$, then

  (a) $W^{j+m,p}(\mathcal{O}) \overset{c}{\hookrightarrow} W^{j,q}(\mathcal{O}_0)$ for $mp < N$ and $q \in [1, Np/(N - mp))$.

  (b) $W^{j+m,p}(\mathcal{O}) \overset{c}{\hookrightarrow} W^{j,q}(\mathcal{O}_0)$ for $mp = N$ and $q \in [1, \infty)$.

(ii) If $mp > N$, then

  (a) $W^{j+m,p}(\mathcal{O}) \overset{c}{\hookrightarrow} W^{j,q}(\mathcal{O}_0)$, $q \in [1, \infty)$.

  (b) $W^{j+m,p}(\mathcal{O}) \overset{c}{\hookrightarrow} C^j(\overline{\mathcal{O}}_0)$.

  (c) $W^{j+m,p}(\mathcal{O}) \overset{c}{\hookrightarrow} C^{j,\alpha}(\overline{\mathcal{O}}_0)$, $\alpha \in (0, m - n/p)$.

REMARK 3.7.2 If $\mathcal{O}$ is an arbitrary domain in $\mathbb{R}^N$, all the above embeddings are compact, provided that $W^{j+m,p}(\mathcal{O})$ is replaced by $W_0^{j+m,p}(\mathcal{O})$. In the case where $\mathcal{O}$ is an unbounded domain, compact embeddings of $W_0^{m,p}(\mathcal{O})$ are studied in [1].

COROLLARY 3.7.3 *For $N = 3$, it holds that $H^1(\mathcal{O}) \overset{c}{\hookrightarrow} L^q(\mathcal{O})$ for $q \in [1, 6]$.*

Regarding fractional order spaces, one of the many existing compact embeddings results is [309].

THEOREM 3.7.4 *Let $\mathcal{O} \subset \mathbb{R}^N$ be a bounded domain with Lipschitz boundary Γ. If $0 \leq \mathsf{s}_1 < \mathsf{s}_2 < \infty$, then $H^{\mathsf{s}_2}(\mathcal{O}) \overset{c}{\hookrightarrow} H^{\mathsf{s}_1}(\mathcal{O})$.*

One of the main differences between the spaces $(H^1(\mathcal{O}))^3$ and $H(\text{curl}, \mathcal{O})$, for regular bounded $\mathcal{O}$, is that the embeddings into $(L^2(\mathcal{O}))^3$ of $H(\text{curl}, \mathcal{O})$, $H(\text{div}, \mathcal{O})$, $H(\text{curl}, \text{div}, \mathcal{O})$, and their intersections with ker curl or ker div are *not* compact, since all these spaces contain the space $\mathsf{H}(\mathcal{O}) := \{u \in (L^2(\mathcal{O}))^3 \,:\, \text{curl}\, u = 0, \text{div}\, u = 0\}$, which is an infinite-dimensional closed subspace of $(L^2(\mathcal{O}))^3$. Nevertheless, there are related compact embeddings (see [91]):

THEOREM 3.7.5 *Let $\mathcal{O} \subset \mathbb{R}^3$ be a bounded domain with Lipschitz boundary Γ. Then*

(*i*) $\{u \in H(\text{curl}, \text{div}, \mathcal{O}) : n \times u|_\Gamma \in (L^2(\Gamma))^3\} \overset{c}{\hookrightarrow} (L^2(\mathcal{O}))^3$,

(*ii*) $\{u \in H(\text{curl}, \text{div}, \mathcal{O}) : n \cdot u|_\Gamma \in L^2(\Gamma)\} \overset{c}{\hookrightarrow} (L^2(\mathcal{O}))^3$,

(*iii*) $\{u \in H(\text{curl}, \mathcal{O}) : \text{div}(\zeta\, u) \in L^2(\mathcal{O}),\, n \times u|_\Gamma \in (L^2(\Gamma))^3\} \overset{c}{\hookrightarrow} (L^2(\mathcal{O}))^3$,

(*iv*) $\{u \in H(\text{curl}, \mathcal{O}) : \text{div}(\zeta\, u) \in L^2(\mathcal{O}),\, n \cdot u|_\Gamma \in L^2(\Gamma)\} \overset{c}{\hookrightarrow} (L^2(\mathcal{O}))^3$,

(*v*) $\{u \in H(\text{curl}, \mathcal{O}) : \int_{\mathcal{O}} (\zeta\, u \cdot \text{grad}\, \psi) = 0\,,\ \forall \psi \in H^1(\mathcal{O})\} \overset{c}{\hookrightarrow} (L^2(\mathcal{O}))^3$,

(*vi*) $\{u \in H_0(\text{curl}, \mathcal{O}) : \text{div}(\zeta\, u) = \text{grad}(\zeta\, \widetilde{\zeta}) \cdot \text{curl}\, u \in L^2(\mathcal{O})\} \overset{c}{\hookrightarrow} (L^2(\mathcal{O}))^3$,

where ζ in cases (iii), (iv), (v), (vi) and $\widetilde{\zeta}$ in case (vi), are positive, bounded away from 0, real-valued functions in $L^\infty(\mathcal{O})$.

3.8 THE OPERATORS OF VECTOR ANALYSIS REVISITED

3.8.1 The operators on function spaces

The action of the grad, curl, and div, operators in function spaces is summarised in the following de Rham diagrams:

$$H^1(\mathcal{O}) \overset{\text{grad}}{\longrightarrow} H(\text{curl}, \mathcal{O}) \overset{\text{curl}}{\longrightarrow} H(\text{div}, \mathcal{O}) \overset{\text{div}}{\longrightarrow} L^2(\mathcal{O}).$$

$$H_0^1(\mathcal{O}) \overset{\text{grad}}{\longrightarrow} H_0(\text{curl}, \mathcal{O}) \overset{\text{curl}}{\longrightarrow} H_0(\text{div}, \mathcal{O}) \overset{\text{div}}{\longrightarrow} L^2(\mathcal{O}).$$

The ranges of the operators grad, curl, and div are closed in the corresponding $L^2(\mathcal{O})$ spaces.

3.8.2 Regularity

Let $\mathcal{O}$ be open and bounded in $\mathbb{R}^3$, with Γ being $C^{k+1,1}$ regular. The following regularity results (see [126]) hold.

PROPOSITION 3.8.1 *If,* $u \in (L^2(\mathcal{O}))^3, \mathrm{curl} u \in (H^k(\mathcal{O}))^3, \mathrm{div} u \in H^k(\mathcal{O})$, *and either* $\gamma_n(u) \in H^{k+1/2}(\Gamma)$ *or* $\gamma_\tau(u) \in (H^{k+1/2}(\Gamma))^3$, *then* $u \in (H^{k+1}(\mathcal{O}))^3$.

3.8.3 Integral formulae revisited

Green's integral formulae can be generalised (see, e.g., [91], [324]). Some of the mostly used generalisations are:

$$\int_{\mathcal{O}} \mathfrak{u}\cdot \mathrm{grad}\, \phi \, dx + \int_{\mathcal{O}} \mathrm{div}\, \mathfrak{u}\, \phi \, dx = \langle \gamma_n(\mathfrak{u})\, \phi \rangle \,, \, \forall \mathfrak{u} \in H(\mathrm{div}, \mathcal{O})\,, \phi \in H^1(\mathcal{O}).$$

$$\int_{\mathcal{O}} \mathrm{curl}\, \mathfrak{v} \cdot \mathfrak{w} \, dx - \int_{\mathcal{O}} \mathfrak{v} \cdot \mathrm{curl}\, \mathfrak{w} \, dx = \langle \gamma_\tau(\mathfrak{v}), \pi_\Gamma(\mathfrak{w}) \rangle \,, \, \forall \mathfrak{v}, \mathfrak{w} \in H(\mathrm{curl}, \mathcal{O})\,.$$

3.8.4 Jump relations

The jump relations of Section 3.3.3.3 also hold in the following setting (see [91]), $\phi|_{\mathcal{O}} \in H^1(\mathcal{O})$ and $\phi|_{\mathcal{O}_e} \in H^1_{loc}(\mathcal{O}_e)$, or $\mathfrak{u}|_{\mathcal{O}} \in H(\mathrm{div}, \mathcal{O})$ and $\mathfrak{u}|_{\mathcal{O}_e} \in H_{loc}(\mathrm{div}, \mathcal{O}_e)$, or $\mathfrak{u}|_{\mathcal{O}} \in H(\mathrm{curl}, \mathcal{O})$ and $\mathfrak{u}|_{\mathcal{O}_e} \in H_{loc}(\mathrm{curl}, \mathcal{O}_e)$.

3.8.5 The kernels of curl and div

Let ker and ran denote the kernel and range of an operator, respectively. The kernels of curl and div are characterised by the celebrated Poincaré lemma:

LEMMA 3.8.2 (POINCARÉ LEMMA)

$$\ker \mathrm{curl} = \mathit{ran}\ \mathrm{grad} = \mathrm{grad}\, H^1(\mathcal{O})\,.$$

$$\ker \mathrm{div} = \mathit{ran}\ \mathrm{curl} = \mathrm{curl}\, (H^1(\mathcal{O}))^3 = \mathrm{curl}\, H(\mathrm{curl}, \mathcal{O})\,.$$

Related results can be proved when boundary conditions are involved. In particular

$$H_0(\mathrm{curl} 0, \mathcal{O}) = \mathrm{grad}\, H_0^1(\mathcal{O}).$$

$$H_0(\mathrm{div} 0, \mathcal{O}) = \mathrm{curl}\, (H_0^1(\mathcal{O}))^3 = \mathrm{curl}\, H_0(\mathrm{curl}, \mathcal{O}).$$

3.8.6 Properties of tangential differential operators

The definitions of Section 3.3.4 are formal; they become precise in the framework of Sobolev spaces [91].

PROPOSITION 3.8.3 *The following mappings are continuous:*

(*i*) Grad : $H^{1/2}(\Gamma) \to H_t^{-1/2}(\Gamma)$.

(*ii*) Div : $H_t^{1/2}(\Gamma) \to H^{-1/2}(\Gamma)$.

(*iii*) $\text{Curl} : H_t^{1/2}(\Gamma) \to H^{-1/2}(\Gamma)$.

(*iv*) $\overrightarrow{\text{Curl}} : H^{1/2}(\Gamma) \to H_t^{-1/2}(\Gamma)$.

3.9 THE MAXWELL OPERATOR

We collect in this section some key results concerning the Maxwell operator

$$\mathsf{M} := \begin{pmatrix} 0 & \text{curl} \\ -\text{curl} & 0 \end{pmatrix}$$

in various function space settings.

3.9.1 The spaces $\mathbb{X}$, $\mathbb{X}_{\mathsf{M}}$, $\mathfrak{X}_{\mathsf{M}}$

We will extensively use the following abbreviations:

$$\mathsf{X} := (L^2(\mathcal{O}))^3.$$

$$\mathbb{X} := \mathsf{X} \times \mathsf{X} = (L^2(\mathcal{O}))^3 \times (L^2(\mathcal{O}))^3.$$

These are Hilbert spaces when equipped with the inner products

$$(\phi_i, \psi_i)_{\mathsf{X}} = \int_{\mathcal{O}} \phi_i \cdot \overline{\psi_i}\, dx, \quad i = 1, 2.$$

$$(\phi, \psi)_{\mathbb{X}} = \int_{\mathcal{O}} (\phi_1 \cdot \overline{\psi_1} + \phi_2 \cdot \overline{\psi_2}) \;\; dx,$$

respectively, where $\phi = (\phi_1, \phi_2)^{tr}$, $\psi = (\psi_1, \psi_2)^{tr}$, and the overbar denotes complex conjugation.

The weighted versions of the above Hilbert spaces are often useful; the inner products are then defined as

$$(\phi_i, \psi_i)_{\mathsf{X}, \mathfrak{w}_i} = \int_{\mathcal{O}} \mathfrak{w}_i \phi_i \cdot \overline{\psi_i}\, dx, \quad i = 1, 2\,,$$

$$(\phi, \psi)_{\mathbb{X}, \mathfrak{w}} = \int_{\mathcal{O}} (\mathfrak{w}_1\, \phi_1 \cdot \overline{\psi_1} + \mathfrak{w}_2\, \phi_2 \cdot \overline{\psi_2})\, dx\,,$$

where $\mathfrak{w}_1$ and $\mathfrak{w}_2$ are appropriately chosen regular-weight functions. A convenient choice for these weights in electromagnetics is $\mathfrak{w}_1 = \varepsilon$, $\mathfrak{w}_2 = \mu$.

THEOREM 3.9.1 *$H(\text{curl}, \mathcal{O})$ and $H_0(\text{curl}, \mathcal{O})$ are dense subspaces of* X.

Additionally, we consider

$$\mathbb{X}_{\mathsf{M}} := H_0(\text{curl}, \mathcal{O}) \times H(\text{curl}, \mathcal{O}),$$

$$\mathfrak{X}_1 := H_0(\text{curl}, \mathcal{O}) \cap H(\text{div}0, \mathcal{O}),$$

$$\mathfrak{X}_2 := H(\text{curl}, \mathcal{O}) \cap H_0(\text{div}0, \mathcal{O}),$$

$$\mathfrak{X}_{\mathsf{M}} := \mathfrak{X}_1 \times \mathfrak{X}_2 \,.$$

REMARK 3.9.2 Observe that the perfect conductor boundary condition is absorbed in the first component of the above Cartesian products.

In the case of spatially dependent coefficients one may consider weighted variants of the above spaces, e.g., using $H(\mathrm{div}_\varepsilon 0, \mathcal{O}) = \{v \in \mathsf{X} : \mathrm{div}(\varepsilon v) = 0\}$ instead of $H(\mathrm{div}0, \mathcal{O})$.

3.9.2 Properties of M

The simplest case to consider is electromagnetic fields of finite energy and work in an L^2 setting. The proper definition of the Maxwell operator in this case is $\mathsf{M} : D(\mathsf{M}) = \mathbb{X}_{\mathsf{M}} \to \mathbb{X}$.

The following results are well known.

THEOREM 3.9.3 *Consider the operator* $\mathsf{M} : D(\mathsf{M}) = \mathbb{X}_{\mathsf{M}} \to \mathbb{X}$. *Then*

(*i*) *The domain* $D(\mathsf{M})$ *is dense in* $\mathbb{X}$ *and* M *is closed.*

(*ii*) *The operator* M *is skew adjoint, i.e.,* $\mathsf{M}^* = -\mathsf{M}$ *and* $D(\mathsf{M}) = D(\mathsf{M}^*)$.

Consequently, Stone's theorem (see Theorem A.8.10 in Appendix A) implies the following:

THEOREM 3.9.4 *The operator* $\mathsf{M} : \mathbb{X}_{\mathsf{M}} \to \mathbb{X}$ *is the generator of a unitary group* $\{\mathsf{T}_{\mathsf{M}}(t)\}_{t\in\mathbb{R}}$ *on* $\mathbb{X}$.

This equivalently means that both M and $-\mathsf{M}$ are generators of strongly continuous semigroups. More precisely, M generates $(\mathsf{T}_{\mathsf{M}}(t))_{t\geq 0}$, whereas $-\mathsf{M}$ generates $(\mathsf{T}_{\mathsf{M}}(t)^*)_{t\geq 0}$. Observe, incidentally, that $\mathsf{T}_{\mathsf{M}}(t)^* = \mathsf{T}_{\mathsf{M}}(t)^{-1}$.

REMARK 3.9.5 It is important to notice that $-\mathsf{M}$ actually appears in the abstract evolution equation (2.10) after the change of variable $t \mapsto -t$, which corresponds to an inversion of time. Since we are studying causal models, dealing with fields that vanish for $t < 0$, we actually use only the semigroup $\mathsf{T}_{\mathsf{M}}(t)$ for $t \geq 0$. However, the semigroup $\mathsf{T}_{\mathsf{M}}(t)$ for $t < 0$ is very useful in certain settings, particularly in problems related to controllability (see Chapter 8) and scattering theory (see Chapter 10).

Other choices are possible. For instance, when we are interested in considering divergence-free solutions, then we may define $\mathsf{M} : D(\mathsf{M}) = \mathfrak{X}_{\mathsf{M}} \to \mathbb{X}$. In this case we have the following theorem.

THEOREM 3.9.6 *Consider the operator* $\mathsf{M} : D(\mathsf{M}) = \mathfrak{X}_{\mathsf{M}} \to \mathbb{X}$. *Then*

(*i*) *The domain* $D(\mathsf{M})$ *is dense in* $\mathbb{X}$ *and* M *is closed.*

(*ii*) *The operator* M *is skew adjoint, i.e.,* $\mathsf{M}^* = -\mathsf{M}$ *and* $D(\mathsf{M}) = D(\mathsf{M}^*)$.

Using the same arguments as above, we obtain (see, e.g., [138]) Theorem 3.9.7.

THEOREM 3.9.7 *The operator* $\mathsf{M} : \mathfrak{X}_{\mathsf{M}} \to \mathbb{X}$ *is the generator of a unitary group* $\{\mathsf{T}_{\mathsf{M}}(t)\}_{t\in\mathbb{R}}$ *on* $\mathbb{X}$.

REMARK 3.9.8 The properties of the Maxwell operator, as described in Theorems 3.9.4 or 3.9.7, are also valid in more general functional settings; for instance, for $0 < p, q < \infty$ the above theorem still holds (see, e.g., [426]) if M is defined on

$$D(\mathsf{M}) = \mathbb{U}_1 \times \mathbb{U}_2\,,$$
$$\mathbb{U}_1 = \{u_1 \in (L^p(\mathcal{O}))^3,\ \mathrm{curl} u_1 \in (L^q(\mathcal{O}))^3,\ n \times u_1|_{\partial\mathcal{O}} = 0\}\,,$$
$$\mathbb{U}_2 = \{u_2 \in (L^2(\mathcal{O}))^3,\ \mathrm{curl} u_2 \in (L^{\frac{p}{p-1}}(\mathcal{O}))^3\}\,.$$

3.9.3 Modified Maxwell operators

It is often convenient to work with a modified version of the Maxwell operator M, namely, with the operator M_{A} defined by

$$\mathsf{M}_{\mathrm{A}} = \mathsf{A}_{\mathrm{or}}^{-1}\mathsf{M},$$

where A_{or} is the 6×6 matrix (see (2.12)):

$$\mathsf{A}_{\mathrm{or}}(x) = \begin{pmatrix} \varepsilon(x) & \xi(x) \\ \zeta(x) & \mu(x) \end{pmatrix},$$

ε, ξ, ζ and μ being 3×3 matrices whose entries are measurable and essentially bounded functions for $x \in \mathcal{O}$. Under sufficient conditions on these matrices, the modified Maxwell operator M_{A} enjoys analogous properties as those of M. In particular, the following result holds [205].

THEOREM 3.9.9 *Under Assumption 2.3.5 for* $\mathsf{A}_{or}(x)$, $\mathsf{M}_{A} : \mathbb{X}_{\mathsf{M}} \to \mathbb{X}$ *is the generator of a strongly continuous group* $\{\mathsf{T}_{\mathsf{M}_A}(t)\}_{t\in\mathbb{R}}$ *on* $\mathbb{X}$.

Further, the above result (under additional regularity assumptions on A_{or}) holds also in the following setting (see, e.g., [138]).

THEOREM 3.9.10 *The operator* $\mathsf{M}_{A} : \mathfrak{X}_{\mathsf{M}} \to \mathbb{X}$ *is the generator of a strongly continuous group* $\{\mathsf{T}_{\mathsf{M}_A}(t)\}_{t\in\mathbb{R}}$ *on* $\mathbb{X}$.

REMARK 3.9.11 In a number of applications, instead of A_{or} we use its diagonal version A_0 (see (2.13)). It is obvious how the above results can be applied to the arising modified Maxwell operator $\mathsf{A}_0^{-1}\mathsf{M}$.

REMARK 3.9.12 For an introduction to operator theory for electromagnetics, see [126], [180], [278].

PART 2

Time-Harmonic Deterministic Problems

Chapter Four

Well Posedness

4.1 INTRODUCTION

Time-harmonic problems constitute an important class within the theory of electromagnetics in complex media. In treating such problems we assume that the temporal evolution of these fields is periodic, with a fixed and prescribed period ϖ. This class covers a variety of important applications, such as wave guide problems, scattering problems, etc. Under the assumption of temporally periodic fields we may use the constitutive relations introduced in Section 2.3.3, which do not contain the convolution terms. The absence of the convolution terms simplifies the treatment of the Maxwell equations, (PDEs) which in this case become partial differential equations depending only on the spatial variables. As we shall see in this chapter, these differential equations, after proper reductions, can lead to problems that can be treated using the theory of variational PDEs, and methods akin to those used in elliptic problems.

In this chapter we present rigorous mathematical results concerning the solvability and well posedness of time-harmonic problems for complex electromagnetic media, with a special emphasis on chiral media. The structure of the chapter is as follows: In Section 4.2 we treat the solvability and well posedness of the interior time-harmonic problem and in Section 4.3 we present some results concerning eigenvalue problems in cavities filled with complex electromagnetic materials. In Section 4.4 we study the behaviour of the interior domain problem for a chiral medium in the limit of low chirality, and we show that this problem converges in the appropriate sense to the relevant problem for an achiral medium. This result may be interpreted as a continuity result (with respect to the data). In Section 4.5 we present some comments related to the well posedness and solvability of exterior problems[1]. Finally, in Section 4.6, using an appropriate finite-dimensional space (namely, the one corresponding to finite element methods) and the variational formulation of the discretised version of the original boundary value problem, we obtain numerical methods for the solution of the Maxwell equations for chiral media. We do not present the actual numerical methods but furnish the analytic framework necessary for their implementation.

[1] These problems will be treated in detail in Chapters 5 and 6, where scattering problems in *homogeneous* chiral media are studied.

4.2 SOLVABILITY OF THE INTERIOR PROBLEM

In this section we consider the solvability of the interior time-harmonic problem, assuming nonconstant coefficients. In view of the Drude-Born-Fedorov constitutive relations (see Section 2.3.3, equation (2.21), in the following we drop the subscript "DBF" from all material parameters), and the Maxwell equations in a domain $\mathcal{O} \subset \mathbb{R}^3$ become

$$\begin{aligned} \operatorname{curl} E &= \beta\gamma^2 E + i\varpi\mu \left(\tfrac{\gamma}{k}\right)^2 H, \\ \operatorname{curl} H &= \beta\gamma^2 H - i\varpi\varepsilon \left(\tfrac{\gamma}{k}\right)^2 E, \end{aligned} \qquad \text{in } \mathcal{O}, \tag{4.1}$$

where $\varpi > 0$ is the angular frequency,

$$k^2 = \varpi^2\varepsilon\mu \quad , \quad \gamma^2 = k^2(1-\beta^2k^2)^{-1}. \tag{4.2}$$

These equations are complemented with the boundary condition

$$n \times E = f, \quad \text{on } \partial\mathcal{O}, \tag{4.3}$$

where[2] $f \in H^{-1/2}(\operatorname{div}, \partial\mathcal{O})$ is a prescribed electric field on $\partial\mathcal{O}$.

If $f = 0$, then we have a homogeneous problem. This problem will have nontrivial solutions for specific values of the frequency ϖ as long as the other parameters of the problem are fixed. This is equivalent to an eigenvalue problem that will reveal the resonant frequencies of $\mathcal{O}$ (the cavity problem).

Assume that we are away from such an eigenvalue. Then we will prove that the interior problem is well posed. Our approach closely follows [11] and [8]. See also [101]. Throughout this chapter we make the following assumption (see [11]).

ASSUMPTION 4.2.1

(*i*) The boundary $\partial\mathcal{O}$ is of class $C^{1,1}$.

(*ii*) The coefficients ε, μ and β are real valued and positive $C^2(\overline{\mathcal{O}})$ functions.

(*iii*) The function $\alpha := \mu^{-1}(1 - \varpi^2\varepsilon\mu\beta^2)$ is positive in $\overline{\mathcal{O}}$.

The assumption $\alpha > 0$ is plausible since the chirality parameter β is usually considered to be small. Furthermore, the above regularity requirements can be relaxed.

THEOREM 4.2.2 *Assume that ϖ is not an eigenvalue of the cavity problem (4.1). Then the boundary value problem (4.1)-(4.3) has a unique solution in $H(\operatorname{curl}, \mathcal{O}) \times H(\operatorname{curl}, \mathcal{O})$ for every $f \in H^{-1/2}(\operatorname{div}, \partial\mathcal{O})$.*

The proof is performed in the following steps.

STEP 1 In order to study the solvability of the interior problem, it is convenient to rewrite the system in terms of the electric field only. To this end,

---

[2]For the definition of the trace space $H^{-1/2}(\operatorname{div}, \partial\mathcal{O})$, see Section 3.4.10.

we apply the curl operator on the first equation (4.1) (after multiplying both sides by $\mu^{-1}k^2\gamma^{-2}$), and using well-known vector identities we find

$$\operatorname{curl}\left(\mu^{-1}(1-\varpi^2\varepsilon\mu\beta^2)\operatorname{curl}E\right) = \varpi^2\operatorname{curl}(\beta\varepsilon E) + i\varpi\operatorname{curl}H. \tag{4.4}$$

In the above, we have used the definition of k^2 given in (4.2).

We now rewrite curl$H$ in terms of the electric field E, using a linear combination of (4.1), as

$$i\varpi\operatorname{curl}H = k^2\beta\mu^{-1}\operatorname{curl}E + \varpi^2\varepsilon E. \tag{4.5}$$

Substituting (4.5) into (4.4), we obtain a single equation for E,

$$\operatorname{curl}(\alpha\operatorname{curl}E) = \varpi^2(\operatorname{curl}(\beta\varepsilon E) + \beta\varepsilon\operatorname{curl}E + \varepsilon E). \tag{4.6}$$

This equation has to be solved with the boundary condition (4.3).

We need to obtain an appropriate weak formulation for the system (4.6) (or its equivalent form, (4.27)). To this end, we take the dot product of (4.3) with a smooth vector function ψ, integrate over the domain $\mathcal{O}$ and use the second Green's formula[3], obtaining

$$\int_{\mathcal{O}} (\alpha\operatorname{curl}E)\cdot\operatorname{curl}\overline{\psi}dx + \int_{\partial\mathcal{O}} (n\times\alpha\operatorname{curl}E)\cdot((n\times\overline{\psi})\times n)ds(x) =$$
$$+\varpi^2\int_{\mathcal{O}} (\varepsilon E + \operatorname{curl}(\varepsilon\beta E) + \varepsilon\beta\operatorname{curl}E)\cdot\overline{\psi}dx.$$

By choosing properly the test function we may eliminate the boundary term. This choice requires that the tangential component of the test function vanish on the boundary $\partial\mathcal{O}$; therefore $\psi \in H_0(\operatorname{curl},\mathcal{O})$.

This argument motivates the introduction of the bilinear form

$$a^{(\beta)}(\cdot,\cdot) : H(\operatorname{curl},\mathcal{O})\times H_0(\operatorname{curl},\mathcal{O}) \to \mathbb{C},$$

defined as

$$\begin{aligned} a^{(\beta)}(E,\psi) := &(\alpha\operatorname{curl}E,\operatorname{curl}\psi) - \varpi^2(\varepsilon E,\psi)\\ &- \varpi^2(\operatorname{curl}(\varepsilon\beta E),\psi) - \varpi^2(\varepsilon\beta\operatorname{curl}E,\psi), \end{aligned} \tag{4.7}$$

where by $(\cdot,\cdot)$ we denote the inner product in the Hilbert space $(L^2(\mathcal{O}))^3$. Our aim is to express the interior problem as a variational problem in terms of this bilinear form and use standard arguments from the theory of partial differential equations (see, e.g., [363]) to provide solvability results.

In doing so, we need to take the following steps:

STEP 2 First reduce problem (4.1) to one with a homogeneous boundary condition. We seek a solution of the form $E = U + E_0$, where $U \in (H^1(\mathcal{O}))^3$ is such that $n\times U = f$ for $x\in\partial\mathcal{O}$. This reduces to the solvability of the operator equation $\gamma_\tau(U) = f$ and is possible for every $f\in H^{-1/2}(\operatorname{div},\partial\mathcal{O})$ by the properties of the tangential trace operator (see Proposition 3.5.3). Substituting this into equation (4.6), we see that E_0 satisfies the equation

$$\operatorname{curl}(\alpha\operatorname{curl}E_0) - \varpi^2(\operatorname{curl}(\beta\varepsilon E_0) + \beta\varepsilon\operatorname{curl}E_0 + \varepsilon E_0) = F, \tag{4.8}$$

---

[3]See Section 3.8.3.

where

$$F := -\text{curl}\,(\alpha \text{curl} U) + \varpi^2(\text{curl}\,(\beta\varepsilon U) + \beta\varepsilon U + \varepsilon U), \tag{4.9}$$

with the homogeneous boundary condition

$$n \times E_0 = 0, \quad x \in \partial\mathcal{O}.$$

Since $f \in H^{-1/2}(\text{div}, \partial\mathcal{O})$, U must be in $H(\text{curl}, \mathcal{O})$, so that in general, $F \in (H^{-1}(\mathcal{O}))^3$. In view of the bilinear form $a^{(\beta)}(\cdot,\cdot)$ defined in (4.7) and the duality pairing $\langle\cdot,\cdot\rangle$ between $H(\text{curl}, \mathcal{O})$ and $(H^{-1}(\mathcal{O}))^3$, the weak form of equation (4.8) can be written as

$$a^{(\beta)}(E_0, \psi) = \langle F, \psi\rangle. \tag{4.10}$$

STEP 3 We now recall the Helmholtz decomposition (see Proposition 3.6.1 (*ii*)) and seek a field E_0 in the form $E_0 = e + \text{grad}\phi$, where $e \in (H_0^1(\mathcal{O}))^3 \cap H(\text{div}, \mathcal{O})$ and ϕ is an electrostatic potential $\phi \in H_0^1(\mathcal{O})$. Substituting this expression into (4.10), we obtain (in terms of the bilinear form $a^{(\beta)}(\cdot,\cdot)$ defined in (4.7))

$$a^{(\beta)}(e, \psi) + a^{(\beta)}(\text{grad}\phi, \psi) = \langle F, \psi\rangle,$$

which, using the vector calculus identities, reduces to

$$a^{(\beta)}(e, \psi) + b^{(\beta)}(\phi, \psi) = \langle F, \psi\rangle,\ \forall\psi \in H_0(\text{curl}, \mathcal{O})\,, \tag{4.11}$$

in terms of the new bilinear form

$$b^{(\beta)} : H_0^1(\mathcal{O}) \times H_0(\text{curl}, \mathcal{O}) \to \mathbb{C}$$

defined by

$$b^{(\beta)}(\phi, \psi) := -\varpi^2 \int_{\mathcal{O}} (\text{curl}(\varepsilon\beta\text{grad}\phi) + \varepsilon\text{grad}\phi) \cdot \overline{\psi} dx. \tag{4.12}$$

This equation contains two unknown entities, the electromagnetic field e and the electrostatic potential ϕ.

STEP 4 We eliminate first the electrostatic potential ϕ. We choose the test function ψ as $\psi = \text{grad}\xi$, $\xi \in H_0^1(\mathcal{O})$ and substitute it into (4.11). For this choice, $\text{curl}\psi = 0$, so that we obtain a vanishing contribution from the first term in the bilinear form $a^{(\beta)}(e, \text{grad}\xi)$. We now integrate the remaining terms by parts using the divergence formula, as well as the fact that $\text{div}\,\text{curl}\, v = 0$. After some algebra, the remaining terms in (4.11) are

$$\varpi^2\left(\int_{\mathcal{O}} \text{div}(\varepsilon\beta\text{curl}e + \varepsilon e)\cdot\xi dx + \int_{\mathcal{O}} \varepsilon\text{grad}\phi\cdot\text{grad}\xi\, dx\right) = \int_{\mathcal{O}} F\cdot\text{grad}\xi dx. \tag{4.13}$$

If we choose e such that $\text{div}(\varepsilon\beta\text{curl}e + \varepsilon e) = 0$, or its equivalent weak form

$$\int_{\mathcal{O}} (\,\varepsilon\beta\text{curl}e + \varepsilon e\,) \cdot \text{grad}\,\xi\, dx = 0, \ \ \forall\xi \in H_0^1(\mathcal{O}),$$

then equation (4.13) reduces to

$$\varpi^2 \int_{\mathcal{O}} \varepsilon\text{grad}\phi \cdot \text{grad}\xi\, dx = \int_{\mathcal{O}} F \cdot \text{grad}\xi\, dx, \tag{4.14}$$

which is the weak form for an elliptic equation for ϕ.

REMARK 4.2.3 The above procedure is equivalent to invoking a Helmholtz-type decomposition of $H_0(\mathrm{curl}, \mathcal{O})$ as

$\{u \in H(\mathrm{curl}, \mathcal{O}) \, : \, (\varepsilon\beta\mathrm{curl}u + \varepsilon u, \, \mathrm{grad}\,\xi) = 0 \, , \, \forall \xi \in H_0^1(\mathcal{O})\} \oplus \mathrm{grad}H_0^1(\mathcal{O})$; see Lemma 4.5 in [324].

The solvability of the elliptic equation (4.14) for the electrostatic potential is obtained by standard arguments based on the Lax-Milgram lemma (see Section A.10 in Appendix A).

LEMMA 4.2.4 *Assume that $\varpi^2 > 0$ and that Assumption 4.2.1 is valid. Then there exists a unique solution $\phi \in H_0^1(\mathcal{O})$ of (4.14) satisfying*

$$||\mathrm{grad}\phi||_{(L^2(\mathcal{O}))^3} \leq c\,||F||_{(H^{-1}(\mathcal{O}))^3} \leq C\,||f||_{H^{-1/2}(\mathrm{div},\partial\mathcal{O})} \, .$$

The first inequality is an immediate consequence of the Lax-Milgram lemma, while the second follows from (4.9) and the properties of the tangential trace operator.

STEP 5 We now turn to the determination of e. Having determined the electrostatic potential ϕ with the use of Lemma 4.2.4, we substitute that back into equation (4.8), or its equivalent weak form (4.10), and now determine e by solving the problem

$$\begin{aligned} \mathrm{curl}(\alpha\mathrm{curl}e) - \varpi^2(\mathrm{curl}(\varepsilon\beta e) + \varepsilon\beta\mathrm{curl}e + \varepsilon e) &= \\ \mathcal{F} + \varpi^2(\mathrm{curl}(\varepsilon\beta\mathrm{grad}\phi) + \varepsilon\mathrm{grad}\phi) \, , \end{aligned} \tag{4.15}$$

which in variational form can be expressed as

$$a^{(\beta)}(e, \psi) = (\mathcal{F}, \psi) + \varpi^2(\mathrm{curl}(\varepsilon\beta\mathrm{grad}\phi), \psi) + \varpi^2(\varepsilon\mathrm{grad}\phi, \psi) \, , \tag{4.16}$$

for all $\psi \in H_0(\mathrm{curl}, \mathcal{O})$. In the above problem, we have included the terms related to U in the new source term $\mathcal{F}$ to ease notation.

STEP 6 The major technical problem we need to address is related to the coercivity of the bilinear form $a^{(\beta)}$. To resolve this difficulty we need to treat $a^{(\beta)}$ as the perturbation of a coercive form and then use the theory of Fredholm operators to show the existence of a solution (see Section A.6 in Appendix A).

Consider the bilinear form

$$a_+^{(\beta)}(e, \psi) := (\alpha\,\mathrm{curl}e, \mathrm{curl}\psi) + \varpi^2(\varepsilon\beta\,\mathrm{curl}e, \psi) + \varpi^2(\varepsilon\, e, \psi) \, , \tag{4.17}$$

which by Assumption 4.2.1 is a coercive continuous form on $H(\mathrm{curl}, \mathcal{O}) \times H_0(\mathrm{curl}, \mathcal{O})$. Therefore, by the Lax-Milgram lemma, for any $\mathfrak{F} \in (H^{-1}(\mathcal{O}))^3$ the variational problem

$$a_+^{(\beta)}(e, \psi) = \langle \mathfrak{F}, \psi \rangle \tag{4.18}$$

has a unique solution.

STEP 7 We will show that the variational problem (4.16) can be treated as a perturbation of the variational problem (4.18). We rewrite the variational problem (4.16) in terms of the bilinear form $a_+^{(\beta)}$ as

$$\begin{aligned} a_+^{(\beta)}(e, \psi) - \varpi^2(\mathrm{curl}(\varepsilon\beta e), \psi) - 2\varpi^2(\varepsilon\beta\mathrm{curl}e, \psi) - 2\varpi^2(\varepsilon e, \psi) &= \\ (\mathcal{F}, \psi) + \varpi^2(\mathrm{curl}(\varepsilon\beta\mathrm{grad}\phi), \psi) + \varpi^2(\varepsilon\mathrm{grad}\phi, \psi) \, . \end{aligned} \tag{4.19}$$

Define the function space $\mathcal{N}^{(\beta)}$ as

$$\mathcal{N}^{(\beta)} = \{u \in H(\mathrm{curl}, \mathcal{O}) : (\varepsilon\beta\mathrm{curl}u + \varepsilon u, \mathrm{grad}\,\xi) = 0,\ \forall \xi \in H_0^1(\mathcal{O})\} \quad (4.20)$$

and the operator $\mathsf{K} : (L^2(\mathcal{O}))^3 \to (L^2(\mathcal{O}))^3$ by $\mathsf{K}e := \varphi$ where φ is the solution of the variational problem

$$a_+^{(\beta)}(\varphi, \psi) = -\varpi^2\{(\mathrm{curl}(\varepsilon\beta e), \psi) + 2(\varepsilon\beta\mathrm{curl}e, \psi) + 2(\varepsilon e, \psi)\} \quad (4.21)$$

for all $\psi \in \mathcal{N}^{(\beta)}$.

LEMMA 4.2.5 *The operator* K *is compact.*

Proof. First of all we notice that by the Lax-Milgram lemma, the operator K is well defined. Then we notice that if $e \in \mathcal{N}^{(\beta)}$, then $\mathsf{K}e \in \mathcal{N}^{(\beta)}$. This can be seen easily, formally at first, by taking the divergence of the above problem. We notice first that if $e \in \mathcal{N}^{(\beta)}$, then the divergence of the right-hand side vanishes. The divergence of the left-hand side equals $\mathrm{div}(\varepsilon\mathsf{K}e + \varepsilon\beta\mathrm{curl}\mathsf{K}e)$, so we conclude that $\mathsf{K}e \in \mathcal{N}^{(\beta)}$. By the regularity of the solution of (4.21) we see that for every $e \in H_0(\mathrm{curl}, \mathcal{O})$ and hence, the above formal argument becomes rigorous.

By Theorem 3.7.5 the embedding of $\mathcal{N}^{(\beta)}$ into $(L^2(\mathcal{O}))^3$ is compact[4]. By the properties of the solution of (4.21) the operator K maps bounded subsets of $(L^2(\mathcal{O}))^3$ into bounded subsets of $\mathcal{N}^{(\beta)}$; therefore, by the compactness of the embedding it follows that it maps bounded sets into precompact sets. Therefore the operator K is compact. □

STEP 8 We now define the map $\mathscr{G}$ as follows: given ϕ, let $\mathscr{G}\phi$ be the solution of the variational system

$$a_+^{(\beta)}(\mathscr{G}, \psi) = (\mathcal{F}, \psi) + \varpi^2(\mathrm{curl}(\varepsilon\beta\mathrm{grad}\phi), \psi) + \varpi^2(\varepsilon\mathrm{grad}\phi, \psi) \quad (4.22)$$

for all $\psi \in \mathcal{N}^{(\beta)}$. Again, by the coercivity of $a_+^{(\beta)}$, $\mathscr{G}$ is well defined by the Lax-Milgram lemma.

STEP 9 Combining (4.21) and (4.22), we see that we may rewrite (4.19) in the form

$$a_+^{(\beta)}(e, \psi) + a_+^{(\beta)}(\mathsf{K}e, \psi) = a_+^{(\beta)}(\mathscr{G}, \psi), \ \ \forall \psi \in H_0(\mathrm{curl}, \mathcal{O}),$$

and by the bilinear property of $a_+^{(\beta)}$ this is equivalent to the operator equation

$$(I + \mathsf{K})\, e = \mathscr{G}\,. \quad (4.23)$$

Since K is a compact operator, this equation is solvable by the standard arguments of the Fredholm theory. According to the Fredholm alternative (see Section A.6 in Appendix A), if the homogeneous equation $\mathsf{K}e = 0$ has only the trivial solution, then for every $\mathscr{G}$ the inhomogeneous equation $(I + \mathsf{K})\, e = \mathscr{G}$ has a unique solution depending continuously on $\mathscr{G}$.

Thus the proof of Theorem 4.2.2 is complete. □

[4] In particular, for $\beta = 0$ this is statement 5 of Theorem 3.7.5, while for $\beta \neq 0$, it follows from statement 4 of the same theorem due to the assumed regularity on ε.

REMARK 4.2.6 In this section we have employed the Drude-Born-Fedorov constitutive relations (2.21). As we have mentioned, in the frequency domain, (2.21) are equivalent to the Tellegen constitutive relations (2.20). Let us consider the more general than (2.20) but of similar structure bianisotropic constitutive relations (2.16). Adopting, as usual, the time convention $e^{-i\varpi t}$, the Maxwell system (in six-vector notation) becomes

$$\mathsf{M}\widetilde{u} = \widetilde{\mathsf{A}_{\rm or}}\widetilde{u} + \mathsf{J}_{\rm A}\,, \tag{4.24}$$

where $\widetilde{\mathsf{A}_{\rm or}}$ is the matrix given in (2.17). Under Assumption 2.3.12, for a fixed frequency, the appropriate sesquilinear form is continuous, bounded and coercive (with suitable coercivity constants depending on $\mathcal{O}$ and on the L^∞ norms of ξ and ζ) in $H(\text{curl}, \mathcal{O})$. Hence the corresponding interior problem has a unique solution. Let us note that the coercivity proof ([261]) in this framework is simpler than the one presented in this section for the case when the Drude-Born-Fedorov constitutive relations (2.21) are considered.

REMARK 4.2.7 In the paper of Ammari and Nédélec [11], the solvability result for the interior problem is proved in a slightly different way, using a mixed variational formulation[5] and the generalisation of Babuška and Brezzi of the Lax-Milgram lemma (see Appendix A). To follow this line, we must rewrite the problem in a suitable weak form using properly selected bilinear forms. As before, we assume that E has a decomposition in the form

$$E = U + e + \text{grad}\phi\,,$$

where $U \in (H^1(\mathcal{O}))^3$ and $n \times U = f$ for $x \in \partial\mathcal{O}$, $e \in H_0(\text{curl}, \mathcal{O}) \cap H(\text{div}, \mathcal{O})$ and $\phi \in H_0^1(\mathcal{O})$. In terms of the bilinear forms a, b defined in (4.7) and (4.12), respectively, the boundary value problem (4.1) - (4.2) can be expressed in weak form using the mixed formulation

$$\begin{aligned} &a^{(\beta)}(e, w) + b^{(\beta)}(\phi, w) = a^{(\beta)}(U, w), \quad \forall w \in H_0(\text{curl}, \mathcal{O})\,, \\ &b^{(\beta)}(\psi, e) = 0 \quad \forall \psi \in H_0^1(\mathcal{O})\,. \end{aligned}$$

The bilinear form $a^{(\beta)}$ satisfies a weaker coercivity property expressed by the inequality

$$|a^{(\beta)}(e, e)| \geq C_1 ||\text{curl}e||_{(L^2(\mathcal{O}))^3} - C_2 ||e||^2_{(L^2(\mathcal{O}))^3}\,,$$

which holds for every e, for some positive constants C_1, C_2. Furthermore, b satisfies the inequality

$$|b^{(\beta)}(\psi, \text{grad}\psi)| \geq C_3 ||\psi||^2_{(H_0^1(\mathcal{O}))^3}$$

for every ψ, for some positive constant C_3, so that the Babuška-Brezzi inf-sup condition (see Appendix A, Section A.10) is satisfied for the linear form $b^{(\beta)}$. By compact embedding theorems we can see that the kernel of the form $b^{(\beta)}$ is compactly embedded[6] in $(L^2(\mathcal{O}))^3$, so that $a^{(\beta)}$ is a compact

[5]Concerning the use of mixed variational formulations of boundary value problems, see [46].

[6]See Section 3.7.

perturbation of a coercive form on the kernel of the form $b^{(\beta)}$. The Babuška-Brezzi generalisation[7] of the Lax-Milgram lemma in conjunction with the Fredholm alternative[8] yields the required result.

4.3 THE EIGENVALUE PROBLEM

We now make a few brief comments regarding the eigenvalue problem, i.e., finding ϖ such that the problem

$$\begin{aligned} &\operatorname{curl}(\alpha \operatorname{curl} E) - \varpi^2(\operatorname{curl}(\beta\varepsilon E) + \beta\varepsilon \operatorname{curl} E + \varepsilon E) = 0, \text{ in } \mathcal{O}, \\ &n \times E = 0, \text{ on } \partial\mathcal{O}, \end{aligned} \tag{4.25}$$

admits nontrivial solutions.

An equivalent formulation of (4.25) is as follows: Let $\mathsf{L}^{(\beta)} : H_0(\operatorname{curl}, \mathcal{O}) \to (H^{-1}(\mathcal{O}))^3$ be the operator[9] that acts on a vector field v as

$$\mathsf{L}^{(\beta)} v := \operatorname{curl}(\alpha \operatorname{curl} v) - \varpi^2(\operatorname{curl}(\beta\varepsilon v) + \beta\varepsilon \operatorname{curl} v + \varepsilon v). \tag{4.26}$$

In terms of this operator, problem (4.25) can be rewritten as

$$\mathsf{L}^{(\beta)} E = 0, \tag{4.27}$$

which in turn can be written as

$$a^{(\beta)}(E, \psi; \varpi^2) = 0, \quad \forall \psi \in H_0(\operatorname{curl}, \mathcal{O})$$

in terms of the bilinear form $a^{(\beta)}$ defined in (4.7).

REMARK 4.3.1 In this section we deviate slightly from the notation of Section 4.2 by keeping explicitly ϖ^2 in the bilinear form, since we are interested in determining those values of ϖ^2 for which nontrivial solutions of the problem exist. The same applies to all the other bilinear forms, e.g., $a_+^{(\beta)}$ and $b^{(\beta)}$.

The full treatment of the spectral properties of the operator $\mathsf{L}^{(\beta)}$ is a rather involved subject that falls outside the scope of the present work. We concentrate here on some specific results that will be useful in the development of the material in this chapter. For simplicity, we assume that β is a non-negative constant. Since chirality effects are usually considered as small effects, we will assume that β is small; therefore, we will adopt a perturbative approach to the eigenvalue problem.

We start by recalling a few facts concerning the operator $\mathsf{L}^{(0)}$. This corresponds to setting $\beta = 0$ in the operator $\mathsf{L}^{(\beta)}$ defined in (4.26) and is in fact the interior achiral Maxwell operator, whose action on a vector field v is given by

$$\mathsf{L}^{(0)} v = \operatorname{curl}\left(\mu^{-1} \operatorname{curl} v\right) - \varpi^2 \varepsilon\, v. \tag{4.28}$$

---

[7]See Appendix A, Section A.10.

[8]See Appendix A, Section A.6.

[9]This operator is related to the bilinear form $a^{(\beta)}$ as follows: $a^{(\beta)}(E, \psi) =: \langle L^{(\beta)} E, \psi \rangle$, for all $\psi \in H_0(\operatorname{curl}, \mathcal{O})$.

Our presentation follows that of [324]. Retracing the steps in Section 4.2 (setting $\beta = 0$) and invoking the Helmholtz decomposition[10], we write any solution as $E = e + \text{grad}\phi$ for $e \in \mathcal{N}^{(0)}$ and $\phi \in H_0^1(\mathcal{O})$, where $\mathcal{N}^{(0)}$ is the function space defined in (4.20) by setting $\beta = 0$. This brings the problem into the variational form

$$a^{(0)}(e, \psi; \varpi^2) = 0, \quad \forall \psi \in \mathcal{N}^{(0)},$$

which, by choosing $\psi = e$, leads by a straightforward application of Friedrichs' inequality[11] to the fact that $\varpi^2 > 0$.

We now rewrite the eigenvalue problem as

$$a_+^{(0)}(e, \psi; 1) = -b^{(0)}(e, \psi; \varpi^2 + 1), \quad \forall \psi \in \mathcal{N}^{(0)},$$

and define the operator $\mathsf{K} : (L^2(\mathcal{O}))^3 \to (L^2(\mathcal{O}))^3$, which acts on a function f, giving the solution of the system

$$a_+^{(0)}(\mathsf{K}f, \psi; 1) = -b^{(0)}(f; 1), \quad \forall \psi \in \mathcal{N}^{(0)}.$$

Similar arguments to the ones applied in the relevant discussion of Section 4.2 lead us to the result that the operator K is compact. Furthermore, using an equivalent norm of $(L^2(\mathcal{O}))^3$, weighted by ε, this operator becomes a self-adjoint operator.

In terms of the operator K, the eigenvalue problem assumes the form

$$\mathsf{K}e = \frac{1}{1 + \varpi^2}\, e,$$

and because of the properties of the operator K, the powerful results of the Hilbert-Schmidt theory (see Appendix A, Section A.6) apply, to yield the following:

THEOREM 4.3.2 *Let $\beta = 0$. There is an infinite discrete set of eigenvalues, ϖ_j^2, $j \in \mathbb{N}$, with the property $\lim_{j\to\infty} \varpi_j^2 = \infty$.*

We now consider the case where $\beta \neq 0$. The following result holds (see, e.g., [14]):

LEMMA 4.3.3 *There exist $\varpi > 0$ and $\beta_0 > 0$ such that for any $\beta \in [0, \beta_0]$, 0 is not an eigenvalue of $\mathsf{L}^{(\beta)}$ in $\mathcal{O}$ with the boundary condition $n \times \cdot = 0$ on $\partial\mathcal{O}$.*

Proof. It is straightforward to see that

$$\left((\mathsf{L}^{(\beta)} - \mathsf{L}^{(0)})v, v\right) = -\beta^2\varpi^2 \int_{\mathcal{O}} \mu\, |\text{curl} v|^2\, dx$$

$$-\beta\, \varpi^2 \int_{\mathcal{O}} \varepsilon\, \text{curl} v \cdot \overline{v}\, dx + \beta\varpi^2 \int_{\mathcal{O}} \varepsilon\, v\, \text{curl}\overline{v}\, dx,$$

---

[10]See Section 3.6.1.

[11]See Theorem 3.2.1 in Section 3.2.2.

from which it follows readily that

$$|((\mathsf{L}^{(\beta)} - \mathsf{L}^{(0)})v, v)| \leq C\,(\beta + \beta^2)\,||\mathrm{curl} v||_{(L^2(\mathcal{O}))^3} + c\,\beta\,||v||_{(L^2(\mathcal{O}))^3},$$

for appropriate constants c, C. Recalling the definition of the norm of the space $H_0(\mathrm{curl}, \mathcal{O})$ we see that the above estimate guarantees that the norm of the operator $\mathsf{L}^{(\beta)} - \mathsf{L}^{(0)}$ in $H_0(\mathrm{curl}, \mathcal{O})$ tends to 0 as $\beta \to 0$. The results for the eigenvalue problem for the nonchiral interior Maxwell operator $\mathsf{L}^{(0)}$, in conjunction with the above observation, lead us to the stated result. □

4.4 LOW CHIRALITY BEHAVIOUR

An interesting modelling question concerns the effect of chirality on the electromagnetic fields as predicted by the solutions of the Maxwell equation. In particular, if we assume that the chirality effects as modelled by the chirality parameter β are considered to be extremely weak, do we recover the behaviour of the electromagnetic fields for the achiral medium? Mathematically, this corresponds to studying the limit of the solutions of the chiral problem, which of course depends on the value of the parameter β as $\beta \to 0$, and finding out whether this limit coincides with the solution of the Maxwell equations for the achiral medium. In this section we formulate this problem and provide a positive answer to it. This problem has been studied for scattering problems in [14]; our approach is an adaptation of the methods of [14] for the interior problem. As in the previous section, we assume for simplicity that β is a non-negative constant.

Consider the interior boundary value problem for the DBF-equations

$$\begin{aligned} \mathrm{curl}\, E^{(\beta)} &= \beta\gamma^2 E^{(\beta)} + i\frac{\varpi\mu}{1-\beta^2\varpi^2\varepsilon\mu} H^{(\beta)}, \\ \mathrm{curl}\, H^{(\beta)} &= \beta\gamma^2 H^{(\beta)} - i\frac{\varpi\varepsilon}{1-\beta^2\varpi^2\varepsilon\mu} E^{(\beta)}, \end{aligned} \quad \text{in}\, \mathcal{O}, \tag{4.29}$$

with the boundary condition

$$n \times E^{(\beta)} = f^{(\beta)}\,, \ \text{on}\, \partial\mathcal{O}\,, \tag{4.30}$$

where $f^{(\beta)} \in H^{-1/2}(\mathrm{div}, \partial\mathcal{O})$ is a known vector field. Let us denote the above problem (consisting of (4.29), (4.30)) by $\mathscr{P}^{(\beta)}$.

Consider, in addition, the interior boundary value problem for the Maxwell equation

$$\mathrm{curl}\left(\mu^{-1}\,\mathrm{curl}\, E^{(0)}\right) - \varpi^2\varepsilon\, E^{(0)} = 0\,,\ \text{in}\, \mathcal{O}, \tag{4.31}$$

with boundary condition

$$n \times E^{(0)} = f^{(0)}\,,\ \text{on}\, \partial\mathcal{O}\,. \tag{4.32}$$

Let us denote the above problem (consisting of (4.31), (4.32)) by $\mathscr{P}^{(0)}$.

The question we want to address can now be expressed in the following way:

If we assume that $\lim_{\beta\to 0} f^{(\beta)} = f^{(0)}$, *in* $H^{-1/2}(\mathrm{div}, \partial\mathcal{O})$, *does the solution of problem* $\mathscr{P}^{(\beta)}$ *tend to the solution of* $\mathscr{P}^{(0)}$ *in some appropriate function space?*

To answer this question we rewrite the problems $\mathscr{P}^{(\beta)}$ and $\mathscr{P}^{(0)}$ in terms of the electric field only. Then, using the interior DBF operator (defined in (4.26)), and the interior achiral Maxwell operator (defined in (4.28)) the chiral and achiral problem become

$$\mathsf{L}^{(\beta)}\, E^{(\beta)} = 0 \tag{4.33}$$

and

$$\mathsf{L}^{(0)}\, E^{(0)} = 0, \tag{4.34}$$

respectively. It is our aim to compare the solutions of problem (4.33) and (4.34) and to assess the effects of chirality on the electromagnetic field.

From (4.33) it follows that

$$\beta\,\mathrm{div}\,(\varepsilon\,\mathrm{curl}\, E^{(\beta)}) + \mathrm{div}\,(\varepsilon\, E^{(\beta)}) = 0 \ . \tag{4.35}$$

We now decompose the electric field $E^{(\beta)}$ as:

$$E^{(\beta)} = e^{(\beta)} - v^{(\beta)} - \mathrm{grad}\,\phi^{(\beta)}$$

where

(*i*) $e^{(\beta)}$ is such that

$$\mathrm{div}\,(\varepsilon\, e^{(\beta)}) = 0 \text{ in } \mathcal{O},$$
$$n \times e^{(\beta)} = 0 \text{ on } \partial\mathcal{O}\,.$$

(*ii*) $\phi^{(\beta)}$ is a scalar function in $H_0^1(\mathcal{O})$ defined as the unique solution to the problem

$$\mathrm{div}\,(\varepsilon\,\mathrm{grad}\,\phi^{(\beta)}) = \beta\,\mathrm{div}\,(\varepsilon\,\mathrm{curl}\, E^{(\beta)}) \ \text{ in } \mathcal{O}\,,$$
$$\phi^{(\beta)} = 0 \text{ on } \partial\mathcal{O}\,.$$

(*iii*) $v^{(\beta)}$ is defined as the unique solution to the problem

$$\mathrm{curl}\,\mathrm{curl}\, v^{(\beta)} + \varpi^2 \varepsilon\, v^{(\beta)} = 0 \ \text{ in } \mathcal{O}\ ,$$

$$n \times v^{(\beta)} = -f^{(\beta)} \ \text{ on } \partial\mathcal{O}\,.$$

(*iv*) $v^{(0)}$ is defined as the unique solution to the same problem that $v^{(\beta)}$ satisfies when $v^{(\beta)}$, $w^{(\beta)}$ and $f^{(\beta)}$ are, respectively, replaced by $v^{(0)}$, $w^{(0)}$ and $f^{(0)}$.

Regarding $v^{(\beta)}$ and $v^{(0)}$, we have the following lemma.

LEMMA 4.4.1 *There exists* $\beta_0 > 0$ *such that for* $0 < \beta \le \beta_0$, *we have the estimate*

$$\| v^{(\beta)} - v^{(0)} \|^2_{H(\mathrm{curl},\mathcal{O})} \le C \, \| f^{(\beta)} - f^{(0)} \|^2_{H^{-1/2}(\mathrm{div},\partial\mathcal{O})},$$

where the constant C *is independent of* β.

Proof. Subtract the equations for $v^{(\beta)}$ and $v^{(0)}$ to obtain an equation for the difference $v^\diamond := v^{(\beta)} - v^{(0)}$, of the form

$$\begin{aligned} \operatorname{curl}\operatorname{curl} v^\diamond - \varepsilon\, v^\diamond &= 0, \quad x \in \mathcal{O}, \\ n \times v^\diamond &= -(\, f^{(\beta)} - f^{(0)}\,), \quad x \in \partial\mathcal{O}. \end{aligned}$$

Using the variational formulation and the estimates for the solution from the Lax-Milgram lemma, we obtain the stated result. □

The proof of the convergence of $E^{(\beta)}$ to $E^{(0)}$ as $\beta \to 0$ is performed in three steps.

LEMMA 4.4.2 (STEP 1) *There exists $\beta_0 > 0$ such that the estimate*

$$\| E^{(\beta)} \|_{H(\mathrm{curl},\mathcal{O})} \leq C$$

holds for any $0 < \beta < \beta_0$, where the constant C is independent of β and $E^{(\beta)}$ is the unique solution of $\mathscr{P}^{(\beta)}$.

Proof. (Sketch) The proof can be obtained following the steps in [14], where a more complicated problem is treated. First, observe that equation (4.33) can be expressed in terms of the bilinear form a (defined in (4.7)), but now taking into account that β is a constant). Using this equation, we obtain the identity (expressed in terms of a)

$$a(E^{(\beta)} + v^{(\beta)}, E^{(\beta)} + v^{(\beta)}) = a(v^{(\beta)}, E^{(\beta)} + v^{(\beta)})\,.$$

We now apply the elementary algebraic inequality

$$s_1\, s_2 \leq \rho\, s_1^2 + \rho^{-1}\, s_2^2\,, \quad \text{for } s_1, s_2 \in \mathbb{R} \text{ and any } \rho > 0$$

to the above identity twice to obtain

$$||\mathrm{curl}\,(\, E^{(\beta)} + v^{(\beta)}\,)||_{(L^2(\mathcal{O}))^3} \leq C\, ||E^{(\beta)} + v^{(\beta)}||_{(L^2(\mathcal{O}))^3} + c||v^{(\beta)}||_{H(\mathrm{curl},\mathcal{O})}\,,$$

for suitable constants c, C. The second term on the right-hand side of the above estimate is uniformly bounded with respect to β by Lemma 4.4.1. It remains to show that the first term enjoys the same property. This claim can be proved using reductio ad absurdum: consider a sequence $\beta_n \to 0$, and let $r_n := \frac{E^{(\beta_n)} + v^{(\beta_n)}}{||E^{(\beta_n)} + v^{(\beta_n)}||}$, where $||\cdot||$ is the $(L^2(\mathcal{O}))^3$-norm, and assume the contrary, i.e., the existence of an unbounded subsequence of r_n. Using the same notation for the unbounded subsequence, we observe that r_n satisfies the equation

$$a(r_n, \psi) = \frac{1}{||E^{(\beta_n)} + v^{(\beta_n)}||_{(L^2(\mathcal{O}))^3}}\, a(v^{(\beta_n)}, \psi), \quad \forall \psi \in H_0(\mathrm{curl}, \mathcal{O}).$$

Therefore, r_n solves the nonhomogeneous interior chiral Maxwell problem with the nonhomogeneous term $v^{(\beta_n)}$, and by the unique solvability of this problem we see that since $||E^{(\beta_n)} + v^{(\beta_n)}||_{(L^2(\mathcal{O}))^3} \to \infty$, $r_n \to 0$ in $H(\mathrm{curl}, \mathcal{O})$, at least up to a subsequence. By the regularity assumption on ε, $\mathrm{div}(\varepsilon\, r_n)$ is uniformly bounded in $(L^2(\mathcal{O}))^3$ with respect to n. By Theorem 3.7.5 (*iii*) we deduce that $r_n \to 0$ in $(L^2(\mathcal{O}))^3$, which is of course a contradiction of $||r_n|| = 1$. □

LEMMA 4.4.3 (STEP 2) *The unique solution* $\phi^{(\beta)}$ *of the problem*

$$\operatorname{div}(\varepsilon \operatorname{grad} \phi^{(\beta)}) = \beta \operatorname{div}(\varepsilon \operatorname{curl} E^{(\beta)}) \ \text{in } \mathcal{O}$$
$$\phi^{(\beta)} = 0 \quad \text{on } \partial\mathcal{O}$$

converges to zero, $H_0^1(\mathcal{O})$*-strongly. Furthermore, there exists* $\beta_0 > 0$ *such that we have*

$$\| \phi^{(\beta)} \|_{H^1(\mathcal{O})} \leq C\,\beta$$

for any $0 < \beta < \beta_0$*, where the constant* C *is independent of* β*.*

Proof. The proof follows by multiplying $\operatorname{div}(\varepsilon \operatorname{grad} \phi^{(\beta)}) = \beta \operatorname{div}(\varepsilon \operatorname{curl} E^{(\beta)})$ by $\overline{\phi^{(\beta)}}$, then integrating by parts over $\mathcal{O}$ and employing Lemma 4.4.2. □

LEMMA 4.4.4 (STEP 3) $e^{(\beta)} \to E^{(0)} + v^{(0)}$ in $H(\text{curl}, \mathcal{O})$, and the rate of convergence is $O(\beta)$.

Proof. It is convenient to further decompose $e^{(\beta)} \in H(\text{curl}, \mathcal{O})$ as

$$e^{(\beta)} = z^{(\beta)} + \operatorname{grad} \varphi^{(\beta)}\ , \tag{4.36}$$

where $z^{(\beta)} \in H(\text{curl}, \mathcal{O})$ satisfies

$$\operatorname{div}(\varepsilon\, z^{(\beta)}) = 0 \ \text{ in } \mathcal{O} \ \ , \ \ z^{(\beta)} \times n = 0 \ \text{ on } \partial\mathcal{O}\,, \tag{4.37}$$

while $\varphi^{(\beta)} \in H_0^1(\mathcal{O})$.

Using the decomposition $E^{(\beta)} = e^{(\beta)} - v^{(\beta)} - \operatorname{grad} \phi^{(\beta)}$ for $e^{(\beta)}$ in (4.33), multiplying the arising equation by a test field in $H_0(\text{curl}, \mathcal{O})$, and integrating over $\mathcal{O}$, we can obtain a variational formulation for (4.33) as

$$\begin{aligned} &a^{(\beta)}(z^{(\beta)}, v) + b^{(\beta)}(v, \varphi^{(\beta)}) = -a^{(\beta)}(v^{(\beta)} + \operatorname{grad} \phi^{(\beta)}, v), \\ &b^{(0)}(z^{(\beta)}, \psi) = 0, \end{aligned} \tag{4.38}$$

for all $(v, \psi) \in H_0(\text{curl}, \mathcal{O}) \times H_0^1(\mathcal{O})$, where we use the forms $a^{(\beta)}$, $b^{(\beta)}$ (see Remark 4.2.7). Note that the second equation is the weak form of the identity (4.37). Further, we introduce $(z^{(0)}, \varphi^{(0)}) \in H_0(\text{curl}, \mathcal{O}) \times H_0^1(\mathcal{O})$ as the unique solution of the variational problem

$$\begin{aligned} &a^{(0)}(z^{(0)}, v) + b^{(0)}(v, \varphi^{(0)}) = -a^{(0)}(v^{(0)}, v), \\ &b^{(0)}(z^{(0)}, \psi) = 0, \end{aligned} \tag{4.39}$$

for all $(v, \psi) \in H_0(\text{curl}, \mathcal{O}) \times H_0^1(\mathcal{O})$. It is known that the variational system (4.39) is a *saddle point formulation* of the interior boundary value problem for the Maxwell equations $\mathscr{P}^{(0)}$. The form $b^{(0)}$ satisfies the *inf-sup condition*[12], and $a^{(0)}$ is a compact perturbation of the coercive form $a_+^{(0)}$ on the kernel $\mathcal{N}^{(0)}$ of $b^{(0)}$ given by

$$\mathcal{N}^{(0)} = \{v \in H(\text{curl}, \mathcal{O}) : \operatorname{div}(\varepsilon v) = 0 \text{ in } \mathcal{O}\ ,\ v \times n = 0 \text{ on } \partial\mathcal{O}\}\,,$$

which is compactly embedded[13] in $(L^2(\mathcal{O}))^3$. □

A lengthy procedure (the details can be found in [14]) based on steps 1, 2 and 3 above leads to the desired result.

---

[12]See Appendix A, Section A.10.

[13]See Section 3.7.

THEOREM 4.4.5 *Let $E^{(\beta)}$ and $E^{(0)}$ be the solutions of problems $\mathscr{P}^{(\beta)}$ and $\mathscr{P}^{(0)}$, respectively. There exists $\beta_0 > 0$ such that for $0 < \beta \leq \beta_0$, the following estimate holds:*

$$\| E^{(\beta)} - E^{(0)} \|_{H(\mathrm{curl},\mathcal{O})} \leq C_1\beta + C_2 \| f^{(\beta)} - f^{(0)} \|_{H^{-1/2}(\mathrm{div},\partial\mathcal{O})},$$

where the two strictly positive constants C_1 and C_2 are independent of β.

Proof. (Sketch) Define $z^\diamond := z^{(\beta)} - z^{(0)}$, $\varphi^\diamond := \varphi^{(\beta)} - \varphi^{(0)}$ as the differences between components of the two solutions and $a^\diamond = a^{(\beta)} - a^{(0)}$, $b^\diamond = b^{(\beta)} - b^{(0)}$ as the differences between the forms. Let

$$\mathscr{R} := a^{(0)}(v^{(0)}, v) - a^{(\beta)}(v^{(\beta)} + \mathrm{grad}\phi^{(\beta)}, v) + a^\diamond(z^{(\beta)}, v) + b^\diamond(v, \varphi^{(\beta)}).$$

Straightforward algebraic manipulations show that $z^\diamond$, $\varphi^\diamond$ satisfy the mixed problem

$$\begin{aligned} a^{(0)}(v^\diamond, v) + b^{(0)}(v, \varphi^\diamond) &= \mathscr{R}, \\ b^{(0)}(z^\diamond, \psi) &= 0, \end{aligned} \tag{4.40}$$

for all $(v, \psi) \in H_0(\mathrm{curl}, \mathcal{O}) \times H_0^1(\mathcal{O})$. Lemmata 4.4.2 and 4.4.3 guarantee that $z^{(\beta)}$ and $\varphi^{(\beta)}$ are uniformly bounded in β, and this gives us an estimate of the order of magnitude of the right-hand side of the first equation of (4.40) in terms of β and $||v^{(\beta)} - v^{(0)}||_{H(\mathrm{curl},\mathcal{O})}$. Based on Lax-Milgram estimates for the variational systems for the forms $a^{(0)}$, $b^{(0)}$, we obtain similar estimates for $z^\diamond$. Then the proof of the theorem is concluded using reductio ad absurdum as before, i.e.; assume that the conclusion does not hold and $||z^\diamond||_{H(\mathrm{curl},\mathcal{O})}$ is not $O\big(\beta + ||v^{(\beta)} - v^{(0)}||_{H(\mathrm{curl},\mathcal{O})}\big)$ and has a subsequence converging to ∞ as $\beta \to 0$. Then the $H(\mathrm{curl}, \mathcal{O})$-normalised sequences $z^\diamond/||z^\diamond||$, $\varphi^\diamond/||\varphi^\diamond||$ satisfy variational problems the solutions of which guarantee that they are uniformly bounded in β, which in turn guarantees that $\left(\frac{z^\diamond}{||z^\diamond||}, \frac{\varphi^\diamond}{||\varphi^\diamond||}\right) \rightharpoonup (z_*^\diamond, \varphi_*^\diamond)$ in $H(\mathrm{curl}, \mathcal{O}) \times H^1(\mathcal{O})$. The limit solves the homogeneous Maxwell problem for $\beta = 0$, and by the uniqueness of this problem, it is identified with $(0, 0)$. Finally, as above, we exploit the compact embedding of $\mathcal{N}^{(0)}$ in $(L^2(\mathcal{O}))^3$ to pass from weak to strong convergence. □

4.5 COMMENTS ON EXTERIOR DOMAIN PROBLEMS

Throughout this section we consider that Assumption 4.2.1 holds appropriately modified for $\mathcal{O}_e$. The solvability of the exterior problem is settled by the following result.

THEOREM 4.5.1 *The exterior chiral problem admits a unique solution.*

This theorem can be established by two different approaches, sketched below.

▷ APPROACH BASED ON THE CALDERÓN OPERATOR

Consider an exterior domain $\mathcal{O}_e$ with boundary $\partial\mathcal{O}_e$. Using the DBF constitutive relations, we see that the electromagnetic fields satisfy the Maxwell equations in the form

$$\begin{aligned}\operatorname{curl} E &= \beta\gamma^2 E + i\varpi\mu\left(\tfrac{\gamma}{k}\right)^2 H,\\ \operatorname{curl} H &= \beta\gamma^2 H - i\varpi\varepsilon\left(\tfrac{\gamma}{k}\right)^2 E,\end{aligned} \quad \text{in } \mathcal{O}_e, \tag{4.41}$$

with boundary condition

$$n \times E = f, \quad \text{on } \partial\mathcal{O}_e$$

and with one of the two Silver-Müller radiation conditions (written in terms of $E - E^{\text{inc}}$ and $H - H^{\text{inc}}$, where E^{inc}, H^{inc} are known fields). Noting that since from the Maxwell equations, H can be expressed in terms of curl$E$, the radiation condition may be re-expressed as

$$\left|\widehat{x} \times \operatorname{curl}(E - E^{\text{inc}})(x) - \beta\gamma^2\, \widehat{x} \times (E - E^{\text{inc}})(x) + \mathrm{i}\,\frac{\gamma^2}{k}(E - E^{\text{inc}})(x)\right| \le \frac{C}{|x|^2},$$

as $|x| \to \infty$ uniformly in all directions $\widehat{x}$. Furthermore, without loss of generality we will assume $f = 0$; if $f \neq 0$ we may always incorporate it in the nonhomogeneous part of the boundary condition at infinity by a simple reformulation of the problem.

Working in a similar manner as for the interior problem, we may eliminate H from the above equations and obtain an equation for E only, of the form

$$\mathsf{L}^{(\beta)}\, E = 0\,, \tag{4.42}$$

with the above boundary condition on $\partial\mathcal{O}_e$ and at infinity.

Equation (4.42) can be treated by considering a sequence of approximate problems. Let $\mathcal{O}_R := \mathcal{O}_e \cap B_R$, where B_R is a ball of radius R centred at the origin. To obtain a variational formulation of this problem we take the dot product with a test function ψ and integrate over the whole domain using the integration by parts formula to obtain

$$a^{(\beta)}(E, \psi) = (n \times (\alpha\, \operatorname{curl} E), (n \times \psi) \times \psi)_{\partial\mathcal{O}_R}, \tag{4.43}$$

where $a^{(\beta)}$ is the bilinear form introduced in equation (4.7), and by $(\,\cdot\,,\cdot\,)_{\partial\mathcal{O}_R}$ we denote the L^2 inner product on the surface $\partial\mathcal{O}_R$.

The boundary $\partial\mathcal{O}_R$ consists of two components: $\partial\mathcal{O}$ and the surface of the sphere B_R, denoted by S_R hereafter. We may consider test functions ψ vanishing on $\partial\mathcal{O}$, and this leaves us with the problem of connecting $n \times \operatorname{curl} E$ (or $n \times H$) on S_R, with E in $\mathcal{O}_R$. This can be accomplished, for instance, by the use of the *exterior Calderón operator* (or *boundary component*, or *admittance* operator); for the Maxwell equations this is the analogue of the celebrated *Dirichlet-to-Neumann map* (for details on the Calderón operator, see, e.g., [91], [106], [132], [324], [334]). Let us recall here the definition and main properties of the exterior[14] Calderón operator.

[14] The interior Calderón operator can be analogously defined ([91]); special care is needed in this case regarding whether $\varpi^2\,\varepsilon\,\mu$ is an eigenvalue of a corresponding selfadjoint operator.

Consider the following auxiliary exterior boundary value problem: let ε_0, μ_0 be, respectively, the permittivity and permeability of the free space. Let m be a fixed vector in $H^{-1/2}(\mathrm{div}, \partial\mathcal{O})$; m can be interpreted as a magnetic current density. We want to find $(E^{\mathrm{e}}, H^{\mathrm{e}}) \in H_{loc}(\mathrm{curl}, \overline{\mathcal{O}_e}) \times H_{loc}(\mathrm{curl}, \overline{\mathcal{O}_e})$ satisfying the equations

$$\mathrm{curl}\, E^{\mathrm{e}}(x) = i\,\varpi\,\mu_0\, H^{\mathrm{e}}(x)\,, \ \mathrm{curl}\, H^{\mathrm{e}}(x) = -i\,\varpi\,\varepsilon_0\, E^{\mathrm{e}}(x)\,, \ x \in \mathcal{O}_e\,,$$

the boundary condition

$$\gamma_\tau^+(E^{\mathrm{e}}) = m\,, \ \mathrm{on}\, \partial\mathcal{O}\,,$$

and one of the Silver-Müller radiation conditions (2.40) or (2.41), as $|x| \to \infty$ uniformly over all directions $x/|x|$. Here γ_τ^+ denotes the trace operator (see Chapter 3) from the exterior of $\partial\mathcal{O}$. It is known ([91], [324]) that this achiral problem has a unique solution.

DEFINITION 4.5.2 *Let $E^{\mathrm{e}}, H^{\mathrm{e}}, m$ be as above. Then the exterior Calderón operator is defined as*

$$\mathcal{C}^e : H^{-1/2}(\mathrm{div}, \partial\mathcal{O}) \to H^{-1/2}(\mathrm{div}, \partial\mathcal{O})\,, \ m = n \times E^{\mathrm{e}} \mapsto \gamma_\tau^+(H^{\mathrm{e}}) = n \times H^{\mathrm{e}}\,.$$

THEOREM 4.5.3 ([91]) *The exterior Calderón operator has the following properties:*

(*i*) *$\mathcal{C}^e$ is an isomorphism in $H^{-1/2}(\mathrm{div}, \partial\mathcal{O})$, with $(\mathcal{C}^e)^2 = -I$.*

(*ii*) *$\mathcal{C}^e$ is an isomorphism in $H^{1/2}(\mathrm{div}, \partial\mathcal{O})$ if $\partial\mathcal{O}$ is C^∞.*

(*iii*) *$\mathcal{R}e \int_{\partial\mathcal{O}} (\mathcal{C}^e\, m) \cdot (n \times \overline{m})\, ds > 0$, for all $m \in H^{-1/2}(\mathrm{div}, \partial\mathcal{O})\,, m \neq 0$.*

(*v*) *$\mathcal{C}^e$ is independent of the material properties inside $\mathcal{O}$.*

Finally, let us define the *chiral exterior Calderón operator* as

$$\mathcal{C}^e_\beta = i\varpi\mu(1 - \varpi^2\varepsilon\mu\beta^2)\left[\mathcal{C}^e - i\varpi\varepsilon\beta\gamma_\tau^+\right]\,.$$

The variational problem (4.43) thus becomes

$$a^{(\beta)}(E, \psi) = \langle n \times \mathcal{C}^e_\beta(E - E^{\mathrm{inc}}), (n \times \psi) \times n \rangle_{\partial\mathcal{O}_R}.$$

Using a generalisation of the variational analysis of Section 4.2, we can show the unique solvability of this problem; see also [11], [12] for a similar approach.

▷ A BOUNDARY INTEGRAL OPERATORS APPROACH

An alternative approach would be to reformulate the problem in terms of boundary integral equations. These equations are in general complicated, and may simplify if the boundary data assume special forms. One such case is when $(E^{\mathrm{inc}}, H^{\mathrm{inc}})$ is assumed to be a known incident electromagnetic wave. This important case corresponds to the scattering problem that will be considered in detail in the next two chapters. One way to tackle this problem is to express the fields as a solution of the boundary integral equation

$$\begin{pmatrix} E \\ H \end{pmatrix} = \begin{pmatrix} E^{\mathrm{inc}} \\ H^{\mathrm{inc}} \end{pmatrix} + \operatorname{curl} \mathsf{K}_1 \begin{pmatrix} n \times H \\ n \times H \end{pmatrix} + \operatorname{grad} \mathsf{K}_2 \begin{pmatrix} n \times H \\ n \times H \end{pmatrix} + \mathsf{K}_3 \begin{pmatrix} n \times H \\ n \times H \end{pmatrix},$$

where K_j , $j = 1, 2, 3$ are appropriate (matrix) boundary integral operators, the exact form of which for the general case of spatially dependent coefficients can be found in, e.g., [10]. For the special case of constant coefficients, explicit forms for the above operators are given in Chapter 5. Using this integral equation, the problem is once more brought into variational form, which may be treated using the Lax-Milgram lemma and its generalisations.

Let us note that the low chirality behaviour of the solution of the exterior problem can also be studied and leads to similar results as for the interior problem (for details, one can consult [12]).

A detailed study of the exterior problem for the achiral Maxwell equations in the space of bounded energy solutions (which, quoting Nédélec, is the "poorest" space where the exterior problem has a unique solution) can be found in the monograph [334].

4.6 TOWARDS NUMERICS

The variational formulation presented in this chapter for the study of the problem of well posedness of the Maxwell equations may form the basis for the numerical treatment of such problems. Since numerical approximations of the solutions are essentially discretisations of the original problem and may present discontinuities in the derivatives, we may no longer consider them classical solutions of the equations; therefore, the (weak) variational treatment is called for.

4.6.1 Discretised version of Lax-Milgram lemma

The starting point for numerical analysis is the weak formulation of the problem as

$$a(u, v) = \langle f, v \rangle,$$

where a is a bilinear continuous and coercive form $a : \mathbb{H} \times \mathbb{H} \to \mathbb{C}$ and $\mathbb{H}$ is an appropriate Hilbert space and $\langle \cdot, \cdot \rangle$ denotes the duality pairing between $\mathbb{H}'$ and $\mathbb{H}$. The existence of solutions is guaranteed by the Lax-Milgram lemma. The numerical approximation consists in working with this problem using a finite-dimensional subspace, $V_h \subset \mathbb{H}$. We will return to this point in more detail; however, for the time being consider the finite-dimensional subspace V_h as containing discretisations of the functions in $\mathbb{H}$ and, as $h \to 0$, the functions in V_h approximate the functions in $\mathbb{H}$ as closely as possible. Since $V_h \subset \mathbb{H}$, we consider the variational problem

$$a(u_h, v_h) = \langle f, v_h \rangle, \quad \forall\, v_h \in V_h,$$

which has a unique solution, $u_h \in V_h$, by arguments similar to the ones employed in the Lax-Milgram lemma. In fact, the finite-dimensional (or discrete) version of this result is known as *Céa's lemma* and provides the simplest convergence estimate for the discretised problem to the original problem in terms of

$$||u - u_h||_{\mathbb{H}} \leq C \inf_{v_h \in V_h} ||u - v_h||_{\mathbb{H}}$$

(see, e.g., [324]). Céa's lemma provides a quasi-optimal error estimate, since the error is (up to the constant C) the best approximation error of u by an element of the finite-dimensional subspace V_h.

4.6.2 Discretised version of mixed variational problems

The discretised version of the Lax-Milgram lemma has a generalised version (see Appendix A, Section A.10) that allows for the numerical treatment of the mixed variational problems. This has the advantage of allowing the relaxation of the various coercivity assumptions. Assume that $V_h \subset \mathbb{H}$ and $S_h \subset \mathbb{S}$ are two finite-dimensional subspaces of the Hilbert spaces $\mathbb{H}$, $\mathbb{S}$, and consider the mixed variational problem:
Given $f \in \mathbb{H}'$ and $g \in \mathbb{S}'$, find $u \in \mathbb{H}$ and $p \in \mathbb{S}$ such that

$$\begin{aligned} &a(u, \phi) + b(\phi, p) = \langle f, \phi \rangle, \quad \forall\, \phi \in \mathbb{H}, \\ &b(u, \psi) = \langle g, \psi \rangle, \quad \forall\, \psi \in \mathbb{S}. \end{aligned} \tag{4.44}$$

The discretised version of this problem is:
Given $f \in \mathbb{H}'$ and $g \in \mathbb{S}'$, find $u \in \mathbb{H}$ and $p \in \mathbb{S}$ such that

$$\begin{aligned} &a(u_h, \phi_h) + b(\phi_h, p_h) = \langle f, \phi_h \rangle, \quad \forall\, \phi \in V_h, \\ &b(u_h, \psi_h) = \langle g, \psi_h \rangle, \quad \forall\, \psi_h \in S_h. \end{aligned} \tag{4.45}$$

Assuming that a is uniformly coercive on a subset of $V_h \times V_h$,

$$|a(u_h, u_h| \geq C_1 ||u_h||^2_{\mathbb{H}}, \quad \forall\, u_h \in \mathcal{Z}_h,$$

where $\mathcal{Z}_h = \{u_h \in V_h, b(u_h, w_h) = 0, \forall\, w_h \in S_h\}$, and that a discrete Babuška-Brezzi condition,

$$\sup_{\phi_h \in V_h} \frac{|b(\phi_h, p_h)|}{||\phi_h||_{\mathbb{H}}} \geq C_2 ||p_h||_{\mathcal{S}},$$

is satisfied, then as long as there exists a $u_h \in V_h$ such that $b(u_h, \psi_h) = \langle g, \psi_h \rangle$ for all $\psi_h \in S_h$, the discretised mixed variational problem (4.45) admits a unique solution. Furthermore, one may obtain error estimates for the discretised problem in a form similar to that obtained by Céa's lemma,

$$||u - u_h||_{\mathbb{H}} \leq C \{ \inf_{v_h \in \mathcal{Z}_h(g)} ||u - v_h||_{\mathbb{H}} + \inf_{q_h \in S_h} ||p - q_h||_{\mathbb{S}} \},$$

where

$$\mathcal{Z}_h(g) = \{u_h \in V_h, : \ b(u_h, \psi_h) = \langle g, \psi_h \rangle, \forall \psi_h \in S_h\}.$$

4.6.3 Discretisation of integral equations

Furthermore, the compact integral equation

$$(I + \mathsf{K})u = \mathscr{G}$$

has a discretisation as

$$(I + \mathsf{K}_h)u_h = \mathscr{G}_h,$$

which has a unique solution and admits the error estimate

$$||u - u_h||_{\mathbb{H}} \leq C\left\{||\mathscr{G} - \mathscr{G}_h||_{\mathbb{H}} + ||(\mathsf{K} - \mathsf{K}_h)u||_{\mathbb{H}}\right\}$$

as long as $\{\mathsf{K}_h\}$ is a collectively compact set[15] of bounded operators converging uniformly on K (see, e.g., [324]).

4.6.4 Finite elements

Having settled the point of the convergence of discretised versions of variational and integral equation problems on their continuous versions, it is important to consider next the nature of the finite-dimensional spaces V_h. One choice is to use as $V_h =: V_n$ the span of an increasing number of linearly independent functions $\{\phi_m \in \mathbb{H},\ m = 1, \cdots, n\}$. Then as $n \to \infty$ the sequence of spaces V_n approximates $\mathbb{H}$. This approach corresponds to the Galerkin expansion, which will be treated in detail in later chapters. Another choice is the one corresponding to finite element methods. This choice is often very useful for the numerical treatment of the Maxwell equations in domains with complicated geometries, which arise often in applications. We choose to present it very briefly here, as the numerical analysis of the Maxwell equations is not among the goals of this book. The interested reader is referred to the excellent books of Bossavit [68] and Monk [324].

A finite element approximation consists of breaking up the domain $\mathcal{O}$ into the union of a finite number of *simplices*, and then approximating the fields with polynomials on the approximation of the domain. A finite element space consists then of the triplet (K, P_K, Σ_K), where $K \subset \mathcal{O}$ is the simplex approximation of the domain (in dimension 3 the simplices are tetrahedra), P_K is a set of polynomials, and Σ_K is a set of linear functionals on P_K ($\Sigma_K : P_K \to \mathbb{R}$), which is needed for assessing the properties of the functions in P_K on various points of the domain, or quantities such as the mean value, etc. For instance, a class of elements of Σ_K may be $\ell_i : P_K \to \mathbb{R}$, defined by $\ell_i(u) = u(k_i)$ for any $u \in P_K$, and a (prescribed) number of points $k_i \in K \subset \mathcal{O}$. In practice, these functionals may be considered as providing the value of the function u at prescribed (mesh) points $\{k_i\}$, $k_i \in K \subset \mathcal{O}$, $i = 1, \cdots, n$. The value of the function u at intermediate points of $\mathcal{O}$ may be provided by an interpolant (see, e.g., [324]).

---

[15]The set $\{\mathsf{K}_h\}$ of bounded linear operators, $\mathsf{K}_h : \mathbb{H} \to \mathbb{H}$, is called collectively compact if for each bounded set $U \subset \mathbb{H}$, the image set $\{\mathsf{K}_h U\}$ has compact closure (is relatively compact).

The numerical treatment of the Maxwell equations requires finite element approximations of the space $H(\mathrm{curl}, \mathcal{O})$, which is the natural functional setting for the variational treatment of such problems. This requires the so-called *curl-conforming edge elements* proposed by Nédélec [335]. The definition of this set requires the definition of a subspace of homogeneous vector polynomials of degree k, $\mathcal{S}_k = \{\pi \in (\widetilde{P}_k)^3, : x \cdot \pi(x) = 0\}$, where by $\widetilde{P}_k$ we denote the set of polynomials of degree exactly k. We then define the space

$$R_k = (P_{k-1})^3 \oplus \mathcal{S}_k, \tag{4.46}$$

where P_{k-1} is the set of polynomials of order at most $k-1$. This is a space of vector polynomials of dimension $\dim(R_k) = \frac{1}{2}k(k+2)(k+3)$.

We have the following definition ([324], Definition 5.30).

DEFINITION 4.6.1 *The space of curl-conforming finite elements is defined as the triplet (K, R_k, Σ_K), where K is the reference tetrahedron, R_k is the space of vector polynomial functions defined in* (4.46)*, and Σ_K is the space of linear functionals on R_k, associated with*

(*i*) *integrals of the fields along the edges of the element $\int_{\wp} u \cdot \hat{e}\, q\, ds$, where $\wp$ is an edge of K, $\hat{e}$ is the unit vector in the direction of $\wp$, and $q \in P_{k-1}(\wp)$,*

(*ii*) *integrals of the fields on the faces of the elements $\int_F u \cdot q\, d\sigma$, where F is a face of K, $q \in (P_{k-2}(F))^3$, and $q \cdot n = 0$,*

(*iii*) *volume integrals of the fields, $\int_K u \cdot q\, dV$, $q \in (P_{k-3}(K))^3$.*

The following theorem ([324], Theorem 5.37) guarantees that the finite elements defined in Definition 4.6.1 provide a good approximation of the elements of $H(\mathrm{curl}, \mathcal{O})$:

THEOREM 4.6.2 *The finite element defined in Definition 4.6.1 is*

(*i*) *$H(\mathrm{curl}, \mathcal{O})$-conforming, i.e., the global finite element space is a subspace of $H(\mathrm{curl}, \mathcal{O})$, and*

(*ii*) *unisolvent, in the sense that specifying a value for each of the functionals in Σ_K uniquely determines a function in P_K.*

Even though the approximating functions, being polynomials, possess high regularity in each constituent component of $\mathcal{O}$, in the whole of $\mathcal{O}$ these functions are less regular, and this is the regularity described by the term "global", in the above theorem.

The space V_h can then be defined as

$$V_h = \{u \in H(\mathrm{curl}, \mathcal{O}) : u|_K \in R_k \, \forall\, K \in \tau_h\},$$

where τ_h is a mesh approximating $\mathcal{O}$.

4.6.5 Discretised versions of the Maxwell equations

Using the finite-dimensional space V_h and the variational formulation of the problems of this chapter, we may obtain numerical methods for the solution of the Maxwell equations.

The numerical solution of the interior problem, for instance, starts from the weak formulation of the problem in equation (4.10). In fact, we may retrace the steps that were followed for the analytic treatment of this equation in Section 4.2, using the discretised versions of the respective problems arising in each step, employing properly selected finite element spaces. Upon invoking the Helmholtz decomposition (see step 3), we express the problem in the form (4.11), which contains the electrostatic potential ϕ. As shown in step 4, this is obtained through the weak formulation (4.14), which can be treated numerically using the discretisation of the Lax-Milgram lemma (Céa's lemma), yielding an approximation ϕ_h for the electrostatic potential. This step requires a finite element space that provides a satisfactory approximation for the elements in $H_0^1(\mathcal{O})$.

We then proceed to the definition of the discretised compact operators K_h through the solution of the discretised version of (4.21) (see step 7), and the discretised maps $\mathscr{G}_h$ through the solution of the discretised version of (4.22) (see step 8) setting $\psi = \psi_h$, where ψ_h belongs to the $H_0(\text{curl}, \mathcal{O})$ conforming finite elements space (see Section 4.6.4) and $\phi = \phi_h$ is the discretised approximation to the electrostatic potential. Finally, we solve the discretised version of the integral operator equation (4.23) (see step 9)

$$(I + \mathsf{K}_h)\, e_h = \mathscr{G}_h$$

(see Section 4.6.3), for the approximation e_h of the electric field.

Alternatively, one may use the mixed variational formulation of the Maxwell system (see Remark 4.2.7) along with the discretised version of the Babuška-Brezzi lemma (see Section 4.6.2). The eigenvalue problems may be treated similarly. At every level the discretised versions of the equations are reduced to linear algrebraic systems of equations, which may be treated using standard methods of numerical linear algebra. We refer to Chapter 7 in [324] for a detailed treatment of various problems of the achiral interior Maxwell system based on the variational formulation; these methods can be extended to chiral media following the guidelines of the present section.

In [94], a weighted regularisation method for the time-harmonic Maxwell equations with a perfectly conducting or impedance boundary condition in composite materials is presented. The computational domain $\mathcal{O}$ is the union of polygonal or polyhedral subdomains made of different materials. As a result, the electromagnetic field presents singularities near geometric singularities, which are the interior and exterior edges and corners. The variational formulation of the weighted regularised problem is given on the subspace of $H(\text{curl}, \mathcal{O})$ whose fields u satisfy $w^\alpha \, \text{div}(\varepsilon u) \in L^2(\mathcal{O})$ and have vanishing tangential trace or tangential trace in $L^2(\partial\mathcal{O})$. The weight function $w(x)$ is equivalent to the distance of x to the geometric singularities, and the minimal weight parameter α is given in terms of the singular exponents of a scalar transmission problem. A density result is proved that guarantees the approximability of the solution field by piecewise regular fields. Numerical results for the discretisation of the source problem by means of Lagrange finite elements of type P_1 and P_2 are given on uniform and appropriately refined two-dimensional meshes. The performance of the method in the case of eigenvalue problems is also addressed in that paper.

Chapter Five

Scattering Problems: Beltrami Fields and Solvability

5.1 INTRODUCTION

This chapter deals with the solvability of time-harmonic electromagnetic wave scattering by an obstacle: either the obstacle or the environment in which it is embedded, or both, is (are) occupied by a chiral material. Regarding general references for scattering theory, we refer to the books [103], [106], [218], [260], [303], [358], [410], [411], [412] and the book chapter [258]. The corresponding theory for an achiral obstacle and the surrounding environment is well known and established: see [103], [106] and the references therein.

We assume that the scatterer and its surrounding space are homogeneous: this allows us to use boundary integral equation (we will use the standard abbreviation BIEs for boundary integral equations) methods for the study of the considered problems. We consider two kinds of problems: first, the scattering of plane electromagnetic waves propagating in chiral space by a perfectly conducting obstacle, and second, the scattering of plane electromagnetic waves by a penetrable obstacle; either the scatterer or the surrounding space, or both, may be filled with a chiral material. Scattering by a (chiral or achiral) obstacle in a chiral environment may seem somehow exotic at a first glance, but it constitutes an attractive problem (from the point of view of both theory and applications): some illustrative examples are turbid chiral media, different classes of contrasting chiral media, the Bruggeman homogenisation of chiral composites, and the method of moments for scattering by a chiral obstacle in a chiral environment; see the references in [21].

In Section 5.2 we introduce the standard terminology regarding the types of polarisation of electromagnetic waves, while in Section 5.3 we introduce the notions of Beltrami fields and the celebrated Bohren decomposition. The formulation of scattering problems in terms of both the electric and the magnetic fields, as well as of the corresponding Beltrami fields, is presented in Section 5.4. Then we address the problem of solvability of the boundary value problems introduced in Section 5.4. This will be established in Section 5.7 by a BIE method; an introduction to BIEs is presented in Section 5.5. In view of the Bohren decomposition, the electromagnetic field is decomposed in terms of suitable Beltrami fields (Section 5.3); some of their properties are studied in Section 5.6. In Section 5.7 it is proved that the perfect conductor in a chiral environment problem has a unique weak solution, which is then shown (given that the boundary of the scatterer is sufficiently smooth) to be a classical one. A similar approach ascertains unique solvability (except for a discrete set of electromagnetic parameters of the scatterer) of the chiral obstacle in a chiral environment problem. In Section 5.8 we present a generalisation of Claus Müller's uniquely solvable set of BIEs to the chiral scatterer in an achiral environment problem. In Section 5.9 we study low chirality approximations for the solutions to the problem considered in the previous section; these approximations are the basis for the possibility of numerical solution of the (complicated) chiral problem with the use of existing sophisticated codes for the solution of the (simpler) achiral problem.

5.2 ELLIPTIC, CIRCULAR AND LINEAR POLARISATION OF WAVES

In this section (based on [187] and [392]) we recall standard notions regarding the types of polarization of electromagnetic waves.

An *elliptically polarised* electromagnetic wave has the form

$$\mathcal{E}(x_3,t) = \widehat{e}_1 E_{0,x_1}\, \mathrm{e}^{\mathrm{i}\,((kx_3-\varpi t)+\varphi_1)} + \widehat{e}_2 E_{0,x_2}\, \mathrm{e}^{\mathrm{i}\,((kx_3-\varpi t)+\varphi_2)}\,,$$

and is described by three independent parameters, the magnitudes E_{0,x_1}, E_{0,x_2} and the phase difference $\varphi = \varphi_1 - \varphi_2$.

We shall see that depending on the choice of the parameters, the end point of $\mathcal{R}e\mathcal{E}$ traces out an ellipse or a circle or a straight line in a fixed plane perpendicular to $\widehat{k}$, the propagation unit vector in the direction of motion. This justifies the terminology elliptic, circular and linear polarisation of the wave, respectively.

Let us select one of the phase factors for $x_3 = t = 0$ so that the wave depends only on the phase difference φ. To simplify the algebra, we set $\varphi_1 = 0$ and $\varphi_2 = \phi$ and consider the resulting form of $\mathcal{E}(x_3,t)$ in terms of its real and imaginary parts,

$$\mathcal{E}(x_3,t) = \widehat{e}_1\, E_{0,x_1}\, \cos(kx_3 - \varpi t) + \widehat{e}_2\, E_{0,x_2}\, \cos(kx_3 - \varpi t + \phi) +$$

$$\mathrm{i}\left(\widehat{e}_1\, E_{0,x_1}\, \sin(kx_3 - \varpi t) + \widehat{e}_2\, E_{0,x_2}\, \sin(kx_3 - \varpi t + \phi)\right).$$

Let

$$E_{x_1}(x_3,t) := \widehat{e}_1\, E_{0,x_1}\, \cos(kx_3 - \varpi t)$$

and

$$E_{x_2}(x_3,t) := \widehat{e}_2\, E_{0,x_2} \cos(kx_3 - \varpi t + \phi)\,.$$

The equation of the curve we are looking for should depend neither on position nor on time, so the $(kx_3 - \varpi t)$ dependence must be eliminated. Using

$$\mathbb{E}_2 := \frac{E_{x_2}}{E_{0,x_2}} = \cos(kx_3 - \varpi t)\, \cos\phi - \sin(kx_3 - \varpi t)\, \sin\phi\,,$$

$$\mathbb{E}_1 := \frac{E_{x_1}}{E_{0,x_1}} = \cos(kx_3 - \varpi t)\,,$$

we obtain

$$\mathbb{E}_2 - \mathbb{E}_1 \cos\phi = -\sin(kx_3 - \varpi t)\, \sin\phi\,.$$

Hence,

$$\sin(kx_3 - \varpi t) = (1 - \mathbb{E}_1^2)^{1/2}\,,$$

whereby

$$\mathbb{E}_2 - \mathbb{E}_1 \cos\phi = (1 - \mathbb{E}_1^2)\sin^2\phi\,.$$

So we end up with

$$\mathbb{E}_2^2 + \mathbb{E}_1^2 - 2\mathbb{E}_2\mathbb{E}_1 \cos\phi = \sin\phi\,,$$

which is the equation of an ellipse rotated with respect to the E_{x_1}, E_{x_2} coordinate system by an angle ϑ given by

$$\tan 2\vartheta = \frac{2E_{0,x_1}E_{0,x_2}}{E_{0,x_1}^2 - E_{0,x_2}^2}\cos\phi\,.$$

If the principal axes of this ellipse are aligned with the coordinate axes, i.e., if $\vartheta = 0$, or equivalently, if $\phi = \pm\pi/2, \pm 3\pi/2, \pm 5\pi/2, \ldots$, then we obtain $\mathbb{E}_2^2 + \mathbb{E}_1^2 = 1$.

Additionally, if

$$E_{0,x_1} = E_{0,x_2} = E_0\,,$$

we get

$$E_{x_1}^2 + E_{x_2}^2 = E_0^2\,,$$

which is a circle. If $\phi = -\pi/2, -3\pi/2, -5\pi/2, \ldots$, then

$$E_{x_1} = \widehat{e}_1\, E_0\, \cos(kx_3 - \varpi t)\,, E_{x_2} = \widehat{e}_2\, E_0 \sin(kx_3 - \varpi t)\,.$$

Let z_0 be an arbitrary point on the positive x_3-axis. At $t = 0$ we have

$$E_{x_1} = \widehat{e}_1\, E_0\, \cos kz_0\,,\;\; E_{x_2} = \widehat{e}_2\, E_0 \sin kz_0\,,$$

while at $t = kz_0/\varpi$ we have

$$E_{x_1} = \widehat{e}_1\, E_0\,,\;\; E_{x_2} = 0\,.$$

So $\mathcal{E}$ is rotating clockwise, as seen by an observer looking back at the source, at an angular frequency ϖ. This wave is *right circularly polarised* (RCP). On the contrary, if $\phi = \pi/2, 3\pi/2, 5\pi/2, \ldots$, then $\mathcal{E}$ rotates counterclockwise and the wave is *left circularly polarised* (LCP).

Further, if $\phi = 2m\pi$, $m \in \mathbb{Z}$, then

$$E_{x_2} = \frac{E_{0,x_2}}{E_{0,x_1}} E_{x_1}\,,$$

while, if $\phi = 2(m+1)\pi$, $m \in \mathbb{Z}$, then

$$E_{x_2} = -\frac{E_{0,x_2}}{E_{0,x_1}} E_{x_1}\,,$$

which are both straight lines with slopes $\pm\frac{E_{0,x_2}}{E_{0,x_1}}$, respectively. In this case, $\mathcal{E}$ is *linearly polarised*.

REMARK 5.2.1 An alternative form of an elliptically polarised wave is

$$\mathcal{E}(x_3,t) = \widehat{e}_{\rm L} E_{0,{\rm L}}\, e^{{\rm i}\,((kx_3-\varpi t)+\varphi_{\rm L})} + \widehat{e}_{\rm R} E_{0,{\rm R}}\, e^{{\rm i}\,((kx_3-\varpi t)+{\rm i}\varphi_{\rm R})} ,$$

where

$$\widehat{e}_{\rm L} = \frac{\sqrt{2}}{2}\,(\widehat{e}_1 - {\rm i}\widehat{e}_2)\ ,\ \widehat{e}_{\rm R} = -\frac{\sqrt{2}}{2}\,(\widehat{e}_1 + {\rm i}\widehat{e}_2)\ .$$

While the linear polarisation unit vectors $\widehat{e}_1, \widehat{e}_2$ are real, the circular unit vectors $\widehat{e}_{\rm L}, \widehat{e}_{\rm R}$ are complex.

5.3 BELTRAMI FIELDS - THE BOHREN DECOMPOSITION

5.3.1 Beltrami fields

A *Beltrami field* is a field that is everywhere (with the possible exception of the region occupied by its source) parallel to its own rotation. So, such a field U is a solution of the equation

$$\operatorname{curl} U = \lambda\, U, \tag{5.1}$$

where the proportionality factor $\lambda \neq 0$ is, in general, a spatially varying function. This concept arose early in the nineteenth century, and it naturally appears in many branches of physics and mathematical physics. Beltrami first introduced it in the study of hydrodynamics. An example of such a flow is the Arnold-Beltrami-Childress (ABC) flow, which is used in chaotic advection in fluid dynamics (see [343]). A Beltrami field can be thought of as a specific solution of the three-dimensional Navier-Stokes (Euler) equations for incompressible, viscous (inviscid) fluid flow with vorticity parallel to the velocity field [1]. Beltrami fields arise naturally in such diverse areas as quark physics, gravitation research, and thermoacoustics; for related references, see [268]. In electromagnetics, Beltrami fields appear in two frameworks: first, in the early 1820s, in studies related to circularly polarised plane waves, and more recently in the study of natural optical activity and the description of the fields' behaviour in chiral media. For the rôle of Beltrami fields in this branch of electromagnetics and related references, see [268]. The second is related to the concept of the *force-free magnetic field* which originated in the early theories of superconductivity (mid-1930s) and appears in of astrophysics and plasma physics. For some comments and related references, see the monograph [268], and [72], [153], [249], [254], [269], [267], [441], [443], [418]. Time-dependent Beltrami fields (where λ appearing in (5.1) is now a function varying not only spatially but temporally as well) are also used in the framework of the time domain analysis of electromagnetic fields; see [38], [151], [269] and the references therein. Finally, Beltrami fields are related to eigenvalue problems for the operator curl, which are important in various applications and are of great mathematical interest; see [349], [442] and the references therein.

[1] In this context, flows defined by Beltrami fields are often called *Trkalian flows*; see [270].

5.3.2 The Bohren decomposition

Let $k^2 = \varpi^2 \varepsilon\mu$ and $\gamma^2 = k^2(1 - k^2\beta^2)^{-1}$. Consider the equations

$$\begin{aligned} \operatorname{curl} E - \mathrm{i}\,\varpi\mu\tfrac{\gamma^2}{k^2}\, H - \beta\gamma^2\, E = 0, \\ \operatorname{curl} H + \mathrm{i}\,\varpi\varepsilon\tfrac{\gamma^2}{k^2}\, E - \beta\gamma^2\, H = 0, \end{aligned}$$

and write them as

$$\begin{pmatrix} \operatorname{curl} E \\ \operatorname{curl} H \end{pmatrix} = \frac{\gamma^2}{k^2} \begin{pmatrix} \beta k^2 & \mathrm{i}\,\varpi\mu \\ -\mathrm{i}\,\varpi\varepsilon & \beta k^2 \end{pmatrix} \begin{pmatrix} E \\ H \end{pmatrix}. \tag{5.2}$$

Diagonalising the matrix in (5.2), we obtain

$$\begin{pmatrix} \operatorname{curl}(\mathrm{i}\,\eta^{-1} E + H) \\ \operatorname{curl}(E + \mathrm{i}\,\eta\, H) \end{pmatrix} = \begin{pmatrix} \frac{k}{1-k\beta} & 0 \\ 0 & -\frac{k}{1+k\beta} \end{pmatrix} \begin{pmatrix} \mathrm{i}\,\eta^{-1}\, E + H \\ E + \mathrm{i}\,\eta\, H \end{pmatrix}, \tag{5.3}$$

where $\eta = \mu^{1/2}\varepsilon^{-1/2}$ is the intrinsic impedance of the medium. Introducing the fields

$$Q_{\mathrm{L}} := \mathrm{i}\,\eta^{-1}\, E + H\,,\ Q_{\mathrm{R}} := E + \mathrm{i}\,\eta\, H\,,$$

we note that (5.3) is written as

$$\begin{aligned} \operatorname{curl} Q_{\mathrm{L}} &= \gamma_{\mathrm{L}}\, Q_{\mathrm{L}}, \\ \operatorname{curl} Q_{\mathrm{R}} &= -\gamma_{\mathrm{R}}\, Q_{\mathrm{R}}, \end{aligned} \tag{5.4}$$

where

$$\gamma_{\mathrm{L}} := k(1 - k\beta)^{-1}\,,\ \gamma_{\mathrm{R}} := k(1 + k\beta)^{-1}\,.$$

Note that $\gamma^2 = \gamma_{\mathrm{L}}\gamma_{\mathrm{R}}$. From (5.4) we clearly have that both Q_{L} and Q_{R} satisfy the vector Helmholtz equation

$$\Delta Q_\lambda + \gamma_\lambda^2 Q_\lambda = 0,\ \lambda = L, R\,,$$

and we see that $\gamma_{\mathrm{L}}, \gamma_{\mathrm{R}}$ are the wave numbers of the Beltrami fields Q_{L}, Q_{R}, respectively. Let us additionally mention that if E, H are divergence free, then Q_{L}, Q_{R} also satisfy this property.

Thus we obtain the *Bohren decomposition* of E, H into Q_{L}, Q_{R}

$$\begin{aligned} E &= Q_{\mathrm{L}} - \mathrm{i}\,\eta\, Q_{\mathrm{R}}, \\ H &= Q_{\mathrm{R}} - \mathrm{i}\,\eta^{-1}\, Q_{\mathrm{L}}. \end{aligned} \tag{5.5}$$

REMARK 5.3.1 It is obvious that when $\beta = 0$ then $\gamma_{\mathrm{L}} = \gamma_{\mathrm{R}} = k$. From (5.4), and if the two complex-valued wave numbers γ_{L} and γ_{R} have positive real parts, we note that Q_{L} is a left-handed Beltrami field and Q_{R} is a right-handed one, and when $\beta \neq 0$ they propagate with different speeds. This is a manifestation of handedness. For details on the physical problem and the appearing physical parameters, we refer to the monographs [273], [268], [289] and the papers [142], [272].

5.4 SCATTERING PROBLEMS: FORMULATION

Consider a bounded domain $\mathcal{O} \subset \mathbb{R}^3$ with smooth boundary $\partial\mathcal{O}$ and its exterior $\mathcal{O}_e = \mathbb{R}^3 \setminus \overline{\mathcal{O}}$, which is assumed to be simply connected. In scattering problems $\mathcal{O}$ will be referred to as the *scatterer* or the *obstacle*. $\mathcal{O}$, $\mathcal{O}_e$ are supposed to be filled with different media whose physical parameters will be denoted by the subscripts i, e, respectively; so we use ε, μ, β, k, η, γ_{L} and γ_{R} with the appropriate subscripts.

ASSUMPTION 5.4.1 ([268]) As far as the physical parameters are concerned, we assume that

(*i*) $\varepsilon_i, \mu_i, \beta_i, \varepsilon_e, \mu_e, \beta_e \in \mathbb{C}$.

(*ii*) $\mathcal{Re}\,\eta_{\mathrm{i}}, \mathcal{Re}\,\eta_{\mathrm{e}}, \mathcal{Re}\,\gamma_{\mathrm{iL}}, \mathcal{Re}\,\gamma_{\mathrm{iR}}, \mathcal{Re}\,\gamma_{\mathrm{eL}}, \mathcal{Re}\,\gamma_{\mathrm{eR}} > 0$.

(*iii*) $\mathcal{Im}\,\eta_{\mathrm{i}}, \mathcal{Im}\,\eta_{\mathrm{e}}, \mathcal{Im}\,\gamma_{\mathrm{iL}}, \mathcal{Im}\,\gamma_{\mathrm{iR}}, \mathcal{Im}\,\gamma_{\mathrm{eL}}, \mathcal{Im}\,\gamma_{\mathrm{eR}} \geq 0$.

(*iv*) $|k_e\beta_e| < 1, |k_i\beta_i| < 1$.

5.4.1 Incident fields

A given electromagnetic field E^{inc}, H^{inc} is incident on the obstacle $\mathcal{O}$. The form of the incident field depends on the type of material with which the complement of $\mathcal{O}$, $\mathcal{O}_e$, is filled. In particular:

▷ If $\mathcal{O}_e$ is filled with an achiral medium, the incident wave has the form

$$E^{\mathrm{inc}}(x) = q\, e^{\,\mathrm{i}k_{\mathrm{e}}\, p\cdot x}, \quad H^{\mathrm{inc}}(x) = -\mathrm{i}\,\eta_{\mathrm{e}}^{-1} q\, e^{\,\mathrm{i}k_{\mathrm{e}}\, p\cdot x}, \tag{5.6}$$

where q is the polarisation vector, p is the propagation unit vector ($p\cdot q = 0$), η_{e} is the impedance, and k_e is the wave number.

▷ If $\mathcal{O}_e$ is filled with a chiral medium, then the incident wave has the form

$$\begin{aligned} E^{\mathrm{inc}}(x) &= q_{\mathrm{L}}\, e^{\mathrm{i}\gamma_{\mathrm{eL}}\, p_{\mathrm{L}}\cdot x} + q_{\mathrm{R}}\, e^{\mathrm{i}\gamma_{\mathrm{eR}}\, p_{\mathrm{R}}\cdot x}, \\ H^{\mathrm{inc}}(x) &= -\mathrm{i}\,\eta_{\mathrm{e}}^{-1}\left(q_{\mathrm{L}}\, e^{\mathrm{i}\gamma_{\mathrm{eL}}\, p_{\mathrm{L}}\cdot x} + q_{\mathrm{R}}\, e^{\mathrm{i}\gamma_{\mathrm{eR}}\, p_{\mathrm{R}}\cdot x}\right), \end{aligned} \tag{5.7}$$

with $p_{\mathrm{L}}\cdot q_{\mathrm{L}} = 0$, $p_{\mathrm{L}} \times q_{\mathrm{L}} = -\mathrm{i}\,q_{\mathrm{L}}$, $p_{\mathrm{R}}\cdot q_{\mathrm{R}} = 0$, and $p_{\mathrm{R}} \times q_{\mathrm{R}} = \mathrm{i}\,q_{\mathrm{R}}$ ([10], [268]).

Under the above assumptions on p_{L}, q_{L}, it follows that

$$\operatorname{curl}\left(q_{\mathrm{L}}\, e^{\mathrm{i}\gamma_{\mathrm{eL}}\, p_{\mathrm{L}}\cdot x}\right) = \gamma_{\mathrm{eL}}\, q_{\mathrm{L}}\, e^{\mathrm{i}\gamma_{\mathrm{eL}}\, p_{\mathrm{L}}\cdot x},$$

$$\operatorname{curl}\left(q_{\mathrm{R}}\, e^{\mathrm{i}\gamma_{\mathrm{eR}}\, p_{\mathrm{R}}\cdot x}\right) = -\gamma_{\mathrm{eR}}\, q_{\mathrm{R}}\, e^{\mathrm{i}\gamma_{\mathrm{eR}}\, p_{\mathrm{R}}\cdot x},$$

i.e., that the incident fields E^{inc}, H^{inc} are combinations of an LCP plane wave and an RCP plane wave, and they satisfy the $\mathcal{O}_{\mathrm{e}}$ version of the chiral Maxwell equations (see (5.11)) below). q_{L} (resp. q_{R}) is the polarisation vector, p_{L} (resp. p_{R}) is the propagation unit vector, and γ_{eL} (resp. γ_{eR}) is the wave number of this left (resp. right) circularly plane wave.

It is convenient to introduce the following notation:

$$\text{If } \lambda = L \text{ then } \mathfrak{m}_{\mathrm{L}} = 1, \text{while, if } \lambda = R \text{ then } \mathfrak{m}_{\mathrm{R}} = -1. \tag{5.8}$$

We shall consider two kinds of scattering problems.

5.4.2 The perfect conductor problem

In this case we assume that $\mathcal{O}_e$ is filled with a chiral material and that $\mathcal{O}$ is a perfect conductor. The scattering problem reads: a given electromagnetic wave of the form (5.7) propagating in $\mathcal{O}_e$ is incident on $\mathcal{O}$. The total electromagnetic field $E^{\rm t}$, $H^{\rm t}$ in $\mathcal{O}_e$ is given by

$$E^{\rm t} = E^{\rm e} + E^{\rm inc}, \quad H^{\rm t} = H^{\rm e} + H^{\rm inc}, \text{ in } \mathcal{O}_e, \tag{5.9}$$

where $E^{\rm e}$, $H^{\rm e}$ is the scattered field that is assumed to satisfy *one* of the Silver-Müller radiation conditions

$$\begin{aligned} \widehat{x} \times H^{\rm e}(x) + \eta_{\rm e}^{-1} E^{\rm e}(x) = o(|x|^{-1}), \quad |x| \to \infty, \\ \widehat{x} \times E^{\rm e}(x) - \eta_{\rm e} H^{\rm e}(x) = o(|x|^{-1}), \quad |x| \to \infty, \end{aligned} \tag{5.10}$$

uniformly for all directions $\widehat{x} = \frac{x}{|x|}$.

The total electromagnetic field $E^{\rm t}$, $H^{\rm t}$ satisfies the following modified form of the Maxwell equations:

$$\left.\begin{aligned} {\rm curl} E^{\rm t} - {\rm i}\,\varpi\mu_e \tfrac{\gamma_{\rm e}^2}{k_e^2} H^{\rm t} - \beta_e\gamma_{\rm e}^2 E^{\rm t} = 0, \\ {\rm curl} H^{\rm t} + {\rm i}\,\varpi\varepsilon_e \tfrac{\gamma_{\rm e}^2}{k_e^2} E^{\rm t} - \beta_e\gamma_{\rm e}^2 H^{\rm t} = 0, \end{aligned}\right\} \text{ in } \mathcal{O}_e, \tag{5.11}$$

where $\gamma_{\rm e}^2 = \gamma_{\rm eL}\gamma_{\rm eR}$ and $k_e^2 = \varpi^2\varepsilon_e\mu_e$. Let us note that k_e is just shorthand notation and not a wave number. In the present case, as for the standard Maxwell equations, $E^{\rm t}$ and $H^{\rm t}$ are divergence free. Since $\mathcal{O}$ is assumed to be perfectly conducting, the following boundary condition must be satisfied:

$$n \times E^{\rm t} = 0, \quad \text{on } \partial\mathcal{O}, \tag{5.12}$$

where n is the unit outward normal vector to $\partial\mathcal{O}$.

The problem (5.9)–(5.12) will be referred to as the *perfect conductor problem* in what follows. It is related to receiving as well as transmitting antennas in various types of sheaths; see [24] and the references therein.

It is often useful to work with the electric field only; after eliminating the magnetic field the above problem becomes

$${\rm curl\, curl}\, E^{\rm t} - 2\beta_e\gamma_e^2 {\rm curl} E^{\rm t} - \gamma_e^2 E^{\rm t} = 0, \quad \text{in } \mathcal{O}_e, \tag{5.13}$$

$$n \times E^{\rm t} = 0, \quad \text{on } \partial\mathcal{O}, \tag{5.14}$$

$$\widehat{x} \times {\rm curl} E^{\rm e}(x) - \beta_e\gamma_e^2 \widehat{x} \times E^{\rm e}(x) + {\rm i}\,\frac{\gamma_e^2}{k_e} E^{\rm e}(x) = o(|x|^{-1}), \tag{5.15}$$

as $|x| \to \infty$, uniformly in all directions $\widehat{x}$.

In view of Bohren's decomposition (5.5) applied in $\mathcal{O}_e$, the perfect conductor problem can be expressed in terms of Beltrami fields as: find $Q^{\rm e}_{\rm L}$, $Q^{\rm e}_{\rm R}$ satisfying the equations

$$\left.\begin{aligned} {\rm curl} Q^{\rm e}_{\rm L} &= \gamma_{\rm eL}\, Q^{\rm e}_{\rm L}, \\ {\rm curl} Q^{\rm e}_{\rm R} &= -\gamma_{\rm eR}\, Q^{\rm e}_{\rm R}, \end{aligned}\right\} \text{ in } \mathcal{O}_e\,, \tag{5.16}$$

the radiation conditions

$$\left.\begin{aligned} \widehat{x} \times Q^{\rm e}_{\rm L}(x) + {\rm i}\, Q^{\rm e}_{\rm L}(x) = o(|x|^{-1}), \\ \widehat{x} \times Q^{\rm e}_{\rm R}(x) - {\rm i}\, Q^{\rm e}_{\rm R}(x) = o(|x|^{-1}), \end{aligned}\right\} |x| \to \infty, \tag{5.17}$$

and the boundary condition

$$n \times Q^{\rm e}_{\rm L} - {\rm i}\,\eta_{\rm e}\, n \times Q^{\rm e}_{\rm R} = n \times E^{\rm inc}, \quad \text{on } \partial\mathcal{O}. \tag{5.18}$$

5.4.3 The transmission problem

Find electric fields E^{i} and E^{e} and magnetic fields H^{i} and H^{e} that satisfy the following equations:

$$\begin{aligned} \mathrm{curl} E^{\mathrm{i}} &= \gamma_{\mathrm{i}}^2 \beta_i \, E^{\mathrm{i}} + \mathrm{i}\, \varpi\, \mu_i \left(\frac{\gamma_{\mathrm{i}}}{k_i} \right)^2 H^{\mathrm{i}}, \\ \mathrm{curl} H^{\mathrm{i}} &= \gamma_{\mathrm{i}}^2 \beta_i \, H^{\mathrm{i}} - \mathrm{i}\, \varpi\, \varepsilon_i \left(\frac{\gamma_{\mathrm{i}}}{k_i} \right)^2 E^{\mathrm{i}}, \end{aligned} \qquad \text{in } \mathcal{O}, \tag{5.19}$$

$$\begin{aligned} \mathrm{curl} E^{\mathrm{e}} &= \gamma_{\mathrm{e}}^2 \beta_e \, E^{\mathrm{e}} + \mathrm{i}\, \varpi\, \mu_e \left(\frac{\gamma_{\mathrm{e}}}{k_e} \right)^2 H^{\mathrm{e}}, \\ \mathrm{curl} H^{\mathrm{e}} &= \gamma_{\mathrm{e}}^2 \beta_e \, H^{\mathrm{e}} - \mathrm{i}\, \varpi\, \varepsilon_e \left(\frac{\gamma_{\mathrm{e}}}{k_e} \right)^2 E^{\mathrm{e}}, \end{aligned} \qquad \text{in } \mathcal{O}_e, \tag{5.20}$$

where $\gamma_{\mathrm{i}}^2 = \gamma_{\mathrm{iL}} \gamma_{\mathrm{iR}}$ and $k_i^2 = \varpi^2 \varepsilon_i \mu_i$, and the transmission conditions ([268])

$$n \times E^{\mathrm{t}} = n \times E^{\mathrm{i}} \text{ and}, \; n \times H^{\mathrm{t}} = n \times H^{\mathrm{i}}, \text{ on } \partial\mathcal{O}, \tag{5.21}$$

where n is the unit outward normal to $\partial\mathcal{O}$. The total fields in $\mathcal{O}_e$ are given by

$$E^{\mathrm{t}} = E^{\mathrm{e}} + E^{\mathrm{inc}}, \quad H^{\mathrm{t}} = H^{\mathrm{e}} + H^{\mathrm{inc}}, \text{ in } \mathcal{O}_e,$$

while the scattered fields E^{e}, H^{e} must satisfy *one* of the Silver-Müller radiation conditions (5.10).

In view of Bohren's decomposition (5.5) applied in $\mathcal{O}$, the transmission problem takes the form: given $E^{\mathrm{inc}}, H^{\mathrm{inc}}$, find $Q^{\mathrm{i}}_{\mathrm{L}}, Q^{\mathrm{i}}_{\mathrm{R}}$ and $Q^{\mathrm{e}}_{\mathrm{L}}, Q^{\mathrm{e}}_{\mathrm{R}}$ satisfying the equations

$$\begin{aligned} \mathrm{curl} Q^{\mathrm{i}}_{\mathrm{L}} &= \gamma_{\mathrm{iL}}\, Q^{\mathrm{i}}_{\mathrm{L}}, \\ \mathrm{curl} Q^{\mathrm{i}}_{\mathrm{R}} &= -\gamma_{\mathrm{iR}}\, Q^{\mathrm{i}}_{\mathrm{R}}, \end{aligned} \qquad \text{in } \mathcal{O}, \tag{5.22}$$

and

$$\begin{aligned} \mathrm{curl} Q^{\mathrm{e}}_{\mathrm{L}} &= \gamma_{\mathrm{eL}}\, Q^{\mathrm{e}}_{\mathrm{L}}, \\ \mathrm{curl} Q^{\mathrm{e}}_{\mathrm{R}} &= -\gamma_{\mathrm{eR}}\, Q^{\mathrm{e}}_{\mathrm{R}}, \end{aligned} \qquad \text{in } \mathcal{O}_e, \tag{5.23}$$

with the radiation conditions

$$\begin{aligned} \widehat{x} \times Q^{\mathrm{e}}_{\mathrm{L}}(x) + \mathrm{i}\, Q^{\mathrm{e}}_{\mathrm{L}}(x) &= o(|x|^{-1}), \\ \widehat{x} \times Q^{\mathrm{e}}_{\mathrm{R}}(x) - \mathrm{i}\, Q^{\mathrm{e}}_{\mathrm{R}}(x) &= o(|x|^{-1}), \end{aligned} \qquad |x| \to \infty, \tag{5.24}$$

and the transmission conditions on $\partial\mathcal{O}$

$$\begin{aligned} n \times (Q^{\mathrm{i}}_{\mathrm{L}} - Q^{\mathrm{e}}_{\mathrm{L}}) + n \times (-\mathrm{i}\, \eta_i\, Q^{\mathrm{i}}_{\mathrm{R}} + \mathrm{i}\, \eta_e\, Q^{\mathrm{e}}_{\mathrm{R}}) &= n \times E^{\mathrm{inc}}, \\ n \times (Q^{\mathrm{i}}_{\mathrm{R}} - Q^{\mathrm{e}}_{\mathrm{R}}) + n \times (-\mathrm{i}\, \eta_i^{-1}\, Q^{\mathrm{i}}_{\mathrm{L}} + \mathrm{i}\, \eta_e^{-1}\, Q^{\mathrm{e}}_{\mathrm{L}}) &= n \times H^{\mathrm{inc}}. \end{aligned} \tag{5.25}$$

Remark 5.4.2 A simple calculation (see, e.g., [273]) shows that when the materials are chiral, no matter if the incident wave is only LCP (resp. RCP), the scattered wave develops both LCP and RCP components. This applies to both the perfect conductor problem and the transmission problem.

REMARK 5.4.3 In both the perfect conductor problem and the transmission problems, stated in terms of the electromagnetic field (E, H), the governing equations are coupled and the boundary or the transmission conditions are uncoupled. On the other hand, when these problems are stated in terms of the Beltrami fields $(Q_{\mathrm{L}}, Q_{\mathrm{R}})$, the equations for the field are uncoupled outside $\partial\mathcal{O}$, but coupling occurs on $\partial\mathcal{O}$ through the boundary conditions. This allows the statement of the perfect conductor problem and the transmission problem as boundary integral equations.

REMARK 5.4.4 The above transmission problem obviously covers the cases of all possible combinations of (chiral or not) scatterer / (chiral or not) environment. These transmission problems will be respectively referred to as the *achiral scatterer in an achiral environment transmission problem* (which will not be considered at all in this book, since it does not refer to chiral media)[2], the *chiral scatterer in an achiral environment transmission problem*, the *achiral scatterer in a chiral environment transmission problem*, and the *chiral scatterer in a chiral environment transmission problem*. The spectrum of applications of the chiral scatterer in an achiral environment transmission problem is wide; see [273], [268], [289] and the references therein. In this work we shall refer often to the chiral scatterer in an achiral environment transmission problem but not to the achiral scatterer in a chiral environment transmission problem, which deals with *turbid* chiral media, i.e., chiral media with achiral particulate inclusions. Nevertheless, this problem has applications in clinical medicine[3] and planetary science[4]; see [23] and the references therein. As for the chiral scatterer in a chiral environment transmission problem, it is the basis for the Bruggeman homogenisation theory of chiral composites and the study of heterogeneous systems constituted by chiral particles in chiral fluids in physical chemistry; see [21] and the references therein.

5.5 AN INTRODUCTION TO BIES

Since we are going to study the solvability of the above scattering problems using BIE methods, in this section, for illustration purposes, we give an outline of this method for the scalar Helmholtz equation. The corresponding theory for the vector case, that is directly related to the Maxwell equations, is very similar; see, e.g., [103].

The consideration of integral equations in relation to boundary value problems for elliptic differential equations has a long history. We very briefly present the basic idea. For a concise yet much more complete introduction, see, e.g., [127], [197], [309]. In particular, a short historical survey is

[2]The theory of the achiral scatterer in an achiral environment transmission problem is well developed and known; see, e.g., [103], [106].

[3]In determining the concentration of blood glucose.

[4]The atmosphere of Titan (the biggest satellite of Saturn) is expected to be characterised by a chiral turbid medium.

contained in [309]. Consider the Helmholtz equation

$$\Delta u + k^2 u = 0 \tag{5.26}$$

in either a bounded, simply connected domain $\mathcal{O} \subset \mathbb{R}^3$, with boundary $\partial\mathcal{O} \in C^2$, or in its complement $\mathcal{O}_e$ in $\mathbb{R}^3$. It is well known that a classical solution $u \in C^2(\mathcal{O}) \cap C^1(\overline{\mathcal{O}})$ (respectively, $u \in C^2(\mathcal{O}_e) \cap C^1(\overline{\mathcal{O}_e})$ can be represented in terms of *surface potentials* via the first Green identity for the Laplacian and the *fundamental solution*

$$\Phi(x, y; k) = \frac{e^{\mathrm{i}\,k|x-y|}}{4\pi\,|x-y|}, \quad x \neq y, \tag{5.27}$$

of the Helmholtz equation. As usual, $k \in \mathbb{C}$ with $\mathcal{I}\mathrm{m}\, k \geq 0$.
In the case of exterior problems the behaviour at infinity is required to satisfy the Sommerfeld radiation conditions

$$u(x) = O(|x|^{-1}),\ \frac{\partial u}{\partial |x|}(x) - iku(x) = o(|x|^{-1}),\ |x| \to \infty,$$

or their weaker version

$$\lim_{R\to\infty} \int_{|x|=R} \left| \frac{\partial u}{\partial |x|}(x) - iku(x) \right|^2 ds = 0\,.$$

The latter is due to Rellich and is used in the variational formulation of exterior problems.
The aforementioned representation reads

$$u(x) = \pm \left(\int_{\partial\mathcal{O}} \Phi(x, y; k) \frac{\partial u}{\partial n}(y) ds(y) - \int_{\partial\mathcal{O}} u(y) \frac{\partial \Phi(x, y; k)}{\partial n_y} ds(y) \right), \tag{5.28}$$

with "+" for $x \in \mathcal{O}$ and "–" for $x \in \mathcal{O}_e$, where n_y denotes the exterior normal to $\partial\mathcal{O}$ at $y \in \partial\mathcal{O}$.

For given *Cauchy data* $u|_{\partial\mathcal{O}}$ and $\frac{\partial u}{\partial n}|_{\partial\mathcal{O}}$, the above *representation formula* defines the solution of the Helmholtz equation everywhere in $\mathcal{O}$, or in $\mathcal{O}_e$, respectively.
The surface potentials appearing in the representation formula are the *single-layer potential*

$$\mathsf{V_k}\phi(x) := \int_{\partial\mathcal{O}} \Phi(x, y; k)\phi(y) ds(y)\,,\ x \in \mathcal{O} \cup \mathcal{O}_e\,,$$

and the *double-layer potential*

$$\mathsf{W_k}\varphi(x) := \int_{\partial\mathcal{O}} \frac{\partial \Phi(x, y; k)}{\partial n_y} \varphi(y) ds(y)\,,\ x \in \mathcal{O} \cup \mathcal{O}_e\,,$$

where ϕ and φ are the corresponding *densities.*
Provided the corresponding limits exist, the *boundary integral operators* we are going to employ are related to the limits of the boundary potentials from either $\mathcal{O}$ on $\partial\mathcal{O}$ ($x \in \partial\mathcal{O}$, $z \in \mathcal{O}$),

$$\mathsf{V_k}\phi(x) := \lim_{z\to x} \mathsf{V_k}\phi(z),$$

$$\mathsf{K_k}(x)\psi(x) := \lim_{z\to x} \mathsf{W_k}\psi(z) + \tfrac{1}{2}\psi(x),$$

$$\mathsf{K_k^*}(x)\phi(x) := \lim_{z\to x} \mathrm{grad}_z \mathsf{V_k}\phi(z)\cdot n_x - \tfrac{1}{2}\phi(x),$$

$$\mathsf{D_k}(x)\psi(x) := -\lim_{z\to x} \mathrm{grad}_z \mathsf{W_k}\psi(z)\cdot n_x,$$

or from $\mathcal{O}_e$ on $\partial\mathcal{O}$ ($x \in \partial\mathcal{O}$, $z \in \mathcal{O}_e$),

$$\mathsf{V_k}\phi(x) := \lim_{z\to x} \mathsf{V_k}\phi(z),$$

$$\mathsf{K_k}(x)\psi(x) := \lim_{z\to x} \mathsf{W_k}\psi(z) - \tfrac{1}{2}\psi(x),$$

$$\mathsf{K_k^*}(x)\phi(x) := \lim_{z\to x} \mathrm{grad}_z \mathsf{V_k}\phi(z)\cdot n_x + \tfrac{1}{2}\phi(x),$$

$$\mathsf{D_k}(x)\psi(x) := -\lim_{z\to x} \mathrm{grad}_z \mathsf{W_k}\psi(z)\cdot n_x,$$

The following standard expressions of these boundary integral operators are essential. Let $\partial\mathcal{O} \in C^2$ and ϕ and ψ be continuous. Then the above limits exist uniformly with respect to all $x \in \partial\mathcal{O}$ and all ϕ and ψ with $\sup_{x\in\partial\mathcal{O}} |\phi(x)| \leq 1\,, \sup_{x\in\partial\mathcal{O}} |\psi(x)| \leq 1$. Further, these limits can be expressed as

$$(\mathsf{V_k}\phi)(x) = \int_{y\in\partial\mathcal{O}\setminus\{x\}} \Phi(x,y;k)\phi(y)ds(y)\,,\ x \in \partial\mathcal{O}\,,$$

$$(\mathsf{K_k}\psi)(x) = \int_{y\in\partial\mathcal{O}\setminus\{x\}} \frac{\partial\Phi(x,y;k)}{\partial n_y}\psi(y)ds(y)\,,\ x \in \partial\mathcal{O}\,,$$

$\mathsf{K_k^*}$ is the adjoint of $\mathsf{K_k}$, i.e.,

$$(\mathsf{K_k^*}\phi)(x) = \int_{y\in\partial\mathcal{O}\setminus\{x\}} \frac{\partial\Phi(x,y;k)}{\partial n_x}\phi(y)ds(y)\,,\ x \in \partial\mathcal{O}\,,$$

while, for ϕ Hölder continuously differentiable with $||\phi||_{C^{1,\theta}} \leq 1\,, \theta \in (0,1)$,

$$(\mathsf{D_k}\phi)(x) = \mathrm{pv}\int_{\partial\mathcal{O}} \frac{\partial^2\Phi(x,y;k)}{\partial n_x \partial n_y}(\phi(y)-\phi(x))ds(y)\,,$$

where "pv" denotes the Cauchy principal value integral[5]. It is not hard to see that the kernels of $\mathsf{V_k}, \mathsf{K_k}$ and $\mathsf{K_k^*}$ are *weakly singular*[6] (with $\alpha = 1$).

[5]Because of the strong singularity of the kernel the integral in the definition of D_k has to be interpreted as a Cauchy principal value integral. Let us briefly recall (see, e.g., [305]) the definition in a simple case: let $f \in C^{0,\theta}([a,b])$ and $x \in (a,b)$. The Cauchy principal value integral is defined as $\mathrm{pv}\int_a^b \frac{f(t)}{x-t}dt := \lim_{\epsilon\to 0}\big(\int_a^{x-\epsilon} \frac{f(t)}{x-t}dt + \int_{x+\epsilon}^b \frac{f(t)}{x-t}dt\big)$. If $f \in C^1([a,b])$, we have that $\mathrm{pv}\int_a^b \frac{f(t)}{x-t}dt = f(a)\ln(x-a) - f(b)\ln(b-x) + \int_a^b f'(t)\ln|x-t|dt$. Further, if $f \in C^{1,\theta}([a,b])$, we have $\frac{d}{dx}\big(\mathrm{pv}\int_a^b \frac{f(t)}{x-t}dt\big) = \frac{f(a)}{x-a} + \frac{f(b)}{b-x} + \mathrm{pv}\int_a^b \frac{f'(t)}{x-t}dt$.

[6]Consider the Banach space $C(\partial\mathcal{O})$ of complex-valued functions on $\partial\mathcal{O}$ equipped with the maximum norm, and the boundary integral operator $\mathsf{C} : C(\partial\mathcal{O}) \to C(\partial\mathcal{O})$ defined by $(\mathsf{C}\phi)(x) := \int_{\partial\mathcal{O}} \mathsf{k}(x,y)\,\phi(y)\,ds(y)\,,\ x \in \partial\mathcal{O}$. The *kernel* k is said to be weakly singular if it is defined and continuous for all $x, y \in \partial\mathcal{O}, x \neq y$, and there exist positive constants c and $\alpha \in (0,2]$ such that for all $x, y \in \partial\mathcal{O}, x \neq y$, we have $|\mathsf{k}(x,y)| \leq c|x-y|^{\alpha-2}$. Then it is well known (see, e.g., Theorem 2.6 in [103]) that, under the assumption that the kernel k is *either* continuous *or* weakly singular, the operator C is compact.

Various continuous extensions of V_k, K_k, K_k^*, D_k between suitable spaces of Hölder continuous or Hölder differentiable functions have been studied, see, e.g., [103], [197], [319].

We are now in a position to consider boundary value problems; we discuss the *interior* Dirichlet problem, namely

$$\Delta u + k^2 u = 0 \text{ in } \mathcal{O}\,, \text{ with } u = f \text{ on } \partial\mathcal{O}\,.$$

In this case only one of the two Cauchy data is given in (5.28), namely $u(x) = f(x)\,, x \in \partial\mathcal{O}$. The "missing" datum is $\frac{\partial u}{\partial n}(x) = g(x)$.

The boundary integral equation for the determination of g then reads

$$\Big(\mathsf{V}_\mathsf{k} g\Big)(x) = \Big(\Big(\frac{1}{2}I + \mathsf{K}_\mathsf{k}\Big)f\Big)(x)\,,\ x \in \partial\mathcal{O}\,,$$

which is a *Fredholm integral equation of the first kind.* This equation, despite the fact that it is ill posed, has proved to be very important both from the analytic as well as from the numerical point of view. It is known that for $k \neq 0$ and $f \in C^{1,\alpha}(\partial\mathcal{O})\,, \alpha \in (0,1)$, the above boundary integral equation is uniquely solvable with $g \in C^\alpha(\partial\mathcal{O})$ except for certain values of $k \in \mathbb{C}$ which are the *exceptional* or *irregular frequencies* of the boundary integral operator V_k. For any irregular frequency κ the operator V_κ has a nontrivial null-space spanned by eigensolutions ϱ related to the eigensolutions ρ of the interior Dirichlet problem

$$-\Delta\rho = \kappa^2\rho \text{ in } \mathcal{O},\ \ \rho = 0 \text{ on } \partial\mathcal{O}\,,$$

by

$$\varrho = \frac{\partial\rho}{\partial n}|_{\partial\mathcal{O}}\,.$$

In this case, for $f \in C^{1,\alpha}(\partial\mathcal{O})$ the boundary integral equation $\mathsf{V}_\mathsf{k} g = \frac{1}{2}f + \mathsf{K}_\mathsf{k} f$ has solutions $g \in C^\alpha(\partial\mathcal{O})$ if and only if the orthogonality condition

$$\int_{\partial\mathcal{O}} f\,\varrho = 0\,,\ \ \forall \varrho \in \ker \mathsf{V}_\kappa\,,$$

is satisfied, see, e.g., [103], [197].

The interior Dirichlet problem for the Helmholtz equation can alternatively be expressed as

$$\Big(\Big(\frac{1}{2}I - \mathsf{K}_\mathsf{k}^*\Big)g\Big)(x) = \Big(\mathsf{D}_\mathsf{k} f\Big)(x)\,,\ x \in \partial\mathcal{O}\,,$$

which is a *Fredholm integral equation of the second kind.* This boundary integral equation has been and still is extensively studied, both analytically and numerically.

The exterior Dirichlet problem can be similarly expressed as a Fredholm integral equation of either the first or the second kind, respectively as

$$\Big(\mathsf{V}_\mathsf{k} g\Big)(x) = \Big(\Big(-\frac{1}{2}I + \mathsf{K}_\mathsf{k}\Big)f\Big)(x)\,,\ x \in \partial\mathcal{O}\,,$$

and

$$\Big(\Big(\frac{1}{2}I + \mathsf{K}_\mathsf{k}^*\Big)g\Big)(x) = -\Big(\mathsf{D}_\mathsf{k} f\Big)(x)\,,\ x \in \partial\mathcal{O}\,.$$

We notice that we may use different boundary integral equations for the same boundary value problem; further, these boundary integral equations may be uniquely or non-uniquely solvable. This is an important issue: for example, the exterior Dirichlet problem for the Helmholtz equation is known, see, e.g., [103], to have a unique solution for all k with $\mathcal{I}m\, k \geq 0$; so the complication of non-uniqueness for the boundary integral equation at the irregular frequencies arises from the selected method of solution rather than from the nature of the problem itself. It is therefore desirable to develop methods leading to boundary integral equations that are uniquely solvable for all values of k. This is done by resorting to the so-called *modified integral equations*. The discussion of this topic lies outside the scope of the present section; one can consult, e.g., [103].

REMARK 5.5.1 In some cases it is not possible to interchange the normal derivative with the integral sign (because this would lead to a nonintegrable integrand). Then one has to resort to *Hadamard finite part integrals*. For $f \in C^{1,\theta}([a,b])$ a two-sided Hadamard finite part integral of order two is defined as

$$\text{pf}\int_a^b \frac{f(t)}{(x-t)^2}dt := \lim_{\epsilon\to 0}\left(\int_a^{x-\epsilon} \frac{f(t)}{(x-t)^2}dt + \int_{x+\epsilon}^b \frac{f(t)}{(x-t)^2}dt - \frac{2f(x)}{\epsilon}\right).$$

It can be shown that

$$\text{pf}\int_a^b \frac{f(t)}{(x-t)^2}dt = -\frac{f(a)}{x-a} - \frac{f(b)}{b-x} - \text{pv}\int_a^b \frac{f'(t)}{x-t}dt,$$

whereby

$$\frac{d}{dx}\left(\text{pv}\int_a^b \frac{f(t)}{x-t}dt\right) = -\text{pf}\int_a^b \frac{f(t)}{(x-t)^2}dt.$$

Therefore, differentiation can be interchanged with integration. For more details one can see, e.g., [197].

So we see that the Dirichlet (and Neumann) boundary value problems for the Helmholtz equation in $\mathcal{O}$ (or in $\mathcal{O}_e$) are reduced (in view of appropriate integral representations of the solutions) to problems defined on a bounded domain of lower dimension, namely, on the boundary $\partial\mathcal{O}$. Such reductions shift the setting from partial differential (unbounded) operators to boundary integral (compact) operators. The very rich Riesz-Fredholm theory for compact operators is then an indispensable arsenal that, combined with potential theory[7], provides a powerful mathematical framework. Boundary integral equation methods are also closely related to constructive techniques and are well suited for numerical computations. One can refer in general to the excellent monographs [9], [197], [309], [398], and for acoustic and electromagnetic scattering problems to [103], [106].

[7]For an introduction to potential theory see [124].

This is the course we follow to deal with boundary value problems and scattering problems for the Maxwell equations supplemented with the Drude-Born-Fedorov constitutive relations. We consider only the perfect conductor boundary condition and transmission boundary conditions, not more general and in particular nonlinear boundary conditions.

Boundary integral operators can also be studied as special cases of pseudodifferential operators. Although the bulk of publications treat the case of L^2-based spaces, the case of spaces based on L^p, $p \neq 2$, is well studied[8]. Further, in the extremely important (e.g., in real engineering problems) case of nonsmooth domains, the behaviour of the dominant singularities of the solution at corners or edges or interior cuts of the domain is the subject of important research. We do not consider any of these aspects.

5.6 PROPERTIES OF BELTRAMI FIELDS

In this section we provide some properties of Beltrami fields that will be useful in what follows.

5.6.1 Representations in terms of spherical wave functions

5.6.1.1 Scalar and vector spherical wave functions

For the reader's convenience, we recall here the definitions of the spherical wave functions that are employed in the following discussion. A classical reference is [325]; in what follows we employ the contemporary condensed notation for these functions (see, e.g., [71]).

Let $\ell = 0, 1, \ldots$, and let J_ℓ, Y_ℓ be the Bessel functions of the first and second kind, respectively. The spherical Bessel functions of the first and second kind, are defined, for $z \in \mathbb{C}$, as

$$j_\ell(z) = \left(\frac{\pi}{2\,z}\right)^{1/2} J_{\ell+\frac{1}{2}}(z)\ ,\ y_\ell(z) = \left(\frac{\pi}{2\,z}\right)^{1/2} Y_{\ell+\frac{1}{2}}(z)\,,$$

respectively, while the spherical Hankel function of the first kind is

$$h_\ell^{(1)}(z) = j_\ell(z) + i\,y_\ell(z)\,.$$

Let $m = 0, \ldots, \ell$, and P_ℓ^m be the associated Legendre function. For the definitions of the Bessel functions and the associated Legendre function, see, e.g., [259], [325].

Let (r, θ, φ) be the spherical coordinates of the vector $x \in \mathbb{R}^3$, and $\widehat{x} = \frac{x}{r}$. The *even* and *odd normalised real scalar spherical harmonic functions* are, respectively, defined as

$$\mathcal{Y}_{\mathsf{e}\,m\,\ell}(\widehat{x}) = \frac{2-\delta_{m0}}{2\pi}\,\frac{2\ell+1}{2}\,\frac{(\ell-m)!}{(\ell+m)!}\,P_\ell^m(cos\theta)\,cos\,m\varphi\,,$$

[8] In [323], the chiral scatterer in achiral environment transmission problem is studied for a scatterer of less smooth boundary than the one considered here, and in L^p-based Sobolev spaces.

$$\mathcal{Y}_{\mathsf{o}\,m\,\ell}(\widehat{x}) = \frac{2-\delta_{m0}}{2\pi}\,\frac{2\ell+1}{2}\,\frac{(\ell-m)!}{(\ell+m)!}\,P_\ell^m(cos\theta)\,sin\,m\varphi\,,$$

where δ_{mn} denotes the Kronecker delta. Let $\ell_0 := [\ell(\ell+1)]^{-1/2}$. The *normalised real vector spherical harmonic functions*, $\mathbf{A}_{j\,\sigma\,m\,\ell}$, $j = 1,2,3$, are defined by

$$\mathbf{A}_{1\,\sigma\,m\,\ell}(\widehat{x}) := \ell_0 r\,\mathrm{curl}(x\,\mathcal{Y}_{\sigma\,m\,\ell}(\widehat{x})) = \ell_0\left[\widehat{\theta}\,\frac{1}{\sin\theta}\,\frac{\partial\mathcal{Y}_{\sigma\,m\,\ell}(\widehat{x})}{\partial\varphi} - \widehat{\varphi}\,\frac{\partial\mathcal{Y}_{\sigma\,m\,\ell}(\widehat{x})}{\partial\theta}\right],$$

$$\mathbf{A}_{2\,\sigma\,m\,\ell}(\widehat{x}) := \ell_0 r\,\mathrm{grad}\,\mathcal{Y}_{\sigma\,m\,\ell}(\widehat{x}) = \ell_0\left[\widehat{\theta}\,\frac{\partial\mathcal{Y}_{\sigma\,m\,\ell}(\widehat{x})}{\partial\theta} + \widehat{\varphi}\,\frac{1}{sin\theta}\,\frac{\partial\mathcal{Y}_{\sigma\,m\,\ell}(\widehat{x})}{\partial\varphi}\right],$$

$$\mathbf{A}_{3\,\sigma\,m\,\ell}(\widehat{x}) := \widehat{x}\,\mathcal{Y}_{\sigma\,m\,\ell}(\widehat{x})\,,$$

for $\sigma = \mathsf{e}, \mathsf{o}$, respectively. They constitute a complete orthonormal set on the unit sphere[9]. Finally, the *normalised outgoing spherical vector wave functions* are[10]

$$\begin{aligned}\boldsymbol{\psi}_{1\,\sigma\,m\,\ell}(x) &:= \ell_0\mathrm{curl}(x\,h_\ell^{(1)}(kr)\,\mathcal{Y}_{\sigma\,m\,\ell}(\widehat{x})) = h_\ell^{(1)}(kr)\,\mathbf{A}_{1\,\sigma\,m\,\ell}(\widehat{x})\,,\\ \boldsymbol{\psi}_{2\,\sigma\,m\,\ell}(x) &:= \ell_0\frac{1}{k}\,\mathrm{curl}\,\mathrm{curl}(x\,h_\ell^{(1)}(kr)\,\mathcal{Y}_{\sigma\,m\,\ell}(\widehat{x}))\\ &= \frac{1}{kr}\left[\left(kr\,h_\ell^{(1)}(kr)\right)'\mathbf{A}_{2\,\sigma\,m\,\ell}(\widehat{x}) + \ell_0^{-1}h_\ell^{(1)}(kr)\,\mathbf{A}_{3\,\sigma\,m\,\ell}(\widehat{x})\right],\end{aligned}$$

where k is the wave number of the vector Helmholtz equation $(\Delta + k^2)\boldsymbol{\psi} = 0$.

Since $\boldsymbol{\psi}_{1\,\sigma\,m\,\ell}(x), \boldsymbol{\psi}_{2\,\sigma\,m\,\ell}(x)$ are defined via $\mathcal{Y}_{\sigma\,m\,\ell}(\widehat{x})$ and $h_\ell^{(1)}(kr)$, asymptotic properties for the former (as $r \to \infty$) can be deduced from well-known corresponding properties of the latter; see [28].

5.6.1.2 Beltrami fields in spherical coordinates

Three-dimensional representations for Beltrami fields follow from the concept of toroidal and poloidal field (see, e.g., [268]). In particular, the following representations hold:

$$Q_\lambda(x) = \mathrm{curl}\,[x\tau_\lambda(x)] + \mathrm{curl}\,\mathrm{curl}\,[x\vartheta_\lambda(x)]\,,$$

where $\tau_\lambda(x)$, $\vartheta_\lambda(x)$, for $\lambda = \mathrm{L}, \mathrm{R}$, are scalar functions of the position vector x that satisfy the Helmholtz equation and are related by

$$\tau_\lambda(x) = \gamma_\lambda\vartheta_\lambda(x).$$

---

[9]Let us note that these functions are closely related to the vector wave functions introduced by Hansen which in [325], are denoted by $\mathbf{C}, \mathbf{B}, \mathbf{P}$, respectively.

[10]They are closely related to the vector wave functions which are, respectively, denoted by $\mathbf{M}, \mathbf{N}$, in [325]. In fact, there is also a third normalised outgoing spherical vector wave function, $\boldsymbol{\psi}_{3\,\sigma\,m\,\ell}$ (or $\mathbf{L}$), but since it is mainly useful in elastodynamics we omit its definition here and instead refer to [71], [325].

Using the Helmholtz decomposition, we obtain the following representation in the spherical coordinate system for Q_{L} and Q_{R} ([28], [268]):

$$Q_{\mathrm{L}}(x) = \sum_{\ell=0}^{\infty}\sum_{m=0}^{\ell} a_{m\ell}\left[\boldsymbol{\psi}_{1\,\mathrm{e}\,m\,\ell}(x) + \boldsymbol{\psi}_{2\,\mathrm{e}\,m\,\ell}(x) - i\left(\boldsymbol{\psi}_{1\,\mathrm{o}\,m\,\ell}(x) + \boldsymbol{\psi}_{2\,\mathrm{o}\,m\,\ell}(x)\right)\right],$$

$$Q_{\mathrm{R}}(x) = \sum_{\ell=0}^{\infty}\sum_{m=0}^{\ell} b_{m\ell}\left[\boldsymbol{\psi}_{1\,\mathrm{e}\,m\,\ell}(x) - \boldsymbol{\psi}_{2\,\mathrm{e}\,m\,\ell}(x) - i\left(\boldsymbol{\psi}_{1\,\mathrm{o}\,m\,\ell}(x) - \boldsymbol{\psi}_{2\,\mathrm{o}\,m\,\ell}(x)\right)\right],$$

where $a_{m\ell}, b_{m\ell}$ are appropriate constants.

5.6.2 Integral representations

In this section we establish integral representations for Beltrami fields, which will be used in Section 5.7 to present a generalisation of the Stratton-Chu representation formula of achiral electromagnetics to the chiral case.

Since we work in source-free regions, the divergence of the Beltrami fields appearing in our study is equal to zero, and in view of the identity

$$\operatorname{curl}\operatorname{curl} u = \operatorname{grad}\operatorname{div} u - \Delta\, u\,,$$

these fields are solutions of a vector Helmholtz equation of the form

$$\Delta\, u + \kappa^2 u = 0\,,$$

so we may use classical layer potentials in terms of the fundamental solution (5.27) of the Helmholtz equation.

5.6.2.1 Interior integral representation

We start with an interior integral representation theorem ([22]) for the solution of the following problem:

$$\begin{array}{l} \operatorname{curl} Q^{\mathrm{i}}_{\mathrm{L}} = \gamma_{\mathrm{iL}}\, Q^{\mathrm{i}}_{\mathrm{L}}, \\ \operatorname{curl} Q^{\mathrm{i}}_{\mathrm{R}} = -\gamma_{\mathrm{iR}}\, Q^{\mathrm{i}}_{\mathrm{R}}, \end{array} \quad \text{in } \mathcal{O}\,. \tag{5.29}$$

THEOREM 5.6.1 *Let Q^{i}_{L}, $Q^{\mathrm{i}}_{R} \in C^1(\mathcal{O}) \cap C(\overline{\mathcal{O}})$ solve the system of equations* (5.29). *Then, for λ =L,R,*

$$-\mathbf{1}_{\mathcal{O}}(x)\, Q^{\mathrm{i}}_{\lambda}(x) = \operatorname{curl}\int_{\partial\mathcal{O}} (\, n\,(y) \times Q^{\mathrm{i}}_{\lambda}(y))\, \Phi(x, y; \gamma_{\mathrm{i}\lambda})\, ds(y)$$
$$+ \frac{\mathfrak{m}_{\lambda}}{\gamma_{\mathrm{i}\lambda}} \operatorname{curl}\operatorname{curl}\int_{\partial\mathcal{O}} (\, n\,(y) \times Q^{\mathrm{i}}_{\lambda}(y))\, \Phi(x, y; \gamma_{\mathrm{i}\lambda})\, ds(y), \tag{5.30}$$

where $\mathbf{1}_{\mathcal{O}}$ is the characteristic (indicator) function of $\mathcal{O}$ and $\mathfrak{m}_{\lambda}$ is defined in (5.8).

5.6.2.2 Exterior integral representation

We now consider integral representations for the corresponding exterior problem (see [22]),

$$\begin{array}{lcl} \operatorname{curl} Q^{\mathrm{e}}_{\mathrm{L}} &=& \gamma_{\mathrm{eL}}\, Q^{\mathrm{e}}_{\mathrm{L}}, \\ \operatorname{curl} Q^{\mathrm{e}}_{\mathrm{R}} &=& -\gamma_{\mathrm{eR}}\, Q^{\mathrm{e}}_{\mathrm{R}}, \end{array} \quad \text{in } \mathcal{O}_e\,. \tag{5.31}$$

with the radiation conditions,

$$\begin{aligned} \widehat{x} \times Q_{\rm L}^{\rm e}(x) + {\rm i}\, Q_{\rm L}^{\rm e}(x) = o(|x|^{-1}), \\ \widehat{x} \times Q_{\rm R}^{\rm e}(x) - {\rm i}\, Q_{\rm R}^{\rm e}(x) = o(|x|^{-1}), \end{aligned} \qquad |x| \to \infty, \tag{5.32}$$

uniformly for all directions $\widehat{x}$.

THEOREM 5.6.2 *Let $Q_L^{\rm e}$, $Q_R^{\rm e} \in C^1(\mathcal{O}_e) \cap C(\overline{\mathcal{O}_e})$ solve the system of equations (5.16) - (5.32). Then, for λ =L,R,*

$$\begin{aligned} \mathbf{1}_{\mathcal{O}_{\rm e}}(x)\, Q_\lambda^{\rm e}(x) = {\rm curl} \int_{\partial\mathcal{O}} (n\,(y) \times Q_\lambda^{\rm e}(y))\, \Phi(x, y; \gamma_{{\rm e}\lambda})\, ds(y) \\ + \frac{\mathfrak{m}_\lambda}{\gamma_{{\rm e}\lambda}} {\rm curlcurl} \int_{\partial\mathcal{O}} (n\,(y) \times Q_\lambda^{\rm e}(y))\, \Phi(x, y; \gamma_{{\rm e}\lambda})\, ds(y), \end{aligned} \tag{5.33}$$

where $\mathbf{1}_{\mathcal{O}_{\rm e}}$ is the characteristic (indicator) function of $\mathcal{O}_{\rm e}$ and $\mathfrak{m}_\lambda$ is defined in (5.8).

5.7 SOLVABILITY

In this section we establish the existence and uniqueness of solutions of the perfect conductor problem, the chiral scatterer in an achiral environment transmission problem, and the chiral scatterer in a chiral environment transmission problem. We present first weak and then classical solvability. We follow a BIE approach.

5.7.1 Integral representation

Using Bohren's decomposition (5.5) and the integral representations (5.30) and (5.33), a straightforward calculation leads to the following generalisation of the *Stratton-Chu integral representation formulae* for chiral media. We introduce, for convenience, fields U and $\mathbb{U}$ (the "dual" field of U) as:

$$\text{If } U = E \text{ then } \mathbb{U} = {\rm i}\, H\,, \text{ while if } U = H \text{ then } \mathbb{U} = -{\rm i}\, E. \tag{5.34}$$

THEOREM 5.7.1 ([22]) *For x in a chiral body embedded in an achiral space, and y on its boundary, we have the Stratton-Chu integral representation formula for chiral media:*

$$\begin{aligned} -8U^{\rm i}(x) = & \sum_{\lambda=L,R} {\rm curl} \int_{\partial\mathcal{O}} \big(n\,(y) \times U^{\rm i}(y)\big)\ \Phi(x, y; \gamma_{{\rm i}\lambda})\, ds(y) \\ & + \sum_{\lambda=L,R} \frac{\mathfrak{m}_\lambda}{\gamma_{{\rm i}L}} {\rm curlcurl} \int_{\partial\mathcal{O}} \big(n\,(y) \times U^{\rm i}(y)\big)\ \Phi(x, y; \gamma_{{\rm i}\lambda})\, ds(y) \\ & + \sum_{\lambda=L,R} \mathfrak{m}_\lambda {\rm curl} \int_{\partial\mathcal{O}} \big(n\,(y) \times \mathbb{U}^{\rm i}(y)\big)\ \Phi(x, y; \gamma_{{\rm i}\lambda})\, ds(y) \\ & + \sum_{\lambda=L,R} \frac{1}{\gamma_{{\rm i}L}} {\rm curlcurl} \int_{\partial\mathcal{O}} \big(n\,(y) \times \mathbb{U}^{\rm i}(y)\big)\ \Phi(x, y; \gamma_{{\rm i}\lambda})\, ds(y), \end{aligned} \tag{5.35}$$

where $x \in \mathcal{O}$, $y \in \partial\mathcal{O}$ and the superscript "i" denotes the interior fields and $\mathfrak{m}_\lambda$ is defined in (5.8).

REMARK 5.7.2 An alternative representation, using a dyadic Green function, can be found in [34].

REMARK 5.7.3 If $\beta = 0$, the usual (achiral) Stratton-Chu integral representation formula is obtained (see [103]).

5.7.2 Layer potentials and boundary integral operators

DEFINITION 5.7.4 *Let $y \in \partial\mathcal{O}$. Given a continuous density function $v(y)$, the single-layer potential of v is*

$$(\mathsf{V}^{\mathrm{i}}(k)v)(x) := \int_{\partial\mathcal{O}} v(y)\Phi(x,y;k)\,ds(y)\,,\ x \in \mathcal{O}.$$

DEFINITION 5.7.5 *Given a tangential vector density, $a \in C_t(\partial\mathcal{O})$, the double-layer potential of a is*

$$(\mathsf{M}^{\mathrm{i}}(k)\,a)(x) := \mathrm{curl}\{\mathsf{V}^{\mathrm{i}}(k)\,a\}\,.$$

Further, define

$$(\mathsf{N}^{\mathrm{i}}(k)\,a)(x) := \mathrm{curl}\{\mathsf{M}^{\mathrm{i}}(k)\,a\}\,.$$

We shall use the superscript "e" to indicate that $x \in \mathcal{O}_e$ in the above definitions.

It is known ([342]) that there are continuous extensions

$$\mathsf{M}^{\mathrm{i}}(k), \mathsf{N}^{\mathrm{i}}(k):\ H^{1/2}(\mathrm{div},\partial\mathcal{O}) \to H^1_{\mathrm{Div}}(\mathcal{O}),$$

$$\mathsf{M}^{\mathrm{e}}(k), \mathsf{N}^{\mathrm{e}}(k):\ \ H^{1/2}(\mathrm{div},\partial\mathcal{O}) \to H^1_{\mathrm{Div},\,loc}(\mathcal{O}_e).$$

Let us now introduce appropriate boundary integral operators. For $x \in \partial\mathcal{O}$, they are defined as

$$\mathsf{M}(k)\,v(x) := n \times \mathrm{pv}\int_{\partial\mathcal{O}} (\mathrm{grad}_x\Phi(x,y;k)) \times v(y)\,ds(y),$$

and

$$\mathsf{N}(k)v(x) := n \times (\mathrm{grad}\int_{\partial\mathcal{O}} \Phi(x,y;k)\mathrm{Div}\,v(y)ds(y) + k^2\int_{\partial\mathcal{O}} \Phi(x,y;k)v(y)ds(y)).$$

It is known ([178], [257], [341], [342]) that there are continuous extensions

$$\mathsf{M}(k), \mathsf{N}(k) : H^{1/2}(\mathrm{div},\partial\mathcal{O}) \to H^{1/2}(\mathrm{div},\partial\mathcal{O}),$$

and that $\mathsf{M}(k)$ is compact. Furthermore, the mappings

$$\mathsf{M}(k) : H^r_t(\partial\mathcal{O}) \to H^{r+1}_t(\partial\mathcal{O}), r \in \mathbb{R},$$

$$\mathsf{N}(k) : H^r_t(\partial\mathcal{O}) \to H^{r-1}_t(\partial\mathcal{O}), r \in \mathbb{R},$$

are continuous.

Additionally, if $v \in H^{1/2}(\mathrm{div},\partial\mathcal{O})$, we have for the traces on $\partial\mathcal{O}$

$$\mathsf{M}(k)v = n \times \mathsf{M}^{\mathrm{i}}(k)\,v + \frac{1}{2}v = n \times \mathsf{M}^{\mathrm{e}}(k)\,v - \frac{1}{2}v,$$

$$\mathsf{N}(k)v = n \times \mathsf{N}^{\mathrm{i}}(k)\,v = n \times \mathsf{N}^{\mathrm{e}}(k)\,v.$$

5.7.3 Boundary integral equations

5.7.3.1 The perfect conductor problem

It is not hard to show that in terms of the above boundary integral operators, the perfect conductor problem (5.16), (5.17), (5.18) admits (formally) a solution given by the following ansatz:

$$Q^{\mathrm{e}}_{\mathrm{L}} := \mathrm{i}\,\eta_{\mathrm{e}}\,[\gamma_{\mathrm{eL}}\mathsf{M}(\gamma_{\mathrm{eL}}) + \mathsf{N}(\gamma_{\mathrm{eL}})]\,\phi, \quad \phi \in H^{1/2}(\mathrm{div}, \partial\mathcal{O}),$$
$$Q^{\mathrm{e}}_{\mathrm{R}} := [-\gamma_{\mathrm{eR}}\mathsf{M}(\gamma_{\mathrm{eR}}) + \mathsf{N}(\gamma_{\mathrm{eR}})]\,\phi, \quad \phi \in H^{1/2}(\mathrm{div}, \partial\mathcal{O}).$$

Let

$$\mathsf{A}_{\mathrm{L,R}} := \gamma_{\mathrm{eL}}\mathsf{M}(\gamma_{\mathrm{eL}}) + \gamma_{\mathrm{eR}}\mathsf{M}(\gamma_{\mathrm{eR}}) + \mathsf{N}(\gamma_{\mathrm{eL}}) - \mathsf{N}(\gamma_{\mathrm{eR}}).$$

$Q^{\mathrm{e}}_{\mathrm{L}}$ and $Q^{\mathrm{e}}_{\mathrm{R}}$ will satisfy (5.18) if and only if

$$\mathrm{i}\,\eta_{\mathrm{e}}\left[\frac{1}{2}\gamma_{\mathrm{eL}} I + \frac{1}{2}\gamma_{\mathrm{eR}} I + \mathsf{A}_{\mathrm{L,R}}\right]\phi = f, \tag{5.36}$$

where I is the identity operator. The operator $\gamma_{\mathrm{eL}}\mathsf{M}(\gamma_{\mathrm{eL}})+\gamma_{\mathrm{eR}}\mathsf{M}(\gamma_{\mathrm{eR}})$ is compact on $H^{1/2}_t(\partial\mathcal{O})$ and on $H^{1/2}(\mathrm{div}, \partial\mathcal{O})$, while $\mathsf{N}(\gamma_{\mathrm{eL}}) - \mathsf{N}(\gamma_{\mathrm{eR}})$ is compact on $H^{1/2}_t(\partial\mathcal{O})$ (but not on $H^{1/2}(\mathrm{div}, \partial\mathcal{O})$) (see [304]).

It is our purpose to show that the operator

$$\frac{1}{2}\gamma_{\mathrm{eL}} I + \frac{1}{2}\gamma_{\mathrm{eR}} I + \mathsf{A}_{\mathrm{L,R}} : H^{1/2}_t(\partial\mathcal{O}) \to H^{1/2}_t(\partial\mathcal{O})$$

is invertible in $H^{1/2}_t(\partial\mathcal{O})$. Then (5.36) will be solvable, and the perfect conductor problem in its Beltrami fields formulation will be as well. Next, by a regularity argument, we shall show that the established solution is in $H^{1/2}(\mathrm{div}, \partial\mathcal{O})$.

A first step is to prove that the corresponding homogeneous problem ((5.36) with $f = 0$) admits only the trivial solution. This is done using different techniques found in [11] for $\gamma_{\mathrm{eL}} > 0$ and $\gamma_{\mathrm{eR}} > 0$, and in [24] allowing complex γ_{eL} and γ_{eR}, at least one of which has a positive imaginary part:

LEMMA 5.7.6 *Assume that at least one of $\mathcal{I}m\,\gamma_{eL}$, $\mathcal{I}m\,\gamma_{eR}$ is positive. Let $\phi \in H^{1/2}(\mathrm{div}, \partial\mathcal{O})$ satisfy*

$$\left[\frac{1}{2}\gamma_{eL} I + \frac{1}{2}\gamma_{eR} I + \mathsf{A}_{L,R}\right]\phi = 0. \tag{5.37}$$

Then $\phi = 0$.

Based on this lemma, and using the standard Fredholm theory (see Section A.6, Appendix A) we end up ([24]) with the following theorem.

THEOREM 5.7.7 *For $f \in H^{1/2}_t(\partial\mathcal{O})$, the boundary integral equation (5.36) has a unique solution ϕ in $H^{1/2}_t(\partial\mathcal{O})$. Further, if $f \in H^{1/2}(\mathrm{div}, \partial\mathcal{O})$ then, $\phi \in H^{1/2}(\mathrm{div}, \partial\mathcal{O})$).*

Proof. (Sketch) Let us note that since[11] (see [341], [304])

$$\mathsf{A}_{\mathrm{L,R}} : H_t^{1/2}(\partial\mathcal{O}) \to H_t^{3/2}(\partial\mathcal{O}) \overset{c}{\hookrightarrow} H^{1/2}(\mathrm{div}, \partial\mathcal{O}), \tag{5.38}$$

if $\psi \in H_t^{1/2}(\partial\mathcal{O})$ is a solution of (5.37), then $\psi \in H^{1/2}(\mathrm{div}, \partial\mathcal{O})$. But, by Lemma 5.7.6, $\psi = 0$, i.e., the homogeneous problem has only the trivial solution. The operator $\mathsf{A}_{\mathrm{L,R}}$ is compact in $H_t^{1/2}(\partial\mathcal{O})$, and hence, by the Fredholm theory (see Section A.6, Appendix A), (5.36) has a unique solution $\phi \in H_t^{1/2}(\partial\mathcal{O})$. Then

$$\mathrm{i}\,\eta_{\mathrm{e}} \mathsf{A}_{\mathrm{L,R}} \phi + \frac{1}{2}\,\mathrm{i}\,\eta_{\mathrm{e}} (\gamma_{\mathrm{eL}} + \gamma_{\mathrm{eR}}) \phi = f,$$

whereby

$$\phi = -\frac{2\mathrm{i}}{\eta_{\mathrm{e}}(\gamma_{\mathrm{eL}} + \gamma_{\mathrm{eR}})} \left\{ f - \mathrm{i}\,\eta_{\mathrm{e}} \mathsf{A}_{\mathrm{L,R}} \phi \right\}.$$

Since $f \in H^{1/2}(\mathrm{div}, \partial\mathcal{O})$, by (5.38) it finally follows that $\phi \in H^{1/2}(\mathrm{div}, \partial\mathcal{O})$, thus completing the proof. □

REMARK 5.7.8 In the case of the chiral scatterer in an achiral environment transmission problem, Theorem 5.7.7 covers the results of [341], where the parameters ε_i, μ_i, and β_i are assumed to be real (while here we allow complex values as well), and [39], where a variational method involving a volume (and not a surface, as here) formulation is used.

In addition, using regularity arguments in [241] and [103], and recalling that $f = -n \times E^{\mathrm{inc}}$, we have the following result ([356]).

THEOREM 5.7.9 *If the boundary $\partial\mathcal{O}$ is of class $C^{3,\alpha}$, the integral equation (5.36) has a unique solution in $C_t(\partial\mathcal{O})$.*

5.7.3.2 The transmission problem

As above, we derive a pair of coupled BIEs using an ansatz in each region.

In $\mathcal{O}$, assume

$$Q^{\mathrm{i}}_{\mathrm{L}} = [\gamma_{\mathrm{iL}}\, \mathsf{M}^{\mathrm{i}}(\gamma_{\mathrm{iL}}) + \mathsf{N}^{\mathrm{i}}(\gamma_{\mathrm{iL}})]\phi_1, \quad \phi_1 \in H^{1/2}(\mathrm{div}, \partial\mathcal{O}),$$
$$Q^{\mathrm{i}}_{\mathrm{R}} = [-\gamma_{\mathrm{iR}}\, \mathsf{M}^{\mathrm{i}}(\gamma_{\mathrm{iR}}) + \mathsf{N}^{\mathrm{i}}(\gamma_{\mathrm{iR}})]\phi_2, \quad \phi_2 \in H^{1/2}(\mathrm{div}, \partial\mathcal{O}),$$

and in $\mathcal{O}_e$, assume

$$Q^{\mathrm{e}}_{\mathrm{L}} = [\gamma_{\mathrm{eL}}\, \mathsf{M}^{\mathrm{e}}(\gamma_{\mathrm{eL}}) + \mathsf{N}^{\mathrm{e}}(\gamma_{\mathrm{eL}})](\zeta_{11}\phi_1 + \zeta_{12}\phi_2),$$
$$Q^{\mathrm{e}}_{\mathrm{R}} = [-\gamma_{\mathrm{eR}}\, \mathsf{M}^{\mathrm{e}}(\gamma_{\mathrm{eR}}) + \mathsf{N}^{\mathrm{e}}(\gamma_{\mathrm{eR}})](\zeta_{21}\phi_1 + \zeta_{22}\phi_2),$$

for some complex constants ζ_{11}, ζ_{12}, ζ_{21}, ζ_{22}.

This leads to a matrix boundary integral equation of the form

$$(\mathsf{L}_0 + \mathsf{K})\,\breve{\phi} = \breve{f}, \tag{5.39}$$

[11]The compactness of the embedding $H_t^{3/2}(\partial\mathcal{O}) \overset{c}{\hookrightarrow} H^{1/2}(\mathrm{div}, \partial\mathcal{O})$ follows from Theorem 3.7.4 and the definition (see Section 3.4.11) of the space $H^{1/2}(\mathrm{div}, \partial\mathcal{O})$.

where $\breve{\phi} = (\phi_1, \phi_2)^{tr}$, $\breve{f} = (f_1, f_2)^{tr}$,

$$\mathsf{K}: \; H_t^{1/2}(\partial\mathcal{O}) \times H_t^{1/2}(\partial\mathcal{O}) \to H_t^{3/2}(\partial\mathcal{O}) \times H_t^{3/2}(\partial\mathcal{O})$$

is a continuous operator given by

$$\mathsf{K} = \begin{pmatrix} \mathsf{K}_{11} & \mathsf{K}_{12} \\ \mathsf{K}_{21} & \mathsf{K}_{22} \end{pmatrix},$$

where

$$\begin{aligned}
\mathsf{K}_{11} &= \gamma_{\mathrm{iL}}\mathsf{M}(\gamma_{\mathrm{iL}}) - \gamma_{\mathrm{eL}}\zeta_{11}\mathsf{M}(\gamma_{\mathrm{eL}}) - \mathrm{i}\,\eta_{\mathrm{e}}\gamma_{\mathrm{eR}}\zeta_{21}\mathsf{M}(\gamma_{\mathrm{eR}}) - \zeta_{11}\mathsf{N}(\gamma_{\mathrm{eL}}) \\
&\quad + \mathsf{N}(\gamma_{\mathrm{iL}}) + \mathrm{i}\,\eta_{\mathrm{e}}\zeta_{21}\mathsf{N}(\gamma_{\mathrm{eR}}), \\
\mathsf{K}_{12} &= -\mathrm{i}\,\eta_{\mathrm{i}}\gamma_{\mathrm{iR}}\mathsf{M}(\gamma_{\mathrm{iR}}) - \gamma_{\mathrm{eL}}\zeta_{12}\mathsf{M}(\gamma_{\mathrm{eL}}) - \mathrm{i}\,\eta_{\mathrm{e}}\gamma_{\mathrm{eR}}\zeta_{22}\mathsf{M}(\gamma_{\mathrm{eR}}) - \zeta_{12}\mathsf{N}(\gamma_{\mathrm{eL}}) \\
&\quad + \zeta_{22}\mathrm{i}\,\eta_{\mathrm{e}}\mathsf{N}(\gamma_{\mathrm{eR}}) - \mathrm{i}\,\eta_{\mathrm{i}}\mathsf{N}(\gamma_{\mathrm{iR}}), \\
\mathsf{K}_{21} &= -\mathrm{i}\,\eta_{\mathrm{i}}^{-1}\gamma_{\mathrm{iL}}\mathsf{M}(\gamma_{\mathrm{iL}}) + \gamma_{\mathrm{eR}}\zeta_{21}\mathsf{M}(\gamma_{\mathrm{eR}}) + \mathrm{i}\,\eta_{\mathrm{e}}^{-1}\gamma_{\mathrm{eL}}\zeta_{11}\mathsf{M}(\gamma_{\mathrm{eL}}) - \zeta_{21}\mathsf{M}(\gamma_{\mathrm{eR}}) \\
&\quad + \zeta_{11}\mathrm{i}\,\eta_{\mathrm{e}}^{-1}N(\gamma_{\mathrm{eL}}) - \mathrm{i}\,\eta_{\mathrm{i}}^{-1}N(\gamma_{\mathrm{iL}}), \\
\mathsf{K}_{22} &= -\gamma_{\mathrm{iR}}\mathsf{M}(\gamma_{\mathrm{iR}}) + \gamma_{\mathrm{eR}}\zeta_{22}\mathsf{M}(\gamma_{\mathrm{eR}}) + \mathrm{i}\,\eta_{\mathrm{i}}^{-1}\gamma_{\mathrm{eL}}\zeta_{12}\mathsf{M}(\gamma_{\mathrm{eL}}) - \zeta_{22}\mathsf{N}(\gamma_{\mathrm{eR}}) \\
&\quad + \mathsf{N}(\gamma_{\mathrm{iR}}) + \zeta_{12}\mathrm{i}\,\eta_{\mathrm{e}}^{-1}\mathsf{N}(\gamma_{\mathrm{eL}}),
\end{aligned}$$

and

$$\mathsf{L}_0 = \begin{pmatrix} \mathsf{L}_{11} & \mathsf{L}_{12} \\ \mathsf{L}_{21} & \mathsf{L}_{22} \end{pmatrix}$$

is given by

$$\begin{aligned}
\mathsf{L}_{11} &= -\frac{1}{2}(\zeta_{11}\gamma_{\mathrm{eL}} + \mathrm{i}\,\eta_{\mathrm{e}}\zeta_{21}\gamma_{\mathrm{eR}} + \gamma_{\mathrm{iL}})I, \\
\mathsf{L}_{12} &= -\frac{1}{2}(\zeta_{12}\gamma_{\mathrm{eL}} + \mathrm{i}\,\eta_{\mathrm{e}}\zeta_{22}\gamma_{\mathrm{eR}} + \mathrm{i}\,\eta_{\mathrm{i}}\gamma_{\mathrm{iR}})I, \\
\mathsf{L}_{21} &= \frac{1}{2}(\zeta_{21}\gamma_{\mathrm{eR}} + \mathrm{i}\,\eta_{\mathrm{e}}^{-1}\zeta_{11}\gamma_{\mathrm{eL}} + \mathrm{i}\,\eta_{\mathrm{i}}^{-1}\gamma_{\mathrm{iL}})I, \\
\mathsf{L}_{22} &= \frac{1}{2}(\zeta_{22}\gamma_{\mathrm{eR}} + \mathrm{i}\,\eta_{\mathrm{e}}^{-1}\zeta_{12}\gamma_{\mathrm{eL}} - \gamma_{\mathrm{iR}})I,
\end{aligned}$$

where ζ_{11}, ζ_{12}, ζ_{21}, ζ_{22} are chosen so that we get as simple as possible equations, in which hypersingular operators appear in suitable, compact combinations. Such a choice ([21]) is given by

$$\zeta_{11} = \frac{1}{2}\left(1 + \frac{\eta_{\mathrm{e}}}{\eta_{\mathrm{i}}}\right), \; \zeta_{12} = \frac{\mathrm{i}}{2}\left(\eta_{\mathrm{e}} - \eta_{\mathrm{i}}\right), \; \zeta_{21} = \frac{\mathrm{i}}{2}\left(\frac{1}{\eta_{\mathrm{e}}} - \frac{1}{\eta_{\mathrm{i}}}\right), \; \zeta_{22} = \frac{1}{2}\left(1 + \frac{\eta_{\mathrm{i}}}{\eta_{\mathrm{e}}}\right).$$

Again, the first step is to show ([21]) the following lemma.

LEMMA 5.7.10 *Let* $\eta_e\eta_i^{-1} > 0$. *Then the homogeneous version of* (5.39) *has only the trivial solution.*

To establish solvability of the chiral scatterer in a chiral environment transmission problem, we need to use the analytic Fredholm theory (Theorem A.6.8 in Appendix A), (instead of the standard Fredholm theory, as for the perfect conductor problem in Theorem 5.7.7), since now there are values of $\eta_e\eta_i^{-1}$ for which the operator L_0 is *not* invertible. Hence:

THEOREM 5.7.11 *Let μ_i, ε_i, ε_e, β_i, β_e be fixed. For $\breve{f} \in H^{1/2}(\partial\mathcal{O}) \times H_t^{1/2}(\partial\mathcal{O})$ the integral equation (5.39) has a unique solution in $H_t^{1/2}(\partial\mathcal{O}) \times H^{1/2}(\partial\mathcal{O})$, except for a discrete set of values of μ_e. Moreover, if $\breve{\phi} \in H^{1/2}(\partial\mathcal{O}) \times H_t^{1/2}(\partial\mathcal{O})$ is a solution of the integral equation (5.39), and if $\breve{f} \in H^{1/2}(\mathrm{div}, \partial\mathcal{O}) \times H^{1/2}(\mathrm{div}, \partial\mathcal{O})$, then $\breve{\phi} \in H^{1/2}(\mathrm{div}, \partial\mathcal{O}) \times H^{1/2}(\mathrm{div}, \partial\mathcal{O})$.*

In addition, we have the following regularity result ([25]).

THEOREM 5.7.12 *If the boundary $\partial\mathcal{O}$ is of class $C^{3,\alpha}$, then the weak solution established in Theorem 5.7.11 is in $C_t(\partial\mathcal{O}) \times C_t(\partial\mathcal{O})$.*

5.7.4 On radiation conditions

The rôle of a radiation condition is twofold. From the mathematical point of view it ensures uniqueness of solutions. From the physical point of view it ensures that the scattered waves are not incoming at infinity, a physically plausible demand.

Let us now discuss radiation conditions for chiral media of infinite extent. Assume that $\mathcal{O}_e$ is filled with a chiral material of parameters $\mu_e, \varepsilon_e, \beta_e$. A given incident electromagnetic wave $E^{\mathrm{inc}}, H^{\mathrm{inc}}$ of the form (5.7), propagating in $\mathcal{O}_e$, is incident on $\mathcal{O} := \mathbb{R}^3 \setminus \overline{\mathcal{O}_e}$. The total electromagnetic field E^{t}, H^{t} in $\mathcal{O}_e$ is given by (5.9), where E^{e}, H^{e} is the scattered field. The Maxwell equations under the Drude-Born-Fedorov constitutive relations in $\mathcal{O}_e$ are (5.11). Define (see Section 5.7.2):

$$\mathsf{K}_{\mathcal{O}_e} := \mathsf{M}^{\mathrm{e}}(\gamma_{\mathrm{eL}}) + \mathsf{M}^{\mathrm{e}}(\gamma_{\mathrm{eR}}) + \frac{1}{\gamma_{\mathrm{eL}}}\mathsf{N}^{\mathrm{e}}(\gamma_{\mathrm{eL}}) + \frac{1}{\gamma_{\mathrm{eR}}}\mathsf{N}^{\mathrm{e}}(\gamma_{\mathrm{eR}}),$$

$$\mathsf{D}_{\mathcal{O}_e} := \mathsf{M}^{\mathrm{e}}(\gamma_{\mathrm{eL}}) - \mathsf{M}^{\mathrm{e}}(\gamma_{\mathrm{eR}}) + \frac{1}{\gamma_{\mathrm{eL}}}\mathsf{N}^{\mathrm{e}}(\gamma_{\mathrm{eL}}) - \frac{1}{\gamma_{\mathrm{eR}}}\mathsf{N}^{\mathrm{e}}(\gamma_{\mathrm{eR}}),$$

and

$$\mathsf{P}_{\mathcal{O}_e,\,L,\,R} := 2\begin{pmatrix} \mathsf{K}_{\mathcal{O}_e} & \mathrm{i}\,\eta_{\mathrm{e}}\,\mathsf{D}_{\mathcal{O}_e} \\ -\frac{\mathrm{i}}{\eta_{\mathrm{e}}}\mathsf{D}_{\mathcal{O}_e} & \mathsf{K}_{\mathcal{O}_e} \end{pmatrix}.$$

Then, along the lines of proof of (5.35), we have the following exterior representation formula (see also [11]) expressed in terms of the boundary integral operators M^{e} and N^{e}:

$$\begin{pmatrix} E^{\mathrm{t}}(x) \\ H^{\mathrm{t}}(x) \end{pmatrix} = \begin{pmatrix} E^{\mathrm{inc}}(x) \\ H^{\mathrm{inc}}(x) \end{pmatrix} + \mathsf{P}_{\mathcal{O}_e,\,L,\,R}\begin{pmatrix} n \times E^{\mathrm{t}}|_{\partial\mathcal{O}} \\ n \times H^{\mathrm{t}}|_{\partial\mathcal{O}} \end{pmatrix}. \tag{5.40}$$

Let B_{R} be a ball in $\mathbb{R}^3$ such that $\overline{\mathcal{O}} \subset B_{\mathrm{R}}$. The meaning of the symbol $\mathsf{P}_{\partial B_R,\,L,\,R}$ is then obvious. The following result (Theorem 4.3 in [11]) settles the issue of radiation conditions for chiral media.

THEOREM 5.7.13 *Let $u, v \in (C^{1,1}(\partial\mathcal{O}_e))^3$, and define E^{t}, H^{t} by*

$$\begin{pmatrix} E^{\mathrm{t}} - E^{\mathrm{inc}} \\ H^{\mathrm{t}} - H^{\mathrm{inc}} \end{pmatrix} = \mathsf{P}_{\mathcal{O}_e,\,L,\,R}\begin{pmatrix} n \times u \\ n \times v \end{pmatrix}. \tag{5.41}$$

Then E^{t}, H^{t} *satisfy the Maxwell equations (5.11) in* $\mathcal{O}_e$ *and the Silver-Müller radiation condition*

$$|\widehat{x} \times H^{\mathrm{e}}(x) + \eta_e^{-1} E^{\mathrm{e}}(x)| = O(|x|^{-2}), \ |x| \to \infty , \tag{5.42}$$

where $E^{\mathrm{e}} = E^{\mathrm{t}} - E^{\mathrm{inc}}$, $H^{\mathrm{e}} = H^{\mathrm{t}} - H^{\mathrm{inc}}$. *Conversely, if* E^{t}, H^{t} *satisfy (5.11) in* $\mathcal{O}_e$*, and*

$$\lim_{R\to\infty} \mathsf{P}_{\partial B_R,\, L,\, R} \begin{pmatrix} n \times E^{\mathrm{t}}|_{\partial B_R} \\ n \times H^{\mathrm{t}}|_{\partial B_R} \end{pmatrix} = 0 , \tag{5.43}$$

in the sense that the left-hand side, together with all its derivatives, tends uniformly to zero on compact sets as $R \to \infty$*, then* E^{t}, H^{t} *have the representation (5.41), with* $u = E^{\mathrm{t}}$ *and* $v = H^{\mathrm{t}}$.

Remark 5.7.14 The case of the anisotropic Maxwell system is treated in [81]. This system (like its isotropic counterpart) is not strongly elliptic. The determination of appropriate radiation conditions is closely related to the asymptotic properties of the corresponding fundamental solutions, and in this case is very difficult. Recall that in the frequency domain, the Maxwell system takes the form (2.17). In the special case in (2.18) that the constituent matrices $\xi_{\mathfrak{F}} = \zeta_{\mathfrak{F}} = 0$ while $\varepsilon_{\mathfrak{F}}$ and $\mu_{\mathfrak{F}}$ are real-valued, symmetric, positive definite and proportional to each other (there exists a $\varrho > 0$ such that $\varepsilon_{\mathfrak{F}} = \varrho\, \mu_{\mathfrak{F}}$), a fundamental solution is explicitly constructed, a Silver-Müller type of radiation condition is formulated, and uniqueness theorems are proved for exterior boundary value problems.

Remark 5.7.15 More general notions than that of a standard radiation condition have been introduced. The starting point has historically been the scalar Helmholtz equation $\Delta u + k^2 u = 0$, regarded as the steady-state oscillation equation of the dissipative wave equation $\partial_{tt} U + \varrho\, \partial_t U - c^2 \Delta U = 0$ (i.e., $U(x,t) = u(x) e^{-i\varpi t}$, with frequency $\varpi > 0$ and wave number $k := \varpi(\varpi + i\varrho)/c^2$). Recall that the appropriate radiation conditions for the scalar Helmholtz equation in dimension 3 are the Sommerfeld radiation conditions $u(x) = O(|x|^{-1})$, $\frac{\partial u}{\partial |x|}(x) - iku(x) = o(|x|^{-1})$, or the weaker version $\lim_{R\to\infty} \int_{|x|=R} |\frac{\partial u}{\partial |x|}(x) - iku(x)|^2 ds = 0$. The formulation of a general "radiation principle", that does not depend on the form of the unbounded domain in which the solution of the steady-state oscillation problem is sought, would be important. There are two possible approaches.

- ▷ The Limiting Amplitude Principle, according to which the solution of the steady-state oscillation equation is determined uniquely by the requirement that it be the limit as $t \to \infty$ of the amplitude of the solution of the Cauchy problem with zero initial condition for the wave equation with periodic right-hand side. For a generalisation of the Limiting Amplitude Principle to exterior problems for a fairly wide class of differential operators under certain additional conditions on the interior boundary of the unbounded domain (see, e.g., [137], [277]).

- ▷ The Limiting Absorption Principle, according to which the solution of the exterior boundary value problem of steady-state oscillations in a medium without absorption is sought as the limit of the bounded solution of the corresponding boundary value problem in the medium with absorption, as the latter tends to zero. There are generalisations of the Limiting Absorption Principle as uniqueness conditions for the solution of exterior boundary value problems for general elliptic operators and for a fairly wide class of interior boundaries of the unbounded domain (see, e.g., [136], [137], [367]). See also Chapter 10.

5.8 GENERALISED MÜLLER'S BIES

In this section we focus our attention on the chiral scatterer in an achiral environment transmission problem, i.e., the chiral obstacle in an achiral environment transmission problem. As discussed below it is possible to express this transmission problem in many ways as a system of boundary integral equations; for all such systems we have existence but not necessarily uniqueness of solutions. To establish uniqueness we present a generalisation of Müller's treatment of the achiral obstacle in an achiral environment case to the chiral obstacle in an achiral environment problem. It is more convenient to work in terms of (E, H) rather than $(Q_{\mathrm{L}}, Q_{\mathrm{R}})$ and with a dimensionless version of this problem, scaling all lengths using a typical length scale for the chiral obstacle (which is eventually taken equal to 1), setting

$$E^{\mathrm{e}} = \mu_e^{1/2}\check{E}^{\mathrm{e}}, \;\; H^{\mathrm{e}} = \varepsilon_e^{1/2}\check{H}^{\mathrm{e}}, \;\; E^{\mathrm{i}} = \mu_i^{1/2}\check{E}^{\mathrm{i}}, \;\; H^{\mathrm{i}} = \varepsilon_i^{1/2}\check{H}^{\mathrm{i}}$$

and similar scalings for E^{t}, H^{t}, E^{inc} and H^{inc}. These scalings reduce the chiral scatterer in an achiral environment transmission problem to the following dimensionless transmission problem, [34], which will be referred to in the sequel as the dimensionless chiral scatterer in an achiral environment transmission problem:

$$\begin{aligned}
&\mathrm{curl}\check{E}^{\mathrm{e}} - \mathrm{i}\,k_e\check{H}^{\mathrm{e}} = 0, \quad \mathrm{curl}\check{H}^{\mathrm{e}} + \mathrm{i}\,k_e\check{E}^{\mathrm{e}} = 0, \text{ in } \mathcal{O}_e,\\
&\mathrm{curl}\check{E}^{\mathrm{i}} - \mathrm{i}\,\frac{\gamma_{\mathrm{i}}^2}{k_i}\check{H}^{\mathrm{i}} - \beta_i\gamma_{\mathrm{i}}^2\check{E}^{\mathrm{i}} = 0, \; \mathrm{curl}\check{H}^{\mathrm{i}} + \mathrm{i}\,\frac{\gamma_{\mathrm{i}}^2}{k_i}\check{E}^{\mathrm{i}} - \beta_i\gamma_{\mathrm{i}}^2\check{H}^{\mathrm{i}} = 0, \text{ in } \mathcal{O}, \quad (5.44)\\
&\mu_e^{1/2}\mu_i^{-1/2} n\times\check{E}^{\mathrm{t}} = n\times\check{E}^{\mathrm{i}}, \; \varepsilon_e^{1/2}\varepsilon_i^{-1/2} n\times\check{H}^{\mathrm{t}} = n\times\check{H}^{\mathrm{i}}, \text{ on } \partial\mathcal{O},
\end{aligned}$$

where the total fields in $\mathcal{O}_e$ are given by

$$\check{E}^{\mathrm{t}} = \check{E}^{\mathrm{e}} + \check{E}^{\mathrm{inc}}, \quad \check{H}^{\mathrm{t}} = \check{H}^{\mathrm{e}} + \check{H}^{\mathrm{inc}}$$

and the scattered field satisfies a Silver-Müller radiation condition

$$\widehat{x}\times\check{H}^{\mathrm{e}} + \check{E}^{\mathrm{e}} = o\left(|x|^{-1}\right), \quad |x|\to\infty,$$

uniformly in all directions $\widehat{x}$.

Using the Stratton-Chu representation formula (5.35), and computing tangential components, we obtain the following system of BIEs for $n\times\check{H}^{\mathrm{t}}$ and

$n \times \check{E}^{\mathrm{t}}$:

$$\{I - \mathsf{M}(k_e)\}\, J_{\mathrm{H}} - \mathrm{i}\; k_e^{-1}\mathsf{N}(k_e)J_{\mathrm{E}} = 2\, J_{\mathrm{H}}^{\mathrm{inc}}, \tag{5.45}$$

$$\{I - \mathsf{M}(k_e)\}\, J_{\mathrm{E}} + \mathrm{i}\; k_e^{-1}\mathsf{N}(k_e)J_{\mathrm{H}} = 2\, J_{\mathrm{E}}^{\mathrm{inc}}, \tag{5.46}$$

$$\begin{aligned}&\left\{2\,I + (\mathsf{M}(\gamma_{\mathrm{iL}}) + \mathsf{M}(\gamma_{\mathrm{iR}})) + (\gamma_{\mathrm{iL}}^{-1}\mathsf{N}(\gamma_{\mathrm{iL}}) - \gamma_{\mathrm{iR}}^{-1}\mathsf{N}(\gamma_{\mathrm{iR}}))\right\} J_{\mathrm{H}}\\ &+ \mathrm{i}\,\eta_{\mathrm{e}}\eta_{\mathrm{i}}^{-1}\left\{(\mathsf{M}(\gamma_{\mathrm{iL}}) - \mathsf{M}(\gamma_{\mathrm{iR}})) + (\gamma_{\mathrm{iL}}^{-1}\mathsf{N}(\gamma_{\mathrm{iL}}) + \gamma_{\mathrm{iR}}^{-1}\mathsf{N}(\gamma_{\mathrm{iR}}))\right\} J_{\mathrm{E}} = 0,\end{aligned} \tag{5.47}$$

$$\begin{aligned}&\left\{2\,I + (\mathsf{M}(\gamma_{\mathrm{iL}}) + \mathsf{M}(\gamma_{\mathrm{iR}})) + (\gamma_{\mathrm{iL}}^{-1}\mathsf{N}(\gamma_{\mathrm{iL}}) - \gamma_{\mathrm{iR}}^{-1}\mathsf{N}(\gamma_{\mathrm{iR}}))\right\} J_{\mathrm{E}}\\ &- \mathrm{i}\,\eta_{\mathrm{i}}\eta_{\mathrm{e}}^{-1}\left\{(\mathsf{M}(\gamma_{\mathrm{iL}}) - \mathsf{M}(\gamma_{\mathrm{iR}})) + (\gamma_{\mathrm{iL}}^{-1}\mathsf{N}(\gamma_{\mathrm{iL}}) + \gamma_{\mathrm{iR}}^{-1}\mathsf{N}(\gamma_{\mathrm{iR}}))\right\} J_{\mathrm{H}} = 0,\end{aligned} \tag{5.48}$$

where

$$J_{\mathrm{H}} = n \times \check{H}^{\mathrm{t}},\;\; J_{\mathrm{E}} = -n \times \check{E}^{\mathrm{t}},\;\; J_{\mathrm{H}}^{\mathrm{inc}} = n \times \check{H}^{\mathrm{inc}},\;\; J_{\mathrm{E}}^{\mathrm{inc}} = -n \times \check{E}^{\mathrm{inc}}.$$

These are four BIEs for the two unknowns J_{E}, J_{H}. To proceed, we shall choose two linear combinations of these equations, namely,

$$\vartheta_1(5.45) + \vartheta_2(5.46) + \vartheta_3(5.47) + \vartheta_4(5.48)$$

and

$$\eth_1(5.45) + \eth_2(5.46) + \eth_3(5.47) + \eth_4(5.48)\,,$$

where ϑ_j and $\eth_j$, $j = 1, 2, 3, 4$, are constants to be specified.

In the achiral case, i.e., when $\beta_i = 0$, several choices have been investigated, both theoretically and numerically (see [183], [304] for review). For all those choices we have *existence*. However, the question of *uniqueness* is less obvious: one good choice for uniqueness has been introduced by Müller, namely,

$$\vartheta_1^{(0)} = \mu_e,\; \vartheta_3^{(0)} = \mu_i,\; \eth_2^{(0)} = \varepsilon_e,\; \eth_4^{(0)} = \varepsilon_i,\; \vartheta_2^{(0)} = \vartheta_4^{(0)} = \eth_1^{(0)} = \eth_3^{(0)} = 0.$$

The chiral case ($\beta_i \neq 0$) is more complicated. Provided that

$$\left(1 + \mu_e\mu_i^{-1}\right)\left(1 + \varepsilon_e\varepsilon_i^{-1}\right) \neq \beta_i k_e^2\,,$$

by a procedure described in [34], we end up with the choice

$$\begin{aligned}&\vartheta_1^{(\beta)} = \mu_e\mu_i^{-1}\left(1 - k_i^2\beta_i^2\right),\; \vartheta_2^{(\beta)} = 0,\; \vartheta_3^{(\beta)} = 1,\; \vartheta_4^{(\beta)} = \mathrm{i}\,k_i\,\beta_i\,\eta_i^{-1}\,\eta_e,\\ &\eth_1^{(\beta)} = 0,\; \eth_2^{(\beta)} = \varepsilon_e\varepsilon_i^{-1}\left(1 - k_i^2\beta_i^2\right),\; \eth_3^{(\beta)} = -\mathrm{i}\,k_i\,\beta_i\,\eta_i\,\eta_e^{-1},\; \eth_4^{(\beta)} = 1,\end{aligned}$$

which reduces to Müller's choice when $\beta_i = 0$, apart from a constant factor (μ_i for $\vartheta_n^{(\beta)}$ and ε_i for $\eth_n^{(\beta)}, n = 1, 2, 3, 4$).

Thus our generalised chiral Müller's equations are

$$\begin{aligned}&\left\{I + \vartheta_1^{(\beta)}(I - \mathsf{M}(k_e)) + \mathsf{A}\right\} J_{\mathrm{H}} + \frac{\vartheta_4^{(\beta)}}{k_i\beta_i}\mathsf{B}J_{\mathrm{E}} = 2\,\vartheta_1^{(\beta)}\,J_{\mathrm{H}}^{\mathrm{inc}},\\ &\left\{I + \eth_2^{(\beta)}(I - \mathsf{M}(k_e)) + \mathsf{A}\right\} J_{\mathrm{E}} - \frac{\eth_3^{(\beta)}}{k_i\beta_i}\mathsf{B}J_{\mathrm{H}} = 2\,\eth_2^{(\beta)}\,J_{\mathrm{E}}^{\mathrm{inc}},\end{aligned} \tag{5.49}$$

where

$$\mathsf{A} = \frac{1}{2}(1 - k_i^2\beta_i^2)\left(\gamma_{\mathrm{iL}}\mathsf{M}(\gamma_{\mathrm{iL}}) + \gamma_{\mathrm{iR}}\mathsf{M}(\gamma_{\mathrm{iR}}) + \mathsf{N}(\gamma_{\mathrm{iL}}) - \mathsf{N}(\gamma_{\mathrm{iR}})\right),$$

$$\mathsf{B} = \frac{1}{2}(1 - k_i^2\beta_i^2)\left(\gamma_{\mathrm{iL}}\mathsf{M}(\gamma_{\mathrm{iL}}) - \gamma_{\mathrm{iR}}\mathsf{M}(\gamma_{\mathrm{iR}}) + \mathsf{N}(\gamma_{\mathrm{iL}}) + \mathsf{N}(\gamma_{\mathrm{iR}}) - \mathsf{N}(k_e)\right).$$

As we mentioned above, when the obstacle is achiral ($\beta_i = 0$), these equations reduce to Müller's system:

$$\begin{aligned} &\left\{I + \mu_e\mu_i^{-1}(I - \mathsf{M}(k_e)) + \mathsf{A}_0\right\} J_{\mathrm{H}} + \mathrm{i}\,\eta_e\eta_i^{-1}\mathsf{B}_0 J_{\mathrm{E}} = 2\,\mu_e\mu_i^{-1} J_{\mathrm{H}}^{\mathrm{inc}}, \\ &\left\{I + \varepsilon_e\varepsilon_i^{-1}(I - \mathsf{M}(k_e)) + \mathsf{A}_0\right\} J_{\mathrm{E}} - \mathrm{i}\,\eta_i\eta_e^{-1}\mathsf{B}_0 J_{\mathrm{H}} = 2\,\varepsilon_e\varepsilon_i^{-1} J_{\mathrm{E}}^{\mathrm{inc}}, \end{aligned} \tag{5.50}$$

where now

$$\mathsf{A}_0 = \mathsf{M}(k_i) \quad \text{and} \quad \mathsf{B}_0 = k_i^{-1}\left(\mathsf{N}(k_i) - \mathsf{N}(k_e)\right).$$

5.9 LOW CHIRALITY APPROXIMATIONS

It is known ([34]) that the solution of the chiral scatterer in an achiral environment transmission problem is an analytic function of $k_i\beta_i$ for $|k_i\beta_i| < 1$. Rather than seek power series solutions in $k_i\beta_i$ of (5.44) directly, we expand the solutions of the governing integral equations (5.49). Thus, we write

$$J_{\mathrm{H}} \simeq J_{\mathrm{H}}^0 + \mathrm{i}\,k_i\beta_i J_{\mathrm{H}}^1 \quad \text{and} \quad J_{\mathrm{E}} \simeq J_{\mathrm{E}}^0 + \mathrm{i}\,k_i\beta_i J_{\mathrm{E}}^1,$$

where the error is $O((k_i\beta_i)^2)$ as $k_i\beta_i \to 0$.

Let

$$\begin{aligned} (\mathsf{T}_0(k_i)v)(x) &= \frac{k_i}{4\pi}\int_{\partial\mathcal{O}} v(y)e^{\mathrm{i}k_\mathrm{i}|x-y|}\,ds(y), \quad x \in \mathcal{O}, \\ \mathsf{T}_1(k_i)v &= n \times \operatorname{curl}\{\mathsf{T}_0(k_i)v\}, \\ \mathsf{T}_2(k_i)v &= n \times \operatorname{curl}\operatorname{curl}\{\mathsf{T}_0(k_i)v\}. \end{aligned}$$

Then for *small* $k_i\beta_i$ we have

$$\gamma_{\mathrm{iL}} \sim k_i(1 + k_i\beta_i), \quad \gamma_{\mathrm{iR}} \sim k_i(1 - k_i\beta_i),$$

$$e^{\,\mathrm{i}\gamma_{\mathrm{iL}}|x-y|} \sim e^{\,\mathrm{i}k_\mathrm{i}|x-y|}\left(1 + k_i^2\beta_i|x-y|\right),$$

$$\mathsf{V}(\gamma_{\mathrm{iL}})v \sim \mathsf{V}(k_i)v + \mathrm{i}\,k_i\beta_i\mathsf{T}_0(k_i)v, \quad \mathsf{V}(\gamma_{\mathrm{iR}})v \sim \mathsf{V}(k_i)v - \mathrm{i}\,k_i\beta_i\mathsf{T}_0(k_i)v,$$

$$\mathsf{M}(\gamma_{\mathrm{iL}})v \sim \mathsf{M}(k_i)v + \mathrm{i}\,k_i\beta_i\mathsf{T}_1(k_i)v, \quad \mathsf{M}(\gamma_{\mathrm{iR}})v \sim \mathsf{M}(k_i)v - \mathrm{i}\,k_i\beta_i\mathsf{T}_1(k_i)v,$$

$$\mathsf{N}(\gamma_{\mathrm{iL}})v \sim \mathsf{N}(k_i)v + \mathrm{i}\,k_i\beta_i\mathsf{T}_2(k_i)v, \quad \mathsf{N}(\gamma_{\mathrm{iR}})v \sim \mathsf{N}(k_i)v - \mathrm{i}\,k_i\beta_i\mathsf{T}_2(k_i)v,$$

with an error of $O((k_i\beta_i)^2)$ as $k_i\beta_i \to 0$.

Substituting these approximations into the exact boundary integral equations (5.49), we find that $[J_{\mathrm{H}}^0, J_{\mathrm{E}}^0]$ solves Müller's equations (5.50). The first-order correction $[J_{\mathrm{H}}^1, J_{\mathrm{E}}^1]$ also solves Müller's equations, *but* replacing on the right-hand sides:

$$2\mu_e\mu_i^{-1}\left(1 - k_i^2\beta_i^2\right) J_{\mathrm{H}}^{\mathrm{inc}} \quad \text{by} \quad -\,k_i^{-1}\mathsf{T}_2 J_{\mathrm{H}}^0 - \eta_e\eta_i^{-1}(2\mathsf{M} + \mathrm{i}\,\mathsf{T}_1)J_{\mathrm{E}}^0,$$

$$2\varepsilon_e\varepsilon_i^{-1}\left(1 - k_i^2\beta_i^2\right) J_{\mathrm{E}}^{\mathrm{inc}} \quad \text{by} \quad -\,k_i^{-1}\mathsf{T}_2 J_{\mathrm{H}}^0 - \eta_i\eta_e^{-1}(2\mathsf{M} + \mathrm{i}\,\mathsf{T}_1)J_{\mathrm{E}}^0.$$

REMARK 5.9.1 In [99], sophisticated codes have been developed for the numerical solution of Müller's equations: the approach followed in that reference is versatile in dealing with any BIE formulation for calculating the scattered fields due to arbitrarily shaped scatterers with arbitrary orientation. The common major difficulties arising from treating singular integrals in integral equation methods have been overcome by the reduction of the order of kernel singularities with the use of an effective regularisation process. Accurate numerical solutions have been obtained that are in excellent agreement with analytic solutions. By a simple modification, the codes in [99] can be used to compute the correction due to chirality of the obstacle; we can see that computing the first-order correction is fairly straightforward. Note that the approximations obtained for J_{E} and J_{H} can be inserted into (5.45)-(5.48) to generate exact solutions of the governing equations in $\mathcal{O}_e$ and $\mathcal{O}$. Let us note, moreover, that in principle one can also calculate higher-order approximations, but the new right-hand sides will be much more complicated.

5.10 MISCELLANEA

5.10.1 The conductive transmission problem

The solvability of the transmission problem describing the scattering of a plane electromagnetic wave propagating in an achiral environment by a chiral obstacle whose boundary is covered by a thin layer of very high conductivity is studied in [41]. The transmission conditions in this case have the form

$$\begin{aligned} &n \times E^{\mathrm{e}} - n \times E^{\mathrm{i}} = -n \times E^{\mathrm{inc}}, \\ &k_e n \times H^{\mathrm{e}} - k_i n \times H^{\mathrm{i}} - \frac{\kappa_i}{\varpi \varepsilon_i \beta_i} \tau (n \times E^{\mathrm{i}}) \times n = -k_e n \times H^{\mathrm{inc}}, \end{aligned}$$

where now

$$k_e^2 = \mathrm{i}\, \varpi \sigma_e, \;\; k_i^2 = \mathrm{i}\, \varpi \mu \sigma_i (1 - \varpi^2 \varepsilon_i \mu_i \beta_i^2)^{-2},$$

$$\kappa_i = \varpi^2 \varepsilon_i \mu_i \beta_i (1 - \varpi^2 \varepsilon_i \mu_i \beta_i^2)^{-1}, \;\; \tau \in (C^{0,\alpha}(\partial \mathcal{O}))^3.$$

5.10.2 Contrasting chiral media

5.10.2.1 Solvability

Consider a chiral obstacle (of electromagnetic parameters $\varepsilon_1, \mu_1, \beta_1$) in a chiral environment of electromagnetic parameters $\varepsilon_2, \mu_2, \beta_2$. If $\varepsilon_1 \neq \varepsilon_2$, $\mu_1 \neq \mu_2$ and $\beta_1 \neq \beta_2$, the resulting transmission problem is frequently referred to as the transmission problem for fully *contrasting* chiral media. Its solvability is settled by Theorems 5.7.11 and 5.7.12 above.

5.10.2.2 Special cases

There are certain interesting special cases of the above transmission problem. Recall that $k = \varpi(\mu\varepsilon)^{1/2}$ and $\eta = (\frac{\mu}{\varepsilon})^{1/2}$ (the impedance). The most important classes of these special cases are (in the following the rôles of subscripts 1 and 2 can be interchanged):

1. Mirror conjugate chiral media: $k_1 = k_2, \beta_1 = -\beta_2, \eta_1 = \eta_2$.

2. Isoimpedant chiral media: $\eta_1 = \eta_2$.

3. Isorefractive chiral media: $k_1 = k_2, \beta_1 = \beta_2$.

4. Cross-refractive chiral media: $k_1 = k_2, \beta_1 = -\beta_2$.

Let us note that the transmission problems for cases 1 and 2 are uniquely solvable without the limitations of Theorem 5.7.11, i.e. with no exception of a discrete set of electromagnetic parameters; this is established in [25]. Cases 3 and 4 are covered by Theorem 5.7.11.

5.10.3 Biisotropic media

The solvability of a transmission problem for biisotropic media is studied in [44].

5.10.4 "Screen" problems

The problem where a partially coated chiral obstacle is embedded in a homogeneous isotropic chiral medium of infinite extent is studied in [40].

5.10.5 Related approaches

In [372], integral equations are obtained for the electromagnetic scattering by an inhomogeneous, isotropic, three-dimensional chiral body. The chiral body is assumed to be in free space, and it can be attached to a perfect electric conducting body. The integral equations are obtained with the help of vector-dyadic identities and the free space dyadic Green's function. These equations are expressed in terms of a volume integral with the electric field as the unknown and surface integrals where the tangential components of the electric field and its curl are the unknowns. The integral equations are then transformed into a linear system of simultaneous equations by means of the moment method technique. Expressions for the scattered field in the far zone are also obtained by replacing the dyadic Green's function and its curl with their approximations for large arguments. Furthermore, closed form expressions are obtained for the fields and dipole moments induced inside an electrically small, homogeneous chiral sphere where it is assumed that the fields are constant. Finally, closed form expressions and numerical results for the fields scattered by the small chiral sphere and its bistatic echo area are also obtained.

In [239], the transmission problem for a homogeneous chiral scatterer embedded in an achiral environment is considered. The problem is reduced to two single boundary integral equations (arising from the choice of the *ansatz*). The first equation is solved by two different methods: the first method suffers from irregular frequencies because there appear eigenvalues of the interior Maxwell problem. In addition, there is a discrete set of values of chirality for which unique solvability does not hold. The second method gives unique solvability for all frequencies. The other equation is uniquely solvable for all frequencies except those corresponding to eigenvalues of a certain associated interior Maxwell problem.

Chapter Six

Scattering Problems: A Variety of Topics

6.1 INTRODUCTION

In this chapter we continue our study of scattering problems in the case where the considered fields have harmonic time dependence and the involved chiral media are homogeneous. For the reader's convenience we start with a section (Section 6.2) containing various important concepts of scattering theory; for simplicity, the presentation is done for the relatively simple case of the scalar Helmholtz equation, whose vector analogue is strongly related to the Maxwell equations. In the following sections we present a variety of topics. In particular, Section 6.3 deals with the establishment of the reciprocity principle, the general scattering theorem and the optical theorem for the chiral obstacle in an achiral environment problem; moreover, a study of the spectrum of the far-field operator and its relation to that of the T-matrix is included. Next, in Section 6.3.2 we consider the chiral obstacle in an achiral environment scattering problem when the incident electromagnetic wave is spherical, coming from a point source; we derive scattering relations involving two spherical waves, and mixed scattering relations involving one plane and one spherical wave. Section 6.4 has as a general underlying concept that of dyadics; it deals with the generalisation of the Atkinson-Wilcox expansion theorem to the perfect conductor in a chiral environment problem. Further, we generalise the low-frequency theory of classical (achiral) electromagnetism (see, e.g., [121], [360]) to the perfect conductor in a chiral environment problem: we obtain an iterative sequence of potential theory type problems with respect to the low-frequency coefficients and derive the leading term approximation of the electric far-field patterns. In Section 6.5 we present the results of Athanasiadis and Kardasi ([28], [30]) on chiral Herglotz wave functions; we first define the Herglotz wave functions of Beltrami and electromagnetic fields in a chiral medium. We then refer to scalar Herglotz functions and the Herglotz condition. We define the left circularly polarised (LCP) and right circularly polarised (RCP) Beltrami-Herglotz functions, which are shown to satisfy the Herglotz condition. Finally, the notion of a chiral Herglotz pair is introduced. The use of the LCP and RCP Beltrami-Herglotz fields to define a chiral Herglotz pair ascertains the validity of some interesting results for such a pair and its far-field patterns. In Section 6.6 we exhibit infinite Fréchet differentiability of the mapping from the boundary of the scatterer onto the far-field patterns, for the perfect conductor in chiral environment problem and derive a characterisation of the Fréchet derivative as a solution to an appropriate boundary value problem. This can be of use in a variety of applications, including the use of Newton's method for solving inverse problems. Let us note that in the case that the chirality measure (β) is equal to zero all the results of this chapter reduce to the corresponding results of classical (achiral) time-harmonic electromagnetic theory. Finally, Section 6.7 is a potpourri of topics related to chiral media. The aim is just to indicate some of the related problems treated in the chiral bibliography; the list is only indicative and by no means exhaustive. None of these directions is presented in detail. In particular, we start with very brief comments on work done on waveguides, then discuss some results of using the quaternionic approach for studying chiral media. Next we concisely discuss periodic structures. A discussion of inverse problems concludes this section.

6.2 IMPORTANT CONCEPTS OF SCATTERING THEORY

In this section we present some of the basic notions of scattering theory. For simplicity, the presentation is done for the scalar Helmholtz equation and follows [103], [121]. The corresponding notions for the Maxwell equations in chiral media are one of the topics of this chapter.

Consider the Helmholtz equation

$$\Delta v(x) + k^2 v(x) = 0, \ \ k \in \mathbb{C}, \ \ \mathcal{I}\mathrm{m}\, k \geq 0, \tag{6.1}$$

equipped with the Sommerfeld radiation condition

$$\lim_{r\to\infty} r\Big(\frac{\partial v(x)}{\partial r} - ikv(x)\Big) = 0, \tag{6.2}$$

uniformly in all directions $\widehat{x} = x/r$, where $r = |x|$.

We are interested in scattering problems, i.e., in studying the manner in which a bounded obstacle (the *scatterer*) $\mathcal{O}$ perturbs a wave originating in $\mathcal{O}_e$, the unbounded exterior of $\overline{\mathcal{O}}$. The scatterer $\mathcal{O}$ is a (nonempty) bounded open set, for convenience additionally assumed to be simply connected, with a sufficiently smooth boundary $\partial\mathcal{O}$; by n we denote the unit outward normal vector on $\partial\mathcal{O}$. So we consider a field[1] v^{inc} incident on $\mathcal{O}$. The resulting total field v^{t} is the superposition of v^{inc} and the scattered field v^{sc}: $v^{\mathrm{t}}(x) = v^{\mathrm{inc}}(x) + v^{\mathrm{sc}}(x)$; v^{sc} satisfies (6.1) in $\mathcal{O}_e$ and (6.2). In addition, a boundary condition[2] must be satisfied on $\partial\mathcal{O}$. Transmission problems are of course important, too, but we do not deal with them in this introduction. Now consider the fundamental solution $\Phi(x, y; k) = e^{\mathrm{i}\,k|x-y|}/(4\pi\,|x-y|)$, $x \neq y$,

---

[1]The incident field $v^{\mathrm{inc}}(x)$ is assumed to be either plane of the form $\exp(ik\widehat{k}\cdot x)$, where $\widehat{k}$ is the direction of propagation, or spherical due to a point source at x_0 of the form $\exp(\mathrm{i}k|x-x_0|)/(\mathrm{i}k|x-x_0|)$. In the former case the incident wave is defined for all $x \in \mathbb{R}^3$ and satisfies (6.1) for all $x \in \mathbb{R}^3$, but does not satisfy (6.2). In the latter, the incident field satisfies (6.2) but is a solution of (6.1) only in $\mathbb{R}^3 \setminus \{x_0\}$.

[2]This can be either the Dirichlet or the Neumann or the Robin condition.

of (6.1). Let

$$\tau(x) := \begin{cases} 1 & x \in \mathcal{O}_e, \\ 1/2 & x \in \partial\mathcal{O}, \\ 0 & x \in \mathcal{O}. \end{cases}$$

Incorporating the boundary values in (5.28) we obtain the following integral representation for the scattered field

$$\tau(x)\, v^{\rm sc}(x) = \int_{\partial\mathcal{O}} \Big(v^{\rm sc}(y) \frac{\partial \Phi(x,y;k)}{\partial n_y} - \Phi(x,y;k) \frac{\partial v^{\rm sc}(y)}{\partial n} \Big) ds(y)\,, \tag{6.3}$$

where $x \in \mathbb{R}^3, y \in \partial\mathcal{O}$.

A direct consequence of the above representation formula is the following result (Atkinson-Wilcox theorem).

THEOREM 6.2.1 *Let $v \in C^2(\mathcal{O}_e)$ be a solution of (6.1) satisfying (6.2). Let R_0 be such that $\{x \in \mathbb{R}^3 : |x| = R_0\} \subset \mathcal{O}_e$, and let (r, θ, φ) be the spherical coordinates of x. Then v has an expansion of the form*

$$v(x) = \frac{e^{\mathrm{i}\, kr}}{r} \sum_{j=0}^{\infty} \frac{F_j(\theta, \varphi)}{r^j} \tag{6.4}$$

that is valid for all $r \geq R_0$ and that converges absolutely and uniformly with respect to the variables (r, θ, φ). The series can be differentiated term by term with respect to (r, θ, φ) any number of times, and the resulting series all converge absolutely and uniformly.

Further, the coefficients F_j are recursively determined in terms of F_0 by the formula

$$2\mathrm{i}kjF_j = j(j-1)F_{j-1} + BF_{j-1}\,,\; j = 1, 2, \dots\;,$$

where $B := \frac{1}{\sin\theta} \frac{\partial}{\partial\theta}\Big(\sin\theta \frac{\partial}{\partial\theta}\Big) + \frac{1}{\sin^2\theta} \frac{\partial^2}{\partial\varphi^2}$ is Beltrami's operator for the sphere.

Clearly, R_0 can be chosen as the radius of the smallest circumscribable sphere around the scatterer and is thus called the *characteristic dimension* of the scatterer.

Additionally, we see that (6.4) can be considered a replacement of the radiation condition by an exact boundary condition on any sphere surrounding the scatterer, thus being suitable for the numerical evaluation of the scattered field.

Finally, (6.4) provides an explicit realisation of the *Dirichlet-to-Neumann map* relating the Neumann data ($\frac{\partial v}{\partial n}$) to the Dirichlet data (v) on a surface. Indeed, taking this surface to be the smallest circumscribing the scatterer sphere, all the coefficients F_j of (6.4) are known in terms of v^∞ and the Dirichlet data for v are obtained by setting $r = R_0$ in (6.4). The Neumann data can be determined in terms of the same coefficients F_j, since

$$\frac{\partial v(x)}{\partial n} = e^{\mathrm{i}kR_0} \sum_{j=0}^{\infty} F_j(\theta, \varphi) \Big(\frac{\mathrm{i}k}{R_0^{j+1}} - \frac{j+1}{R_0^{j+2}} \Big)\,. \tag{6.5}$$

Asymptotically, the Dirichlet-to-Neumann map on a sphere appears in the radiation condition

$$\frac{\partial v(x)}{\partial n} = \mathrm{i}\, k\, v(x) + O(|x|^{-2}), \quad |x| \to \infty,$$

where the term $O(|x|^{-2})$ can be determined exactly in view of (6.5).

The first results regarding such expansions can be found in [430], [431].

6.2.1 The far field

Let F_0 be the first coefficient in the series of Theorem 6.2.1 and S^2 the unit sphere in $\mathbb{R}^3$. Every solution v of (6.1) satisfying (6.2) has the following asymptotic behaviour:

$$v(x) = \frac{e^{\mathrm{i}\, k|x|}}{|x|} v^\infty(\widehat{x}) + O(|x|^{-2}), \ |x| \to \infty .$$

The function $v^\infty : S^2 \to \mathbb{C}$ defined as $v^\infty(\widehat{x}) = F_0(\theta, \varphi)$, where F_0 is the coefficient of the first term of the expansion (6.4), called the *far-field pattern*[3] of v. Hence, knowledge of the far-field pattern permits reconstruction of the scattered field up to the smallest circumscribing $\partial\mathcal{O}$ sphere. It can be shown that v^∞ is an analytic function of $\widehat{x}$ and of k.

In view of asymptotics, as $|x| \to \infty$, for $\Phi(x, y; k)$ and its gradient we obtain from (6.3) that

$$v^\infty(\widehat{x}) = -\int_{\partial\mathcal{O}} \Big(\frac{\partial v(y)}{\partial n_y} + \mathrm{i}\, k\, (\widehat{x} \cdot \widehat{n}_y)\, v(y)\Big) e^{-\mathrm{i}\, k\, \widehat{x} \cdot y} ds(y) . \tag{6.6}$$

The following comments ([121]) indicate the importance of the far-field pattern: in the case in which the incident wave is planar, the far-field pattern (often denoted in this case by $v^\infty(\widehat{x}, \widehat{k})$ instead of $v^\infty(\widehat{x})$) describes the response of the scatterer in the *direction of observation* $\widehat{x}$ owing to a wave excitation in the *direction of incidence* $\widehat{k}$. If the incident wave is spherical, then v^∞ describes the response of the scatterer in the direction of observation $\widehat{x}$ owing to the excitation provided by the particular incident wave. The far-field pattern contains all the angular characteristics of the scattered far field and does not vary with $|x|$. A variation of either the scatterer's geometry (via $\partial\mathcal{O}$) or its physics (via the boundary conditions) is "felt" by v^∞ alone. Conversely, the determination of whether the observation takes place in the far field is based on the particular form of the scattered field, which is the product of v^∞ (being a scatterer-dependent angular function) and $e^{\mathrm{i}\, k|x|}/|x|$ (being a fixed radial function).

Further, we observe that the integral representation (6.3) for the scattered field is a combination of *monopoles* and *dipoles* on $\partial\mathcal{O}$, while the integral representation (6.6) for the far-field pattern is a weighted distribution of plane waves on the boundary, all of which propagate in the direction $-\widehat{x}$.

---

[3]Or *scattering amplitude*, or *scattering coefficient*, or *radiation pattern*, or *radiation function*.

6.2.2 Cross sections

The *intensity* (*power flux*) of the scattered field due to a plane incident wave in the far field where the medium is lossless ($k \in \mathbb{R}^+$) is ([121]) given by

$$\mathfrak{I}(x) = \alpha\, \mathcal{I}\mathrm{m}\Big(\overline{v^{\mathrm{sc}}(x)}\, \mathrm{grad}\, v^{\mathrm{sc}}(x)\Big)\,,$$

where α is a positive constant. For an incident plane wave

$$v^{\mathrm{inc}}(x) = \exp(\mathrm{i}\, k\, (\widehat{k}\cdot x)),$$

the intensity in the direction of propagation $\widehat{k}$ is given by

$$\widehat{k}\cdot \mathfrak{I}^{\mathrm{inc}}(x) = \alpha\, k\,.$$

The scattered field in the radiation zone (where it propagates in the radial direction) has the radial intensity

$$\widehat{r}\cdot \mathfrak{I}^{\mathrm{sc}}(x) = \alpha\, \mathcal{I}\mathrm{m}\Big(\overline{v^{\mathrm{sc}}(x)}\, \frac{\partial v^{\mathrm{sc}}}{\partial r}(x)\Big)\,, r = |x| \to \infty\,,$$

which, in view of the radiation condition, gives, as $r \to \infty$,

$$\widehat{r}\cdot \mathfrak{I}^{\mathrm{sc}}(x) = \alpha\, k\, |v^{\mathrm{sc}}(x)|^2 + O(r^{-3}) = \alpha\, k\, r^{-2}\, |v^{\infty}(\widehat{x})|^2 + O(r^{-3})\,,$$

whereby, as $r \to \infty$,

$$r^2 \widehat{x}\cdot \mathfrak{I}^{\mathrm{sc}}(x) = \alpha\, k\, |v^{\mathrm{sc}}(x)|^2 + O(r^{-3}) = \alpha\, k\, |v^{\infty}(\widehat{x})|^2 + O(r^{-1})\,.$$

The *differential scattering cross section* is then defined as

$$\sigma_{\mathrm{dif}}(\widehat{x}) := \lim_{r\to\infty} \frac{r^2 \widehat{x}\cdot \mathfrak{I}^{\mathrm{sc}}(x)}{\widehat{k}\cdot \mathfrak{I}^{\mathrm{inc}}(x)} = |v^{\infty}(\widehat{x})|^2\,,$$

and specifies the amount of power scattered in the observation direction $\widehat{x}$ relative to the incident power flux in the direction of propagation.

The *scattering cross section* or *total cross section* is defined as the average of $\sigma_{\mathrm{dif}}(\widehat{x})$ over all directions:

$$\sigma_{\mathrm{sc}} := \frac{1}{4\pi}\int_{S^2} \sigma_{\mathrm{dif}}(\widehat{x}) ds(\widehat{x}) = \int_{S^2} |v^{\infty}(\widehat{x})|^2 ds(\widehat{x})\,,$$

which by Green's formula yields

$$\sigma_{\mathrm{sc}} = \mathcal{I}\mathrm{m}\Big(\overline{v^{\mathrm{sc}}}\ \frac{\partial v^{\mathrm{sc}}}{\partial n}\Big) ds\,.$$

So the scattering cross section is defined as the ratio of the time average rate (over a period) at which energy is scattered by the obstacle to the corresponding time average at which the energy of the incident wave crosses a unit area normal to the direction of propagation. The scattering cross section has the dimensions of area and is a measure of the disturbance caused by the obstacle to the incident wave.

The corresponding expression for the total field defines the *absorption cross section*

$$\sigma_{\mathrm{abs}} = -\mathcal{I}\mathrm{m}\Big(\overline{v^{\mathrm{t}}}\ \frac{\partial v^{\mathrm{t}}}{\partial n}\Big) ds\,,$$

where the minus sign indicates that the power flux is inward. The absorption cross section defines the total energy absorbed by the scatterer if it is penetrable and lossy. For lossless scatterers, as for scatterers with either the Dirichlet or the Neumann condition on their boundary, we have that $\sigma_{\mathrm{abs}} = 0$. In the case that the Robin condition, $\frac{\partial v(x)}{\partial n} + ik\nu v(x) = 0$ on $\partial\mathcal{O}$ ($\nu > 0$), is satisfied on the scatterer's boundary, the scatterer absorbs energy, and $\sigma_{\mathrm{abs}} = \nu \int_{\partial\mathcal{O}} |v^{\mathrm{t}}|^2 ds > 0$.

Finally, the *extinction cross section*, defined as

$$\sigma_{\mathrm{ext}} := \sigma_{\mathrm{sc}} + \sigma_{\mathrm{abs}} ,$$

describes the total energy that the scatterer extracts from the incident wave either by radiation or by absorption.

6.2.3 Basic scattering theorems

THEOREM 6.2.2 (RECIPROCITY RELATIONS)

(*i*) PLANE WAVES *The far-field pattern in the direction $\widehat{x}$ due to a plane wave in the direction $\widehat{k}$ is equal to the far-field pattern in the direction $-\widehat{k}$ due to a plane wave in the direction $-\widehat{x}$, i.e.,*

$$v^{\infty}(\widehat{x}, \widehat{k}) = v^{\infty}(-\widehat{k}, -\widehat{x}) .$$

(*ii*) SPHERICAL WAVES *The scattered field, $v^{\mathrm{sc}}(x_1, x_2)$, at x_1 due to a point source*[4] *at x_2 in $\mathcal{O}_{\mathrm{e}}$ is equal to the scattered field, $v^{\mathrm{sc}}(x_2, x_1)$, at x_2 due to a point source at x_1. The same relation holds for the total fields, as well. So,*

$$v^{\mathrm{sc}}(x_1, x_2) = v^{\mathrm{sc}}(x_2, x_1) \;\; \textit{and} \;\; v^{\mathrm{t}}(x_1, x_2) = v^{\mathrm{t}}(x_2, x_1) .$$

THEOREM 6.2.3 (THE GENERAL SCATTERING THEOREM)

$$v^{\infty}(\widehat{k}_1, \widehat{k}_2) - \overline{v^{\infty}}(\widehat{k}_2, \widehat{k}_1) = \frac{ik}{2\pi} \int_{S^2} \overline{v^{\infty}}(\widehat{x}, \widehat{k}_1) \; v^{\infty}(\widehat{x}, \widehat{k}_2) \, ds(\widehat{x})$$

$$+\frac{1}{4\pi} \int_{\partial\mathcal{O}} \left(v^{\mathrm{t}}(x, \widehat{k}_2) \, \frac{\partial \overline{v^{\mathrm{t}}}(x, \widehat{k}_1)}{\partial n} - \overline{v^{\mathrm{t}}}(x, \widehat{k}_1) \, \frac{\partial \, v^{\mathrm{t}}(x, \widehat{k}_2)}{\partial n} \right) ds(x).$$

THEOREM 6.2.4 (THE OPTICAL THEOREM[5]) *The total energy that the scatterer removes from the incident field is proportional to the value of the far-field pattern in the forward direction $\widehat{k}$, i.e.,*

$$\sigma_{\mathrm{ext}} = \frac{4\pi}{k} \, \mathcal{I}m(v^{\infty}(\widehat{k}, \widehat{k})) ,$$

a relation that easily follows from the general scattering theorem by setting $\widehat{k}_1 = \widehat{k}_2 = \widehat{k}$.

[4] Recall that the field due to a point source at x_0 is given by $e^{ik|x-x_0|}/(ik|x-x_0|)$.

[5] Also known as the *forward scattering theorem* or the *fundamental extinction theorem*.

A corollary of this result is that the power scattered in the forward direction can never be zero.

The relation between the forward far-field pattern and the total energy scattered, given by the optical theorem, can be interpreted as a relation between the incident and the scattered wave that establishes the mechanism of energy transfer.

REMARK 6.2.5 The optical theorem actually states that the total energy the scatterer removes from the incident field is proportional to the value of the far-field pattern in the direction in which the incident wave is travelling (forward scattering).

6.3 BACK TO CHIRAL MEDIA: SCATTERING RELATIONS AND THE FAR-FIELD OPERATOR

Here again, we consider the dimensionless chiral scatterer in an achiral environment transmission problem treated in Section 5.8, and study scattering relations and related issues.

6.3.1 Plane waves

The electromagnetic *far-field pattern* $(\check{E}^{\infty}(\widehat{x}), \check{H}^{\infty}(\widehat{x}))$ is defined in terms of the scattered electromagnetic field $(\check{E}^{\mathrm{e}}(x), \check{H}^{\mathrm{e}}(x))$ by the relations ([106])

$$\check{E}^{\mathrm{e}}(x) = \frac{e^{\,\mathrm{i}\,k_e|x|}}{|x|}\check{E}^{\infty}(\widehat{x}) + O(|x|^{-2}), \quad |x| \to \infty,$$

$$\check{H}^{\mathrm{e}}(x) = \frac{e^{\,\mathrm{i}\,k_e|x|}}{|x|}\check{H}^{\infty}(\widehat{x}) + O(|x|^{-2}), \quad |x| \to \infty,$$

uniformly in all directions $\widehat{x}$. For the (dimensionless) incident electromagnetic field, we take

$$\begin{aligned} \check{E}^{\mathrm{inc}}(x;\widehat{d},p) &= \mathrm{i}\,(k_e\varpi)p\, e^{\,\mathrm{i}\,(k_e\varpi)\widehat{d}\cdot\widehat{x}}, \\ \check{H}^{\mathrm{inc}}(x;\widehat{d},p) &= \widehat{d}\times E^{\mathrm{inc}}(x;\widehat{d},p) \end{aligned}$$

(with $\varpi = 1$; see the comment at the beginning of Section 5.8), where the unit vector $\widehat{d}$ describes the direction of propagation and the complex vector p gives the polarisation and satisfies $\widehat{d}\cdot p = 0$. We indicate the dependence of the scattered field, of the total exterior and interior fields, and of the far-field pattern on the incident direction $\widehat{d}$ and the polarisation p by writing $(\check{E}^{\mathrm{e}}(x;\widehat{d},p), \check{H}^{\mathrm{e}}(x;\widehat{d},p))$, $(\check{E}^{\mathrm{t}}(x;\widehat{d},p), \check{H}^{\mathrm{t}}(x;\widehat{d},p))$, and $(\check{E}^{\infty}(x;\widehat{d},p), \check{H}^{\infty}(x;\widehat{d},p))$, respectively. In what follows we again use the "U-$\mathbb{U}$ notation" (5.34). The meaning of the symbols $\check{U}^{\mathrm{e}}$, $\check{U}^{\mathrm{inc}}$, $\check{U}^{\mathrm{t}}$ and $\check{U}^{\infty}$

is now clear. Moreover, we shall employ the Twersky [417] notation[6],

$$\{\check{U}_1, \check{U}_2\}_{\partial\mathcal{O}} := \int_{\partial\mathcal{O}} [(n \times \check{U}_1) \cdot \check{\mathbb{U}}_2 - (n \times \check{U}_2) \cdot \check{\mathbb{U}}_1] \, ds.$$

Let S^2 denote the unit sphere in $\mathbb{R}^3$. In [35] the following result is established:

THEOREM 6.3.1 *The far-field pattern* $\check{U}^\infty$ *satisfies the reciprocity principle*

$$q \cdot \check{U}^\infty(\widehat{x}; \widehat{d}, p) = p \cdot \check{U}^\infty(-\widehat{d}; -\widehat{x}, q),$$

for all $\widehat{d}, \widehat{x} \in S^2$ *and* $p, q \in (\mathbb{C} \setminus \mathbb{R})^3$ *with* $p \cdot \widehat{d} = q \cdot \widehat{x} = 0$.

We thus observe that the standard reciprocity relation for the achiral case, described in [106], is also valid for the chiral case. For a discussion of the basic reciprocity theorems for electromagnetic wave fields in time-invariant configurations we refer the reader to [131].

We then have the following basic scattering theorem ([35]).

THEOREM 6.3.2 *The far-field pattern satisfies the relation*

$$\begin{aligned} & q \cdot \overline{\check{U}^\infty(\widehat{x}; \widehat{d}, p)} + \overline{p} \cdot \check{U}^\infty(\widehat{d}; \widehat{x}, q) \\ &= -\frac{1}{2\pi} \int_{S^2} \overline{\check{U}^\infty(\widehat{y}; \widehat{d}, p)} \cdot \check{U}^\infty(\widehat{y}; \widehat{x}, q) \, ds(q) - \frac{i}{4\pi} \left\{ \overline{\check{U}^{\mathrm{t}}(\cdot\, ; \widehat{d}, p)}, \check{U}^{\mathrm{t}}(\cdot\, ; \widehat{x}, q) \right\}_{\partial\mathcal{O}}, \end{aligned}$$

for all $\widehat{d}, \widehat{x} \in S^2$ *and* $p, q \in (\mathbb{C} \setminus \mathbb{R})^3$ *with* $\widehat{d} \cdot p = \widehat{x} \cdot q = 0$, *where*

$$\begin{aligned} & \frac{\mathrm{i}}{2} \left\{ \overline{\check{U}^{\mathrm{t}}(\cdot\, ; \widehat{d}, p)}, \check{U}^{\mathrm{t}}(\cdot\, ; \widehat{x}, q) \right\}_{\partial\mathcal{O}} = \mathcal{I}m \left(\frac{\gamma_i^2}{\mu_\star} \overline{\varepsilon_\star k_i} \right) (\check{\mathbb{U}}_1, \check{\mathbb{U}}_2) \\ & + \mathcal{I}m \left(\frac{\gamma_i^2}{\overline{\mu_\star} \varepsilon_\star k_i} \right) (\check{U}_1, \check{U}_2) + \mathcal{I}m\,(\beta_i \gamma_i^2) \left[\frac{1}{\mu_\star \overline{\varepsilon_\star}} (\check{\mathbb{U}}_1, \check{U}_2) + \frac{1}{\overline{\mu_\star} \varepsilon_\star} (\check{U}_1, \check{\mathbb{U}}_2) \right], \end{aligned}$$

$\check{U}_1 = \check{U}(\cdot\ ; \widehat{x}, q)$, $\check{U}_2 = U(\cdot\ ; \widehat{d}, p)$, (u, v) *is the usual inner product in* $(L^2(\mathcal{O}))^3$, *and* $\mu_\star = \mu_e^{1/2} \mu_i^{-1/2}$ $\varepsilon_\star = \varepsilon_e^{1/2} \varepsilon_i^{-1/2}$, $\gamma_i^2 = k_i^2 (1 - k_i^2 \beta_i^2)^{-1}$.

REMARK 6.3.3 If $\varepsilon_i, \mu_i, \beta_i \in \mathbb{R}$, then $\frac{\mathrm{i}}{2} \left\{ \overline{\check{U}^{\mathrm{t}}(\cdot\, ; \widehat{d}, p)}, \check{U}^{\mathrm{t}}(\cdot\, ; \widehat{x}, q) \right\}_{\partial\mathcal{O}} \equiv 0$.

THEOREM 6.3.4 *The following relation holds:*

$$\sigma_{sc} = -4\pi\, \mathcal{R}e \left(\overline{p} \cdot \check{U}^\infty(\widehat{d}; \widehat{d}, p) \right) - \frac{\mathrm{i}}{2} \left\{ \overline{\check{U}^{\mathrm{t}}(\cdot\, ; \widehat{d}, p)}, \check{U}^{\mathrm{t}}(\cdot\, ; \widehat{x}, q) \right\}_{\partial\mathcal{O}}.$$

REMARK 6.3.5 If ε_i and μ_i are positive and β_i is real, we obtain that

$$\sigma_{sc} = -4\pi\, \mathcal{R}e\, (\overline{p} \cdot \check{U}^\infty(\widehat{d}; \widehat{d}, p)),$$

which coincides with the classical *optical theorem* ([131], [417]).

---

[6] Let us note that $\{\cdot, \cdot\}_{\partial\mathcal{O}}$ is related to the *reciprocity gap functional* which has recently found many applications in inverse problems in acoustics, electromagnetics and linear elasticity in connection with the *linear sampling method* (see Section 6.7.4).

To conclude this section, we introduce the far-field operator $\mathsf{F}_U^\infty : L_t^2(S^2) \to L_t^2(S^2)$, corresponding to the far-field pattern $\check{U}^\infty$, defined (see [106]) by

$$(\mathsf{F}_U^\infty h)(\widehat{x}) = \int_{S^2} \check{U}^\infty(\widehat{x}; \widehat{q}, h(\widehat{q}))\, ds(\widehat{q}).$$

We have the following result ([35]).

THEOREM 6.3.6 *Suppose that* $\varepsilon_i, \mu_i > 0\,,\ \beta_i \in \mathbb{R}$. *Then* F_U^∞ *has a countable number of eigenvalues, all lying on the circle* $|\zeta|^2 + 4\pi\, \mathcal{R}e\, \zeta = 0$.

There is an interesting relation between the spectra of the far-field operator and the T-matrix. For the sake of brevity we just recall here the definition of the T-matrix, and refer to [397] for its properties.

Surround an obstacle by a sphere of radius d and centre O. The T-matrix connects the incident field with the scattered field, and is defined by

$$f_\varrho = \sum_\varsigma T_{\varrho\varsigma}\, a_\varsigma\,, \quad \text{or} \quad f = T\, a\,,$$

where we assume that the given incident field can be expanded as

$$U^{\text{inc}}(x) = \sum_\varrho a_\varrho \mathcal{R}e\boldsymbol{\psi}_\varrho(x)\,,\ |x| < d\,,$$

and the scattered field as

$$U^{\text{sc}}(x) = \sum_\varrho f_\varrho \boldsymbol{\psi}_\varrho(x)\,,\ |x| > d\,,$$

in terms of $\boldsymbol{\psi}_\varrho$ that is used as an abbreviation for $\boldsymbol{\psi}_{1\,\sigma\,m\,\ell}, \boldsymbol{\psi}_{2\,\sigma\,m\,\ell}$, the normalised outgoing spherical vector wave functions (see Section 5.6.1.1). Therefore, the T-matrix connects the known coefficients a_ϱ with the unknown coefficients f_ϱ. In the above formulae, ϱ and ς are multi-indices. The composition of the obstacle (chiral or achiral, homogeneous or inhomogeneous, perfectly conducting or otherwise) enters through its T-matrix.

The final result of this section is the following theorem ([35]).

THEOREM 6.3.7 *The number* λ *is an eigenvalue of the far-field operator* F_U^∞ *if and only if* $\dfrac{\lambda}{4\pi}$ *is an eigenvalue of the* T*-matrix.*

6.3.2 Spherical waves

We consider an incident spherical electromagnetic wave due to a point source located at a point with position vector a with respect to the origin; this incident wave $(E_a^{\text{inc}}, H_a^{\text{inc}})$ has the form ([36], [33]):

$$\begin{aligned} E_a^{\text{inc}}(x;\widehat{p}) &= (\mathrm{i}\, k_e)^{-1} \operatorname{curl}\left(\frac{h_0(k_e\,|x-a|)}{h_0(k_e\,|a|)}\, \widehat{a} \times \widehat{p} \right), \\ H_a^{\text{inc}}(x;\widehat{p}) &= (\mathrm{i}\, k_e)^{-1} \eta_e^{-1} \operatorname{curl} E_a^{\text{inc}}(x;\widehat{p}), \end{aligned} \tag{6.7}$$

where $\widehat{p}$ is a constant unit vector with $\widehat{p} \cdot a = 0$, and $h_0(y) = \mathrm{i}\, y^{-1}\, e^{\mathrm{i}\, y}$ is the spherical Hankel function of the first kind and order zero. Physically,

$(E_a^{\rm inc}, H_a^{\rm inc})$ represents the field generated by a magnetic dipole with dipole moment $\widehat{a} \times \widehat{p}$ (see [106], or [121]). The above form of $E_a^{\rm inc}$ ensures that when the point source tends to infinity, the spherical wave reduces to a plane electric wave with direction of propagation $-\widehat{a}$ and polarisation $\widehat{p}$. The total exterior electric field E_a^t is given by

$$E_a^t(x;\widehat{p}) = E_a^{\rm inc}(x;\widehat{p}) + E_a^{\rm e}(x;\widehat{p}), \quad x \in \mathcal{O}_e \setminus \{a\},$$

where $E_a^{\rm e}(x;\widehat{p})$ is the scattered electric field, which is assumed to satisfy the radiation condition

$$\widehat{x} \times {\rm curl} E_a^{\rm e} + {\rm i}\, k_e |x| E_a^{\rm e} = o(|x|^{-2}), \ |x| \to \infty, \tag{6.8}$$

uniformly in all directions $\widehat{x}$.

The behaviour of $E_a^{\rm e}$ in the radiation zone is given by

$$E_a^{\rm e}(x) = h_0(k_e|x|)\, g_{\rm A}(\widehat{x}) + O(|x|^{-2}), \ |x| \to \infty,$$

where $g_{\rm A}(\widehat{x})$ is the electric far-field pattern. We use this notation for the far-field pattern (instead of E^∞) only in this section because of its relation to the spherical far-field pattern generator $G_{\rm A}$, which is defined below. The total exterior electric field solves the equation

$$ {\rm curl}\,{\rm curl} E_a^t = k_e^2 E_a^t \quad \text{in } \mathcal{O}_e.$$

We note that the incident electric field satisfies the radiation condition (6.8), and hence the total electric field also satisfies (6.8).

The incident electromagnetic waves are transmitted into the chiral scatterer. Let $E_a^{\rm i}$ be the total electric field in the interior. Then $E_a^{\rm i}$ satisfies

$$ {\rm curl}\,{\rm curl} E_a^{\rm i} - 2\beta_i \gamma_i^2 {\rm curl} E_a^{\rm i} - \gamma_i^2 E_a^{\rm i} = 0 \quad \text{in } \mathcal{O}\ ,$$

where $\gamma_i^2 = k_i^2(1 - k_i^2\beta_i^2)^{-1}$, $k_i^2 = \varpi^2 \varepsilon_i \mu_i$.

On the surface $\partial\mathcal{O}$ of the scatterer we have the following transmission conditions:

$$\begin{aligned} n \times E_a^t &= n \times E_a^{\rm i} \\ n \times {\rm curl} E_a^t &= \frac{\mu_e}{\mu_i}\frac{k_i^2}{\gamma_i^2} n \times {\rm curl} E_a^{\rm i} - \frac{\mu_e}{\mu_i} k_i^2 \beta_i n \times E_a^{\rm i}. \end{aligned}$$

In the following for an incident time-harmonic spherical wave $E_a^{\rm inc}(x;\widehat{p})$ due to a point source located at a, we will denote the total field in $\mathcal{O}_e$, the scattered field and the far-field pattern by writing $E_a^t(x;\widehat{p})$, $E_a^{\rm e}(x;\widehat{p})$ and $g_{\rm A}(\widehat{x};\widehat{p})$, respectively, indicating the dependence on the position a of the point source and the polarisation $\widehat{p}$. Also, the total electric field in $\mathcal{O}$ will be denoted by $E_a^{\rm i}(x;\widehat{p})$.

We are interested in relations between these fields. We consider a point source at a with polarisation $\widehat{p}_1$ and another point source at b with polarisation $\widehat{p}_2$.

We define the *spherical far-field pattern generator* ([33]) as

$$G_b(a;\widehat{p}_2) = e^{\,{\rm i}\, k_e |a|} a \times \left[{\rm curl} E_b^{\rm e}(a;\widehat{p}_2) - \frac{{\rm i}\, k_e}{2\pi} \int_{S^2} \widehat{x} \times g_b(\widehat{x};\widehat{p}_2)\, e^{\,{\rm i}\, k_e \widehat{x}\cdot a} ds(\widehat{x}) \right].$$

Now the general scattering theorem for spherical electric waves scattered by a chiral obstacle is formulated as follows.

THEOREM 6.3.8 *For any two point source locations in $\mathcal{O}_e$, a and b, and for any polarisations, $\widehat{p}_1$ and $\widehat{p}_2$, we have*

$$\widehat{p}_1 \cdot G_b(a;\widehat{p}_2) + \widehat{p}_2 \cdot \overline{(G_A(b;\widehat{p}_1))} + \frac{1}{2\pi}\int_{S^2} g_b(\widehat{x};\widehat{p}_2)\cdot\overline{(g_A(\widehat{x};\widehat{p}_1))}\,ds(\widehat{x}) = \mathcal{E}_{a,b}(\widehat{p}_1;\widehat{p}_2),$$

where

$$\mathcal{E}_{a,b}(\widehat{p}_1;\widehat{p}_2) = \int_{\mathcal{O}} (\mathfrak{r}_1 \mathrm{curl}\overline{(E^{\mathrm{i}}_a(x;\widehat{p}_1))} \cdot \mathrm{curl} E^{\mathrm{i}}_b(x;\widehat{p}_2) \mathfrak{r}_2 \overline{(E^{\mathrm{i}}_a(x;\widehat{p}_1))} \cdot E^{\mathrm{i}}_b(x;\widehat{p}_2))dx$$

$$+ \mathfrak{r}_3 \int_{\mathcal{O}} (\overline{(E^{\mathrm{i}}_a(x;\widehat{p}_1))} \cdot \mathrm{curl} E^{\mathrm{i}}_b(x;\widehat{p}_2) - \mathrm{curl}\overline{(E^{\mathrm{i}}_a(x;\widehat{p}_1))} \cdot E^{\mathrm{i}}_b(x;\widehat{p}_2))dx,$$

and

$$\mathfrak{r}_1 = \frac{k_e}{2\pi}\mathcal{I}m\,(\frac{\mu_e}{\mu_i}\frac{k_i^2}{\gamma_i^2}),\ \mathfrak{r}_2 = -\frac{k_e}{2\pi}\mathcal{I}m\,(\frac{\mu_e}{\mu_i}k_i^2),\ \mathfrak{r}_3 = \frac{k_e}{2\pi}\mathcal{I}m\,(\frac{\mu_e}{\mu_i}k_i^2\beta_i).$$

In [121] a reciprocity relation for spherical waves scattered by an achiral obstacle has been proved. The same relation also holds for a penetrable chiral scatterer; see [20].

THEOREM 6.3.9 *For any two point source locations in $\mathcal{O}_e$, a and b, for any polarisations, $\widehat{p}_1$ and $\widehat{p}_2$, and for a penetrable chiral scatterer, we have*

$$h_0(k_e|a|)\,(\widehat{b}\times\widehat{p}_2)\cdot \mathrm{curl} E^{\mathrm{e}}_a(\widehat{b};\widehat{p}_1) = h_0(k_e|b|)\,(\widehat{a}\times\widehat{p}_1)\cdot \mathrm{curl} E^{\mathrm{e}}_b(\widehat{a};\widehat{p}_2).$$

The *scattering cross section* due to a point source at a is defined ([121]) as

$$\sigma_a^{\mathrm{sc}} = \frac{1}{k_e^2}\int_{S^2} |g_{\mathrm{A}}(\widehat{x};\widehat{p})|^2\,ds(\widehat{x}),$$

the *absorption cross section* is defined as

$$\sigma_a^{\mathrm{abs}} = \frac{1}{k_e}\mathcal{I}\mathrm{m}\int_{\partial\mathcal{O}} n\cdot(E^t_a \times \mathrm{curl}\overline{E^t_a})\,ds,$$

and the *extinction cross section*, σ_a^{ext}, is defined by

$$\sigma_a^{\mathrm{ext}} = \sigma_a^{\mathrm{sc}} + \sigma_a^{\mathrm{abs}}.$$

It is not hard to see first that

$$\sigma_a^{\mathrm{abs}} = -\frac{2\pi}{k_e^2}\mathcal{E}_{a,a}(\widehat{p};\widehat{p}),$$

and then that

$$\sigma_a^{\mathrm{ext}} = -\frac{4\pi}{k_e^2}\,\mathcal{R}e\,[\widehat{p}\cdot G_{\mathrm{A}}(a;\widehat{p})]\,.$$

Now let

$$E^{\mathrm{inc}}_{\mathrm{pl}}(x;\widehat{d},\widehat{p}) = \widehat{p}\,\exp\{\mathrm{i}\,k_e\widehat{d}\cdot x\}$$

be an incident time-harmonic *plane* electric wave, where the unit vector $\widehat{d}$ describes the direction of propagation and the unit vector $\widehat{p}$ gives the polarisation. We will indicate the dependence of the total field in $\mathcal{O}_e$, the total field in $\mathcal{O}$, the scattered field and the electric far-field pattern on the incident direction $\widehat{d}$, and the polarisation $\widehat{p}$ by writing $E^{\mathrm{t}}(x;\widehat{d},\widehat{p})$, $E^{\mathrm{i}}(x;\widehat{d},\widehat{p})$, $E^{\mathrm{e}}(x;\widehat{d},\widehat{p})$, and $g(\widehat{x};\widehat{d},\widehat{p})$, respectively.

6.3.3 Mixture of plane and spherical waves

Here we consider *mixed* situations, i.e., relate fields due to one spherical electric wave $E_a^{\mathrm{inc}}(x;\widehat{p}_1)$ and one plane electric wave $E_{\mathrm{pl}}^{\mathrm{inc}}(x;-\widehat{b},\widehat{p}_2)$; we do this by letting $|b|\to\infty$ in our previous results.

Using some asymptotics, we can easily see that for the spherical electric wave (6.7) we have

$$\lim_{|b|\to\infty} E_b^{\mathrm{inc}}(x;\widehat{p}) = E_{\mathrm{pl}}^{\mathrm{inc}}(x;-\widehat{q},\widehat{p}). \tag{6.9}$$

THEOREM 6.3.10 *For two incident point source electric waves, $E_a^{\mathrm{inc}}(x;\widehat{p}_1)$ and $E_b^{\mathrm{inc}}(x;\widehat{p}_2)$, we have*

$$\lim_{|a|\to\infty} G_b(a;\widehat{p}_2) = g_b(-\widehat{a};\widehat{p}_2)\,, \quad \lim_{|a|\to\infty} G(a;-\widehat{q},\widehat{p}_2) = g(-\widehat{a};-\widehat{q},\widehat{p}_2).$$

We can now let $|b|\to\infty$ in the general scattering theorem (Theorem 6.3.8).

THEOREM 6.3.11 *Let $E_a^{\mathrm{inc}}(x;\widehat{p}_1)$ be an incident spherical electric wave and let $E_{\mathrm{pl}}^{\mathrm{inc}}(x;-\widehat{q},\widehat{p}_2)$ be an incident plane electric wave. Then*

$$\begin{aligned}&\widehat{p}_1\cdot G(a;-\widehat{q},\widehat{p}_2) + \widehat{p}_2\cdot\overline{g_A(\widehat{q};\widehat{p}_1)}\\ &+\frac{1}{2\pi}\int_{S^2} g(\widehat{x};-\widehat{q},\widehat{p}_2)\cdot\overline{g_A(\widehat{x};\widehat{p}_1)}\,ds(\widehat{x}) = \lim_{|b|\to\infty}\mathcal{E}_{a,b}(\widehat{p}_1;\widehat{p}_2).\end{aligned}$$

To conclude, we note that we also have

$$\lim_{|a|\to\infty}\lim_{|b|\to\infty} G_b(a;\widehat{p}_2) = \lim_{|b|\to\infty}\lim_{|a|\to\infty} G_b(a;\widehat{p}_2) = g\,(-\widehat{a};-\widehat{q},\widehat{p}_2).$$

This can be used to verify that the known scattering relations for plane wave incidence ([106], [121]) are recovered when $|a|\to\infty$ and $|b|\to\infty$. Furthermore, (6.9) and the reciprocity principle for plane waves ([121]) give the following limiting property:

$$\lim_{|a|\to\infty}\widehat{p}_1\cdot G(a;-\widehat{q},\widehat{p}_2) = \lim_{|b|\to\infty}\widehat{p}_2\cdot G(-b;\widehat{a},\widehat{p}_1).$$

REMARK 6.3.12 The problem of the scattering of a spherical electromagnetic wave propagating in a chiral medium by a bounded chiral obstacle is studied in [20], as far as the derivation of reciprocity and general scattering theorems (relating the scattered fields due to the presence of a point source located at two different points) is concerned. A forward scattering theorem and mixed (plane/spherical) scattering relations are also established. In [45] certain radiation integral relations ("radiation principles") are established that relate the fields and far-field patterns created by spherical wave excitation due to a dipole immersed in the interior of a layered chiral obstacle. A reciprocity principle and a general radiation theorem (as well as an optical theorem resulting from the latter) are proved, relating the total, primary and secondary Beltrami fields to the respective far-field patterns. Mixed (plane/spherical) radiation-scattering theorems are also derived.

6.4 USING DYADICS

Dyadics may provide a very useful formulation for some key results in scattering theory. Dyadic Green functions play a key rôle in electromagnetics; see [400]. An introduction to dyadic analysis can be found in Appendix D.

6.4.1 An expansion result of the Atkinson-Wilcox type

We deal again with the scattering problem for an obstacle in a chiral environment. In this section, *any kind* of boundary condition on $\partial\mathcal{O}$ that ensures well posedness of the corresponding boundary value problem can be considered (and not necessarily the perfect conductor condition).

Letting U^{e} be equal either to E^{e} or H^{e}, we have

$$\operatorname{curl}\operatorname{curl} U^{\mathrm{e}} - 2\beta_e \gamma_{\mathrm{e}}^2 \operatorname{curl} U^{\mathrm{e}} - \gamma_{\mathrm{e}}^2 U^{\mathrm{e}} = 0, \quad \text{in } \mathcal{O}_e, \tag{6.10}$$

$$\widehat{x} \times \operatorname{curl} U^{\mathrm{e}}(x) - \beta_e \gamma_{\mathrm{e}}^2 \widehat{x} \times U^{\mathrm{e}}(x) + \mathrm{i}\,\frac{\gamma_{\mathrm{e}}^2}{k_e} U^{\mathrm{e}}(x) = o(|x|^{-1}), \tag{6.11}$$

as $|x| \to \infty$ uniformly in all directions $\widehat{x}$.

Let us denote by $\widetilde{\widetilde{B}}(x,x')$ the infinite medium Green dyadic. Chirality imposes the appearance of two components, a left-handed one and a right-handed one, in $\widetilde{\widetilde{B}}$; as in [268] we can write

$$\widetilde{\widetilde{B}}(x,x') = \widetilde{\widetilde{B}}_{\mathrm{L}}(x,x') + \widetilde{\widetilde{B}}_{\mathrm{R}}(x,x'), \tag{6.12}$$

where, for $\lambda = \mathrm{L}$, R,

$$\widetilde{\widetilde{B}}_\lambda(x,x') = \frac{k_e \gamma_{e\lambda}}{2\gamma_{\mathrm{e}}^2}\left(\widetilde{\widetilde{I}} + \frac{1}{{\gamma_{\mathrm{e}\lambda}}^2}\operatorname{grad}\operatorname{grad} + \frac{\mathfrak{m}_\lambda}{\gamma_{\mathrm{e}\lambda}}\operatorname{curl}_x \widetilde{\widetilde{I}}\right)\Phi(x,x';\gamma_{\mathrm{e}\lambda}).$$

Here $\widetilde{\widetilde{I}}$ is the identity dyadic, $\Phi(x,x';k)$ is the fundamental solution of the Helmholtz equation and $\mathfrak{m}_\lambda$ is defined in (5.8). This decomposition is reflected in an Atkinson-Wilcox type result for chiral media stated below, based on the following integral representation of the scattered field, ([268])

$$\begin{aligned} U^{\mathrm{e}}(x) &= -2\beta_e \gamma_{\mathrm{e}}^2 \int_{\partial\mathcal{O}} \widetilde{\widetilde{B}}(x,x') \cdot (\, n \times U^{\mathrm{e}}(x'))\, ds(x') \\ &\quad + \int_{\partial\mathcal{O}} \widetilde{\widetilde{B}}(x,x') \cdot (\, n \times \operatorname{curl} U^{\mathrm{e}}(x'))\, ds(x') \\ &\quad + \int_{\partial\mathcal{O}} \operatorname{curl}_x \widetilde{\widetilde{B}}(x,x') \cdot (\, n \times U^{\mathrm{e}}(x'))\, ds(x'), \quad x \in \mathcal{O}_e. \end{aligned} \tag{6.13}$$

Let ρ_0 be the radius of the smallest sphere circumscribable around the scatterer. Then we have an Atkinson-Wilcox type of expansion result for chiral media.

THEOREM 6.4.1 *Let U^{e} be a twice continuously differentiable field satisfying (6.10) and (6.11) in the domain $r > \rho_0$, where (r,θ,φ) are the spherical coordinates of the observation point x. Then*

$$U^{\mathrm{e}}(x) = \frac{e^{i\gamma_{eL} r}}{r}\sum_{j=0}^{\infty}\frac{F_j^L(\theta,\varphi)}{r^j} + \frac{e^{i\gamma_{eR} r}}{r}\sum_{j=0}^{\infty}\frac{F_j^R(\theta,\varphi)}{r^j}\,, \tag{6.14}$$

which converges for $r > \rho_0$. *The series in (6.14), as well as those obtained by term-by-term differentiation of any order, converge absolutely and uniformly in the closed domain* $r \geq r_0 > \rho_0$, $\theta \in [0, \pi]$, $\varphi \in [0, 2\pi]$.

A consequence of the above expansion theorem is the following proposition by which we recurrently obtain all the coefficients in the series (6.14) from the leading coefficient $F_0^\lambda(\theta, \varphi)$. In complete analogy with the standard Atkinson-Wilcox expansion for the Helmholtz equation, the leading coefficient has the interpretation of a far-field.

THEOREM 6.4.2 *The coefficients* $F_j^\lambda(\theta, \varphi)$, $\lambda = L, R$, *of the series (6.14) can be determined from* $F_0^\lambda(\theta, \varphi)$ *by the recurrence relations*

$$(\gamma_e^2 - \gamma_{e\lambda}^2) F_1^\lambda + 2\mathrm{i}\, \gamma_e^2 \beta_e \gamma_{e\lambda} \widehat{r} \times F_1^\lambda = 2\gamma_e^2 \beta_e (\widehat{r} - \mathsf{D}) \times F_0^\lambda \ ,$$

and, for $j \geq 1$,

$$\begin{aligned}&(\gamma_e^2 - \gamma_{e\lambda}^2) F_{j+1}^\lambda + 2\mathrm{i}\, \gamma_e^2 \beta_e \gamma_{e\lambda} \widehat{r} \times F_{j+1}^\lambda = \\ &2j\mathrm{i}\, \gamma_{e\lambda} F_j^\lambda + 2\gamma_e^2 \beta_e \left((j+1)\widehat{r} - \mathsf{D} \right) \times F_j^\lambda - [\mathsf{B} + j(j-1)] F_{j-1}^\lambda,\end{aligned} \quad (6.15)$$

where[7] $\mathsf{D} = \widehat{\theta} \frac{\partial}{\partial \theta} + \frac{1}{\sin \theta} \widehat{\varphi} \frac{\partial}{\partial \varphi}$ *and* $\mathsf{B} = \mathsf{D} \cdot \mathsf{D}$ *is the Beltrami operator.*

One can easily obtain from the above recursion relations that if the radiation patterns $F_0^\lambda(\theta, \varphi)$, $\lambda = L, R$, are zero, then the scattered field is also zero (a useful property in inverse scattering problems).

The proofs of the results in this section may be found in [26].

6.4.2 Low-frequency theory

In this section we consider the perfect conductor problem, in terms of the electric field only. We make extensive use of dyadic Green functions (see [400]).

Using (5.14), the fact that E^{inc} satisfies (5.13), and the following reciprocity properties of the dyadic Green function

$$\widetilde{\widetilde{B}}(x', x)^{tr} = \widetilde{\widetilde{B}}(x, x') \quad \text{and} \quad \left(\mathrm{curl}_{x'} \widetilde{\widetilde{B}}(x', x) \right)^{tr} = \mathrm{curl}_x \widetilde{\widetilde{B}}(x, x'),$$

we obtain from the integral representation (6.13) that

$$E^{\mathrm{t}}(x) = E^{\mathrm{inc}}(x) + \int_{\partial \mathcal{O}} \widetilde{\widetilde{B}}(x, x') \cdot (\, n \times \mathrm{curl} E^{\mathrm{t}}(x'))\, ds(x'), \quad x \in \mathcal{O}_e. \quad (6.16)$$

Using some asymptotics ([24], [121]) we have the following expression for the asymptotic behaviour of $\widetilde{\widetilde{B}}_\lambda$, $\lambda = L, R$:

$$\widetilde{\widetilde{B}}_\lambda(x, x') = \frac{k_e \gamma_{e\lambda}}{8\pi \alpha_e^2} \frac{e^{\,\mathrm{i}\gamma_{e\lambda}|x|}}{|x|} \, e^{-\mathrm{i}\, \gamma_{e\lambda} \widehat{x} \cdot x'} \widetilde{\widetilde{\mathfrak{J}}}_\lambda + O(|x|^{-2}), \quad |x| \to \infty, \quad (6.17)$$

---

[7] As usual, $\widehat{r}, \widehat{\theta}, \widehat{\varphi}$ are the unit vectors of the spherical coordinate system.

uniformly for all $x' \in \partial \mathcal{O}$, where

$$\widetilde{\widetilde{\mathfrak{J}}}_\lambda = \widetilde{\widetilde{I}} - \widehat{x} \otimes \widehat{x} + \mathrm{i}\, \mathfrak{m}_\lambda \widehat{x} \otimes \widetilde{\widetilde{I}},$$

and $\mathfrak{m}_\lambda$ is defined in (5.8). Using (6.17) and the integral representation (6.16), we end up with

$$E^{\mathrm{e}}(x) = \frac{e^{\,\mathrm{i}\gamma_{\mathrm{eL}}|x|}}{|x|} E_{\mathrm{L}}^{\infty}(x) + \frac{e^{\,\mathrm{i}\gamma_{\mathrm{eR}}|x|}}{|x|} E_{\mathrm{R}}^{\infty}(x) + O(|x|^{-2}), \quad |x| \to \infty, \tag{6.18}$$

where for $x \in \mathcal{O}_e, x' \in \partial \mathcal{O}$,

$$E_\lambda^{\infty}(x) = \frac{k_e \gamma_{e\lambda}}{8\pi \alpha_e^2} \int_{\partial \mathcal{O}} e^{\,-\mathrm{i}\,\gamma_{e\lambda}\widehat{x}\cdot x'} \widetilde{\widetilde{\mathfrak{J}}}_\lambda \cdot [\,n \times \mathrm{curl} E^{\mathrm{t}}(x')]\, ds(x'). \tag{6.19}$$

Let us remark that in the case that $\mathcal{O}_e$ is achiral space ($\beta_e = 0$), (6.18) and (6.19) yield the familiar electric far-field pattern ([106]). The appearance of the left and the right electric far-field patterns is due to the chiral nature of the host medium; (see also [268]).

We are now in a position to present the low-frequency theory for the perfect conductor problem.

The solution E^{t} of the scattering problem (5.13) - (5.15) considered as a function of ϖ is analytic in a neighbourhood of zero ([34]). Thus, E^{t} can be expressed as a convergent power series of ϖ in the low-frequency region

$$E^{\mathrm{t}}(x) = \sum_{j=0}^{\infty} \frac{\varpi^j}{j!} \Theta_j(x), \quad x \in \mathcal{O}_e, \tag{6.20}$$

where the low-frequency coefficients $\Theta_j(x)$ are independent of ϖ. Inserting the expansion (6.20) into (5.13) and equating equal powers of ϖ, the following iterative sequence of PDEs is obtained:

$$\begin{aligned}
\mathrm{curl}\,\mathrm{curl}\,\Theta_j(x) &= 0, \quad x \in \mathcal{O}_e, \quad j = 0, 1, \\
\mathrm{curl}\,\mathrm{curl}\,\Theta_j(x) &= j(j-1)\varepsilon_e \mu_e \beta_e^2 \mathrm{curl}\,\mathrm{curl}\,\Theta_{j-2}(x) \\
&\quad + 2j(j-1)\varepsilon_e \mu_e \beta_e\ \mathrm{curl}\,\Theta_{j-2}(x) \\
&\quad + j(j-1)\varepsilon_e \mu_e\, \Theta_{j-2}(x), \quad x \in \mathcal{O}_e, \quad j = 2, 3, \ldots.
\end{aligned}$$

Moreover Θ_j, $j = 0, 1, 2, \ldots$, are divergence free in $\mathcal{O}_e$. Hence, the low-frequency coefficients Θ_j for $j = 0, 1$, are solutions of the vector Laplace equation, while Θ_j for $j \geq 2$ are solutions of the vector Poisson equation, since the low-frequency coefficients of the right-hand sides are known from previous steps. The boundary condition (5.14) is transformed into

$$n \times \Theta_j(x) = 0, \quad x \in \partial \mathcal{O}.$$

We note that a similar expansion can be obtained for the magnetic field H. In particular, if we consider the low-frequency expansion

$$H^{\mathrm{t}}(x) = \sum_{j=0}^{\infty} \frac{\varpi^j}{j!} \Psi_j(x), \quad x \in \mathcal{O}_e,$$

with $\Psi_n(x)$ independent of ϖ, then the coefficients Θ_j, Ψ_j are connected by the relations

$$\begin{aligned}
&\operatorname{curl}\Theta_0 = 0,\\
&\operatorname{curl}\Theta_1 = \mathrm{i}\,\mu\,\Psi_0, \qquad (6.21)\\
&\operatorname{curl}\Theta_j - (j-1)\varepsilon_e\mu_e\beta_e(\beta_e \operatorname{curl}\Theta_{j-2} - j\,\Theta_{j-2}) = \mathrm{i}\,j\,\Psi_{j-1}, \quad j \geq 2.
\end{aligned}$$

To derive a suitable form of the radiation condition for the low-frequency coefficients, we use the integral representation (6.16), where all the fields are expanded in power series with respect to ϖ. The incident electric plane wave (5.7) assumes the following form, established in the Appendix of [24]:

$$E^{\mathrm{inc}}(x) = \sum_{j=0}^{\infty} \frac{\varpi^j}{j!} \left[q_{\mathrm{L}} f_{\mathrm{L},j}(p_{\mathrm{L}} \cdot x) + q_{\mathrm{R}} f_{\mathrm{R},j}(p_{\mathrm{R}} \cdot x)\right]. \qquad (6.22)$$

The functions $f_{\lambda,j}$, $\lambda = \mathrm{L}, \mathrm{R}$, have the form

$$f_{\lambda,j}(\tau) = \sum_{m=0}^{j} \mathfrak{d}_{\lambda,j,m}\tau^m,$$

where the coefficients $\mathfrak{d}_{\lambda,j,m}$ are dependent only on the physical parameters ε_e, μ_e, β_e.

The infinite medium Green dyadic $\tilde{\tilde{B}}(x, x')$ given by (6.12) has the following expansion:

$$\tilde{\tilde{B}}(x, x') = \sum_{j=0}^{\infty} \frac{\varpi^j}{j!}\tilde{\tilde{\Gamma}}_j(x, x') + \frac{\tilde{\tilde{T}}_0(x, x')}{4\pi\varepsilon_e\mu_e\varpi^2}, \qquad (6.23)$$

where

$$\tilde{\tilde{\Gamma}}_j(x, x') = \tilde{\tilde{\Gamma}}_{\mathrm{L},j}(x, x') + \tilde{\tilde{\Gamma}}_{\mathrm{R},j}(x, x'),$$

with

$$\tilde{\tilde{\Gamma}}_{\lambda,0}(x, x') = \frac{1}{16\pi|x - x'|}\left[\left(1 + 2\beta_e^2\right)\tilde{\tilde{I}} + \left(1 - 6\beta_e^2\right)\frac{(x - x') \otimes (x - x')}{|x - x'|^2}\right]$$

and, for $j = 1, 2, \ldots$,

$$\begin{aligned}
8\pi\tilde{\tilde{\Gamma}}_{\lambda,j}(x, x') &= \frac{\tilde{\tilde{I}}}{|x - x'|}\left[f_{\lambda,j}(|x - x'|) + j\beta_e\sqrt{\varepsilon_e\mu_e}\, f_{\lambda,j-1}(|x - x'|)\right]\\
&+ \frac{1}{(j+1)(j+2)\varepsilon_e\mu_e}\tilde{\tilde{S}}_{\lambda,j+2}(x, x') - \frac{\beta_e\,\mathfrak{m}_\lambda}{(j+1)\sqrt{\varepsilon_e\mu_e}}\tilde{\tilde{S}}_{\lambda,j+1}(x, x')\\
&+ \frac{\mathfrak{m}_\lambda}{(j+1)\sqrt{\varepsilon_e\mu_e}}(x - x') \times \tilde{\tilde{S}}_{\lambda,j+1}(x, x') + j\,\mathfrak{m}_\lambda\beta_e^2\sqrt{\varepsilon_e\mu_e}\tilde{\tilde{S}}_{\lambda,j-1}(x, x')\\
&- j\,\mathfrak{m}_\lambda\beta_e^2\sqrt{\varepsilon_e\mu_e}\,(x - x') \times \tilde{\tilde{S}}_{\lambda,j-1}(x, x') - \beta_e^2\tilde{\tilde{S}}_{\lambda,j}(x, x'),
\end{aligned}$$

with

$$\widetilde{\widetilde{S}}_{\lambda,j}(x,x') = \sum_{m=0}^{j} (m-1)\mathfrak{d}_{\lambda,j,m}\widetilde{\widetilde{T}}_m(x,x')$$

and

$$\widetilde{\widetilde{T}}_m(x,x') = \frac{1}{|x-x'|^3}\left[\widetilde{\widetilde{I}} + (m-3)\frac{(x-x')\otimes(x-x')}{|x-x'|^2}\right].$$

Since

$$\widetilde{\widetilde{T}}_0(x,x') = O\left(|x|^{-3}\right), \quad |x| \to \infty,$$

from (6.16), (6.22) and (6.23) we obtain the following asymptotic representation for the jth low-frequency coefficient:

$$\begin{aligned}\Theta_j(x) = {} & q_{\mathrm{L}} f_{\mathrm{L},j}(p_{\mathrm{L}}\cdot x) + q_{\mathrm{R}} f_{\mathrm{R},j}(p_{\mathrm{R}}\cdot x)\\ & + \sum_{\rho=0}^{j-1}\binom{j}{\rho}\int_{\partial\mathcal{O}} \widetilde{\widetilde{\Gamma}}_{j-\rho}(x-x')\cdot[n \times \mathrm{curl}\Theta_\rho(x')]\, ds(x')\\ & + O\left(|x|^{-1}\right), \quad |x|\to\infty.\end{aligned}$$

The far-field behaviour of the electric field at low-frequencies can be derived from (6.19), using (6.20) and the expansion

$$e^{-\mathrm{i}\gamma_{\mathrm{e}\lambda}\widehat{x}\cdot x'} = \sum_{j=0}^{\infty}\frac{\varpi^j}{j!} f_{\lambda,j}(-\widehat{x}\cdot x').$$

So, we have

$$E_\lambda^\infty(x) = \frac{\widetilde{\widetilde{\mathfrak{J}}}_\lambda}{8\pi}\cdot\Sigma_\infty,$$

where

$$\begin{aligned}\Sigma_\infty := {} & \sum_{j=0}^{\infty}\frac{\varpi^j}{j!}\sum_{\rho=0}^{j}\binom{j}{\rho}\int_{\partial\mathcal{O}} f_{\lambda,j}(-\widehat{x}\cdot x')\; n \times \mathrm{curl}\Theta_{j-\rho}(x')\, ds(x')\\ & + \beta_e\sqrt{\varepsilon_e\mu_e}\sum_{j=0}^{\infty}\frac{\varpi^j}{j!}\sum_{\rho=0}^{j-1}\binom{j-1}{\rho}\int_{\partial\mathcal{O}} f_{\lambda,j}(-\widehat{x}\cdot x')\; n \times \mathrm{curl}\Theta_{j-\rho-1}(x')\, ds(x').\end{aligned}$$

From this relation, taking into account (6.21) and the fact that for any closed surface surrounding a free-charge region the corresponding surface integrals of Θ_0 or Ψ_0 are equal to zero, we admit the leading term approximation as $\varpi \to 0$:

$$\begin{aligned}E_\lambda^\infty(x) = {} & \frac{\varpi^2}{8\pi}\widetilde{\widetilde{\mathfrak{J}}}_\lambda\cdot\left[\int_{\partial\mathcal{O}} n \times \mathrm{curl}\Theta_2(x')\, ds(x')\right.\\ & \left. -2\mathrm{i}\int_{\partial\mathcal{O}} (x\cdot x')\; n \times \mathrm{curl}\Theta_1(x')\, ds(x')\right] + O(\varpi^3), \quad \varpi\to 0.\end{aligned}$$

REMARK 6.4.3 As far as the development of low-frequency theory for transmission problems is concerned, let us mention that the chiral scatterer in an achiral environment transmission problem (to be more precise, the more general problem where the chiral scatterer consists of an "onion-like" structure of a finite number of nested homogeneous chiral layers of different electromagnetic material parameters in each layer, with a perfectly conducting or impedant core) is treated in [43]. The case of an achiral scatterer in a chiral environment is studied in [23].

REMARK 6.4.4 Some other very interesting issues of low-frequency theory for chiral media are considered in [10].

REMARK 6.4.5 In [27], scattering theorems for dyadic chiral fields are studied, while in [29] the authors' results (see Section 6.5) for the perfect conductor scattering problem on Beltrami-Herglotz fields, Herglotz pairs and density theorems in the vector case are established in the dyadic case, as well.

6.5 HERGLOTZ WAVE FUNCTIONS

The asymptotic properties of metaharmonic functions (i.e., solutions of the Helmholtz equation) have been the subject of extensive study over the years. The modern interest in the field, initiated by a lecture by Herglotz in the mid-1940s, was further advanced by Müller [327], Hartman [184], and Hartman and Wilcox [185] in the 1950s and 1960s, and was used extensively in the context of multiple scattering theory by Twersky in the 1960s [415], [416], [417]. In addition to the sine qua non rôle that Herglotz wave functions play in multiple scattering, they proved to be very useful in inverse scattering theory; Colton and Kirsch [102], Colton and Kress [104], [105], [107] and Colton and Monk [108], [109] used Herglotz wave functions to develop an effective method for shape reconstruction in acoustic and electromagnetic inverse scattering theory. An excellent source for this approach is [106]. Density results for the vector Helmholtz equation may be found in [433]. Dassios and Rigou [122] extended to linear elasticity the basic properties of the acoustic Herglotz functions. In this section we present the related results of Athanasiadis and Kardasi [28], [30] concerning chiral media.

6.5.1 Herglotz functions

A *Herglotz scalar wave fuction* is a solution U in $\mathbb{R}^3$ of the Helmholtz equation $\Delta U + k^2 U = 0$, $k \in \mathbb{R}$, satisfying the growth condition

$$\|U\|_H^2 := \lim_{\varrho \to \infty} \frac{1}{\varrho} \int_{B(0,\varrho)} |U(x)|^2 dx < \infty,$$

where, as proved in [185], the above quantity exists and defines a norm, called the *Herglotz norm*. It is known ([106]) that a function is a Herglotz function if and only if there exists a function $g \in L^2(S^2)$, called the *Herglotz kernel*, such that

$$U(x) = \int_{S^2} g(\widehat{d}) e^{\mathrm{i}\, k \widehat{d} \cdot x} ds(\widehat{d}).$$

Furthermore, U and g are related by

$$\|U\|_H^2 = \frac{8\pi^2}{k^2} \|g\|^2_{L^2(S^2)}.$$

In analogy with the scalar case, let us now consider the vector Herglotz wave function

$$q(x) = \int_{S^2} b(\widehat{d}) e^{\mathrm{i}\, \gamma \widehat{d} \cdot x} ds(\widehat{d}), \tag{6.24}$$

with a vector Herglotz kernel $b \in (L^2(S^2))^3$, and $\gamma > 0$; i.e., the Cartesian components of q are scalar Herglotz wave functions. These represent solutions of the vector Helmholtz equation $\Delta q + \gamma^2 q = 0$ in $\mathbb{R}^3$. The definition of the Herglotz norm for the vector case is obvious.

REMARK 6.5.1 The term *Herglotz function* is also used for a holomorphic function $h : \mathbb{C}^+ \to \mathbb{C}^+ \cup \mathbb{R}$, where $\mathbb{C}^+ := \{z : \mathcal{I}m\, z > 0\}$. These functions are closely related to the so-called Nevanlinna functions, or Pick functions, or R-functions, and have been thoroughly studied because of their relation to positive harmonic functions and Hardy spaces, the theory of continued fractions and the problem of moments, and the spectral theory of self-adjoint operators. These functions turned to be very important recently in the following sense: it is known that they can be represented by positive measures on the real line, a representation that can be interpreted as a dispersion relation for passive systems. A set of integral identities (called "sum rules") for physical systems has been derived ([62]) from this representation. These sum rules constitute a very promising setting for deriving physical limitations for passive systems.

6.5.2 Beltrami-Herglotz functions

Chirality imposes the appearance of two components, a left-handed one and a right-handed one, in Herglotz functions. For q given by (6.24), it can easily be shown that

$$\operatorname{div} q(x) = i\gamma \int_{S^2} e^{\mathrm{i}\, \gamma \widehat{d} \cdot x} \widehat{d} \cdot b\,(\widehat{d}) ds(\widehat{d})\,,$$

$$\operatorname{curl} q(x) = i\gamma \int_{S^2} e^{i\gamma \widehat{d} \cdot x} \widehat{d} \times\, b\,(\widehat{d}) ds(\widehat{d})\,.$$

Thus, we see that a vector Herglotz wave function $q(x)$ is a Beltrami field if and only if $\widehat{d} \cdot b\,(\widehat{d}) = 0$ and for LCP fields $\widehat{d} \times b\,(\widehat{d}) = -i\, b\,(\widehat{d})$, while $\widehat{d} \times b\,(\widehat{d}) = i\, b\,(\widehat{d})$ for RCP fields. This observation leads to the definition of the following subspaces of $(L^2(S^2))^3$:

$$\mathbb{T}^2_{\mathrm{L}}(S^2) = \{b_{\mathrm{L}} : S^2 \to \mathbb{C}^3 : b_{\mathrm{L}} \in (L^2(S^2))^3,\, n \cdot b_{\mathrm{L}} = 0,\, n \times b_{\mathrm{L}} = -ib_{\mathrm{L}}\},$$

$$\mathbb{T}^2_{\mathrm{R}}(S^2) = \{b_{\mathrm{R}} : S^2 \to \mathbb{C}^3 : b_{\mathrm{R}} \in (L^2(S^2))^3,\ n \cdot b_{\mathrm{R}} = 0,\ n \times b_{\mathrm{R}} = i b_{\mathrm{R}}\}.$$

A vector Herglotz wave function that is also a Beltrami field is called a *Beltrami-Herglotz function.* Athanasiadis and Kardasi [28] introduced this concept for the study of the Maxwell equations (5.2) in chiral media, which in view of Bohren's decomposition (5.5) take the form (5.4) in terms of Beltrami fields. In particular:

DEFINITION 6.5.2 *Let $b_\lambda \in \mathbb{T}^2_\lambda(S^2)$. A λ-CP [8] Beltrami-Herglotz function, $\lambda = L, R$, is a function of the form*

$$q_\lambda(x) = \int_{S^2} b_\lambda(\widehat{d}_\lambda) e^{i\gamma_\lambda \widehat{d}_\lambda \cdot x} ds(\widehat{d}_\lambda). \tag{6.25}$$

The function b_λ is called the λ-CP Beltrami-Herglotz kernel.

The Beltrami-Herglotz functions are characterised by a growth condition:

THEOREM 6.5.3 *The vector q_λ is a λ-CP, $\lambda = L, R$, Beltrami-Herglotz function if and only if it is a λ-CP Beltrami field satisfying $||q_\lambda||^2_H < \infty$.*

Beltrami-Herglotz functions can be expanded in terms of spherical vector wave functions (see Section 5.6.1). The inversion of (6.25) is given in the following result ([28]) which provides an expression for the Herglotz kernels b_λ directly from q_λ without the need to resort to such series expansions.

THEOREM 6.5.4 *If q_λ, $\lambda = L, R$, is a Beltrami-Herglotz function and b_λ is the corresponding Herglotz kernel, then*

$$b_\lambda(\widehat{x}) = \lim_{\varrho \to \infty} \frac{1}{2\varrho} \int_0^\varrho r e^{-i\gamma_\lambda r} \left[\frac{\partial\, q_\lambda(x)}{\partial r} + i\gamma_\lambda\, q_\lambda(x) \right] dr,$$

with $r = |x|$ and the limit interpreted in the $(L^2(S^2))^3$ sense.

Beltrami-Herglotz functions are important because of the following density result ([28]).

THEOREM 6.5.5 *Let $\mathcal{O}$ be a bounded domain in $\mathbb{R}^3$ with C^2 boundary. For every Beltrami field $Q_\lambda \in C^{2,\alpha}(\overline{\mathcal{O}})$, $\alpha \in (0, 1)$, and for every $\epsilon > 0$ there exists a Beltrami-Herglotz function q_λ, λ=L,R, such that*

$$\max_{x \in \overline{\mathcal{O}}} |Q_\lambda(x) - q_\lambda(x)| \le \epsilon.$$

Similar density results have been proved by Colton and Monk [108] for the Helmholtz equation in starlike domains in $\mathbb{R}^2$, and by Dassios and Rigou [123] for the reduced Navier equation for three-dimensional linear elasticity.

[8] CP stands for "circularly polarised".

6.5.3 Chiral Herglotz pairs

Using the LCP and the RCP Beltrami-Herglotz functions, we introduce the concept of a chiral Herglotz pair.

DEFINITION 6.5.6 *A chiral Herglotz pair is a pair of vector fields of the form*

$$\mathcal{E}(x) = q_{L}(x) - i\eta q_{R}(x), \quad \mathcal{H}(x) = i\eta^{-1} q_{L}(x) + q_{R}(x), \tag{6.26}$$

where q_λ, $\lambda = L, R$ are λ-CP Beltrami-Herglotz functions.

Chiral Herglotz pairs are generated by linear combinations of Beltrami-Herglotz kernels.

DEFINITION 6.5.7 *The vector field $b_L + b_R$, where $b_\lambda \in \mathbb{T}^2_\lambda(S^2)$, $\lambda = L, R$, is called the electric Herglotz kernel for the electric Herglotz field $\mathcal{E}$.*

Let us denote the set of all electric Herglotz kernels by

$$\mathbb{T}^2_{\mathrm{LR}}(S^2) = \{b : S^2 \to \mathbb{C}^3 : \ b = b_{\mathrm{L}} + b_{\mathrm{R}}, \text{ where } b_\lambda \in \mathbb{T}^2_\lambda(S^2), \ \lambda = L, R\},$$

and note that it is a subset[9] of $\mathbb{T}^2_{\mathrm{L}}(S^2) \oplus \mathbb{T}^2_{\mathrm{R}}(S^2)$.

Chiral Herglotz pairs obviously represent solutions to (5.19) in $\mathbb{R}^3$. Using Theorem 6.5.3, we characterise chiral Herglotz pairs.

THEOREM 6.5.8 *A solution $(\mathcal{E}, \mathcal{H})$ to equations (5.19) in $\mathbb{R}^3$ satisfies*

$$||\mathcal{E}||^2_H + ||\mathcal{H}||^2_H < \infty, \tag{6.27}$$

if and only if it is a chiral Herglotz pair.

Furthermore, the following density result holds:

THEOREM 6.5.9 *Let $\mathcal{O}$ be an open, bounded and connected subset of $\mathbb{R}^3$ with C^2 boundary $\partial\mathcal{O}$, and suppose that E and H are $C^{2,\alpha}(\overline{\mathcal{O}})$, $\alpha \in (0,1)$, solutions of (5.19). Then, for every $\epsilon > 0$ there exists a chiral Herglotz pair $(\mathcal{E}, \mathcal{H})$ such that*

$$\max_{x\in\overline{\mathcal{O}}} |E(x) - \mathcal{E}(x)| \le \epsilon, \quad \max_{x\in\overline{\mathcal{O}}} |H(x) - \mathcal{H}(x)| \le \epsilon.$$

The proof can be found in [28].

The following asymptotic forms, as $r = |x| \to \infty$, for $\mathcal{E}$ and $\mathcal{H}$ hold

$$\begin{aligned}
\mathcal{E}(x) &= \Big[b_{\mathrm{L}}(\widehat{x}) h_0^{(1)}(\gamma_{\mathrm{L}} r) - i\eta b_{\mathrm{R}}(\widehat{x}) h_0^{(1)}(\gamma_{\mathrm{R}} r)\Big] \\
&\quad + \Big[b_{\mathrm{L}}(-\widehat{x}) h_0^{(1)}(-\gamma_{\mathrm{L}} r) - i\eta b_{\mathrm{R}}(-\widehat{x}) h_0^{(1)}(-\gamma_{\mathrm{R}} r)\Big] + O\left(\frac{1}{r^2}\right), \\
\mathcal{H}(x) &= \left[\frac{1}{i\eta} b_{\mathrm{L}}(\widehat{x}) h_0^{(1)}(\gamma_{\mathrm{L}} r) + b_{\mathrm{R}}(\widehat{x}) h_0^{(1)}(\gamma_{\mathrm{R}} r)\right] \\
&\quad + \left[\frac{1}{i\eta} b_{\mathrm{L}}(-\widehat{x}) h_0^{(1)}(-\gamma_{\mathrm{L}} r) + b_{\mathrm{R}}(-\widehat{x}) h_0^{(1)}(-\gamma_{\mathrm{R}} r)\right] + O\left(\frac{1}{r^2}\right),
\end{aligned} \tag{6.28}$$

where $h_0^{(1)}$ is the spherical Hankel function of the first kind.

---

[9] $\mathbb{T}^2_{\mathrm{L}}(S^2) + \mathbb{T}^2_{\mathrm{R}}(S^2)$ is shown to be a direct sum in Lemma 3.2 of [30].

6.5.4 Superposition of incident electric fields

In [108], [109], Colton and Monk applied a superposition technique[10] to solve the inverse acoustic scattering problem, while in [106], Colton and Kress applied this method to the electromagnetic case. Dassios and Rigou [122], [123] studied the superposition of incident dyadic fields for the case of linear elasticity. Here we present the extension of the superposition technique to the case of electromagnetic fields in chiral media by Athanasiadis and Kardasi [30].

It is now easy to prove the following[11].

THEOREM 6.5.10 *For given densities* $b_\lambda \in \mathbb{T}^2_\lambda(S^2)$, $\lambda = L, R$, *the solution to the perfect conductor scattering problem for the incident wave*

$$\mathcal{E}^{\mathrm{inc}}(x) = \int_{S^2} b_L(\widehat{d}_L) e^{i\gamma_L \widehat{d}_L \cdot x} ds(\widehat{d}_L) + \int_{S^2} b_R(\widehat{d}_R) e^{i\gamma_R \widehat{d}_R \cdot x} ds(\widehat{d}_R)$$

is given by the relation

$$\begin{aligned} \mathcal{E}^{\mathrm{e}}(x) = &\int_{S^2} \{E^{\mathrm{e}}_L(x; \widehat{d}_L, b_L(\widehat{d}_L)) + E^{\mathrm{e}}_R(x; \widehat{d}_L, b_L(\widehat{d}_L))\} ds(\widehat{d}_L) \\ &+ \int_{S^2} \{E^{\mathrm{e}}_L(x; \widehat{d}_R, b_R(\widehat{d}_R)) + E^{\mathrm{e}}_R(x; \widehat{d}_R, b_R(\widehat{d}_R))\} ds(\widehat{d}_R), \end{aligned} \tag{6.29}$$

and has the far-field pattern

$$\begin{aligned} \mathcal{E}^{\infty}(\widehat{x}) = &\int_{S^2} \{E^{\infty}_L(\widehat{x}; \widehat{d}_L, b_L(\widehat{d}_L)) + E^{\infty}_R(\widehat{x}; \widehat{d}_L, b_L(\widehat{d}_L))\} ds(\widehat{d}_L) \\ &+ \int_{S^2} \{E^{\infty}_L(\widehat{x}; \widehat{d}_R, b_R(\widehat{d}_R)) + E^{\infty}_R(\widehat{x}; \widehat{d}_R, b_R(\widehat{d}_R))\} ds(\widehat{d}_R), \end{aligned} \tag{6.30}$$

where E^{∞}_λ, $\lambda = L, R$, *is the far-field pattern corresponding to scattered field* E^{e}_λ.

REMARK 6.5.11 If the incident wave is either LCP or RCP, then the above theorem reduces to a simpler form; see Corollary 3.1 in [30].

Let us recall the reciprocity principle for chiral media ([22]).

THEOREM 6.5.12 *Consider two time-harmonic plane electric waves*

$$E^{\mathrm{inc}}_j(x) = E^{\mathrm{inc}}_L(x; \widehat{d}_{L,j}, p_{L,j}) + E^{\mathrm{inc}}_R(x; \widehat{d}_{R,j}, p_{R,j}), \quad j = 1, 2,$$

incident on the scatterer $\mathcal{O}$. *Then the corresponding far-field patterns* $E^{\infty}_{\lambda,j}$,

---

[10] This is a constrained optimisation method for solving the inverse problem seeking an optimal solution in the orthogonal complement of the closure of the set of far-field patterns. This leads to reducing the problem of solving an exterior scattering problem to that of an interior boundary value problem, in view of Herglotz wave functions.

[11] We use the notation of Section 6.3.

$\lambda = L, R$, satisfy the reciprocity relation

$$\begin{aligned}
&\frac{1}{\gamma_L^2}\, p_{L2} \cdot [E_{L1}^{\infty}(-\widehat{d}_{L2}; \widehat{d}_{L1}, p_{L1}) + E_{L1}^{\infty}(-\widehat{d}_{L2}; \widehat{d}_{R1}, p_{R1})] \\
&+ \frac{1}{\gamma_R^2} p_{R2} \cdot [E_{R1}^{\infty}(-\widehat{d}_{R2}; \widehat{d}_{L1}, p_{L1}) + E_{R1}^{\infty}(-\widehat{d}_{R2}; \widehat{d}_{R1}, p_{R1})] \\
&= \frac{1}{\gamma_L^2} p_{L1} \cdot [E_{L2}^{\infty}(-\widehat{d}_{L1}; \widehat{d}_{L2}, p_{L2}) + E_{L2}^{\infty}(-\widehat{d}_{L1}; \widehat{d}_{R2}, p_{R2})] \\
&+ \frac{1}{\gamma_R^2} p_{R1} \cdot [E_{R2}^{\infty}(-\widehat{d}_{R1}; \widehat{d}_{L2}, p_{L2}) + E_{R2}^{\infty}(-\widehat{d}_{R1}; \widehat{d}_{R2}, p_{R2})],
\end{aligned} \tag{6.31}$$

for all $\widehat{d}_{\lambda,j} \in S^2$, $p_{\lambda,j} \in \mathbb{C}^3$, $\lambda = L, R$, $j = 1, 2$, and $p_{\lambda,j} \cdot \widehat{d}_{\lambda,j} = q \cdot \widehat{x} = 0$.

We note that for either LCP or RCP incidence the reciprocity principle reduces to the more familiar form of Corollary 6.5.13.

COROLLARY 6.5.13 *If both $E_{\lambda 1}^{\text{inc}}$ and $E_{\lambda 2}^{\text{inc}}$ are either LCP or RCP incident plane waves on the scatterer $\mathcal{O}$, then the corresponding far-field patterns $E_{\lambda,j}^{\infty}$, $\lambda = L, R$, $j = 1, 2$, satisfy the reciprocity relation*

$$p_{\lambda 2} \cdot E_{\lambda}^{\infty}(-\widehat{d}_{\lambda 2}; \widehat{d}_{\lambda 1}, p_{\lambda 1}) = p_{\lambda 1} \cdot E_{\lambda}^{\infty}(-\widehat{d}_{\lambda 1}; \widehat{d}_{\lambda 2}, p_{\lambda 2}),$$

for all $\widehat{d}_{\lambda j} \in S^2$, $p_{\lambda j} \in \mathbb{C}^3$, $\lambda = L, R$, $j = 1, 2$ and $p_{\lambda j} \cdot \widehat{d}_{\lambda j} = q \cdot \widehat{x} = 0$.

The following result also holds; its proof follows from an argument analogous to that of [102], p. 157.

THEOREM 6.5.14 *Let $E^{\text{e}} \in C^2(\mathbb{R}^3 \setminus \overline{\mathcal{O}})$ be a solution of (5.13)-(5.15) for which the electric far-field pattern vanishes identically. Then $E^{\text{e}} = 0$ in $\mathbb{R}^3 \setminus \overline{\mathcal{O}}$.*

In the following result (whose proof can be found in [30]) provides a necessary and sufficient condition such that the set of far-field patterns is complete in $\mathbb{T}_{\lambda}^2(S^2)$, $\lambda = L, R$.

LEMMA 6.5.15 *Let $\{\widehat{d}_{\lambda,n}\}$, $\lambda = L, R$, be a sequence of unit vectors that is dense in S^2. We define the sets of LCP and RCP electric far-field patterns by*

$$\mathcal{F}_{\lambda} := \{E_{\lambda}^{\infty}(\cdot; \widehat{d}_{\lambda,n}, \widehat{e}_j) : n = 1, 2, \ldots, \textit{ and } j = 1, 2, 3\}, \quad \lambda = L, R,$$

where $\widehat{e}_j$ are the Cartesian unit vectors. Then $\mathcal{F}_{\lambda}$ is complete in $\mathbb{T}_{\lambda}^2(S^2)$, $\lambda = L, R$, if and only if there does not exist a nontrivial electric Herglotz field $\mathcal{E}$ such that $n \times \mathcal{E} = 0$ on $\partial\mathcal{O}$.

The nontrivial solution may be interpreted as an eigenfunction of the operator form of the equation (5.13)-(5.15).

6.5.5 The far-field operators

Using the far-field equation derived above, we define related far-field operators and study some of their properties.

DEFINITION 6.5.16 *The operators*

$$\mathsf{F}^\infty_\lambda : \mathbb{T}^2_\lambda(S^2) \longrightarrow \mathbb{T}^2_\lambda(S^2), \quad \lambda = L, R,$$

defined by

$$(\mathsf{F}^\infty_\lambda b_\lambda)(x) := \int_{S^2} \{E^\infty_L(\widehat{x}; \widehat{d}_\lambda, b_\lambda(\widehat{d}_\lambda)) + E^\infty_R(\widehat{x}; \widehat{d}_\lambda, b_\lambda(\widehat{d}_\lambda))\} ds(\widehat{d}_\lambda) \tag{6.32}$$

will be respectively called the LCP and the RCP far-field operators related to the far-field equation (6.30).

In the following result we furnish some properties of the operators defined in (6.32).

PROPOSITION 6.5.17 *For the bounded integral operator* $\mathsf{F}^\infty_\lambda$, $\lambda = L, R$, *we have*

(*i*) $\mathsf{F}^\infty_\lambda(\mathbb{T}^2_\lambda(S^2))^\perp = \ker((\mathsf{F}^\infty_\lambda)^*) \quad , \quad \ker((\mathsf{F}^\infty_\lambda)^*)^\perp = \overline{\mathsf{F}^\infty_\lambda(\mathbb{T}^2_\lambda(S^2))}.$

(*ii*) $\mathsf{F}^\infty_\lambda$ *is injective and has dense range if and only if there does not exist a nontrivial electric Herglotz field* $\mathcal{E}$ *such that* $n \times \mathcal{E} = 0$ *on* $\partial\mathcal{O}$.

The proof of (*i*) is based on Theorem 4.6 of [102], while for the proof of (*ii*) see Lemma 4.2 of [30].

Let us now consider an electric field given by a superposition of the form

$$\mathcal{E}^{\text{inc}}(x) = \int_{S^2} b_{\text{L}}(\widehat{d}_{\text{L}}) e^{i\gamma_{\text{L}} \widehat{d}_{\text{L}} \cdot x} ds(\widehat{d}_{\text{L}}) + \int_{S^2} b_{\text{R}}(\widehat{d}_{\text{R}}) e^{i\gamma_{\text{R}} \widehat{d}_{\text{R}} \cdot x} ds(\widehat{d}_{\text{R}}),$$

where $b_\lambda \in T^2_\lambda(S^2)$, $\lambda = L, R$, are weight functions, i.e., the incident wave is an electric Herglotz field. Then, according to Theorem 6.5.10, the corresponding far-field pattern is given by

$$\begin{aligned} \mathcal{E}^\infty(\widehat{x}) = &\int_{S^2} \{E^\infty_{\text{L}}(\widehat{x}; \widehat{d}_{\text{L}}, b_{\text{L}}(\widehat{d}_{\text{L}})) + E^\infty_{\text{R}}(\widehat{x}; \widehat{d}_{\text{L}}, b_{\text{L}}(\widehat{d}_{\text{L}}))\} ds(\widehat{d}_{\text{L}}) \\ &+ \int_{S^2} \{E^\infty_{\text{L}}(\widehat{x}; \widehat{d}_{\text{R}}, b_{\text{R}}(\widehat{d}_{\text{R}})) + E^\infty_{\text{R}}(\widehat{x}; \widehat{d}_{\text{R}}, b_{\text{R}}(\widehat{d}_{\text{R}}))\} ds(\widehat{d}_{\text{R}}). \end{aligned} \tag{6.33}$$

In order that the scattered field be a solution $\mathcal{E}^{\text{e}}$ to (5.13)-(5.15) with corresponding far-field pattern $\mathcal{E}^\infty$, we have to solve the integral equation

$$\mathsf{F}^\infty_{\text{LR}}(b_{\text{L}} + b_{\text{R}}) = \mathcal{E}^\infty, \tag{6.34}$$

where the integral operator

$$\mathsf{F}^\infty_{\text{LR}} : \mathbb{T}^2_{\text{LR}}(S^2) \longrightarrow \mathbb{T}^2_{\text{LR}}(S^2) \tag{6.35}$$

is defined by

$$\mathsf{F}^\infty_{\text{LR}}(b_{\text{L}} + b_{\text{R}}) = \mathsf{F}^\infty_{\text{L}} b_{\text{L}} + \mathsf{F}^\infty_{\text{R}} b_{\text{R}},$$

and the far-field patterns $E^\infty(\cdot;\widehat{d})$ for all incident directions $\widehat{d}$ are assumed to be known. An interesting discussion of far-field patterns can be found in [19]. The integral operator defined in (6.35) is called the *electric far-field operator* related to the far-field equation (6.30).

As a consequence of Proposition 6.5.17(*ii*) (considering the adjoint operator of $\mathsf{F}^\infty_{\mathrm{LR}}$), we can easily obtain the following corollary.

COROLLARY 6.5.18 *The integral operator F^∞_{LR} is injective and has dense range if and only if there does not exist a nontrivial electric Herglotz field $\mathcal{E}$ such that $n \times \mathcal{E} = 0$ on $\partial\mathcal{O}$.*

Finally, in the following result we address the question of the possibility of finding a superposition of incident plane electric waves such that the resulting far-field patterns coincide with some prescribed far-field patterns. Its proof is based directly on the previous discussion and the results of this section and is therefore omitted; it can be found in [30]. Let us note that it is the existence and not the uniqueness of a solution that is of interest in this setting.

THEOREM 6.5.19 *Let E^{e}, H^{e} be a radiating solution to (5.19) in $\mathcal{O}_e$ that creates electric far-field patterns E^∞_λ, $\lambda = L, R$. Then the linear integral equation (6.34) possesses a solution $b = (b_L + b_R) \in \mathbb{T}^2_{LR}(S^2)$ if and only if $E^{\mathrm{e}}, H^{\mathrm{e}}$ are defined in $(C(\mathbb{R}^3 \setminus \overline{\mathcal{O}_e}))^3$ and belong to $(C(\mathbb{R}^3 \setminus \mathcal{O}_e))^3$, and the interior perfect conductor boundary value problem*

$$\begin{aligned}
\operatorname{curl} E^{\mathrm{int}} &= \beta\gamma^2 E^{\mathrm{int}} + i\varpi\mu\left(\frac{\gamma}{k}\right)^2 H^{\mathrm{int}}, \quad \text{in } \mathcal{O},\\
\operatorname{curl} H^{\mathrm{int}} &= \beta\gamma^2 H^{\mathrm{int}} - i\varpi\varepsilon\left(\frac{\gamma}{k}\right)^2 E^{\mathrm{int}}, \quad \text{in } \mathcal{O},\\
n \times E^{\mathrm{int}} &= -n \times E^{\mathrm{e}}, \quad \text{on } \partial\mathcal{O},
\end{aligned} \tag{6.36}$$

has a solution $\mathcal{E}^{\mathrm{int}}, \mathcal{H}^{\mathrm{int}}$ that is an electromagnetic Herglotz pair.

6.6 DOMAIN DERIVATIVE

We now consider the perfect conductor problem. It is known (see [24]) that a scattered electromagnetic wave behaves like the sum of two outgoing spherical waves:

$$E^{\mathrm{e}}(x) = \frac{e^{\,\mathrm{i}\gamma_{\mathrm{eL}}|x|}}{|x|}E^\infty_{\mathrm{L}} + \frac{e^{\,\mathrm{i}\gamma_{\mathrm{eR}}|x|}}{|x|}E^\infty_{\mathrm{R}} + O\left(|x|^{-2}\right), \quad |x| \to \infty.$$

The fields E^∞_{L} and E^∞_{R} are referred to as the left-handed and the right-handed electric far-field patterns of E^{e}; the notion of far-field pattern is discussed further in Sections 6.2 and 6.3.

The main purpose of this section is to study the properties of the mapping from the boundary $\partial\mathcal{O}$ of $\mathcal{O}$ onto the far-field pattern $(E^\infty_{\mathrm{L}}, E^\infty_{\mathrm{R}})$. We will exhibit infinite Fréchet differentiability (see Section A.3.1 in Appendix A) for this mapping in appropriate spaces and derive a characterisation of the Fréchet derivative as a solution to a chiral boundary value problem.

Consider perturbations $\mathcal{O}(\mathfrak{r})$ of a reference domain[12] $\mathcal{O}_0$, with $C^{3,\alpha}$ boundary $\partial\mathcal{O}_0$, such that $\Gamma(\mathfrak{r}) := \partial\mathcal{O}(\mathfrak{r}) := \{x + \mathfrak{r}(x) : x \in \partial\mathcal{O}_0\}$ for some $C^{3,\alpha}$ vector fields $\mathfrak{r} : \partial\mathcal{O}_0 \to \mathbb{R}^3$. Note that $\partial\mathcal{O}(0) = \partial\mathcal{O}_0$, and for $\|\mathfrak{r}\|_{C^{3,\alpha}}$ sufficiently small, each set $\Gamma(\mathfrak{r})$ is the boundary of some domain $\mathcal{O}(\mathfrak{r})$ with boundary of class $C^{3,\alpha}$. For a given $\mathfrak{r} \in C^{3,\alpha}$ and a given vector field $a : \mathbb{R}^3 \to \mathbb{R}^3$, we consider the (invertible) operator $\mathcal{T}(\mathfrak{r})$ whose action on a is

$$(\mathcal{T}(\mathfrak{r})\, a\,)(x) := \, a\,(x + \mathfrak{r}(x)), \quad x \in \partial\mathcal{O}_0.$$

Using this operator we may transform all functions and boundary integral operators on $\partial\mathcal{O}(\mathfrak{r})$ into corresponding quantities on $\partial\mathcal{O}_0$. It is important to note that $\mathcal{T}(\mathfrak{r})$ does not necessarily relate tangential fields to tangential fields between the two domains. To handle the particular spaces of tangential fields we use the projection operator $\mathsf{P}_1 : C(\partial\mathcal{O}_0) \to C_t(\partial\mathcal{O}_0)$, defined by $\mathsf{P}_1 a = \pi_{\partial\mathcal{O}_0}(a) = (n_0 \times a|_{\partial\mathcal{O}_0}) \times n_0$, where n_0 is the outward unit normal to $\partial\mathcal{O}_0$. By $\mathsf{P}_2(\mathfrak{r})$ we denote the inverse of the restriction, $\mathsf{P}_1(\mathfrak{r})$, of P_1 to the space $C_t(\partial\mathcal{O}(\mathfrak{r}))$, given by

$$\mathsf{P}_2(\mathfrak{r})\, a \, := \, a \, - \left(\frac{n\,(\mathfrak{r}) \cdot a}{n\,(\mathfrak{r}) \cdot n_0} \right) n_0,$$

where $n(\mathfrak{r})$ is the outward unit normal on $\partial\mathcal{O}(\mathfrak{r})$.

If $\mathcal{A}(\mathfrak{r}) : C_t(\partial\mathcal{O}(\mathfrak{r})) \to C_t(\partial\mathcal{O}(\mathfrak{r}))$ is an operator and c a constant such that $cI + \mathcal{A}(\mathfrak{r})$ is invertible, where c is a constant, then we have

$$\mathsf{P}_1(\mathfrak{r})(cI{+}\mathcal{A}(\mathfrak{r}))^{-1}\mathsf{P}_2(\mathfrak{r}) = (\mathsf{P}_1(\mathfrak{r})(cI{+}\mathcal{A}(\mathfrak{r}))\mathsf{P}_2(\mathfrak{r}))^{-1} = (cI{+}\mathsf{P}_1(\mathfrak{r})\mathcal{A}(\mathfrak{r})\mathsf{P}_2(\mathfrak{r}))^{-1},$$

on $C_t(\partial\mathcal{O}(\mathfrak{r}))$. Using this fact, along with the identity $I = \mathsf{P}_2(\mathfrak{r})\mathsf{P}_1(\mathfrak{r})$ on $C_t(\partial\mathcal{O}(\mathfrak{r}))$, we obtain ([356]) that the solution of the perfect conductor problem in $\mathcal{O}(\mathfrak{r})$ can be expressed as

$$E^{\mathrm{e}} = -2\mathsf{A}(\mathfrak{r})\mathsf{P}_2(\mathfrak{r})(cI + \mathsf{P}_1(\mathfrak{r})\mathcal{A}(\mathfrak{r})\mathsf{P}_2(\mathfrak{r}))^{-1}\mathsf{P}_1(\mathfrak{r})(\,n(\mathfrak{r}) \times \mathcal{T}(\mathfrak{r})E^{\mathrm{inc}}), \quad (6.37)$$

where[13]

$$c := \mathrm{i}\,\eta_{\mathrm{e}}(\gamma_{\mathrm{eL}} + \gamma_{\mathrm{eR}})\,, \;\; \mathsf{A}(\mathfrak{r}) := \mathrm{i}\,\eta_{\mathrm{e}}\mathsf{A}_{\mathrm{L,R}}(\mathfrak{r})\,, \;\; \mathcal{A}(\mathfrak{r}) := \frac{c}{2}I + \mathsf{A}(\mathfrak{r})\,.$$

Let

$$\mathfrak{A} := \left\{\mathfrak{r} \in (C^{3,\alpha}(\partial\mathcal{O}_0))^3 : \; \|\mathfrak{r}\| \le d\right\},$$

for appropriately chosen $d > 0$. If we consider the restriction of the solution to the scattering problem to some (arbitrary) bounded set $G \subset \mathcal{O}_e$, then the representation (6.37) defines the mapping

$$F(\mathfrak{r}) : C_t(\partial\mathcal{O}_0) \to (C(G))^3, \; F(\mathfrak{r})E^{\mathrm{inc}} := E^{\mathrm{e}}(\mathfrak{r}), \quad (6.38)$$

---

[12]That is, we consider a family of domains $\mathcal{O}(\mathfrak{r})$, in the vicinity of the original domain $\mathcal{O}_0$ all of which are parametrised in terms of $\partial\mathcal{O}_0$ using $\Gamma(\mathfrak{r}) := \partial\mathcal{O}(\mathfrak{r}) := \{x + \mathfrak{r}(x) : x \in \partial\mathcal{O}_0\}$ for various choices of $\mathfrak{r}$.

[13]Recall that $\mathsf{A}_{\mathrm{L,R}} := \gamma_{\mathrm{eL}}\mathsf{M}(\gamma_{\mathrm{eL}}) + \gamma_{\mathrm{eR}}\mathsf{M}(\gamma_{\mathrm{eR}}) + \mathsf{N}(\gamma_{\mathrm{eL}}) - \mathsf{N}(\gamma_{\mathrm{eR}})$, which clearly depends on the choice of $\partial\mathcal{O}(\mathfrak{r})$.

where by $E^{\mathrm{e}}(\mathfrak{r}) := E^{\mathrm{e}}$, as given by (6.37), we denote the solution to the problem for the choice $\partial\mathcal{O}(\mathfrak{r})$. The meaning of the above is the following: consider a fixed function $E^{\mathrm{inc}} : \mathbb{R}^3 \to \mathbb{C}^3$, such that E^{inc} is continuous on the boundary of the reference domain $\mathcal{O}_0$. Choosing $\mathfrak{r}$, we obtain a perturbation $\mathcal{O}(\mathfrak{r})$ of the reference domain, and the boundary data are derived by evaluating the function E^{inc} on $\partial\mathcal{O}(\mathfrak{r})$ using $(\mathcal{T}(\mathfrak{r})E^{\mathrm{inc}})(x)$, $x \in \partial\mathcal{O}_0$. By letting $\mathfrak{r}$ vary, therefore letting $\partial\mathcal{O}(\mathfrak{r})$ vary, but keeping the function E^{inc} fixed, we obtain different boundary data[14], and therefore we obtain different solutions to the problem. It is one of the aims of the present section to quantify the variation of these solutions as the domain changes. This can be accomplished in terms of the Fréchet derivative of the solution with respect to $\mathfrak{r}$ (see, e.g., [60], [3], [98]).

To this end, we use the results of [354] applied to the integral operators which build up the mapping $F(\mathfrak{r})$. For fixed $\mathfrak{r}$, the operator $F(\mathfrak{r})$ can be decomposed as $F(\mathfrak{r}) = F_1(\mathfrak{r}) \circ F_2(\mathfrak{r}) \circ F_3(\mathfrak{r})$, where

$$F_1(\mathfrak{r}) : C_t(\partial\mathcal{O}_0) \to (C(G))^3, \ F_1(\mathfrak{r})u := \mathsf{A}(\mathfrak{r})\mathsf{P}_2(\mathfrak{r})u,$$
$$F_2(\mathfrak{r}) : C_t(\partial\mathcal{O}_0) \to C_t(\partial\mathcal{O}_0), \ F_2(\mathfrak{r})v := \left(cI + \mathsf{P}_1(\mathfrak{r})\mathcal{A}(\mathfrak{r})\mathsf{P}_2(\mathfrak{r})\right)^{-1} v,$$
$$F_3(\mathfrak{r}) : C_t(\partial\mathcal{O}_0) \to C_t(\partial\mathcal{O}_0), \ F_3(\mathfrak{r})w := -2\mathsf{P}_1(\mathfrak{r})\left(n(\mathfrak{r}) \times \mathcal{T}(\mathfrak{r})w\right).$$

We now consider an operator from $\mathfrak{A}$ to the space of operators from $C_t(\partial\mathcal{O}_0) \to C(G)$, which for every choice of $\mathfrak{r} \in \mathfrak{A}$ renders the operator $F(\mathfrak{r}) : C_t(\partial\mathcal{O}_0) \to C(G)$ mapping the boundary data on the $\partial\mathcal{O}(\mathfrak{r})$ to the field in G. This can be visualised as follows: Fix $w \in C_t(\mathcal{O}_0)$, and consider

$$\breve{F}_1 : \mathfrak{A} \to \mathcal{L}(C_t(\partial\mathcal{O}_0), C(G)), \ \breve{F}_1 := \mathsf{A}(\mathfrak{r})\mathsf{P}_2(\mathfrak{r}),$$
$$\breve{F}_2 : \mathfrak{A} \to \mathcal{L}(C_t(\mathcal{O}_0), C_t(\partial\mathcal{O}_0)), \ \breve{F}_2(\mathfrak{r}) = \left(cI + \mathsf{P}_1(\mathfrak{r})\mathcal{A}(\mathfrak{r})\mathsf{P}_2(\mathfrak{r})\right)^{-1}, \qquad (6.39)$$
$$\breve{F}_3 : \mathfrak{A} \to C_t(\mathcal{O}_0), \breve{F}_3(\mathfrak{r}) := F_3(\mathfrak{r})w,$$

where by $\mathcal{L}(X_1, X_2)$ we denote the set of linear and bounded operators between the Banach spaces X_1, X_2. Note that $\breve{F}_1$, $\breve{F}_2$ are operator valued. Since for each $\mathfrak{r}$ we obtain $F(\mathfrak{r})$, an operator $C_t(\partial\mathcal{O}_0) \to C(G)$, we may consider the mapping $\breve{F} : \mathfrak{A} \to \mathcal{L}(C_t(\partial\mathcal{O}_0), C(G))$ with action $\breve{F}(\mathfrak{r})w = F(\mathfrak{r})w$, for every $w \in C_t(\partial\mathcal{O}_0)$.

We summarise the infinite Fréchet differentiability of the above operators in the following result (see [356]).

PROPOSITION 6.6.1 *The operators $\breve{F}_1$, $\breve{F}_2$ and $\breve{F}_3$, defined in* (6.39) *are infinitely Fréchet differentiable with respect to* $\mathfrak{r}$.

Using the product rule (Theorem A.2 of [98]) applied to the decomposition of $\breve{F}$ using the operators (6.39) we obtain, in view of Proposition 6.6.1, the required differentiability of $\breve{F}$. This leads to the following.

THEOREM 6.6.2 *The mapping $F : \mathfrak{A} \to C(G)$ is infinitely Fréchet differentiable with respect to* $\mathfrak{r}$.

[14] As the given function E^{inc} is calculated on different surfaces $\partial\mathcal{O}(\mathfrak{r})$.

This means that the solution to the chiral scattering problem is a continuous function on G that depends infinitely differentiably on the boundary of the scatterer, as long as it is $C^{3,\alpha}$ and such that $\|\mathfrak{r}\| \leq d$ for a suitably chosen $d > 0$. In particular, this guarantees that the solution of the problem depends continuously on the boundary of the scatterer.

Above we established the Fréchet differentiability of the scattered field for the chiral scattering problem. In principle, the derivative can be calculated by differentiation of the kernels of the involved integral operators and an application of the product rule as stated in [98]. However, to avoid the complexity of this procedure both from the viewpoint of implementation time and from the viewpoint of computational costs, the characterisation of the derivative of E^{e} as a solution to some corresponding boundary value problem has become very popular over the last years (see, e.g., [106], [98], [114], and the references therein).

We conclude this section with a characterisation of the chiral medium scattering problem, established in [356].

THEOREM 6.6.3 *The action of the Fréchet derivative at* $\mathfrak{r} = 0$ *of the chiral scattering problem,* $\frac{\partial E^{\mathrm{e}}}{\partial \mathfrak{r}}(0)$*, on a function* $h = (h_1, h_2, h_3) \in (C^{3,\alpha}(\partial \mathcal{O}_0))^3$ *solves the exterior chiral problem with the boundary values given by*

$$n_0 \times \left.\frac{\partial E^{\mathrm{e}}}{\partial \mathfrak{r}}\right|_{\mathfrak{r}=0} h = -\left.\frac{\partial n}{\partial \mathfrak{r}}\right|_{\mathfrak{r}=0} \times \left.E^{\mathrm{t}}\right|_{\mathfrak{r}=0} - n_0 \times \sum_{j=1}^{3} \partial_{x_j} \left.E^{\mathrm{t}}\right|_{\mathfrak{r}=0} h_j.$$

Note that the derivatives with respect to $\mathfrak{r}$ are Fréchet derivatives, while the derivatives with respect to x_j are the usual partial derivatives. A coordinate-free expression of the above expression can be found in [356].

We would like to briefly sketch two important applications of the differentiability properties and their characterisation. First, the differentiability properties may be used for perturbation methods to calculate the scattered field or its far-field patterns from the solution of the corresponding boundary value problems for simple geometries. Second, consider the operator $\mathfrak{S}_D$ mapping the surfaces $\Gamma = \partial \mathcal{O}_\star$, of some inclusions $\mathcal{O}_\star$, onto the far-field pattern for scattering of a chiral plane wave. We assume that Γ is in an appropriate Banach space, e.g., by using sufficiently smooth vector fields $\mathfrak{r} : \Gamma_0 \to \mathbb{R}^3$ on the boundary of some reference domain Γ_0 and $\mathfrak{r} \mapsto \Gamma(\mathfrak{r}) := \{x + \mathfrak{r}(x) : x \in \Gamma_0\}$. Then, the *Newton method* for finding an unknown set of inclusions Γ from knowledge of the far-field pattern is given by the iterative scheme

$$\Gamma_{n+1} := \Gamma_n - \left(\frac{\partial \mathfrak{S}_D}{\partial \Gamma}(\Gamma_n)\right)^{-1} \mathfrak{S}_D(\Gamma_n).$$

Here, the derivative $\frac{\partial \mathfrak{S}_D}{\partial \Gamma}(\Gamma_n)$ needs to be calculated and inverted. For this, the characterisation of this derivative as a solution to a certain boundary value problem is an important step, since each Newton step is reduced to solving one or several boundary value problems (see, e.g., [355] or [106]). Domain derivatives play an important rôle in the theory of inverse problems, as well as in shape optimisation (see [190] and the references therein).

6.7 MISCELLANEA

6.7.1 Waveguides

Problems regarding chirowaveguides or biisotropic waveguides have been studied by many authors. A representative very short list would include [143], [163], [164], [165], [435], [203], [204].

6.7.2 Quaternionic methods for chiral media

Quaternionic analysis is the most natural and close generalisation of complex analysis that preserves many of the latter's important features.

It is well known that in vector analysis the vector product does not permit the formation of an algebra. In the nineteenth century various attempts were made to construct an algebra in $\mathbb{R}^3$, until eventually Hamilton discovered that it is necessary instead to consider $\mathbb{R}^4$, for which a convenient algebra already existed. Quaternionic analysis was introduced by Hamilton in 1843 and was further developed by G. C. Moisil, N. Teodorescu[15] and K. R. Füter. For an introduction to quaternionic analysis the reader may refer to, e.g., [251].

A quaternion q is an element of $\mathbb{R}^4$, and hence can be represented as $q = \sum_{k=0}^{3} q_k i_k$, where $q_k \in \mathbb{R}$, $k = 0, 1, 2, 3$ are the "components" of q, and i_k, $k = 0, 1, 2, 3$, are the elements of the standard orthonormal basis. The equality and the addition of two quaternia are defined componentwise. The concept of quaternion differs from that of any vector in $\mathbb{R}^4$ through the definition of quaternionic multiplication, based on the proper definition of multiplication of the elements of the basis. All laws of algebra are valid regarding quaternionic arithmetic, with one exception: quaternionic multiplication is not commutative. Thus, in algebraic terms, real quaternia form a skew field. There is an obvious generalisation to complex quaternia (also called biquaternia).

The operator $\mathfrak{D} := \sum_{k=1}^{3} i_k \frac{\partial}{\partial x_k}$ is called the *Moisil-Teodorescu* (or Dirac) *operator*. The quaternionic equation[16] $\mathfrak{D}f = 0$ is equivalent to a system, called the Moisil-Teodorescu system, that can be expressed in terms of the usual operators div, curl and grad; it is regarded as a natural generalisation of the Cauchy-Riemann equations to the three-dimensional space. It is easy to see that $\mathfrak{D}^2 = -\Delta$, where Δ is the Laplacian. Hence, each component of a function f satisfying $\mathfrak{D}f = 0$ is a harmonic function.

In recent years quaternionic methods have been used by many authors in a wide spectrum of problems in mathematical physics. Vladislav V. Kravchenko and his group have extensively employed quaternionic methods in the study of electromagnetic problems in chiral media. We briefly review their work here.

---

[15] His name often appears as either Théodoresco or Theodoresco.

[16] A function f satisfying this equation is called $\mathfrak{D}$-holomorphic, hyperholomorphic, or monogenic.

In [246] it is observed, in the achiral setting, that the Maxwell system in the time-harmonic case for homogeneous isotropic media can be decoupled and written in the form of two separate biquaternionic equations with the aid of the operators $(\mathfrak{D} + \alpha)$ and $(\mathfrak{D} - \alpha)$[17] applied to purely vectorial biquaternia φ and ψ [18]. Here φ and ψ are appropriate linear combinations of the electric and magnetic fields E and H, $\mathfrak{D}$ is the Moisil-Teodorescu operator and $\alpha \in \mathbb{C}$ with $\mathcal{I}m\, \alpha \geq 0$ is the wave number. With the aid of this decoupling procedure, Kravchenko showed that the Stratton-Chu formulae are easily obtained from the Cauchy integral formulae for the biquaternia φ and ψ; moreover, he considered and solved the problem of analytic extendability of the electromagnetic field from the boundary. These results, complemented with some integral representations for other physical quantities (e.g., the impulse of the electromagnetic field), appeared later in [247] and in the book [256]. In [237], similar ideas are applied to the time-harmonic field in a chiral medium. In [248], using the fact that the quaternionic equations to which the Maxwell system reduces are closely related to the Dirac system from relativistic quantum mechanics, the relationship between solutions of both these systems is studied, with a discussion of the physical meaning. In [238], a system (suitable for solving scattering problems for time-harmonic electromagnetic fields in homogeneous chiral media) of biquaternionic fundamental solutions is introduced, with very promising results from numerical tests owing to the correct asymptotic behaviour at infinity, as well as near the boundary of a scatterer. In [253] and [251] the Maxwell equations for time-harmonic fields in chiral media are studied in the case of stratified media. Classes of exact and asymptotic solutions are obtained. In [249] the equation $\text{curl}U + \lambda U = 0$ is studied in the case in which λ is a function of the spatial variables. In particular, when λ depends on one Cartesian variable only, integral representations for solutions are obtained in terms of fundamental solutions for associated stationary Schrödinger equations. This study has been continued in [254], [252] in view of the pseudoanalytic function theory. In this case (i.e., when λ depends on one Cartesian variable), the equation describing Beltrami fields is reduced to specific type of equation called *Vekua equation* for which a complete system of solutions is obtained explicitly. In [250] and [251] the Maxwell system describing *time-dependent* electromagnetic fields in inhomogeneous isotropic media is reformulated in the form of a single biquaternionic equation (in the case of homogeneous media, this result has been known since the early twentieth century). This is a novel result for inhomogeneous chiral media. For homogeneous chiral media the Maxwell system for time-dependent fields is written as a single biquaternionic equation in [172]. This biquaternionic reformulation is used

[17] These operators permit a factorisation of the three-dimensional Helmholtz operator, exactly as the operators ∂ and $\overline{\partial}$ factorise the two-dimensional Laplacian. This observation has played a sine qua non rôle in the development of modern quaternionic analysis. For details, see [174].

[18] These ideas work in higher dimensions, as well: the skew field of quaternia is then replaced by an appropriate Clifford algebra (see, e.g., [175]).

to obtain in explicit form a Green function satisfying the causality principle. A review of this and the preceding results can be found in [236]. See also [255].

6.7.3 Periodic structures

Periodic structures (gratings) have received increasing attention over the years because of their important applications in integrated optics, optical lenses, antireflective structures, holography, lasers, communications and computing. In the achiral case, significant mathematical results have been established by many authors; one can consult the bibliography in [14]. Chiral gratings give rise to new features and applications; e.g., they are capable of converting a linearly polarised incident field into two nearly circularly polarised diffracted modes in different directions. Various physical and computational aspects regarding the propagation of electromagnetic waves inside periodic chiral media have been studied by many authors; again, one can consult the bibliography in [14]. The mathematical theory in the time-harmonic case has been studied in [4], [5], [6], [14]: a time-harmonic electromagnetic plane wave is incident on a *biperiodic* structure (i.e., a three-dimensional structure that is periodic in two orthogonal directions) filled with a chiral inhomogeneous material; this structure is assumed to separate two chiral homogeneous regions. The diffraction problem consists in the study of the propagation of the reflected and transmitted waves away from this structure. A variational formulation of the diffraction problem by chiral gratings is first introduced. The main result refers to the well posedness of the problem: it is established that for all (with the possible exception of a discrete set of) frequencies, the diffraction problem has a unique quasi-periodic weak solution. The approach is based on a variant of the Hodge decomposition and a compact embedding result. The reduction of the problem to a bounded domain problem is a key step; this can be done by two approaches: either in view of a pair of transparent boundary conditions (based on the Bohren decomposition of fields inside homogeneous chiral media) or by deriving exact radiation conditions on the boundary of the heterogeneous chiral medium (based on the coupling of a finite element method in the inhomogeneous chiral medium with an integral equations, or boundary element method on the periodic interfaces). The aforementioned variational approach assumes that the material parameters ε, μ and β are only bounded functions. The incidence angles and grating shapes can be arbitrary, and the geometry can be very general. In [444] the problem of electromagnetic scattering by a periodic chiral structure is considered. The medium is homogeneous and the structure is periodic in one direction and invariant in another direction. The electromagnetic fields inside the chiral medium are governed by the Maxwell equations together with the Drude-Born-Fedorov constitutive relations. The problem is simplified to a two-dimensional scattering problem, and the existence and uniqueness of solutions are treated by an integral equations approach. It is shown that for all (with the possible exception of a discrete set) wave numbers, the corresponding integral equation has a unique solution. In [445], the scattering of time-harmonic electromagnetic waves propagating in a homogeneous chiral environment by a chiral grating is studied. The reduction to a two-dimensional problem is again employed, and the existence and uniqueness of solutions are studied by a variational approach. The diffraction problem is solved by a finite element method with perfectly matched absorbing layers. The corresponding scattering problem for the case of a perfectly conducting grating is studied in [446], by an integral equation approach. An exact boundary value solution for the scattering of a transverse magnetic (TM) wave by an elliptic chiral cylinder is developed and presented in [235]. The solution is based on the separation-of-variables technique in the elliptic cylinder coordinates system and expressed in terms of Mathieu and modified Mathieu functions. The incident, transmitted and scattered electromagnetic waves are expressed in terms of infinite series of wave functions. The matrix form of the expansion coefficients is found by applying the boundary conditions and the orthogonality of the Mathieu functions. Numerical results of the forward- and back-scattered echo widths for co- and cross-polarised waves with TM and TE polarised incident fields are presented and discussed for various cases. In [61] the authors prove the uniqueness of the solution of time-harmonic electromagnetic boundary value problems defined on a smooth bounded domain that is filled with both lossless and lossy media. Further, they obtain the regularity of electromagnetic fields in chiral media and the absence of resonances in cavities loaded with chiral and partly lossy dielectrics. The potential applications of electromagnetic chirality, together with the fact that a general ellipsoid can simulate a variety of common objects, such as discs, needles, etc., constitute the motivation of [228]. Scattering of electromagnetic waves incident on an ellipsoidal object with a chiral property is analysed using Fourier analysis, in conjunction with an integral equations technique. The inner field of the scatterer is described in terms of a superposition of plane waves satisfying the vector Helmholtz equation inside the chiral medium, and a Galerkin procedure is employed to solve a volume integral equation obtained by applying Green's theorem. A highly stable numerical code is obtained and verified by comparison with previously published solutions. Numerical computations have been carried out for various scatterers, and interesting scattering properties of waves from chiral media are observed. In the particular case of an incident linearly polarised plane wave, the subsequent change of its polarisation is investigated and discussed. Wave scattering by a chiral grating is studied in [438]. Numerical results are given and physical properties are discussed, including the influence of frequency, angle of incidence, and aspect ratio. At high frequencies the authors find anomalous coupling regions known as Wood's anomalies, which are explained by the excitation and reradiation of leaky waveguide modes in the periodic layer. The chiral grating can posses both frequency-selection and mode-conversion properties. In [214] the angular scattering from radially stratified spherical chiral objects is investigated. Based on the principles of invariant embedding, a matrix Riccati equation is formulated that can be used to examine basic scattering properties of spherical chiral structures with radial inhomogeneities in permittivity, permeability, and chirality. High- and low-frequency limits as well as weak reflection and constant impedance cases for this equation are examined. Further, it is shown that in the limit of large radii of curvature, this formulation yields the planar result. Some interesting results appear also in [436]. Time-harmonic electromagnetic fields in thin chiral curved layers are considered and studied in [13].

6.7.4 Inverse problems

The inverse scattering problem can be broadly divided into two classes, the *inverse obstacle problem* and the *inverse medium problem*. In the inverse obstacle problem, the scattering object is a homogeneous obstacle with given boundary data and the inverse problem is to determine the obstacle from knowledge of the scattered field at infinity (i.e., the far-field pattern). The inverse medium problem is to determine the constitutive parameters of a (possibly) inhomogeneous medium, from knowledge of the far-field pattern.

The inverse scattering problem has only recently progressed from a collection of ad hoc techniques with little rigorous mathematical basis to an area of intense activity with at least the beginnings of a solid mathematical foundation. The reason for this is that the inverse scattering problem is inherently nonlinear and improperly posed. Nevertheless, the inverse scattering problem is fundamental in areas such as radar, sonar, geophysical exploration, medical imaging and nondestructive testing; for different aspects of the theory and its applications see, [106], [190], [208], [219], [242], [355], [359], [402] and the references therein.

Consider the direct transmission problem describing the scattering of a time-harmonic plane wave propagating in an achiral space by a chiral scatterer. The solvability of this problem in $\mathbb{R}^3$ is discussed in Section 5.7 for a homogeneous scatterer and in [308] for an inhomogeneous scatterer. The homogeneous scatterer case in $\mathbb{R}^2$ is studied in [160]. The study of the corresponding inverse problem has two main areas: the first is the unique determination of the chiral scatterer from knowledge of the far-field pattern, and the second consists in uniquely determining the electromagnetic parameters in the scatterer from knowledge of the fields on its boundary. Without referring to the exact formulation and detailed presentation of the related results, we briefly describe the main ideas.

Consider two sets of data $(\mathcal{O}_j,\ \varepsilon_j,\ \mu_j,\ \beta_j)$, $j = 1, 2$, describing the scattering problem for homogeneous chiral scatterers (ε_j, μ_j, β_j: constants) in a homogeneous chiral environment. If the electric far-field patterns $E_{\infty,1}$ and $E_{\infty,2}$ coincide for *all* incident directions and *all* polarisations, then the scatterers coincide, i.e., $\mathcal{O}_1 = \mathcal{O}_2$. For three dimensions this is proved in [42] based on a weak volume formulation of the transmission problem, in view of Isakov's approach in [106], [208]. For two dimensions it is proved in [160], in view of a simplified version of Isakov's approach introduced by Kirsch and

Kress (see [106]), based on a boundary integral equation method, with the use of a modified Bohren decomposition in which the coupling in the boundary conditions is shifted from Neumann-to-Dirichlet data. In addition, it is established in [393] that if $\mathcal{O}_1 = \mathcal{O}_2$ and the electric far-field patterns coincide for all incident directions and all polarisations, then $\varepsilon_1 = \varepsilon_2$, $\mu_1 = \mu_2$ and $\beta_1 = \beta_2$. In two dimensions, a Newton-type method for the reconstruction of the boundary of a starlike scatterer has been developed in [160]. To cope with the ill posedness of the problem, a regularisation suggested by Hohage and Schormann (see the references in [160]) is implemented. The following question is treated in [307]: from electromagnetic information obtainable at the boundary of a body, can one determine material parameters, and their normal derivatives, at the boundary of the body? The answer to this question is given in two physically distinct situations. The first is when the relationship between the electromagnetic fields depends on the conductivity, the electric permittivity and the magnetic permeability of the body, and these parameters, together with their normal derivatives, are shown to be recoverable at the boundary. The second is when the chirality of the body is taken into account. It is also shown how a layer-stripping algorithm may be derived to estimate the unknown parameters near the boundary in both situations. In [31] the inverse electromagnetic scattering problem by a perfect conductor in a chiral environment is studied in view of the authors' results for Herglotz wave functions for chiral media. The reconstruction of biisotropic material parameters from experimental data is treated in [366]. In [157] the linear sampling method[19] for boundary reconstruction of a chiral obstacle is analysed.

For the corresponding nonhomogeneous problem (ε, μ, β are now position dependent), taking into account as well the (spatially varying) conductivity σ, it is shown in [308] that the knowledge of a boundary map for the electromagnetic fields (Dirichlet-to-Neumann map formalism) uniquely determines the electromagnetic parameters ε, μ, β and σ in the interior of the scatterer. This is done by rewriting the modified chiral Maxwell equations as a first-order perturbation of the Laplacian and constructing exponentially growing solutions; the result is thus obtained in the spirit of complex geometrical optics. In [307] the problem of determining the conductivity, permittivity, permeability and chirality measure (as well as their partial derivatives with respect to the x_3 direction) on the boundary of a scatterer $\mathcal{O}$ from knowledge of the Calderón operator is studied. To this end, two results pertaining to pseudodifferential operators are provided. In [281] the problem of determining inhomogeneous (space-dependent) coefficients in the (time-independent) constitutive relations for the Maxwell equations is explored: the data are taken to be two sets of measurements of the electric flux density and the

[19]The linear sampling method (LSM), introduced by Colton and Kirsch in 1996, is the oldest and most developed of the qualitative methods in inverse scattering theory. It is based on solving a linear integral equation and then using the equation's solution as an indicator function for the determination of the support of the scattering object. For the LSM in electromagnetics, introduced by Haddar and Monk in 2002, see [84].

magnetic flux density obtained at points forming a subset of the spatial domain where the Maxwell equations are considered, for all times from a given time interval. The constitutive relations considered in the paper correspond to the biisotropic model of a medium and involve three scalar parameters (as functions of the space variable): the dielectric permittivity, the magnetic permeability, and a magnetoelectric coupling parameter, which is a measure of nonreciprocity in the model. Under certain a priori bounds on the parameters to be reconstructed (defining an admissible set of parameters), the author establishes a Lipschitz-type stability estimate for the inverse problem. The proof is based on reducing the Maxwell equations to a weakly coupling system of hyperbolic equations and subsequently applying a Carleman-type estimate.

In [192] the transmission problem for a chiral obstacle $\mathcal{O}$ of non-constant ε, μ, β surrounded by free-space ($\varepsilon_e = \mu_e = 1$, $\beta_e = 0$) is studied; the solvability of the corresponding direct scattering problem is settled by a variational technique. Further, the inverse problem of determining the shape of the scatterer from the knowledge of the (fixed, positive) wavenumber k and of the far-field patterns (resulting from a plane incident field $p\,e^{i\,k\,d\cdot x}$) for all $d, x/|x| \in S^2$ and $p \in \mathbb{C}^3$, is studied. This is accomplished with the use of an appropriate factorisation method, involving the far-field operator, the Herglotz operator and some range identities. An application is provided for the case of an infinite chiral cylinder of the form $D \times \mathbb{R}$, where D is a bounded domain in the x_1x_2-plane.

Another problem refers to stratified chiral media. Through an analysis of the scattering matrix, information is obtained about the material parameters at normal or oblique incidence ([73], [74], [75]): uniqueness results and reconstruction formulae are given for these parameters. In particular, in [73] the propagation of a transversely polarised time-harmonic electromagnetic plane wave normally incident on a stratified nonreciprocal chiral (biisotropic) slab is considered. The structure of the scattering matrix is analysed. The inverse problem of reconstruction of material characteristics of a medium is studied. The approach is to calculate an explicit asymptotic expansion for the symbol of a boundary operator that is assumed to be known (from boundary measurements); this expansion is shown in each case to determine the unknown parameters at the boundary. Further, in [76], a Riemann-Hilbert approach is presented to solving the inverse problem of the reconstruction of the parameters of a dispersive stratified chiral medium. The frequency dependence of the medium's parameters is supposed to be given by a single-resonance Lorentz model. It is assumed that a plane harmonic wave impinges normally on the chiral slab from a vacuum region. The scattering produced by the slab is measured and constitutes the input data for the inverse problem. The main result is to show, by treating the inverse problem as an analytic factorisation problem in the complex plane frequencies, that the scattering data as functions of alternating frequency allow reconstruction of three independent combinations of four spatially varying medium parameters. In [379] an inverse problem for a stratified uniaxial bianisotropic slab is considered in the frequency domain. The problem is treated as an analytic factorisation problem in the frequency complex plane. Uniqueness in the parameter reconstruction is discussed and illustrated under a particular choice of a priori information. In [155] the propagation of a transient electromagnetic field in a stratified, dispersive and anisotropic slab and related direct and inverse problems are investigated. The field is generated by a transient external three-dimensional source. The analysis relies on the wave-splitting concept and a two-dimensional Fourier transformation in the transverse spatial coordinates. An investigation of the physical properties of the split fields is undertaken. To solve the direct and inverse scattering problems, wave propagators are used. This method is a generalisation and a unification of the previously used embedding and Green functions methods. The wave propagator approach provides an exact solution of the transmission operator. From this solution it is possible to extract the first precursor (the Sommerfeld forerunner). These results also hold for a bianisotropic slab. An inverse problem is outlined using reflection and transmission data corresponding to four two-dimensional Fourier parameters. Because of the stratification of the medium, the inverse Fourier transformation is not needed in the inverse problem. In [365], time-varying wave propagation and time-harmonic wave propagation in biisotropic materials are reviewed and the connection between the formulations is established via the temporal Fourier transform. An alternative method to determine the dispersive properties of a biisotropic slab from sinusoidal scattering data at normal incidence is presented. A numerical example (realistic, synthetic scattering data) is given to illustrate the theory. Furthermore, experimental data are presented and used to generate the permittivity, permeability, and chirality parameters of a specific man-made chiral slab in the range 3.5-18 GHz. On the basis of the results of inversion, the question of whether the passivity concept is too austere is raised.

PART 3

Time-Dependent Deterministic Problems

Chapter Seven

Well Posedness

7.1 INTRODUCTION

In this chapter we deal with electromagnetic fields in complex media in the time domain. The most general form for the constitutive relations is assumed to be nonlocal in time, in the form of convolutions. The convolution models dispersive effects in the medium. We address questions related to the well posedness of the Maxwell equations in this setting.

The structure of the chapter is as follows. In Section 7.2 we discuss the Maxwell equations in complex media in the time domain and show that they can be expressed as integrodifferential equations. In Section 7.3 we provide a convenient functional setting that allows us to treat the Maxwell equation as an integrodifferential evolution equation, and in Section 7.4 we provide some solvability and well posedness results for this problem based on a semigroup approach. In Section 7.5 we discuss two alternative approaches to the solvability of evolution problems in the time domain, namely, the evolution family approach and an approach using finite-dimensional approximations (Faedo-Galerkin) approach. Finally, in Section 7.6 we present some extensions related to evolution problems. This chapter is partly based on [284], [394].

7.2 THE MAXWELL EQUATIONS IN THE TIME DOMAIN

The starting point for the analysis of this chapter is the Maxwell system for isotropic linear complex media in an interior bounded domain $\mathcal{O}$, in the time domain, as expressed in Section 2.4.6 and in particular in equation (2.46). For convenience, we repeat this initial-boundary value problem here.

$$\begin{aligned} &\partial_t \left(\mathsf{A}_{\mathrm{or}}\, u + \mathsf{G}_{\mathrm{d}} \star u\right) = \mathsf{M}u - j, \quad (t,x) \in (0,T] \times \mathcal{O}, \\ &u(0,x) = u_0(x), \quad x \in \mathcal{O}, \\ &n(x) \times u_1(t,x) = 0, \quad (t,x) \in (0,T) \times \mathcal{O}, \end{aligned} \tag{7.1}$$

where $u = (E,H)^{tr} =: (u_1,u_2)^{tr}$ and M is the Maxwell operator. Further, let G_{d} be the susceptibility matrix and A_{or} be the optical response matrix (see (2.11)). Assuming, additionally, that G_{d} is weakly differentiable with respect to time and that $\mathsf{G}_{\mathrm{d}}(\cdot,0) = 0$, we may rewrite the above equation in the equivalent form

$$\partial_t u = \mathsf{M}_{\mathrm{A}} u + \mathsf{G}_{\mathrm{A}} \star u + \mathsf{J}_{\mathrm{A}}, \tag{7.2}$$

where $\mathsf{G}_{\mathrm{A}} = -\mathsf{A}_{\mathrm{or}}^{-1}\,\partial_t \mathsf{G}_{\mathrm{d}}$, $\mathsf{M}_{\mathrm{A}} = \mathsf{A}_{\mathrm{or}}^{-1}\mathsf{M}$, $\mathsf{J}_{\mathrm{A}} = -\mathsf{A}_{\mathrm{or}}^{-1} j$. By Assumption 2.3.5, equations (7.1) and (7.2) are equivalent.

7.3 FUNCTIONAL FRAMEWORK AND ASSUMPTIONS

In this section we introduce the functional setting that will be needed for the study of equations (7.1) or (7.2), as well as the assumptions on the data of the problem that will be used in proving the well posedness of the above models.

We will use the following function spaces (see Section 3.9.1):

$$\begin{aligned}
&\mathsf{X} := (L^2(\mathcal{O}))^3,\\
&\mathbb{X} := \mathsf{X} \times \mathsf{X},\\
&\mathsf{X}_1 := H_0(\mathrm{curl}, \mathcal{O}), \ \ \mathsf{X}_2 := H(\mathrm{curl}, \mathcal{O}),\\
&\mathbb{X}_{\mathsf{M}} := \mathsf{X}_1 \times \mathsf{X}_2.
\end{aligned}$$

The proper combination of these function spaces provides the right functional environment for the treatment of the problem. Note that if $u \in \mathbb{X}_{\mathsf{M}}$, then u satisfies the perfect conductor boundary condition. Other choices are of course possible.

To keep the discussion as general as posible, we formulate our results in terms of an abstract integrodifferential equation. Let $\mathbb{H}$ and $\mathbb{H}_{\mathsf{M}}$ be two Hilbert spaces and $\mathsf{M}_{\mathrm{A}} : \mathbb{H}_{\mathsf{M}} \to \mathbb{H}$. The space $\mathbb{H}_{\mathsf{M}}$ is endowed with the graph norm of M_{A}. In our setting, $(\mathbb{H}, \mathbb{H}_{\mathsf{M}})$ will be used as a proxy for the pair $(\mathbb{X}, \mathbb{X}_{\mathsf{M}})$ or any other relevant choice.

The following standing assumption on the operator M_{A} is crucial for our approach. As will be seen later, in Section 7.4.2, the state space can be chosen so that this assumption holds true.

ASSUMPTION 7.3.1 The operator $\mathsf{M}_{\mathrm{A}} : \mathbb{H}_{\mathsf{M}} \to \mathbb{H}$ is the generator of a C_0 semigroup $\{\mathsf{T}_{\mathsf{M}_{\mathrm{A}}}(t)\}_{t\in\mathbb{R}}$ on the Hilbert space $\mathbb{H}$.

Regarding the definition and properties of semigroups, see Section A.8 in Appendix A.

We are now able to state (7.2) as an abstract integrodifferential equation of the form

$$u' = \mathsf{M}_{\mathrm{A}} u + \mathsf{G}_{\mathrm{A}} \star u + \mathsf{J}_{\mathrm{A}} \tag{7.3}$$

on the Hilbert space $\mathbb{H}$.

REMARK 7.3.2 ([205]) Consider the Maxwell equation $\mathsf{d}'(t) = (\mathcal{L}u)'(t) = \mathsf{M}u(t) + j(t)$, $t \geq 0$, $u(0) = u_0$, with $\mathcal{L}(u) = \mathsf{A}_{\mathrm{or}} u + \mathsf{G}_{\mathrm{d}} \star u$. If $u : \mathbb{R}^+ \to \mathbb{X}_{\mathsf{M}}$ and $J : \mathbb{R}^+ \to H(\mathrm{div}, \mathcal{O})$ (recall that $j = (-J, 0)^{tr}$), it follows that $\mathsf{d} : \mathbb{R}^+ \to H(\mathrm{div}, \mathcal{O}) \times H(\mathrm{div}, \mathcal{O})$ and $\rho : \mathbb{R}^+ \to L^2(\mathcal{O})$. So, if $\mathrm{div}\, \mathsf{d}(0) = (\rho(0), 0)^{tr} = (\mathcal{L}u)(0) = \mathsf{A}_{\mathrm{or}} u_0$ is given at $t = 0$, then $\mathrm{div} D = \rho$ and $\mathrm{div} B = 0$ for all $t > 0$. Therefore Gauss's laws (2.2) are true, as long as the initial datum

u_0 is such that $\mathsf{A}_{\text{or}} u_0 \in H(\text{div}\rho(0), \mathcal{O}) \times H(\text{div}0, \mathcal{O})$, where $H(\text{div}\rho(0), \mathcal{O}) = \mathfrak{r} + H(\text{div}0, \mathcal{O})$ with $\mathfrak{r} \in L^2(\mathcal{O}, \mathbb{R}^3) : \text{div}\mathfrak{r} = \rho(0)$. So the modelling requires just Ampère's and Faraday's laws (2.1), the equation of continuity (2.7), and the above regularity on $(E_0, H_0)^{tr}$.

7.4 SOLVABILITY

We now consider the problem of well posedness of the mathematical model governing the evolution of electromagnetic fields in complex media in the time domain, as expressed by the integrodifferential equation (7.3). This is a very important issue, since as it stands, this equation is nothing more than a formal expression of the Maxwell equations supplemented with a properly selected set of constitutive relations and boundary conditions. However, unless one manages to show that this evolution equation admits solutions having the right qualitative behaviour, one cannot assume this model is a satisfactory one for the physical situation at hand, and it is also very risky to try and use this model for numerical purposes.

Several alternative approaches to the solvability of system (7.3) can be considered. Our primary approach to this problem is to adopt a semigroup formulation, based on the semigroup generated by the Maxwell operator. Then the convolution terms, which model the chirality and the dispersive effects, are treated as perturbations of this semigroup. Through the use of general fixed point schemes, the well posedness as well as interesting qualitative properties of the model can be shown. This approach is adopted for the following reasons: (a) the semigroup (group, actually) generated by the Maxwell operator is very well studied and (b) the chiral terms are usually small, so that it is plausible to treat them as perturbations. The results of this section are based on [285]; they are related to those in [205]. See also [64], [65], [179].

7.4.1 Different notions of solutions

For the integrodifferential equation (7.3) a variety of different types of solutions can be defined regarding *spatial* or *temporal* regularity.

DEFINITION 7.4.1 *A function $u \in C([0,T];\ \mathbb{H})$ is called a mild solution of (7.3) if for all $t \in [0,T]$,*

$$u(t) = \mathsf{T}_{\mathsf{M}_A}(t)u_0 + \int_0^t \mathsf{T}_{\mathsf{M}_A}(t-s) \int_0^s \mathsf{G}_A(s-r)u(r)dr\, ds + \int_0^t \mathsf{T}_{\mathsf{M}_A}(t-s)\mathsf{J}_A(s)\, ds.$$

DEFINITION 7.4.2 *A function $u \in C([0,T];\ \mathbb{H})$ is called a weak solution of (7.3) if for every $\phi \in D(\mathsf{M}_A^*) = D(\mathsf{M}_A)$:*

(*i*) *The function* $(u(t), \phi)$ *is absolutely continuous on* $[0, T]$, *where* $(\cdot, \cdot)$ *denotes the inner product*[1] *in* $\mathbb{H}$.

(*ii*) *For almost all* $t \in [0, T]$,

$$(u(t), \phi)' = -(u(t), \mathsf{M}_A\phi) + ((\mathsf{G}_A \star u)(t), \phi) + (\mathsf{J}_A(t), \phi).$$

(*iii*) $u(0) = u_0$.

REMARK 7.4.3 Since M_A is closed and densely defined, one can easily prove that a mild solution is a weak solution, and vice versa.

DEFINITION 7.4.4 *An* $\mathbb{H}$*-valued function* u *is called a strong solution of (7.3) if* $u \in W^{1,1}([0, T]; \mathbb{H})$ *and* $u(t)$ *satisfies equation (7.3) and the initial condition* $u(0) = u_0$, *a.e. on* $[0, T]$.

REMARK 7.4.5 By the form of (7.3) it can be seen that for a strong solution we have $u(t) \in D(\mathsf{M}_A)$, a.e. in $[0, T]$, and $\mathsf{M}_A u, \mathsf{G}_A \star u \in L^1([0, T]; \mathbb{H})$. Therefore, a strong solution possesses higher spatial regularity than a mild or a weak solution.

Solutions that assume a higher degree of regularity are classical solutions.

DEFINITION 7.4.6 *A function* $u \in C^1([0, T]; \mathbb{H}) \cap C([0, T]; D(\mathsf{M}_A))$ *that satisfies equation (7.3) and the initial condition* $u(0) = u_0$, *for all* $t \in [0, T]$, *is called a classical solution.*

REMARK 7.4.7 It is easily seen that for a classical solution, it holds that $\mathsf{M}_A u$, and $\mathsf{G}_A \star u \in C([0, T]; \mathbb{H})$.

The following theorem guarantees the well posedness of (7.3) in the weak sense:

THEOREM 7.4.8 *Under the assumptions*

(*i*) $\mathsf{G}_A \in L^2([0, T]; (L^\infty(\mathcal{O}))^{6\times 6})$,

(*ii*) $\mathsf{J}_A \in L^1([0, T]; \mathbb{H})$,

the equation (7.3) is weakly[2] *well posed in* $\mathbb{H}$.

Proof. Choose a positive b, and consider the Banach space $C([0, T]; \mathbb{H})$, with the norm $\|u\|_b = \sup_{t\in[0,T]} e^{-bt}\|u(t)\|_{\mathbb{H}}$. For $T < \infty$, this norm is clearly equivalent to the usual norm of $C([0, T]; \mathbb{H})$. We define the map F on $C([0, T]; \mathbb{H})$ by

$$\begin{aligned}\mathsf{F}(u(t)) := \mathsf{T}_{\mathsf{M}_A}(t)u_0 &+ \int_0^t \mathsf{T}_{\mathsf{M}_A}(t-s)\int_0^s \mathsf{G}_A(s-r)u(r)dr\, ds \\ &+ \int_0^t \mathsf{T}_{\mathsf{M}_A}(t-s)\mathsf{J}_A(s)\, ds,\end{aligned} \tag{7.4}$$

[1] In general, e.g., in nonlinear problems, in place of the inner product the duality pairing between the involved spaces should appear.

[2] In view of Remark 7.4.3, "weakly" can be replaced by "mildly".

with $s \leq t \in [0,T]$. It is not hard to establish that F maps $C([0,T];\mathbb{H})$ into $C([0,T];\mathbb{H})$, and that for an appropriate choice of b, F is a contraction. Therefore, by the Banach contraction theorem (see Theorem A.9.1 in Appendix A), F has a unique fixed point that is the solution of (7.3). The continuous dependence of the solution on the data follows similarly. $\square$

REMARK 7.4.9 The choice of b in the equivalent norm $||\cdot||_b$ so that the map F is a contraction can be thought of in an equivalent manner, as follows: instead of choosing b as large as we wish to render F a contraction, we may take $b = 0$ and study the fixed point equation in an interval $[t_0, t_0 + \Delta t]$ for arbitrary t_0. For Δt small enough, F is a contraction in the original norm, therefore guaranteeing the existence of a solution in this interval. A standard continuation argument may provide solutions in $[0,T]$.

In the assumptions of the three following theorems, $\mathsf{G}_{\mathsf{A}}(t)\mathsf{y}$ denotes the action of the matrix $\mathsf{G}_{\mathsf{A}}(t)$ on the six-vector y locally in t, i.e., *not* as a convolution.

The following result guarantees that equation (7.3) is strongly well posed.

THEOREM 7.4.10 *Assume that* $u_0 \in \mathbb{H}_{\mathsf{M}}$ *and that*

(*i*) $\exists\, C > 0$: $||\mathsf{G}_{A}(t)\mathsf{y}||_{\mathbb{H}_{\mathsf{M}}} < C\,||\mathsf{y}||_{\mathbb{H}_{\mathsf{M}}}\ \forall\, \mathsf{y} \in \mathbb{H}$ *a.e. in* $[0,T]$;

(*ii*) $\mathsf{J}_{A} \in L^1([0,T];\ \mathbb{H}_{\mathsf{M}})$;

Then (7.3) is strongly well posed in $\mathbb{H}$.

Proof. Let $b > 0$. We consider the Banach space $L^1([0,T];\ \mathbb{H}_{\mathsf{M}})$ with norm $||u||_b = \int_0^T e^{-bt}||u(t)||_{\mathbb{H}_{\mathsf{M}}}\,dt$. For $T < \infty$ the norm $||\cdot||_b$ is equivalent to the usual norm of $L^1([0,T];\mathbb{H}_{\mathsf{M}})$. For $s \leq t \in [0,T]$, we define the map F on $L^1([0,T];\ \mathbb{H}_{\mathsf{M}})$ as in (7.4). Based on the equivalence of the norms we can show that F maps $L^1([0,T];\ \mathbb{H}_{\mathsf{M}})$ into $L^1([0,T];\ \mathbb{H}_{\mathsf{M}})$. It is not hard to establish now that F is a contraction on $L^1([0,T];\ \mathbb{H}_{\mathsf{M}})$ for $b > 0$ sufficiently large. It is clear[3] that the resulting fixed point satisfies the properties required by Definition 7.4.4. So, (7.3), under the assumptions (*i*)-(*ii*), has a unique strong solution that is b-exponentially bounded. Remark 7.4.9 applies here as well. $\square$

Working in a similar fashion we may obtain classical solutions of equation (7.1) at the cost of imposing stricter regularity conditions on the data of the problem.

THEOREM 7.4.11 *Assume that* $u_0 \in \mathbb{H}_{\mathsf{M}}$ *and that*

(*i*) $\exists\, C > 0$: $||\mathsf{G}_{A}(t)\mathsf{y}||_{\mathbb{H}} < C\,||\mathsf{y}||_{\mathbb{H}}\ \forall\, t \in [0,T]$, $\mathsf{y} \in \mathbb{H}$.

(*ii*) $\exists\, C > 0$: $||\mathsf{G}_{A}(t)\mathsf{y}||_{\mathbb{H}_{\mathsf{M}}} < C\,||\mathsf{y}||_{\mathbb{H}_{\mathsf{M}}}\ \forall\, t \in [0,T]$, $\mathsf{y} \in \mathbb{H}_{\mathsf{M}}$.

[3]Taking into account as well the absolute continuity of $u(t)$ on $[0,T]$.

(iii_1) $\mathsf{J}_A \in L^1([0,T];\ \mathbb{H}_\mathsf{M}) \cap C([0,T];\mathbb{H})$, *or*

(iii_2) $\mathsf{J}_A \in W^{1,1}([0,T];\ \mathbb{H}) \cap C([0,T];\mathbb{H})$.

Then (7.3) is classically well posed in $\mathbb{H}$.

Proof. For $b > 0$, consider the Banach space $C([0,T];\mathbb{H}_\mathsf{M})$ equipped with the norm $\|u\|_b = \sup_{t\in[0,T]} e^{-bt}\|u(t)\|_{\mathbb{H}_\mathsf{M}}$. For $T < \infty$, this norm is equivalent to the usual norm of $C([0,T];\mathbb{H}_\mathsf{M})$. Define the map F on $C([0,T];\mathbb{H}_\mathsf{M})$ as in (7.4).

Consider the function $u_\mathsf{G} : [0,T] \to \mathbb{H}$, $u_\mathsf{G}(t) := \int_0^t \mathsf{G}_\mathsf{A}(t-s)u(s)\,ds$. Clearly, $u_\mathsf{G} \in C([0,T];\mathbb{H})$. Furthermore, since $u_\mathsf{G}(t) \in\ \mathbb{H}_\mathsf{M} = D(\mathsf{M}_\mathsf{A})$ for every $t \in [0,T]$ and $\mathsf{M}_\mathsf{A} u_\mathsf{G} \in\ L^1([0,T];\ \mathbb{H})$ the functions $u(t) = \int_0^t \mathsf{T}_{\mathsf{M}_\mathsf{A}}(t-s)u_\mathsf{G}(s)\,ds$ and $\mathsf{M}_\mathsf{A} u(t)$ are continuous on $[0,T]$, so $u(t) \in C([0,T];\ \mathbb{H}_\mathsf{M})$.

By a similar argument we find that the functions $v(t) = \int_0^t \mathsf{T}_{\mathsf{M}_\mathsf{A}}(t-s)\mathsf{J}_\mathsf{A}(s)ds$ and $\mathsf{M}_\mathsf{A} v(t)$ are continuous on $[0,T]$; hence, $v(t) \in C([0,T];\mathbb{H}_\mathsf{M})$. So, F maps $C([0,T];\mathbb{H}_\mathsf{M})$ into $C([0,T];\mathbb{H}_\mathsf{M})$.

Furthermore, we may show that for $b > 0$ large enough, F is a contraction on $C([0,T];\mathbb{H}_\mathsf{M})$, and thus has a unique fixed point in $C([0,T];\mathbb{H}_\mathsf{M})$ that satisfies the assertions of Definition 7.4.6. Following the proof of Theorem 2.4, p. 107 in [346], it can be shown that this solution is continuously differentiable. So, (7.1) is classically well posed and the unique classical solution is b-exponentially bounded. Remark 7.4.9 applies here as well. □

Finally, we observe that the classical solution may be obtained by replacing some of the conditions concerning spatial regularity of the kernel with conditions concerning temporal regularity of kernel. In particular:

THEOREM 7.4.12 *Under the assumptions of Theorem 7.4.11 with (ii) replaced by*

$$\exists\ \mathsf{g} \in L^1([0,T];\mathbb{R}):\ ||\mathsf{G}'_A \mathsf{y}||_\mathbb{H} \le \mathsf{g}(t)\,||\mathsf{y}||_{\mathbb{H}_\mathsf{M}},\ \ \forall\ t \in [0,T],\ \mathsf{y} \in \mathbb{H}_\mathsf{M},$$

(7.3) is classically well posed.

Proof. We consider again the map $\mathsf{F} : C([0,T];\mathbb{H}_\mathsf{M}) \to C([0,T];\mathbb{H}_\mathsf{M})$, as defined by (7.4), and prove that under the above assumptions, this map is a contraction.

Consider the continuous function $u_\mathsf{G}(t) = \int_0^t \mathsf{G}_\mathsf{A}(t-s)u(s)\,ds$. Since $u \in C([0,T];\mathbb{H}_\mathsf{M})$, we observe that $u'_\mathsf{G}(t) = \mathsf{G}_\mathsf{A}(0)u(t) + \int_0^t \mathsf{G}'_\mathsf{A}(t-s)u(s)\,ds \in L^1([0,T];\ \mathbb{H})$.

Therefore, the function $u_\mathsf{T}(t) = \int_0^t \mathsf{T}_{\mathsf{M}_\mathsf{A}}(t-s)u_\mathsf{G}(s)\,ds$ is differentiable, and its derivative,

$$u'_\mathsf{T}(t) = \mathsf{T}_{\mathsf{M}_\mathsf{A}}(t)u_\mathsf{G}(0) + \int_0^t \mathsf{T}_{\mathsf{M}_\mathsf{A}}(t-s)u'_\mathsf{G}(s)\,ds = \int_0^t \mathsf{T}_{\mathsf{M}_\mathsf{A}}(t-s)u'_\mathsf{G}(s)\,ds,$$

is continuous on $[0,T]$. So, we obtain that $u_\mathsf{T}(t) \in \mathbb{H}_\mathsf{M} = D(\mathsf{M}_\mathsf{A})$ and $\mathsf{M}_\mathsf{A} u_\mathsf{T}(t)$ is continuous on $[0,T]$, with $\mathsf{M}_\mathsf{A} u_\mathsf{T}(t) = u'_\mathsf{T}(t) - u_\mathsf{G}(t)$. Moreover, since $\{\mathsf{G}_\mathsf{A}(t)\}$

is a family of bounded operators in $\mathbb{H}$ such that $\sup_{t\in[0,T]} \|\mathsf{G}_{\mathsf{A}}\|_{\mathcal{L}(\mathbb{H})} \le C$ for some $C > 0$, for any $y \in \mathbb{H}_{\mathsf{M}}$ we have:

$$\|\mathsf{G}_{\mathsf{A}}(t)y\|_{\mathbb{H}} \le C\|y\|_{\mathbb{H}} \le C\|y\|_{\mathbb{H}_{\mathsf{M}}}.$$

Hence, applying (7.5) for $y = u(t)$, in conjunction with our assumptions we obtain :

$$\|\mathsf{M}_{\mathsf{A}} u_{\mathsf{T}}(t)\|_{\mathbb{H}} \le \|u_{\mathsf{T}}'(t)\|_{\mathbb{H}} + \|G(t)\|_{\mathbb{H}} \le \|\int_0^t \mathsf{T}_{\mathsf{M}_{\mathsf{A}}}(t-s)u_{\mathsf{G}}'(s)\, ds\|_{\mathbb{H}} + \|u_{\mathsf{G}}(t)\|_{\mathbb{H}}$$

$$\le T\left(\|\mathsf{G}_{\mathsf{A}}(0)\|_{\mathcal{L}(\mathbb{H})} + \int_0^t \mathsf{g}(s)\, ds + C\right) \sup_{t\in[0,T]} \|u(t)\|_{\mathbb{H}_{\mathsf{M}}}.$$

Finally we obtain that, for fixed T, there is a constant c such that

$$\|u_{\mathsf{T}}(t)\|_{\mathbb{H}_{\mathsf{M}}} \le c \sup_{t\in[0,T]} \|u(t)\|_{\mathbb{H}_{\mathsf{M}}}\,, \tag{7.5}$$

with $c > T_m(\|\mathsf{G}_{\mathsf{A}}(0)\|_{\mathcal{L}(\mathbb{H})} + \int_0^T \mathsf{g}(s)\, ds + C)$, and $T_m = \max\{T, T^2\}$.

By a similar argument as before, since $\mathsf{J}_{\mathsf{A}} \in W^{1,1}([0,T];\ \mathbb{H})$, we see that for the continuous function $v(t) = \int_0^t \mathsf{T}_{\mathsf{M}_{\mathsf{A}}}(t-s)\mathsf{J}_{\mathsf{A}}(s)ds$ we have that $v(t) \in \mathbb{H}_{\mathsf{M}} = D(\mathsf{M}_{\mathsf{A}})$ and $\mathsf{M}_{\mathsf{A}} v(t)$ is continuous on $[0,T]$. So F maps $C([0,T];\mathbb{H}_{\mathsf{M}})$ into $C([0,T];\mathbb{H}_{\mathsf{M}})$.

Now, using (7.5), we can check that F is a contraction on $C([0,T];\mathbb{H}_{\mathsf{M}})$. Indeed, defining $u^{\diamond} = u^{(1)} - u^{(2)}$, for $u^{(1)}, u^{(2)} \in C([0,T];\mathbb{H}_{\mathsf{M}})$, we have

$$e^{-bt}\|\int_0^t \mathsf{T}_{\mathsf{M}_{\mathsf{A}}}(t-s)\int_0^s \mathsf{G}_{\mathsf{A}}(s-r)u^{\diamond}(r)dr\, ds\|_{\mathbb{H}_{\mathsf{M}}} \le e^{-bt}c \sup_{t\in[0,T]} \|u^{\diamond}(t)\|_{\mathbb{H}_{\mathsf{M}}}\,,$$

so it is clear that for $b > 0$ sufficiently large, the map F is a contraction on $C([0,T];\mathbb{H}_{\mathsf{M}})$. It is easy to check that the unique fixed point $u(t),\ \ t \in [0,T]$, satisfies the assertions of Definition 7.4.6 and consequently equations (7.3). So (7.3) is classically well posed. □

REMARK 7.4.13 In the case that G_{d} is independent of x, assumption (ii) holds, provided that G_{d}' is bounded by an L^1 function. Similar comments apply in the case in which G_{d} may depend on x and $\mathbb{H}$ is an L^2-based Hilbert space, e.g., in $\mathbb{X}$.

REMARK 7.4.14 Along the lines of the above procedure, one may obtain solutions of (7.3) of higher regularity by imposing sufficiently smooth data. The results of this section are by no means optimal. Optimal results can be obtained using more sophisticated methods than the ones employed here. For instance, resorting to other fixed point theorems we may obtain existence results under weaker assumptions on the data. We intentionally avoid such an approach to minimise the necessary technicalities.

7.4.2 Finite energy solutions

Different types of solutions can be obtained by the proper choice of the state space $\mathbb{H}$ and the domain of the operator M_{A}. In particular, the following choices yield important types of finite energy solutions.

▷ Non (necessarily)-divergence-free solutions. The choice $\mathbb{H} = \mathbb{X}$ and $D(\mathsf{M}_{\mathsf{A}}) = \mathbb{X}_{\mathsf{M}}$ provides the proper functional setting for studying finite energy solutions for (7.1). In view of Proposition 3.9.9, Assumption 7.3.1 holds for this choice of function spaces. Therefore, one may apply in a straightforward manner the abstract results of Section 7.4.1 to show the mild, strong and classical well posedness of (7.1) under the appropriate regularity conditions on the data.

▷ Divergence-free solutions. The same choice for $\mathbb{H}$ and $\mathbb{H}_{\mathsf{M}}$ as above, $(\mathbb{H}, \mathbb{H}_{\mathsf{M}}) = (\mathbb{X}, \mathbb{X}_{\mathsf{M}})$ may yield solutions with divergence in $\mathbb{X}$ or even divergence-free solutions. For instance, if $u(0) \in \mathbb{X}$ and $\mathsf{J}_{\mathsf{A}} \in C^1([0,T];\mathbb{X})$, then by the abstract theory $u(t) \in \mathbb{X}_{\mathsf{M}}$ for all t, and taking the divergence of the Maxwell equations (weakly), we obtain in conjunction with the *equation of continuity* that u has well-defined divergence in $\mathbb{X}$, with divergence properties that are determined by the initial data (see Remark 7.3.2).

To obtain solutions for which the divergence can be defined, other choices for the functional setting are possible. For example, one may choose one of the following.

▷ $\mathbb{H} = \mathbb{X}$ and $\mathbb{H}_{\mathsf{M}} = (H_0(\mathrm{curl}, \mathcal{O}) \cap H(\mathrm{div}, \mathcal{O})) \times (H(\mathrm{curl}, \mathcal{O}) \cap H(\mathrm{div}, \mathcal{O}))$. Noting that $\mathsf{M} : \mathbb{H}_{\mathsf{M}} \to \mathbb{X}$ is a semigroup generator, apply the abstract results of Section 7.4.1 to obtain solutions with well-defined divergence without resorting to the charge conservation equation. These results extend to the operator M_{A}.

▷ $\mathbb{H} = \mathbb{X}$ and $\mathbb{H}_{\mathsf{M}} = \mathfrak{X}_{\mathsf{M}} = \big(H_0(\mathrm{curl}, \mathcal{O}) \cap H(\mathrm{div}0, \mathcal{O})\big) \times \big(H(\mathrm{curl}, \mathcal{O}) \cap H_0(\mathrm{div}0, \mathcal{O})\big)$. This choice guarantees that the fields are divergence free. Then $\mathsf{M} : \mathbb{H}_{\mathsf{M}} \to \mathbb{H}$ is a semigroup generator (see Theorem 3.9.7; see also [138]). These results extend to the operator M_{A} in the case where A_{or} or A_0 have constant coefficients, or in the case of spatially dependent but smooth enough coefficients but working in the weighted versions of the $H(\mathrm{div}0, \mathcal{O})$ spaces.

▷ Still other choices are possible; see, e.g., [92], [93], [282], [450].

7.5 OTHER POSSIBLE APPROACHES TO SOLVABILITY

We now turn to the discussion of other possible approaches to the solvability of equation (7.1)

7.5.1 Evolution families approach

More compact expressions of the solution of Volterra-type problems of forms more general than (7.3) can be found in [133], [169], [170], [357] in terms of the theory of resolvent operators. According to this theory, the unique classical solution of (7.3) is given by

$$u(t) = R(t)u_0 + \int_0^t R(t-s)\mathsf{J}_{\mathsf{A}}(s)\, ds, \quad t \in [0,T],$$

where $\{R(t)\}_{t\geq 0}$, is the resolvent operator family that is admitted for (7.3).

We have the following result, along the lines of [133]:

THEOREM 7.5.1 *Under the assumptions of Theorem 7.4.12, (7.3) admits a resolvent operator $\{R(t)\}_{t\geq 0}$ and has a classical unique solution given by*

$$u(t) = R(t)u_0 + \int_0^t R(t-s)\mathsf{J}_{\mathsf{A}}(s)\, ds, \quad t \in [0,T].$$

Proof. (Sketch) Since, by assumption, $\mathsf{G}_{\mathsf{A}}(t)$ is continuous as an operator from $\mathbb{H}_{\mathsf{M}_\mathsf{A}}$ to $\mathbb{H}$ for any $t \geq 0$, we see that Hypothesis H2 of [133] is fulfilled, and the result follows by application of the abstract results of [133]. □

For a similar result in the whole of $\mathbb{R}^3$, see Chapter 10.

In many applications we prefer the expression of the solution of (7.3) using semigroup theory instead of using the theory of resolvent operators, even if in the latter case the solution has a simpler and more compact form. The resolvent operator $\{R(t)\}_{t\geq 0}$ is an abstract mathematical object, whereas the unitary group $(\mathsf{T}_{\mathsf{M}_\mathsf{A}}(t))_{t\in\mathbb{R}}$, generated by the Maxwell operator, is quite well studied in the literature. Generally, in [173], several results concerning the expansion of a solution, which is expressed using semigroup theory, are proved. These results may lead to a numerical scheme.

7.5.2 A Faedo-Galerkin approach

An alternative approach to the well posedeness problem is to employ a Faedo-Galerkin approach. This will result in a finite-dimensional approximation of the evolution equation. The Faedo-Galerkin approach is a very versatile approach for the study of evolution problems that has interesting applications in numerical analysis and can be generalised for the study of nonlinear problems.

The following theorem, which follows the general lines of [70], provides a result concerning the solvability of the Maxwell equations in chiral media based on the Faedo-Galerkin method.

THEOREM 7.5.2 *Under the assumptions of Theorem 7.4.8, equation (7.1) has a unique weak solution $u \in W^{1,\infty}([0,T];\mathbb{X}) \cap L^\infty([0,T];\mathbb{X}_\mathsf{M})$, satisfying*

$$\| u \|_{L^\infty([0,T];\mathbb{X}_\mathsf{M})} + \| u' \|_{L^\infty([0,T];\mathbb{X})} \leq C\,(\| j \|_{W^{1,1}([0,T];\mathbb{X})} + \| u(0) \|_{\mathbb{X}_\mathsf{M}}).$$

Proof. (Sketch) We will work in terms of the weak form of the evolution equation and use the Faedo-Galerkin method to provide an existence result.

Let e_n, $n = 1, 2, \cdots$, be an orthonomal basis of $\mathbb{X}_\mathsf{M}$. We will first look for solutions in $V_m = span(e_1, \cdots, e_m)$, which is a finite-dimensional subspace of $\mathbb{X}_{\mathsf{M}}$. This candidate solution will be of the form $u_m = \sum_{i=1}^{m} u^{(i)}\, e_i$, where $u^{(i)}$ are scalar functions. To avoid introducing cumbersome notation, we will identify $u_m \in V_m$ with the m-dimensional vector $u_m = (u^{(1)}, \cdots, u^{(m)})$ and use the same notation for both. We substitute this expression into the weak form of the equation and then, using as test functions the elements of the basis e_i, one by one, we obtain, after integrating once with respect to time, a system of m integral equations of the Volterra type for the scalar functions, of the form

$$Au_m + K \star u_m - \int_0^t M u_m(s)ds = \int_0^t j_m(s)ds, \tag{7.6}$$

where $u_m = (u^{(1)}, \cdots, u^{(m)}) \in \mathbb{R}^m$, $A, M, K \in \mathbb{R}^{m\times m}$ and j_m is the projection of j on V_m. The matrix functions A, M, K are defined through the relations

$$(Au_m)_i := (\mathsf{A}_{\text{or}}\, u_m, e_i),\ (K \star u_m)_i := (\mathsf{G}_{\text{d}} \star u_m, e_i)\,,\ (Mu_m)_i := (\mathsf{M}\, u_m, e_i)\,,$$

where the subscript i denotes the corresponding coordinate of the vector $\mathfrak{S} u_m$, $\mathfrak{S}$ being a proxy for A, K, M.

The solvability of the finite-dimensional system (7.6) is obtained by the standard theory of the Volterra equations (see, e.g., Lemma 1.1 in [70]). The time regularity of the solution u_m is dictated by the regularity of the kernel function K, as well as by the regularity of the source term[4].

We now obtain a priori estimates for the projections of the solution in V_m, independent of the dimension m. To this end, in the weak form of the equation, and assuming $u \in V_m$ as the solution, we use u as a test function. By the properties of the Maxwell operator we see that

$$\int_{\mathcal{O}} (\mathsf{A}_{\text{or}} u_m + \mathsf{G}_{\text{d}} \star u_m)' \cdot u_m\, dx = \int_{\mathcal{O}} j_m \cdot u_m\, dx.$$

Integrating over $[0,t]$ we get the estimate

$$\begin{aligned} \int_{\mathcal{O}} \mathsf{A}_{\text{or}}\, u_m(t) \cdot u_m(t)\, dx &\le \int_{\mathcal{O}} \mathsf{A}_{\text{or}}\, u_m(0) \cdot u_m(0)\, dx \\ &+ \int_0^t \Theta(s)\, U_m^2(s)\, ds + \|\, j_m \,\|_{L^1([0,t];\mathbb{X})}\, U_m(t)\,, \end{aligned} \tag{7.7}$$

where $U_m(t) = \sup_{s\in[0,t]} \|\, u_m(s) \,\|_{\mathbb{X}}$ and

$$\Theta(s) = \|\, \mathsf{G}_{\text{d}}(0) \,\|_{L^\infty(\mathcal{O})} + \int_0^t \|\, \mathsf{G}_{\text{d}}'(s) \,\|_{L^\infty(\mathcal{O})}\; ds\,.$$

By Gronwall's inequality (see Section A.11 in Appendix A) we obtain the following bound for the finite-dimensional approximation of the solution[5],

$$\|\, u_m \,\|_{L^\infty([0,\,T];\,\mathbb{X})} \le C_1\, (\, \|\, j \,\|_{L^1([0,\,T];\,\mathbb{X})} + \|\, u(0) \,\|_{\mathbb{X}_{\mathsf{M}}})\,,$$

[4]In fact, u_m inherits the regularity of the above-mentioned data of the problem, that is, $u_m \in W^{r,1}((0,T),\mathbb{R}^m)$, as long as $K \in W^{r,1}((0,T),\mathbb{R}^{m\times m})$ and $\int^t j(s)\, ds \in W^{r,1}((0,T),\mathbb{R}^m)$.

[5]And taking into account the obvious estimate $\|\, j_m \,\|_{L^1([0,\,T];\,\mathbb{X})} \le \|\, j \,\|_{L^1([0,\,T];\,\mathbb{X})} < \infty$.

where the bound is uniform in m.

Similar arguments applied to the weak form of the equation, differentiated twice with respect to t, provide the estimate

$$\| u_m' \|_{L^\infty([0,T];\mathbb{X})} \leq C_2 \left(\| j \|_{W^{1,1}([0,T];\mathbb{X})} + \| u(0) \|_{\mathbb{X}_{\mathsf{M}}} \right),$$

where C_2 is a constant depending on the data of the problem but not on the dimension m of the approximation.

We now consider the approximation $u_m = \sum_{i=1}^m u^{(i)} e_i$ as a sequence in $\mathbb{X}$. The above estimates allow us to conclude that if we consider the sequence $\{u_m\}$ of finite-dimensional approximations to the solution of the problem, and allow $m \to \infty$, then the sequences $\{u_m\}$ and $\{u_m'\}$ are bounded sequences in the function space $L^\infty([0,T];\mathbb{X})$. By standard results (see Section A.2 in Appendix A), this guarantees the existence of a subsequence $\{u_{m_k}\}$ that converges in the weak-$*$ topology to a limit u such that $u_{m_k} \stackrel{*}{\rightharpoonup} u$ and $u_{m_k}' \stackrel{*}{\rightharpoonup} u'$ in $L^\infty([0,T];\mathbb{X})$. The finite-dimensional approximations u_m satisfy

$$\int_{\mathcal{O}} (\mathsf{A}_{\text{or}} u_m + K \star u_m)' \cdot \phi_n \, dx = -\int_{\mathcal{O}} u_m \cdot \mathsf{M}\phi_n \, dx - \int_{\mathcal{O}} j \cdot \phi_n \, dx$$

for every $\phi_n \in V_n$. We first take the limit as $n \to \infty$, and by the density of V_n in $\mathbb{X}_{\mathsf{M}}$, we find that

$$\int_{\mathcal{O}} (\mathsf{A}_{\text{or}} u_m + K \star u_m)' \cdot v \, dx = -\int_{\mathcal{O}} u_m \cdot \mathsf{M}v \, dx - \int_{\mathcal{O}} j \cdot v \, dx \tag{7.8}$$

for every $v \in \mathbb{X}_{\mathsf{M}}$. We now take the limit as $m \to \infty$, and by the weak star convergence results of u_m and u_m', we see that the limit u satisfies (7.8) for every $v \in V$. By an integration by parts, we see that this is equivalent to

$$\int_{\mathcal{O}} (\mathsf{A}_{\text{or}} u + \mathsf{G}_{\text{d}} \star u)' \cdot v \, dx = \int_{\mathcal{O}} \mathsf{M}u \cdot v \, dx - \int_{\mathcal{O}} j \cdot v \, dx, \tag{7.9}$$

and by the density of $\mathbb{X}_{\mathsf{M}}$ in $\mathbb{X}$, we deduce that (7.9) is equivalent to the weak form of equation (7.1). Solving the equation for $\mathsf{M}u$ provides a bound for the $L^\infty([0,T];\mathbb{X})$ norm of curl$u$. The Banach-Steinhaus theorem (uniform boundedness principle) (see Section A.7 in Appendix A) is essential in showing that the weak limit of $\mathsf{M}u_m$ is $\mathsf{M}u$. □

The choice of the basis e_i is an important issue. An interesting choice would be to use the eigenfunctions of the curl operator or the eigenfunctions of the Maxwell operator. Such a choice would simplify immensely the form of the finite-dimensional approximations in terms of Volterra integral equations, and this will be important if one wishes to use this approximation for the numerical analysis of the problem. One particular choice may be in terms of the eigenfunctions of the curl operator, proposed by Moses [326]. The use of these eigenfunctions for the study of chiral media has been proposed in [151].

The Faedo-Galerkin approach may be used for the study of more regular solutions, imposing higher regularity on the data. On the other hand, the

Faedo-Galerkin method may be used within a variational formulation in order to provide solutions of the equations under possibly weaker assumptions on the data, and may also prove a very useful tool for numerical approximation of the equations, by generalising the discussion of Section 4.6 to the time-dependent problem.

7.6 MISCELLANEA

7.6.1 More on the semigroup approach

It is possible to employ the semigroup approach directly by extending the phase space of the original problem. Then, in the proper functional setting, (7.3) is equivalent to an abstract ODE,

$$z' = \mathscr{C} z, \tag{7.10}$$

where $z = (w, u)^{tr}$ is an element of the extended phase space (u being the original unknown six-vector function and w being a new variable incorporating the effects of the convolution terms) and $\mathscr{C}$ is the generator of the extended semigroup. This semigroup comprises the right shift semigroup (whose generator is the derivative operator with respect to the time variable) and the semigroup generated by M_{A}. A semigroup approach for (7.10) yields the required well posedness results.

This approach has been employed for general linear integrodifferential equations in Banach spaces ([53], [133], [321]; see also the monograph [141]).

7.6.2 Pseudoparabolic equations

Equations of the general form

$$\mathfrak{L}u' = \mathfrak{M}u + \mathfrak{f} \tag{7.11}$$

(and nonlinear extensions) have been studied for a long time by a variety of methods. They are of interest not only for the sake of generalisations but also because they arise naturally in a vast variety of applications (e.g., in acoustics, electromagnetics, viscoelasticity, heat conduction, etc.). They have been called *Sobolev-type equations*, the term generally denoting equations or systems in which spatial derivatives are mixed with the time derivative of highest order. For a descriptive review of work prior to 1976 and many historic examples, see [89]. In particular, equation (7.11) is called *strongly regular* if $\mathfrak{L}^{-1}\mathfrak{M}$ is continuous, *weakly regular* if $\mathfrak{L}$ is invertible but does not dominate $\mathfrak{M}$, and *degenerate* if $\mathfrak{L}$ is not invertible (see [399]). Strongly regular Sobolev equations are also widely known as *pseudoparabolic*. Using techniques from the theory of Sobolev-type equations, we may consider the low chirality limit for the Maxwell equations (see [283]).

Chapter Eight

Controllability

8.1 INTRODUCTION

In this chapter we present some issues concerning the controllability of the Maxwell equations for complex media in the time domain. Controllability issues constitute an important class of problems that present considerable interest both from the mathematical as well as from the applications point of view.

The structure of this chapter is as follows. In Section 8.2 we formulate the problem and discuss our main strategy for its treatment, using a fixed point approach. This approach is based on the controllability problem for the Maxwell equations and treats the integrodifferential terms in the constitutive relations for the complex media as a perturbation. In Section 8.3 we present some results on the controllability of the Maxwell system that are essential in the study of the controllability problem in complex media. In Section 8.5 we provide details on the solution of the controllability problem for complex media using the fixed point scheme. In this chapter our main approach is a constructive one; however, in Section 8.6.1 we provide an abstract approach to controllability based on semigroup arguments, which allows us to present our results in a unified framework. In Section 8.6.2 we briefly discuss the problem of boundary controllability. In Section 8.6.3 we discuss the problem of optimal control for Maxwell equations. Finally, in Section 8.6.4 we present miscellaneous results concerning controllability for time-harmonic problems, etc.

8.2 FORMULATION

Our starting point is the general control system (2.49) derived in Section 2.4.7 (and restated here for convenience):

$$u' = \mathsf{M}_{\text{A}}\, u + \mathsf{G}_{\text{A}} \star u + \mathsf{J}_{\text{A}} + \mathcal{B}\, v, \tag{8.1}$$

where $u = (E, H)^{tr}$ is the state of the system, $v = (v_1, v_2)^{tr}$ is the control, and $\mathcal{B}$ is an operator quantifying how the control v affects the system's evolution equation. Although $\mathcal{B}$ can in general be an unbounded, or a noninvertible, operator, to avoid technicalities we assume here that it is a bounded and boundedly invertible operator. As usual, we will consider the Maxwell equations as an abstract ODE in the appropriate function setting. For instance, for internal controllability problems we will consider (8.1) an abstract ODE in the function space

$$\mathfrak{X}_{\mathsf{M}} := \big(H_0(\mathrm{curl}, \mathcal{O}) \cap H(\mathrm{div}0, \mathcal{O})\big) \times \big(H(\mathrm{curl}, \mathcal{O}) \cap H_0(\mathrm{div}0, \mathcal{O})\big)$$

which incorporates the boundary conditions, as well as the divergence-free property of the fields[1].

The main object of this chapter is the problem of the controllability of system (8.1), which can now be stated as follows:

> *Given $T > 0$, an initial condition $u(0) = U_0$ and a final condition $u(T) = U_T$, can we find a control procedure $v^\flat(\cdot)$ such that the solution of the system (8.1) with $v(\cdot) = v^\flat(\cdot)$ satisfies $u(0) = U_0$ and $u(T) = U_T$?*

This problem has been treated in detail for a large number of linear equations, with or without convolution terms (see, e.g., [244], [290], [291]). Although extensive work on the subject of controllability of the Maxwell equations exists (see, e.g., [139], [140], [223], [264], [336], [350], [422]), the subject is still "virgin" for the case of complex media (with the exceptions of [116], [117], [118], [195]). The linear problem with the inclusion of a convolution term can be treated exactly; nevertheless, we choose to adopt here a perturbative approach based on the *Hilbert uniqueness method*, mainly because this method can be easily extended to treat the problem of controllability of nonlinear systems. However, for the sake of completeness, we will sketch the alternative approach in a separate section. For detailed presentations of the Hilbert uniqueness method, developed by J.-L. Lions, we refer the reader to [265], [290], [291], [453].

Our approach is based on the following fixed point scheme. Assume that the state space of the system is a Hilbert space $\mathbb{H}$. At this point we do not specify the exact nature of this Hilbert space; we will return to this later. Fix a function $z(\cdot) \in \mathbb{H}$ and consider the linear system

$$u' = \mathsf{M}_{\mathrm{A}}\, u + \mathsf{J}^{\square}_{\mathrm{A}} + \mathcal{B}\, v, \tag{8.2}$$

where $\mathsf{J}^{\square}_{\mathrm{A}} = \mathsf{J}_{\mathrm{A}} + \mathsf{G}_{\mathrm{A}} \star z$. Assuming for the moment the controllability of the linear system (8.2), we define as $v_z(\cdot)$ the control procedure that is needed to drive the system (8.2) from $u(0) = U_0$ to $u(T) = U_T$. Let $u_z(\cdot)$ be the solution of

$$u' = \mathsf{M}_{\mathrm{A}}\, u + \mathsf{J}^{\square}_{\mathrm{A}} + \mathcal{B}\, v_z, \tag{8.3}$$

with initial condition $u(0) = U_0$. By the definition of v_z, we know that $u(T) = U_T$. Define the map $\mathsf{F} : C([0,T];\mathbb{H}) \to C([0,T];\mathbb{H})$ by

$$F(z(\cdot)) = u(\cdot)$$

If this map has a fixed point, i.e., if there exists a function $\mathfrak{u} \in C([0,T];\mathbb{H})$ such that $\mathsf{F}(\mathfrak{u}) = \mathfrak{u}$, then this $\mathfrak{u}$ is the solution of the controlled system (8.1)

[1] This choice is not restrictive; other choices are possible (e.g., the function spaces used in Chapter 7 may be used). However, we make this choice here to show some alternative results under modified assumptions on the properties of the fields.

that connects the states U_0 and U_T in time T, and furthermore v_{u} is the required control $v^{\flat}$.

There are several possible treatments of the problem of showing that the map F has a fixed point. For instance, one may use the Banach contraction mapping theorem for this purpose. This, in general, will work either for small enough initial data, or small enough T, or small enough amplitudes of the chirality coefficients, or a proper combination of these. Another alternative is to use fixed point theorems that are of a more global nature, such as the Schauder fixed point theorem. The application of this theorem may allow us to obtain controllability results that are less restrictive in terms of parameter values or in terms of the initial data. For this reason we choose to adopt this approach here.

8.3 CONTROLLABILITY OF ACHIRAL MEDIA: THE HILBERT UNIQUENESS METHOD

From the above discussion it is clear that the first main step in the study of controllability of the Maxwell equations in complex media is the study of the achiral case, i.e., when $\mathsf{G}_{\mathrm{A}} = 0$. To this end we consider the controllability of the system

$$u' = \mathsf{M}_{\mathrm{A}}\, u + \mathsf{J}_{\mathrm{A}} + \mathcal{B}\, v. \tag{8.4}$$

The question is to construct a control procedure v such that the system is driven from U_0 to U_T in time T.

We will approach this problem using the Hilbert uniqueness method. This requires the introduction and study of some auxiliary problems.

Consider first the backward adjoint problem

$$\begin{aligned} -\Phi' &= \mathsf{M}_{\mathrm{A}}^{*}\Phi, \\ \Phi(T) &= \Phi_T, \end{aligned} \tag{8.5}$$

where by $\mathsf{M}_{\mathrm{A}}^{*}$ we denote the adjoint of M_{A}. By the properties of the Maxwell operator, within the functional setting adopted here, we have that $\mathsf{M}_{\mathrm{A}}^{*} = -\mathsf{M}_{\mathrm{A}}$, so that the backward adjoint problem (8.5) is equivalent to

$$\begin{aligned} \Phi' &= \mathsf{M}_{\mathrm{A}}\Phi, \\ \Phi(T) &= \Phi_T. \end{aligned} \tag{8.6}$$

Assuming temporarily that this problem is well posed (we return to this issue shortly; see Section 8.4.1), we may obtain the function Φ.

For this function Φ we consider the following auxiliary forward problem:

$$\begin{aligned} \Psi' &= \mathsf{M}_{\mathrm{A}}\Psi + \mathcal{G}\,\Phi, \\ \Psi(0) &= 0, \end{aligned} \tag{8.7}$$

where $\mathcal{G}$ is some operator that will be specified in due course. Assuming for the time being the well posedness of this problem, as we did for the adjoint backward problem (see Section 8.4.1), we may find the final condition $\Psi(T)$.

We now define the operator Λ by $\Lambda(\Phi_T) = \Psi(T)$. This map connects the final condition, Φ_T, of the backward adjoint problem (8.6) to the final state, $\Psi(T)$, of the forward nonhomogeneous problem (8.7). We must be careful with the domain and the range of this map; however, for the time being we define the map as $\Lambda : \mathbb{H}_1 \to \mathbb{H}_1$, where $\mathbb{H}_1$ is an appropriate subset of the state space $\mathbb{H}$.

Finally, we consider the auxiliary system

$$\begin{aligned} V' &= \mathsf{M}_{\text{A}} V + \mathsf{J}_{\text{A}}, \\ V(0) &= U_0. \end{aligned} \tag{8.8}$$

This is nothing else but the uncontrolled Maxwell system, starting at U_0. The solution of this system will provide us with $V(T)$, which is the state that the original system would get to in the absence of control.

By the linearity of the equations we observe that if we set $U = V + \Psi$, then the function U will satisfy the system

$$U' = \mathsf{M}_{\text{A}}\, U + \mathsf{J}_{\text{A}} + \mathcal{G}\,\Phi,$$

which, if $\mathcal{G} = \mathcal{B}$, coincides with the original control system. Now let us consider the initial and final states. For the initial state we have

$$U(0) = V(0) + \Psi(0) = U_0.$$

For the final state we have $U(T) = V(T) + \Psi(T)$. Therefore, if $\Psi(T) = U_T - V(T)$, then $U(T) = U_T$, which is the required final state. This would result in the solution of the problem, since if $\Phi^\flat$ is the solution of (8.6) such that the solution of (8.7) satisfies $\Psi(T) = U_T - V(T)$, then $\Phi^\flat$ is the desired control v.

It remains to see whether such a solution $\Phi^\flat$ indeed exists. To this end, we need to use the map Λ, which connects the final states of the adjoint backward and forward systems. Suppose that the operator equation

$$\Lambda(\Phi_T^\flat) = U_T - V(T) \tag{8.9}$$

is satisfied for some $\Phi_T^\flat \in \mathbb{H}_1$. Then this $\Phi_T^\flat$ is the proper final condition for the adjoint system (8.6). The construction of the control then is given by the following procedure:

- ▷ Solve (8.8) with initial condition U_0 to obtain $V(T)$.
- ▷ Solve the operator equation (8.9) to obtain $\Phi_T^\flat$.
- ▷ Solve the backward adjoint equation (8.6) with final condition $\Phi_T^\flat$ to obtain the solution $\Phi^\flat(\cdot)$.
- ▷ The required control is $v^\flat = \Phi^\flat$.
- ▷ The desired path is given by the solution of (8.4) with the control $v^\flat = \Phi^\flat$.

Therefore, the whole problem reduces to a detailed understanding of the mapping Λ and the solvability of the operator equation (8.9). If this equation has a solution for any U_T, then, assuming always the well posedness of the auxiliary problems, it is possible to find a control procedure $v(\cdot)$ that allows us to steer the system (8.4) from any initial state U_0 to any final state U_T within the required time T. In this case we say that the system is *exactly controllable.* The control procedure constructed above is a *minimal-norm control*[2].

From the above discussion, we see that exact controllability of the system is equivalent to the invertibility of the operator Λ on the whole of $\mathbb{H}_1$. The Hilbert uniqueness method essentially consists in trying to define appropriate Hilbert spaces $\mathbb{H}_1$, endowed with the proper norms, so as to guarantee the invertibility of Λ. We treat this problem following [196], in the specific setting of the Maxwell equations, and avoid a general abstract discussion to keep the presentation as clear and as self-contained as possible. For an excellent and extremely clear discussion of the general abstract approach, we refer the interested reader to [112]. See also [414].

8.4 THE FORWARD AND BACKWARD PROBLEMS

8.4.1 Well posedness

We first establish that both the backward and the forward problems (8.6) and (8.7) are well posed. Recall that

$$\begin{aligned}
&\mathsf{X} := (L^2(\mathcal{O}))^3, \ \mathbb{X} := \mathsf{X} \times \mathsf{X}, \\
&\mathfrak{X}_1 := H_0(\mathrm{curl}, \mathcal{O}) \cap H(\mathrm{div}0, \mathcal{O}), \ \ \mathfrak{X}_2 := H(\mathrm{curl}, \mathcal{O}) \cap H_0(\mathrm{div}0, \mathcal{O}), \\
&\mathfrak{X}_{\mathsf{M}} := \mathfrak{X}_1 \times \mathfrak{X}_2 ,
\end{aligned}$$

and define

$$\mathbb{W}^{1,1} := C([0,T];\mathbb{X}) \cap W^{1,1}([0,T],\mathbb{X}) .$$

THEOREM 8.4.1 *(i)Let $\Phi_T \in \mathfrak{X}_{\mathsf{M}}$. Then the backward problem (8.6) (and hence (8.5)) is well posed in $C([0,T];\mathfrak{X}_{\mathsf{M}}) \cap C^1([0,T];\mathbb{X})$.*
(ii)Let $\mathcal{B} \in \mathcal{L}(\mathbb{W}^{1,1},\mathbb{W}^{1,1})$ be invertible. Then the nonhomogeneous forward problem (8.7) with $\mathcal{G} = \mathcal{B}$, namely,

$$\begin{aligned}
&\Psi' = \mathsf{M}_A \Psi + \mathcal{B}\,\Phi, \\
&\Psi(0) = 0,
\end{aligned}$$

is well posed in $C([0,T];\mathfrak{X}_{\mathsf{M}}) \cap C^1([0,T];\mathbb{X})$.

Proof. (Sketch) The proof is based on the facts that within the functional setting employed, we have (a) $\mathsf{M}_{\mathsf{A}}^* = -\mathsf{M}_{\mathsf{A}}$, and (b) M_{A} is the generator of

[2] In general the controllability problem does not have a unique solution unless an extra condition is required, that of choosing the control of minimal norm. This coincides with the control constructed by the Hilbert uniqueness method described above (see, e.g., [291].

a C_0 group. The group property of $e^{t\mathsf{M}_A}$ guarantees that the properties of the forward and the backward problem are equivalent. The rest follows by standard semigroup arguments. □

By the regularity properties of the Maxwell equations (which follow easily from semigroup arguments; see Appendix A, Theorem A.8.14), we see that the map Λ is well defined as a map from $\mathfrak{X}_M \to \mathfrak{X}_M$. It remains to study the invertibility of Λ. We will approach this problem using an application of the Lax-Milgram lemma.

To this end, we need the following auxiliary results:

8.4.2 An estimate for the backward problem

Consider first the backward problem

$$\begin{aligned} -v' &= \mathsf{M}_A^* v, \\ v(T) &= \alpha, \end{aligned} \tag{8.10}$$

which is a final value problem for some $\alpha \in \mathfrak{X}_M$. Here, M_A^* is the adjoint of M_A in $D(\mathsf{M}_A) = \mathfrak{X}_M$.

LEMMA 8.4.2 *Consider any $\tau \in [0, T]$. The solution to the backward problem (8.10) satisfies an estimate of the form*

$$||\alpha||^2_{\mathfrak{X}_M} \leq C\rho_1(\tau) \int_{T-\tau}^{T} ||v(t)||^2_{\mathfrak{X}_M} dt, \tag{8.11}$$

where $\rho_1(\tau)$ blows up as $\tau \to 0$.

Proof. Recall that for the specific choice of $\mathfrak{X}_M$ considered here, the operator M_A has the property $\mathsf{M}_A^* = -\mathsf{M}_A$, so that the backward adjoint system assumes the form

$$\begin{aligned} v' &= \mathsf{M}_A\, v, \\ v(T) &= \alpha. \end{aligned} \tag{8.12}$$

Therefore, we may use the results for the backward continuation of the Maxwell system. In fact, since the Maxwell operator generates a C_0 group, these results are straightforward adaptations of the results for the forward system. Now let us denote $\alpha := (\alpha_1, \alpha_2)^{tr}$ $v := (v_1, v_2)^{tr}$. Assuming for simplicity, and without loss of generality, that $\varepsilon = \mu = 1$, we have

$$\partial_t^2 v_i + \operatorname{curl}\operatorname{curl} v_i = 0, \quad i = 1, 2. \tag{8.13}$$

We will use the following known fact (see, e.g., [334]): there exist $C_1 > 0$ and $C_2 > 0$ that depend only on $\mathcal{O}$, such that for any $w_1 \in \mathfrak{X}_1$, one has

$$||w_1||_{\mathfrak{X}_1} \leq C_1 ||\mathrm{curl} w_1||_{\mathsf{X}}, \tag{8.14}$$

and for any $w_2 \in \mathfrak{X}_2$, one has

$$||w_2||_{\mathfrak{X}_2} \leq C_2 ||\mathrm{curl} w_2||_{\mathsf{X}}. \tag{8.15}$$

Working in a similar fashion with the corresponding problem in [182], we multiply the identity

$$\partial_t^2 v_1 + \operatorname{curl}\operatorname{curl} v_1 = 0 \tag{8.16}$$

by $\rho(t)v_1$, where $\rho(t) := (T-\tau-t)^2(T-t)^2$, and integrate over $(T-\tau, T)\times\mathcal{O}$.

We get

$$\int_{T-\tau}^{T}\int_{\mathcal{O}} \rho' v_1 \partial_t v_1 dx\, dt + \int_{T-\tau}^{T}\int_{\mathcal{O}} \rho|\partial_t v_1|^2 dx\, dt = \int_{T-\tau}^{T}\int_{\mathcal{O}} \rho|\operatorname{curl} v_1|^2 dx\, dt.$$

Noting that

$$\int_{T-\tau}^{T}\int_{\mathcal{O}} \rho' v_1 \partial_t v_1 dx\, dt = \int_{T-\tau}^{T}\int_{\mathcal{O}} \frac{\rho'}{\sqrt{\rho}}\sqrt{\rho} v_1 \partial_t v_1 dx\, dt,$$

and using standard inequalities, we see that for all $\lambda > 0$,

$$\begin{aligned}\int_{T-\tau}^{T}\int_{\mathcal{O}} \rho(t)|\operatorname{curl} v_1|^2 dx\, dt &\le \int_{T-\tau}^{T}\int_{\mathcal{O}} \rho(t)|\partial_t v_1|^2 \\ &+\frac{\lambda}{2}\sup_{T-\tau\le t\le T}\frac{\rho'(t)^2}{\rho(t)}\int_{T-\tau}^{T}\int_{\mathcal{O}} |\partial_t v_1|^2 dx\, dt + \frac{1}{2\lambda}\int_{T-\tau}^{T}\int_{\mathcal{O}} \rho(t)|v_1|^2 dx\, dt.\end{aligned} \tag{8.17}$$

Using then (8.14), which implies that there exists a constant C_3 such that $||v_1|||_{\mathbb{X}} \le C_3\, ||\operatorname{curl} v_1||_{\mathbb{X}}$ (see, e.g., Corollary 3.51 in [324]), by choosing λ large enough,

$$\begin{aligned}\int_{T-\tau}^{T}\int_{\mathcal{O}} \rho(t)|\operatorname{curl} v_1|^2 dx\, dt \le C\Bigg(&\int_{T-\tau}^{T}\int_{\mathcal{O}} \rho(t)|\partial_t v_1|^2 dx\, dt \\ &+ \sup_{T-\tau\le t\le T}\frac{\rho'(t)^2}{\rho(t)}\int_{T-\tau}^{T}\int_{\mathcal{O}} |\partial_t v_1|^2 dx\, dt\Bigg),\end{aligned} \tag{8.18}$$

for some constant C that depends only on $\mathcal{O}$.

Next we establish that, for $i = 1, 2$,

$$\int_{T-\tau}^{T}\int_{\mathcal{O}} \rho(t)(|\partial_t v_i|^2 + |\operatorname{curl} v_i|^2) dx\, dt = ||\operatorname{curl}\alpha||_{\mathbb{X}}^2 \int_{T-\tau}^{T} \rho(t) dt, \tag{8.19}$$

where we use the notation $\operatorname{curl}\alpha = (\operatorname{curl}\alpha_1, \operatorname{curl}\alpha_2)^{tr}$.

Adding $\int_{T-\tau}^{T}\int_{\mathcal{O}} \rho(t)|\operatorname{curl} v_1|^2 dx\, dt$ to both sides of (8.18), using (8.19), we obtain the inequality

$$\begin{aligned}||\operatorname{curl}\alpha||_{\mathbb{X}}^2 \int_{T-\tau}^{T} \rho(s) ds &\le C\int_{T-\tau}^{T}\int_{\mathcal{O}} \Big(\rho(t)|\partial_t v_1|^2 + \rho_1(\tau, T)|\partial_t v_1|^2\Big) dx\, dt \\ &+ \int_{T-\tau}^{T}\int_{\mathcal{O}} \rho(t)|\operatorname{curl} v_1|^2 dx\, dt \le C\int_{T-\tau}^{T}\int_{\mathcal{O}} (|\operatorname{curl} v_1|^2 + |\operatorname{curl} v_2|^2) dx\, dt,\end{aligned} \tag{8.20}$$

where $\rho_1(\tau, T) := \sup_{T-\tau\le t\le T}\frac{\rho'(t)^2}{\rho(t)}$, for some constant C that depends on the suprema of the functions ρ and $\frac{(\rho')^2}{\rho}$. In the above we used the fact that

$|\partial_t v_1|^2 = |\mathrm{curl} v_2|^2$, as dictated by the Maxwell equations, and we estimated (for $i = 1, 2$)

$$\int_{T-\tau}^{T}\int_{\mathcal{O}} \rho(t)\,|\mathrm{curl} v_i|^2\,dx\,dt \le \sup_{T-\tau\le t\le T} \rho(t) \int_{T-\tau}^{T}\int_{\mathcal{O}} |\mathrm{curl} v_i|^2\,dx\,dt\,.$$

Combining (8.20) with (8.14), (8.15) gives the desired result for $\rho_1(\tau) = (\int_{T-\tau}^{T} \rho(s)\,ds)^{-1}$. □

REMARK 8.4.3 Let us note that (8.19) is formally deducible from the fact that the energy of the Maxwell system (written as described above in the form (8.13)) is conserved.

Indeed, by multiplying (8.16) by $\varrho(t)\partial_t v_1$, where $\varrho(t) := \int_{2T-\tau-t}^{T} \rho(s)\,ds$, and integrating over $[T-\tau, T]\times\mathcal{O}$, we get $I_1 + I_2 = 0$, where

$$I_1 = \int_{T-\tau}^{T}\int_{\mathcal{O}} \varrho(t)\,\partial_t v_1\,\partial_t^2 v_1\,dx\,dt,$$

$$I_2 = \int_{T-\tau}^{T}\int_{\mathcal{O}} \varrho(t)\,(\partial_t v_1)\,\mathrm{curl}\mathrm{curl} v_1\,dx\,dt.$$

Observe that $\varrho(t)$ is such that $\varrho(T-\tau) = 0$, $\varrho(T) = \int_{T-\tau}^{T} \rho(s)\,ds$ and $\varrho'(t) = \rho(t)$. In I_1, integrate with respect to t by parts, to obtain

$$\begin{aligned} I_1 &= \int_{\mathcal{O}} \varrho(T)\,|\partial_t v_1|^2_{t=T}\,dx - \int_{T-\tau}^{T}\int_{\mathcal{O}} \varrho'(t)\,|\partial_t v_1|^2\,dx\,dt \\ &\quad - \int_{T-\tau}^{T}\int_{\mathcal{O}} \varrho(t)\,v_1\,\partial_t^2 v_1 dx\,dt \\ &= \varrho(T)\,||\mathrm{curl}\alpha_2||^2_{\mathsf{X}} - \int_{T-\tau}^{T}\int_{\mathcal{O}} \rho(t)\,|\partial_t v_1|^2\,dx\,dt - I_1, \end{aligned}$$

since by the Maxwell equations we have $|\mathrm{curl} v_2|^2 = |\partial_t v_1|^2$. This gives

$$I_1 = \frac{1}{2}\,\varrho(T)\,||\mathrm{curl}\alpha_2||^2_{\mathsf{X}} - \frac{1}{2}\int_{T-\tau}^{T}\int_{\mathcal{O}} \rho(t)\,|\partial_t v_1|^2\,dx\,dt$$

In I_2, integrate by parts with respect to x to obtain

$$I_2 = \int_{T-\tau}^{T}\int_{\mathcal{O}} \varrho(t)\partial_t(\mathrm{curl}\, v_1)\mathrm{curl}\, v_1 dx\,dt = \frac{1}{2}\int_{T-\tau}^{T} \varrho(t)\partial_t(\mathrm{curl} v_1)|\mathrm{curl} v_1|^2\,dxdt,$$

which is then integrated by parts with respect to t to give

$$I_2 = \frac{1}{2}\varrho(T)||\mathrm{curl}\alpha_1||_{\mathsf{X}} - \frac{1}{2}\int_{T-\tau}^{T}\int_{\mathcal{O}} \rho(t)\,|\mathrm{curl} v_1|^2\,dx\,dt.$$

Then $I_1 + I_2 = 0$ gives (8.19).

8.4.3 An estimate for the forward problem

Next we consider the forward problem (8.7), where the nonhomogeneous term is $\mathcal{B}\,v$, with v being the solution of the problem (8.10). We assume that the operator $\mathcal{B}$ is realised as a matrix operator with entries $\mathcal{B}_{ij}$, $i, j = 1, 2$.

We use the notation $\Psi = (\Psi_1, \Psi_2)^{tr}$ where Ψ_i, $i = 1, 2$ are three-vectors. By differentiating with respect to time once and using the divergence-free property of the fields, this is reduced to a vector wave equation of the form

$$\begin{aligned} &\partial_t^2 w + \operatorname{curl}\operatorname{curl} w = \mathcal{B}^{\nabla}\partial_t v + \mathcal{B}^{\triangle} v, \\ &w(T-\tau) = \partial_t w(T-\tau) = 0 \end{aligned} \tag{8.21}$$

where $w = \Psi_1$, $v = (v_1, v_2)^{tr}$ is a solution of the backward problem (8.10) and

$$\begin{aligned} \mathcal{B}^{\nabla}\partial_t v &= (\mathcal{B}_{22} + \mathcal{B}_{11})\,\partial_t v_1 + (\mathcal{B}_{12} - \mathcal{B}_{21})\,\partial_t v_2, \\ \mathcal{B}^{\triangle} v &= \operatorname{grad}\mathcal{B}_{21} \times v_1 + \operatorname{grad}\mathcal{B}_{22} \times v_2. \end{aligned} \tag{8.22}$$

An equivalent formulation in terms of Ψ_2 is possible with different definitions of the operators $\mathcal{B}^{\nabla}$, $\mathcal{B}^{\triangle}$.

The following estimate for (8.21) will be needed.

LEMMA 8.4.4 *Let* $w(T) = (w_1(T), w_2(T))^{tr} =: w_T = (w_{T,1}, w_{T,2})^{tr}$. *Then*

$$||w_T||^2_{\mathfrak{X}_{\mathsf{M}}} \le C \int_{T-\tau}^{T} \int_{\mathcal{O}} \sum_{i=1}^{2} |\partial_t v_i|^2 dx\, dt.$$

Proof. To avoid cumbersome notation, in what follows C is used as a proxy for a constant, the value of which may vary from estimate to estimate. Furthermore, without loss of generality, we assume that $\operatorname{grad}\mathcal{B}_{ij} = 0, i, j = 1, 2$, since otherwise $\mathcal{B}^{\triangle}\,v$ is subordinate[3] to $\mathcal{B}^{\nabla}\,\partial_t v$, as long as $\operatorname{grad}\mathcal{B}_{ij}$ is weakly differentiable.

In a similar fashion as for the wave equation (see, e.g., [144]), multiply the equation by $\partial_t w$ and integrate over $\mathcal{O}$. This gives

$$\begin{aligned} \frac{1}{2}\frac{d}{dt}||\partial_t w_i||^2_{\mathsf{X}} + \frac{1}{2}\frac{d}{dt}||\operatorname{curl} w_i||^2_{\mathsf{X}} &= \int_{\mathcal{O}} \sum_{j=1}^{2} \mathcal{B}_{ij}\partial_t v_j \partial_t w_i dx \\ &\le C\left(||\partial_t w_i||^2_{\mathsf{X}} + \sum_{i=1}^{2} ||\partial_t v_i||^2_{\mathsf{X}}\right). \end{aligned} \tag{8.23}$$

For $i = 1, 2$, define

$$v_i(t) := ||\partial_t w_i||^2_{\mathsf{X}} + ||\operatorname{curl} w_i||^2_{\mathsf{X}}.$$

Then (8.23) implies that

$$v_i'(t) \le C\left(v_i(t) + \sum_{j=1}^{2} ||\partial_t v_j||^2_{\mathsf{X}}\right), \quad T-\tau \le t \le T\,,\ i = 1, 2\,,$$

[3]In the sense that the norm of $\mathcal{B}^{\triangle}\mathsf{v}$ is bounded by the norm $\mathcal{B}^{\nabla}\mathsf{v}$ for any suitable v.

and a straightforward application of Gronwall's inequality gives

$$\sup_{T-\tau\leq t\leq T} v_i(t) \leq C \int_{T-\tau}^{T} \sum_{j=1}^{2} ||\partial_t v_j||_{\mathsf{X}}^2 \, dt$$

By the definition of $v_i(t)$ and the equivalence of the norm $||w||_{\mathfrak{X}_{\mathsf{M}}}$ with $||\mathrm{curl} w||_{\mathsf{X}}$, we obtain the stated result. □

8.4.4 Solvability of the operator equation (8.9)

We consider the map $\Lambda : \mathfrak{X}_{\mathsf{M}} \to \mathfrak{X}_{\mathsf{M}}$ defined by

$$\alpha \mapsto v \mapsto w \mapsto w(T) =: \Lambda(\alpha),$$

where v is the solution of the backward equation (8.10) and w is the solution of the forward equation (8.21).

The following property of Λ is essential for what follows:

LEMMA 8.4.5 *There exists an* $\alpha^\flat$ *such that* $\Lambda(\alpha^\flat) = \mathsf{h}$, *for every* $\mathsf{h} \in \mathfrak{X}_{\mathsf{M}}$.

Proof. We split the proof into four steps.
STEP 1 First, we observe that the operator $\Lambda : \mathfrak{X}_{\mathsf{M}} \to \mathfrak{X}_{\mathsf{M}}$ is a linear operator. This follows easily from the linearity of the adjoint system and the linearity of the forward system.
STEP 2 We next observe that the operator $\Lambda : \mathfrak{X}_{\mathsf{M}} \to \mathfrak{X}_{\mathsf{M}}$ is continuous. Indeed, by definition, $\Lambda(\alpha) = w(T)$. By Lemma 8.4.4, concerning the solutions of the forward problem (8.21), we have

$$||\Lambda(\alpha)||_{\mathfrak{X}_{\mathsf{M}}}^2 \leq \int_{T-\tau}^{T} \int_{\mathcal{O}} \sum_{i=1}^{2} |\partial_t v_i|^2 dx \, dt. \tag{8.24}$$

Working as for the proof of equation (8.19), but choosing $\Psi(t) = t - (T-\tau)$, we obtain for $i = 1, 2$ the identity

$$\int_{T-\tau}^{T} \int_{\mathcal{O}} \left(|\partial_t v_i|^2 + |\mathrm{curl} v_i|^2\right) dx \, dt = \tau \, ||\mathrm{curl}\alpha||_{\mathbb{X}}^2 ,$$

which gives the estimate

$$\int_{T-\tau}^{T} \int_{\mathcal{O}} |\partial_t v|^2 \, dx \, dt \leq C \, ||\alpha||_{\mathfrak{X}_{\mathsf{M}}}^2. \tag{8.25}$$

Combining (8.24) and (8.25), we obtain

$$||\Lambda(\alpha)||_{\mathfrak{X}_{\mathsf{M}}} \leq C \, ||\alpha||_{\mathfrak{X}_{\mathsf{M}}},$$

which proves the continuity of the operator Λ.
STEP 3 We now prove the coercivity of Λ. Multiply the forward equation (8.21) by $\partial_t v_i$, $i = 1, 2$ and integrate over $[T-\tau, T] \times \mathcal{O}$. This gives

$$\begin{aligned} &\int_{T-\tau}^{T} \int_{\mathcal{O}} \partial_t^2 w_1 \partial_t v_1 dx \, dt + \int_{T-\tau}^{T} \int_{\mathcal{O}} \partial_t v_1 \mathrm{curl}\,\mathrm{curl} w_1 dx \, dt = \\ &\mathcal{B}_{11} \int_{T-\tau}^{T} \int_{\mathcal{O}} \mid \partial_t v_1 \mid^2 dx \, dt + \mathcal{B}_{12} \int_{T-\tau}^{T} \int_{\mathcal{O}} \partial_t v_1 \, \partial_t v_2 dx \, dt. \end{aligned} \tag{8.26}$$

Integration of the first term on the left-hand side by parts with respect to t gives

$$I_1 := \int_{T-\tau}^{T}\int_{\mathcal{O}} \partial_t^2 w_1 \partial_t v_1 dx\, dt$$
$$= \int_{\mathcal{O}} \partial_t w_1(T)\partial_t v_1(T) dx - \int_{T-\tau}^{T}\int_{\mathcal{O}} \partial_t^2 v_1\, \partial_t w_1 dx\, dt. \qquad (8.27)$$

The second term on the left-hand side gives

$$I_2 := \int_{T-\tau}^{T}\int_{\mathcal{O}} \partial_t v_1 \mathrm{curl}\,\mathrm{curl} w_1 dx\, dt$$
$$= \int_{\mathcal{O}} v_1 \mathrm{curl}\,\mathrm{curl} w_1 \Big|_{t=T-\tau}^{t=T} dx - \int_{T-\tau}^{T}\int_{\mathcal{O}} v_1 \mathrm{curl}\,\mathrm{curl}\partial_t w_1 dx\, dt$$
$$= \int_{\mathcal{O}} \mathrm{curl} v_1 curl w_1 \Big|_{t=T-\tau}^{t=T} dx - \int_{T-\tau}^{T}\int_{\mathcal{O}} \mathrm{curlcurl} v_1\, \partial_t w_1\, dx\, dt$$
$$= \int_{\mathcal{O}} \mathrm{curl}\alpha_1 \mathrm{curl} w_{\mathrm{T},1} dx - \int_{T-\tau}^{T}\int_{\mathcal{O}} \mathrm{curlcurl} v_1\, \partial_t w_1\, dx\, dt, \qquad (8.28)$$

where we first integrated by parts with respect to t and then integrated by parts with respect to x (once for the first integral and twice for the second).

Combining (8.26)-(8.28) and using the fact the v_1 solves the adjoint equation (8.10), we obtain

$$\int_{\mathcal{O}} \partial_t w_1(T)\partial_t v_1(T) dx + \int_{\mathcal{O}} \mathrm{curl}\alpha_1 \mathrm{curl} w_{\mathrm{T},1} dx =$$
$$\mathcal{B}_{11} \int_{T-\tau}^{T}\int_{\mathcal{O}} |\partial_t v_1|^2 dx\, dt + \mathcal{B}_{12} \int_{T-\tau}^{T}\int_{\mathcal{O}} \partial_t v_1\, \partial_t v_2 dx\, dt. \qquad (8.29)$$

We remark that the first term on the left-hand side of the above relation can be expressed as a linear combination of $\mathrm{curl}\alpha_1 \mathrm{curl} w_{\mathrm{T},1}$, $\mathrm{curl}\alpha_2 \mathrm{curl} w_{\mathrm{T},2}$, which may be recombined to form the inner product $(\alpha, w_{\mathrm{T}})_{\mathfrak{X}_{\mathsf{M}}}$. Indeed, from (8.21), $\partial_t w(t = T)$ is equal to a linear combination of $\mathrm{curl}\alpha_i$, $\mathrm{curl} w_{\mathrm{T,i}}$ and v_i, $\partial_t v_i$, $i = 1, 2$, where v_i are the components of v. Since v solves (8.13) (which is the equivalent form of (8.10)), by the conservation of the X norm for the solution of (8.13) combined with the conservation of the X norm for the solution of (8.10), we see that the $\mathbb{X}$ norm of v is related to the $\mathbb{X}$ norm of α, which in turn is bounded above by the $\mathbb{X}$ norm of $\mathrm{curl}\alpha$.

We now integrate the equation for w_2 by $\partial_t v_2$ and perform the same steps to obtain

$$\int_{\mathcal{O}} \partial_t w_2(T)\partial_t v_2(T) dx + \int_{\mathcal{O}} curl\alpha_2 \mathrm{curl} w_{\mathrm{T},2} dx =$$
$$\mathcal{B}_{22} \int_{T-\tau}^{T}\int_{\mathcal{O}} |\partial_t v_2|^2 dx\, dt + \mathcal{B}_{21} \int_{T-\tau}^{T}\int_{\mathcal{O}} \partial_t v_1\, \partial_t v_2 dx\, dt. \qquad (8.30)$$

Adding (8.29) and (8.30), we get the following estimate:

$$(\alpha, w_{\mathrm{T}})_{\mathfrak{X}_{\mathsf{M}}} \geq C \int_{T-\tau}^{T}\int_{\mathcal{O}} \left(|\partial_t v_1|^2 + |\partial_t v_1|^2\right) dx\, dt. \qquad (8.31)$$

The left-hand side of (8.31) can be estimated by the Cauchy-Schwarz inequality by

$$(\alpha, w_{\mathrm{T}})_{\mathfrak{X}_{\mathsf{M}}} \leq ||\alpha\,||_{\mathfrak{X}_{\mathsf{M}}}\ ||w_{\mathrm{T}}||_{\mathfrak{X}_{\mathsf{M}}}. \tag{8.32}$$

By the backward uniqueness estimate of Lemma 8.4.2, the right-hand side of (8.31) can be estimated by

$$\int_{T-\tau}^{T}\int_{\mathcal{O}} \left(|\partial_t v_1|^2 + |\partial_t v_2|^2\right) dx\,dt = \int_{T-\tau}^{T}\int_{\mathcal{O}} \left(|curl v_1|^2 + |curl v_2|^2\right) dx\,dt$$
$$\geq C\,||\alpha||^2_{\mathfrak{X}_{\mathsf{M}}}. \tag{8.33}$$

Combining (8.31), (8.32), and (8.33), we obtain

$$||\alpha||_{\mathfrak{X}_{\mathsf{M}}}\,||w_{\mathrm{T}}||_{\mathfrak{X}_{\mathsf{M}}} \geq C\,||\alpha||^2_{\mathfrak{X}_{\mathsf{M}}},$$

which readily provides the estimate

$$||w_{\mathrm{T}}||_{\mathfrak{X}_{\mathsf{M}}} = ||\Lambda(\alpha)||_{\mathfrak{X}_{\mathsf{M}}} \geq C\,||\alpha||_{\mathfrak{X}_{\mathsf{M}}}, \tag{8.34}$$

for all $\alpha \in \mathfrak{X}_{\mathsf{M}}$.

STEP 4 The inequality (8.34), along with continuity property, guarantees the invertibility of the operator Λ, using the standard arguments of the Lax-Milgram lemma.

Thus the proof is complete. □

8.5 CONTROLLABILITY: COMPLEX MEDIA

In this section we prove the exact controllability of the Maxwell equations in complex media using a fixed point argument based on the Schauder fixed point theorem (see Appendix A, Theorem A.9.3). For the required compactness we need a version of the Arzelà-Ascoli theorem (Theorem A.5.2, Appendix A). Let $\mathfrak{X}_{\mathsf{M}}, \mathbb{W}^{1,1}$ be the spaces defined in Section 8.4. We have the following controllability result for the full chiral system.

THEOREM 8.5.1 *Assume that $u_0 \in \mathfrak{X}_{\mathsf{M}}$ and that $\mathsf{J}_A \in \mathbb{W}^{1,1}$. In addition let $\mathcal{B} \in \mathcal{L}(\mathbb{W}^{1,1}, \mathbb{W}^{1,1})$ be invertible. Then the full Maxwell system (8.1) is exactly controllable.*

Proof. The idea is based on a fixed point argument similar to that sketched in Section 8.2. Consider the problem

$$u' = \mathsf{M}_{\mathrm{A}}\,u + \mathsf{G}_{\mathrm{A}} \star z + \mathsf{J}_{\mathrm{A}} + \mathcal{B}\,v\,. \tag{8.35}$$

Fix $z \in C([0,T];\mathbb{X})$. Let $v_z(\cdot)$ be the control procedure that drives the system (8.35) from U_0 at time $t = 0$ to U_T at time $t = T$. Consider now the problem

$$u' = \mathsf{M}_{\mathrm{A}}\,u + \mathsf{G}_{\mathrm{A}} \star z + \mathsf{J}_{\mathrm{A}} + \mathcal{B}\,v_z\,, \tag{8.36}$$

with initial condition $u(0) = U_0$. By the definition of v_z we know that $u(T) = U_T$.

Define the map $\mathsf{F} : C([0,T];\mathbb{X}) \to C([0,T];\mathbb{X})$ by

$$z \mapsto \mathsf{F}(z(\cdot)) := u(\cdot)\,,$$

where u is the solution to (8.36) with initial condition $u(0) = U_0$.

The establishment of the existence of a fixed point of F clearly settles the exact controllability of the full (chiral) Maxwell system. We shall prove the existence of a (not necessarily unique) fixed point of F by applying the Schauder fixed point theorem (Theorem A.9.3 in Appendix A).

To this end, we must show that F maps an element of $C([0,T];\mathbb{X})$ to a subspace that is compactly embedded in $C([0,T];\mathbb{X})$. Since the map F is defined by the solution of (8.36) with initial condition $u(0) = U_0$, we resort to the regularity theory for this system. By the regularity properties (see Appendix A, Theorem A.8.14) of the mild solutions of abstract nonhomogeneous equations of the form

$$u' = \mathsf{M}_{\mathrm{A}} u + \mathsf{J}_{\mathrm{A}}^{\square},$$

for $\mathsf{M}_{\mathrm{A}} : D(\mathsf{M}_{\mathrm{A}}) = \mathfrak{X}_{\mathsf{M}} \to \mathbb{X}$, we have that

$$u \in C([0,T];\mathfrak{X}_{\mathsf{M}}) \cap C^1([0,T];\mathbb{X}),$$

as long as $U_0 \in \mathfrak{X}_{\mathsf{M}}$ and $\mathsf{J}_{\mathrm{A}}^{\square} \in \mathbb{W}^{1,1}$. Clearly, if $z \in C([0,T];\mathbb{X})$, then $\mathsf{G}_{\mathrm{A}} \star z \in C^1([0,T];\mathbb{X})$. Since $\mathsf{J}_{\mathrm{A}}^{\square} = \mathsf{J}_{\mathrm{A}} + \mathsf{G}_{\mathrm{A}} \star z + \mathcal{B}\, v_z$, it remains to show that $\mathcal{B}\, v_z \in \mathbb{W}^{1,1}$.

Recall from Section 8.3 that v_z is constructed via the following steps:

▷ We first solve the problem

$$\begin{aligned} V' &= \mathsf{M}_{\mathrm{A}} V + \mathsf{J}_{\mathrm{A}}, \\ V(0) &= U_0. \end{aligned} \tag{8.37}$$

By the regularity properties (see Appendix A, Theorem A.8.14) of the mild solutions of abstract inhomogeneous equations, we have that $V \in C([0,T];\mathfrak{X}_{\mathsf{M}})$, since $U_0 \in \mathfrak{X}_{\mathsf{M}}$ and $\mathsf{J}_{\mathrm{A}} \in \mathbb{W}^{1,1}$.

▷ We solve the operator equation $\Lambda(\Phi_T^\flat) = U_T - V(T)$, where U_T is the desired final state and $V(T)$ is the solution of the uncontrolled system (8.37) at $t = T$. This operator is invertible in $\mathfrak{X}_{\mathsf{M}}$, thus establishing the existence of a solution $\Phi_T^\flat \in \mathfrak{X}_{\mathsf{M}}$.

▷ Using the above found $\Phi_T^\flat$ as a final condition, we solve the backward (adjoint) system

$$\begin{aligned} -\Phi' &= \mathsf{M}_{\mathrm{A}}^* \Phi, \\ \Phi(T) &= \Phi_T^\flat. \end{aligned} \tag{8.38}$$

Noting that $\mathsf{M}_{\mathrm{A}}^* = -\mathsf{M}_{\mathrm{A}}$ is the generator of a strongly continuous group, this is equivalent to an initial value problem for the uncontrolled Maxwell equations with initial condition $\Phi_T^\flat$; by the same regularity arguments as above, we obtain $\Phi(t) \in C([0,T];\mathfrak{X}_{\mathsf{M}}) \cap C^1([0,T];\mathbb{X})$.

▷ The solution Φ of (8.38) yields the desired control procedure v_z.

Therefore, by the assumptions on $\mathcal{B}$, $\mathcal{B}v_z \in C([0,T];\mathfrak{X}_{\mathsf{M}}) \cap C^1([0,T];\mathbb{X})$, whereby $\mathcal{B}v_z \in \mathbb{W}^{1,1}$, as desired. Since the image $\mathsf{F}(z(\cdot))$ of $z(\cdot)$ is the mild solution of (8.35) with control v given by v_z, in view of the properties of $\mathsf{J}_{\mathsf{A}}^{\square}$ proved in the above steps, $u \in C^1([0,T];\mathbb{X}) \cap C([0,T];\mathfrak{X}_{\mathsf{M}})$ and is an equicontinuous family. The characterisation of compact subsets of $C([0,T];\mathbb{X})$ can now be performed by the Arzelà-Ascoli theorem, and in particular by its version Theorem A.5.2 (Appendix A), by noting that all its assumptions hold true, in view of the above arguments and by the compact embedding $\mathfrak{X}_{\mathsf{M}} \overset{c}{\hookrightarrow} \mathbb{X}$ (see Theorem 3.7.5, or Corollary 3.49 in [324], p. 71). Since the map F is continuous, the Schauder fixed point theorem (Theorem A.9.3, Appendix A) establishes that F admits a fixed point. □

REMARK 8.5.2 If the control v were exerted in a part $\mathcal{O}_0$ of $\mathcal{O}$ only, then T would have to satisfy $T > T_{\text{cr}}$, where T_{cr} is a critical time related to the time needed for the field activity in $\mathcal{O}_0$ to reach $\mathcal{O} \setminus \mathcal{O}_0$. This is due to the finite speed of propagation of the solutions of the Maxwell equations (by analogy with the corresponding property of solutions to the wave equation). This remark also holds for boundary controllability.

8.6 MISCELLANEA

8.6.1 An abstract approach

The problem of controllability for chiral media may be treated using a general abstract approach. This approach uses the semigroup formulation of the problem. It is inspired by the work of Bensoussan [57].

To make the presentation simpler, we first omit the chiral term, which will be added shortly after.

Consider the control system

$$u' = \mathsf{M}_{\mathsf{A}} u + \mathcal{B} v, \tag{8.39}$$

where $\mathcal{B}$ is a "control-to-state" operator. Let us denote by $\mathbb{H}$ the state space of the system and by $\mathbb{V}$ the control space of the system; both are assumed to be Hilbert spaces. Assume that $\mathsf{M}_{\mathsf{A}} : D(\mathsf{M}_{\mathsf{A}}) \to \mathbb{H}$ genererates a C_0 semigroup, $\mathsf{T}_{\mathsf{M}_{\mathsf{A}}}(t)$, on $\mathbb{H}$. Assume further that $\mathcal{B} \in \mathcal{L}(\mathbb{V}, \mathbb{H})$.

The mild solution of the system (8.39) can be expressed as

$$u(t) = \mathsf{T}_{\mathsf{M}_{\mathsf{A}}}(t) u_0 + \int_0^t \mathsf{T}_{\mathsf{M}_{\mathsf{A}}}(t-s)\, \mathcal{B}\, v(s)\, ds. \tag{8.40}$$

If we consider controls of the general form

$$v(t) = \mathcal{B}^* \, \mathsf{T}_{\mathsf{M}_{\mathsf{A}}}(t)\, \xi,$$

for some $\xi \in \mathbb{H}$, then equation (8.40) becomes

$$u(t) = \mathsf{T}_{\mathsf{M}_{\mathsf{A}}}(t)\, \left(u_0 + \mathsf{L}_{\mathsf{co}}(t)\xi\right),$$

where

$$\mathsf{L}_{\mathsf{co}}(t) := \int_0^t \mathsf{T}^*_{\mathsf{M}_{\mathrm{A}}}(s)\, \mathcal{B}\,\mathcal{B}^*\, \mathsf{T}_{\mathsf{M}_{\mathrm{A}}}(s)\, ds$$

is called the *controllability operator* for the control system (8.39).

Observe that by the special form assumed for the control, ξ uniquely determines the control process v. Equation (8.41) thus connects ξ (and therefore the control process) with the state of the system at time t. If for some $T > 0$ the operator equation

$$u_T = \mathsf{T}_{\mathsf{M}_{\mathrm{A}}}(T)\ (u_0 + \mathsf{L}_{\mathsf{co}}(T)\xi)$$

has a solution $\xi \in \mathbb{H}$, then there exists a control $v(t)$ that can drive the system (8.39) from state u_0 to state u_T in time T.

If the operator $\mathsf{L}_{\mathsf{co}}(T)$ were coercive, then an application of the Lax-Milgram lemma would automatically yield the required result. However, the coercivity of this operator is attained only if we change the functional setting of the equation.

One sees immediately that as long as an inequality of the form

$$(\mathsf{L}_{\mathsf{co}}(T)\xi, \xi) \geq C(\xi, \xi) \tag{8.41}$$

holds (where now we may have to interpret the operator $\mathsf{L}_{\mathsf{co}}(T)$ as an operator between two suitably chosen function spaces $\mathbb{K}_1$ and $\mathbb{K}_2$), then this operator is invertible (by Lax-Milgram-type results). However, recall the definition of the controllability operator. Using this definition, we obtain

$$\int_0^T (\mathsf{T}^*_{\mathsf{M}_{\mathrm{A}}}(s)\mathcal{B}\,\mathcal{B}^*\mathsf{T}_{\mathsf{M}_{\mathrm{A}}}(s)\xi, \xi)ds = \int_0^T ||\mathcal{B}\,\mathsf{T}_{\mathsf{M}_{\mathrm{A}}}(s)\xi||\, ds \geq C||\xi||^2, \tag{8.42}$$

and since $\mathsf{T}_{\mathsf{M}_{\mathrm{A}}}(s)\xi$ can be interpreted as the solution $\phi(s)$ of the uncontrolled system at time s, having started at initial state ξ, we see that (8.42) is an inequality relating the future states of the system to the initial state, i.e.,

$$\int_0^T ||\mathcal{B}\,\phi(s)||^2\, ds \geq ||\phi(0)||^2. \tag{8.43}$$

Such an inequality is called an *observability inequality* and is equivalent to the controllability of the system.

The treatment so far is very general, makes minimal or no reference to the nature of the control system, other than the semigroup property, and reduces the controllability issue to verifying condition (8.41) or its equivalent condition (8.41). However, there is no free lunch! Proving an observability inequality is usually the difficult part of a controllability problem and requires detailed study of the specific problem at hand. A possible choice of $\mathbb{H}$ can be $\mathbb{X}$, while for $D(\mathsf{M}_{\mathrm{A}})$ we may choose $\mathfrak{X}_{\mathsf{M}}$. To obtain the coercivity result we need to define an appropriate Gelfand triple setting; for a detailed treatment; see [57].

Regarding now the inclusion of the chiral terms: they can be treated in a perturbative manner, akin to the fixed point scheme approach we have used in this chapter. We compute the controllability operator for the system including the chiral terms. Repeating the above calculations, we see that the controllability operator for the chiral medium $\mathsf{L}_{\mathsf{co}}^{\mathsf{chi}}(T)$ assumes the form

$$\mathsf{L}_{\mathsf{co}}^{\mathsf{chi}}(T) = \mathsf{L}_{\mathsf{co}}(T) + \mathsf{K}^{\mathsf{chi}}(T),$$

where $\mathsf{K}^{\mathsf{chi}}(T)$ is the operator, including the chirality effects.

An alternative approach would be to use the evolution family for the integrodifferential equation.

8.6.2 Boundary controllability

An important extension of the ideas developed in this chapter applies to the case of boundary controllability, i.e., when the control is exerted at the boundary $\partial\mathcal{O}$ of the domain $\mathcal{O}$, or even at a part $\partial\mathcal{O}_o$ of it. Boundary controllability problems are often of interest, as in a number of practical applications it is easier to exert a control at the boundary of the domain than in its interior, often in the form of electric currents. Boundary controllability results for the achiral Maxwell system are proved in [140], provided that T is greater than a critical value T^* depending on the geometrical properties of the domain $\mathcal{O}$. The main reason for that stems from the fact that the Maxwell equation has wave propagation properties. This dictates that any information from the boundary of the domain has to propagate to the interior of the domain with the wave propagation velocity in the medium. As a result of that, the "communication" between the boundary $\partial\mathcal{O}$ and any point $x \in \mathcal{O}$, has to take place in finite time. The problem for complex media is still open.

8.6.3 Optimal control problems

Another class of important problems comprises optimal control problems. In such problems we are not simply interested in steering the control system to a desired final state, we want to do so in the "cheapest" possible way, where the cost is defined in terms of a cost functional, $\mathcal{J}(u, v)$, involving the control and the possible deviation from the desired final state. The theory of optimal control is a well-studied subject for finite- and infinite-dimensional control systems with many important applications. A particular subclass of problems that has been heavily studied in the past consists of linear quadratic control problems. In this class of problems, the control system is a linear evolution equation, whereas the cost functional is a quadratic functional of the general form

$$\mathcal{J}(u, v) = \int_0^T (||\mathscr{C}_1 u(t) - U(t)||_{\mathbb{H}}^2 + ||\mathscr{C}_2 v(t)||_{\mathbb{V}}^2) dt + ||\mathscr{C}_3(u(T) - U_T)||_{\mathbb{H}}^2,$$

where $\mathbb{H}$ and $\mathbb{V}$ are proxies for the state and the control space of the system, respectively. The operators $\mathscr{C}_1, \mathscr{C}_2, \mathscr{C}_3$ are operators that quantify the relative importance of the intertemporal deviation from a desired path $U(t)$ ending at U_T, the cost of the control procedure and the deviation from the desired final state, respectively, in the overall cost functional. An optimal control problem consists of finding $v \in \mathbb{V}$ such that $\mathcal{J}(u, v)$ is minimised.

Problems of this form can be solved using the infinite-dimensional Riccati equation, or generalisations of the Pontryagin maximum principle (see, e.g., [58], [146], [275]). This theory can be extended in a perturbative fashion, via fixed point arguments, to include the effects of the convolution terms; however, that is beyond the scope of the present work.

8.6.4 Controllability for time-harmonic fields

Suppose that the electromagnetic fields we are interested in are all time-harmonic of a given frequency ϖ. We consider $\mathcal{F}$ to be an electric current that acts on a subset Γ_0 of the boundary $\partial\mathcal{O}$. This electric current is considered a control and is chosen in such a way as to drive the system to a desired state. The problem we have to treat then consists of the following equations in $\mathcal{O}$:

$$i\varepsilon\,\varpi\,(E + \beta\,\mathrm{curl}E) + \mathrm{curl}H = 0\ ,$$
$$i\mu\,\varpi\,(H + \beta\,\mathrm{curl}H) - \mathrm{curl}E = 0\ ,$$

with boundary conditions

$$H \times n = \mathcal{F}\ ,\ \text{on } \Gamma_0\,,$$
$$H \times n = 0\ ,\ \text{on } \partial\mathcal{O} \setminus \Gamma_0\,.$$

Recall that $L^2_t(\partial\mathcal{O} \setminus \Gamma_0)$ is the space of tangential fields that vanish on Γ_0. In [117] the following result concerning controllability of the time-harmonic system is proved.

THEOREM 8.6.1 *There exists an open dense subset $\mathfrak{I}$ of $\mathbb{R}^+$ such that if ε, μ, ϖ satisfy $\frac{1}{\varepsilon\mu\varpi^2} \in \mathfrak{I}$, then, as $\mathcal{F} \in L^2_t(\Gamma_0)$ varies, the set of* curl$H$ *restricted to $\partial\mathcal{O} \setminus \Gamma_0$ is dense in $L^2_t(\partial\mathcal{O} \setminus \Gamma_0)$, for sufficiently small $\beta > 0$.*

This theorem provides an interesting approximate controllability result for time-harmonic fields in chiral media. One may interpret the density of the set of curl$H$ in $L^2_t(\partial\mathcal{O} \setminus \Gamma_0)$, as $\mathcal{F}$ varies, as follows: it is possible to choose $\mathcal{F}$ so as to bring curl$H$ as close as we wish to a chosen field in $L^2_t(\partial\mathcal{O} \setminus \Gamma_0)$; therefore, we may choose the boundary control to drive the state of the system as close as we wish to a given state.

Chapter Nine

Homogenisation

9.1 INTRODUCTION

Composite materials containing finely mixed constituent parts, possibly exhibiting a well-defined structure, are encountered almost everywhere, either in natural forms or as fabricated materials (e.g., bones, wood, metals, rocks, polycrystalline materials, or concrete, carbon fibres, ceramics, foams, etc., respectively; see, e.g., [322]). They are designed to display desirable properties that may not be exhibited by homogeneous media. Complex electromagnetic media are often composite materials.

The evolution of physical phenomena in composite materials may be modelled using boundary value problems with a periodic structure. This periodic structure leads to complications in both the analytic and the numerical treatment of these problems, especially if the period of the structure is small compared to the size of the region in which the system is to be studied. In such cases an asymptotic analysis is often useful in obtaining a simpler, average description of the phenomenon that models the bulk properties of the material. If we define a parameter $\epsilon = \frac{\ell_m}{\ell_M}$ as the ratio of the characteristic length scales ℓ_m and ℓ_M of the microstructure and the size of the material, respectively, then the asymptotic description is valid in the limit as $\epsilon \to 0$. This asymptotic theory, which started in terms of formal asymptotic expansions and gradually developed into a well-founded rigorous mathematical theory, is called *homogenisation theory* and aims at establishing the macroscopic behaviour of such systems. This is effected by proving that in the limit as, $\epsilon \to 0$, the phenomenon may be modelled in terms of a boundary value problem, often similar in form to the original problem but with constant coefficients rather than variable (periodic) ones. This means that the inhomogeneous material is replaced by a homogeneous fictitious one (the "homogenised" material) whose global characteristics are a good approximation of the initial ones.

The theoretical background of homogenisation has developed considerably since the first results were reported by Spagnolo [391] in the early 1960s, using the method of *G-convergence*. Later, the methods of Γ-*convergence* and *H-convergence* were introduced by De Giorgi and Franzoni (see [119], [129], [130], [333]), and Tartar [404], respectively. Murat [329] and Tartar [403] introduced the *compensated compactness theorem* (or div-curl lemma) which is a powerful tool with which to prove convergence; representative of the mathematical work on homogenisation are the monographs [59], [97], [216],

[375]; see also [405] for optimisation problems leading to homogenisation questions. See also [331], [332], [375], [407], [421].

Within the electromagnetic community, homogenisation of composites has a huge literature - see [381] and references therein - the major part of which is devoted to dielectrics. The literature on bianisotropic composites is much smaller. Among the recent developments are the work on Maxwell Garnett and Bruggeman formalisms for different classes of bianisotropic inclusions (see [317], [382] and the references therein) and work on the strong property fluctuation theory for bianisotropic composites (see [296], [317] and the references therein). See also [274], [296], [297], [316].

We work in the time domain and consider dissipative bianisotropic media along the lines of [51] and [395]. For the corresponding problem for isotropic media, see [18], [302], [374], [425] and the references therein.

The structure of this chapter is as follows. In Section 9.2 we formulate our model and the homogenisation problem for a complex electromagnetic medium. In Section 9.3 we present a formal two-scale expansion that motivates our approach and allows us to predict the form of the homogenised system. In Section 9.4 we provide a rigorous approach to the homogenisation problem for media in the optical response region, which is then generalised in Section 9.5 for dispersive media. The convergence of the evolution equations for the fields in periodic media to those for a homogenised medium is proved and the coefficients of the homogenised medium are identified in terms of auxiliary elliptic problems. Finally, in Section 9.6 we collect some general comments and alternative approaches to the homogenisation problem.

9.2 FORMULATION

Let $\mathcal{O}$ be a domain in $\mathbb{R}^3$, filled by a complex electromagnetic medium modelled by constitutive relations of the general form (2.12),

$$\mathsf{d} = \mathsf{A}_{\mathrm{or}} u + \mathsf{G}_{\mathrm{d}} \star u, \tag{9.1}$$

where now the material is spatially inhomogeneous, i.e., $\mathsf{A}_{\mathrm{or}} = \mathsf{A}_{\mathrm{or}}(x)$, $\mathsf{G}_{\mathrm{d}} = \mathsf{G}_{\mathrm{d}}(x)$. The evolution of the field $u = (u_1, u_2)^{tr}$ in $\mathcal{O}$ is governed by the Maxwell equations, which can be expressed in six-vector notation as

$$(\mathsf{A}_{\mathrm{or}} u + \mathsf{G}_{\mathrm{d}} \star u)' = \mathsf{M} u + j, \tag{9.2}$$

which without loss of generality[1] is complemented with the initial condition,

$$u = 0, \; x \in \mathcal{O}, \; t = 0, \tag{9.3}$$

and the perfect conductor boundary condition,

$$n \times u_1 = 0, \; t > 0, \; x \in \partial\mathcal{O}, \tag{9.4}$$

where n is the outward unit normal on $\partial\mathcal{O}$.

We now assume a special type of spatial inhomogeneity for the medium;

[1]The effects of nonzero initial conditions are treated in detail in Section 9.5.2.

ASSUMPTION 9.2.1 The medium exhibits small-scale periodicity, i.e.,

$$\begin{aligned} \mathsf{A}_{\text{or}} &= \mathsf{A}^{\epsilon}_{\text{or}}(x) = \mathsf{A}^{\text{per}}_{\text{or}}\left(\frac{x}{\epsilon}\right), \\ \mathsf{G}_{\text{d}} &= \mathsf{G}^{\epsilon}_{\text{d}}(x) = \mathsf{G}^{\text{per}}_{\text{d}}\left(\frac{x}{\epsilon}\right), \end{aligned} \tag{9.5}$$

where $\mathsf{A}^{\text{per}}_{\text{or}}(\cdot)$, $\mathsf{G}^{\text{per}}_{\text{d}}(\cdot)$ are periodic matrix-valued functions on the set (parallelepided) $Y = [0, \ell_1] \times [0, \ell_2] \times [0, \ell_3] \subset \mathbb{R}^3$ and $0 < \epsilon \ll 1$.

REMARK 9.2.2 Without loss of generality we may assume $\ell_1 = \ell_2 = \ell_3$, so that Y is a cube. The set Y may be considered the *fundamental cell* of the medium; the whole medium structure can be generated by repeating the structure in Y using translations.

REMARK 9.2.3 In what follows, to ease notation, we will often drop the superscript per from $\mathsf{A}^{\text{per}}_{\text{or}}$ and $\mathsf{G}^{\text{per}}_{\text{d}}$ and use the notation $\mathsf{A}^{\epsilon}_{\text{or}}(x) = \mathsf{A}_{\text{or}}\left(\frac{x}{\epsilon}\right)$ and $\mathsf{G}^{\epsilon}_{\text{d}}(x) = \mathsf{G}_{\text{d}}\left(\frac{x}{\epsilon}\right)$ instead.

We will also need the following definition.

DEFINITION 9.2.4 *If $a : Y \to \mathbb{R}$ is a periodic function, then the periodic averaging operator is defined by*

$$\langle a \rangle := \frac{1}{|Y|} \int_Y a(y)\, dy,$$

where $|Y|$ is the Lebesgue measure[2] *of Y.*

To be able to model the small-scale periodic microstructure, we must let ϵ vary over a range of arbitrarily small values. Since $\mathsf{A}^{\text{per}}_{\text{or}}(x)$ is periodic with period Y, it follows that $\mathsf{A}^{\text{per}}_{\text{or}}\left(\frac{x}{\epsilon}\right)$ is periodic with period ϵY. We are therefore led to a sequence of boundary value problems,

$$(\mathsf{A}^{\epsilon}_{\text{or}} u^{\epsilon} + \mathsf{G}^{\epsilon}_{\text{d}} \star u^{\epsilon})' = \mathsf{M} u^{\epsilon} + j, \tag{9.6}$$

with initial condition $u^{\epsilon} = 0$, and the perfect conductor boundary condition

$$n \times u^{\epsilon}_1 = 0, \ \text{ on } \mathcal{O}. \tag{9.7}$$

The explicit t and x dependence is omitted for simplicity.

The solution of the above sequence of boundary value problems exists for all $\epsilon > 0$ (by Theorem 9.2.5). This generates a sequence of functions $u^{\epsilon} = u^{\epsilon}(x, t)$.

THEOREM 9.2.5 (EXISTENCE AND UNIFORM BOUNDS) *Assume that j is locally Hölder continuous, and further that $j \in L^1([0, T], \mathbb{X})$. Then the Maxwell system (9.6)–(9.7) has a unique solution $u^{\epsilon} = (u^{\epsilon}_1, u^{\epsilon}_2)^{tr} = (E^{\epsilon}, H^{\epsilon})^{tr}$ in $C([0, T], \mathbb{X})$ satisfying the uniform bounds $||u^{\epsilon}(t)||_{\mathbb{X}} < C$, $\epsilon > 0$.*

---

[2] Clearly, $|Y| = \ell_1\, \ell_2\, \ell_3$ in the case where $Y = [0, \ell_1] \times [0, \ell_2] \times [0, \ell_3] \subset \mathbb{R}^3$.

Proof. The claim follows either by minor adjustments of the results of Section 7.4 (using semigroup theory) or by minor modifications of the Faedo-Galerkin approach of Section 7.5.2. □

The question of interest to homogenisation theory is what happens in the limit of very small-scale microstructures. This corresponds to considering the limit as $\epsilon \to 0$ in the above mathematical formulation, which in turn corresponds to considering boundary value problems with very rapidly oscillating coefficients and solutions. The presence of fast oscillations requires special attention to the concept of limit that should be adopted, as in general, sequences of fast oscillating functions fail to be convergent, as a whole or even up to subsequences, in the usual (strong) sense. One can escape this situation by "smoothing out" the fast oscillations, achieved by multiplying by suitable test functions, integrating, and then taking the limit. This convergence corresponds to the notion of weak convergence. For certain classes of sequences of functions, even the notion of weak convergence is not adequate; one has to employ the notion of weak-$*$ convergence (see Appendix C). The situation becomes even more complicated if one considers derivatives of such functions.

REMARK 9.2.6 The standard notations "$\rightharpoonup$" and "$\overset{*}{\rightharpoonup}$" for weak and weak-$*$ convergence, respectively, will be used in what follows.

The following example (see, e.g., [97]) is illustrative.

EXAMPLE 9.2.7 Consider the sequence of functions $f^\epsilon(x) = \sin\left(\frac{x}{\epsilon}\right)$. In the limit as $\epsilon \to 0$, these functions present fast oscillations, which may provide a model for the material microstructure. However, the limit $\lim_{\epsilon\to 0} f^\epsilon(x)$ is not defined at almost any point. On the other hand, one can see easily that the sequence $f^\epsilon \rightharpoonup 0$, in $L^2([-\ell, \ell])$ for any $\ell > 0$, but not strongly!

Suppose now that in the limit as $\epsilon \to 0$, the sequence $u^\epsilon(x,t)$ converges to a limit, $u^* = u^*(x,t)$, in some weak sense. Can this limit u^* be considered an electromagnetic field that is physically acceptable? In other words, is there a (fictitious) homogeneous (in space) material, modelled by a constitutive relation

$$\mathsf{d}^* = \mathsf{A}^h_{\text{or}} u^* + \mathsf{G}^h_{\text{d}} \star u^*,$$

with A^h_{or}, G^h_{d} matrices with spatially independent coefficients, such that the limit u^* of the sequence of fields u^ϵ satisfies the Maxwell equation

$$(\mathsf{A}^h_{\text{or}} u^* + \mathsf{G}^h_{\text{d}} \star u^*)' = \mathsf{M} u^* + j,$$

equipped with some appropriate initial and boundary conditions? If the answer to this question is positive, then the remaining tasks are (a) to identify the homogenised coefficients A^h_{or}, G^h_{d} in terms of the parameters of the original medium $\mathsf{A}^\epsilon_{\text{or}}$, $\mathsf{G}^\epsilon_{\text{d}}$ and (b) to identify the initial and boundary conditions for the homogenised problem. None of these tasks is simple. Concerning the former, we need to provide an averaging procedure that will lead to the homogenised medium parameters; this is not simply a straightforward averaging of the original medium coefficients (see Remark 9.2.8) but requires averaging over properly selected periodic "weight" functions that may be obtained as the solution of auxiliary periodic problems called the cell problems. Regarding the latter, the homogenised problem may have to be solved using modified initial or boundary conditions obtained from those of the original problem, once more using some kind of weighted averaging in terms of the solutions of the cell problems.

The physical intuition behind the above procedure is the following: the fine microstructure of the medium is not "felt" by the electromagnetic field created by an external excitation. The field "feels" a simpler medium that is spatially independent, the homogenised medium. This allows a simplified and more efficient approximate description of the material.

REMARK 9.2.8 At this point, let us recall that if $\phi : Y \subset \mathbb{R}^3 \to \mathbb{R}$ is periodic and $\phi^\epsilon(x) = \phi(x/\epsilon)$, then

$$\phi_\epsilon \rightharpoonup \langle \phi \rangle \quad \text{in } L^p(\mathcal{O}_0)\,, \qquad 1 < p < +\infty,$$

for all bounded $\mathcal{O}_0 \subset \mathbb{R}^3$. In this respect it is remarkable (and against intuition) that the homogeneous susceptibility matrices A^h_{or} and $\mathsf{G}^h_{\text{d}}(t)$ are in fact not obtained by simply averaging the functions $\mathsf{A}_{\text{or}}(y)$ and $\mathsf{G}_{\text{d}}(y,t)$ over Y. A very nice intuitive example of this surprising fact is furnished by the simple one-dimensional heat equation $\partial_t u^\epsilon = \partial_x(a(\frac{x}{\epsilon})\partial_x u^\epsilon)$, where $a : [0,\ell] \to \mathbb{R}$ is a periodic function of period ℓ. In this case, one may show explicitly that the sequence of functions $\{u^\epsilon\}$ converges weakly to a function u^*, which solves the constant coefficient heat equation $\partial_t u^* = \partial_x(a^h\,\partial_x u^*)$, but $a^h \neq \langle a \rangle$. The proper constant is instead given by $a^h = \langle a^{-1} \rangle$. This remark shows that special care should be taken when trying to define the parameters of the homogenised medium.

9.3 A FORMAL TWO-SCALE EXPANSION

To provide some insight concerning the structure of the homogenised system for the Maxwell equations in a periodic medium we give some formal arguments using a two-scale expansion. As we shall see, this expansion will give us an idea of what the coefficients of the homogenised system should look like, thus facilitating the rigorous treatment that will follow.

To make the exposition more clear, we abandon the six-vector notation and work in terms of the electromagnetic fields (E^ϵ, H^ϵ). After taking the Laplace transform[3] of the Maxwell equations (with respect to t), and dropping the explicit dependence on the Laplace variable p for simplicity of no-

[3]Recall that the Laplace transform $\widehat{\mathfrak{s}}(p)$, $p \in \mathbb{C}^+ = \{p \in \mathbb{C} : \mathcal{R}e\, p > 0\}$ of a function $\mathfrak{s}(t), t > 0$, is defined as $\widehat{\mathfrak{s}}(p) := \int_0^\infty e^{-pt}\mathfrak{s}(t)dt$. For a complete account of the theory of the Laplace transform, readers are referred to the classical monograph [429].

tation, we obtain

$$
\begin{aligned}
\operatorname{curl}\widehat{H}^{\epsilon} &= p(\varepsilon^{\epsilon}_{\mathfrak{L}}\,\widehat{E}^{\epsilon} + \xi^{\epsilon}_{\mathfrak{L}}\,\widehat{H}^{\epsilon}) + \widehat{J}, \\
-\operatorname{curl}\widehat{E}^{\epsilon} &= p(\mu^{\epsilon}_{\mathfrak{L}}\,\widehat{H}^{\epsilon} + \zeta^{\epsilon}_{\mathfrak{L}}\,\widehat{E}^{\epsilon}),
\end{aligned} \tag{9.8}
$$

where[4] $\varepsilon^{\epsilon}_{\mathfrak{L}}, \xi^{\epsilon}_{\mathfrak{L}}, \mu^{\epsilon}_{\mathfrak{L}}, \zeta^{\epsilon}_{\mathfrak{L}}$ are the Laplace transforms of the matrices composing $\mathsf{A}^{\epsilon}_{\text{or}}(x) + \mathsf{G}^{\epsilon}_{\text{d}}(x, t)$.

We now *assume* that the fields have an expansion in power series in terms of the small coefficient ϵ of the form

$$
\begin{aligned}
\widehat{E}^{\epsilon}(x) &= \widehat{E}^{(0)}(x) + \epsilon\,\widehat{E}^{(1)}(x) + \epsilon^2\,\widehat{E}^{(2)}(x) + \cdots, \\
\widehat{H}^{\epsilon}(x) &= \widehat{H}^{(0)}(x) + \epsilon\,\widehat{H}^{(1)}(x) + \epsilon^2\,\widehat{H}^{(2)}(x) + \cdots,
\end{aligned}
$$

Because of the special choice for the structure of the medium coefficients, as expressed in equation (9.5), the functions $\widehat{E}^{(j)}(x)$, $\widehat{H}^{(j)}(x)$ will be of the form

$$
\widehat{E}^{(j)}(x) = \widehat{E}^{(j)}\left(x, \frac{x}{\epsilon}\right), \quad \widehat{H}^{(j)}(x) = \widehat{H}^{(j)}\left(x, \frac{x}{\epsilon}\right),
$$

so that we may consider them functions of two variables x and $y = \frac{x}{\epsilon}$, which hereafter are considered independent variables. Because of the periodicity of the fields, the functions $\operatorname{grad}_y \widehat{E}^{(j)}(x, y)$, $\operatorname{grad}_y \widehat{H}^{(j)}(x, y)$ will be periodic in y.

Using this two-scale expansion, we see that the curl operator becomes

$$
\operatorname{curl} u = \operatorname{curl}_x u + \frac{1}{\epsilon}\operatorname{curl}_y u\,,
$$

so when acting on $\widehat{E}, \widehat{H}$ and using the power series expansion we obtain to order ϵ^{-1} that

$$
\operatorname{curl}_y \widehat{E}^{(0)} = 0\,, \ \operatorname{curl}_y \widehat{H}^{(0)} = 0\,, \tag{9.9}
$$

a fact that implies the existence of two scalar functions, $\Psi_1(x, y)$, $\Psi_2(x, y)$, such that

$$
\begin{aligned}
\widehat{E}^{(0)}(x, y) &= \widehat{E^{\bowtie}}(x) + \operatorname{grad}_y \Psi_1(x, y)\,, \\
\widehat{H}^{(0)}(x, y) &= \widehat{H^{\bowtie}}(x) + \operatorname{grad}_y \Psi_2(x, y)\,.
\end{aligned} \tag{9.10}
$$

Combining this with the chosen boundary conditions and the periodicity arguments, we see that when averaging over the small-scale dependence, we obtain

$$
\langle \widehat{E}^{(0)}(x, y) \rangle = \widehat{E^{\bowtie}}(x)\,, \ \langle \widehat{H}^{(0)}(x, y) \rangle = \widehat{H^{\bowtie}}(x)\,.
$$

We now proceed to the next order $O(\epsilon^0)$. This gives the equations

$$
\begin{aligned}
+\operatorname{curl}_y \widehat{H}^{(1)} &= p\big(\varepsilon_{\mathfrak{L}} \widehat{E}^{(0)} + \xi_{\mathfrak{L}}\,\widehat{H}^{(0)}\big) - \operatorname{curl}_x \widehat{H}^{(0)}, \\
-\operatorname{curl}_y \widehat{E}^{(1)} &= p\big(\mu_{\mathfrak{L}} \widehat{H}^{(0)} + \zeta_{\mathfrak{L}}\,\widehat{E}^{(0)}\big) + \operatorname{curl}_x \widehat{E}^{(0)}.
\end{aligned} \tag{9.11}
$$

---

[4]The same notation (9.5) is employed for the submatrices $\varepsilon^{\epsilon}(x), \xi^{\epsilon}(x), \mu^{\epsilon}(x), \zeta^{\epsilon}(x)$ and their Laplace transforms $\varepsilon^{\epsilon}_{\mathfrak{L}}, \xi^{\epsilon}_{\mathfrak{L}}, \mu^{\epsilon}_{\mathfrak{L}}, \zeta^{\epsilon}_{\mathfrak{L}}$ with respect to t, as for the matrices $\mathsf{A}^{\epsilon}_{\text{or}}(x)$, $\mathsf{G}^{\epsilon}_{\text{d}}(x)$.

We substitute (9.10) into (9.11) and then take the divergence of (9.11) with respect to y. Noting that

$$\operatorname{div}_y \operatorname{curl}_x \widehat{E}^{(0)} = \operatorname{div}_y \operatorname{curl}_x \widehat{E^{\bowtie}}(x) + \operatorname{div}_y \operatorname{curl}_x \operatorname{grad}_y \Psi_1,$$
$$\operatorname{div}_y \operatorname{curl}_x \widehat{H}^{(0)} = \operatorname{div}_y \operatorname{curl}_x \widehat{H^{\bowtie}}(x) + \operatorname{div}_y \operatorname{curl}_x \operatorname{grad}_y \Psi_2,$$

and since $\operatorname{div}_y \operatorname{curl}_x \Psi_i = 0$, $i = 1, 2$, this gives a consistency condition for the scalar functions $\Psi_1(x, y)$, $\Psi_2(x, y)$ in terms of the elliptic system

$$\begin{aligned} 0 &= \operatorname{div}_y(\varepsilon_{\mathfrak{L}} \operatorname{grad}_y \Psi_1) + \operatorname{div}_y(\xi_{\mathfrak{L}} \operatorname{grad}_y \Psi_2) + \operatorname{div}_y(\varepsilon_{\mathfrak{L}} \widehat{E^{\bowtie}}(x)) + \operatorname{div}_y(\xi_{\mathfrak{L}} \widehat{H^{\bowtie}}(x)), \\ 0 &= \operatorname{div}_y(\zeta_{\mathfrak{L}} \operatorname{grad}_y \Psi_1) + \operatorname{div}_y(\mu_{\mathfrak{L}} \operatorname{grad}_y \Psi_2) + \operatorname{div}_y(\zeta_{\mathfrak{L}} \widehat{E^{\bowtie}}(x)) + \operatorname{div}_y(\mu_{\mathfrak{L}} \widehat{H^{\bowtie}}(x)), \end{aligned} \tag{9.12}$$

where the solution (Ψ_1, Ψ_2) is considered to be periodic in y.

The solvability of this system can be guaranteed by the Lax-Milgram lemma. The solution of the system can be written in the form

$$\begin{aligned} \Psi_1(x, y) &= \Lambda^{(1)}(y) \cdot \widehat{E^{\bowtie}}(x) + \Lambda^{(2)}(y) \cdot \widehat{H^{\bowtie}}(x), \\ \Psi_2(x, y) &= \Lambda^{(3)}(y) \cdot \widehat{E^{\bowtie}}(x) + \Lambda^{(4)}(y) \cdot \widehat{H^{\bowtie}}(x), \end{aligned} \tag{9.13}$$

where $\Lambda^{(i)}(y) = (\Lambda_1^{(i)}, \Lambda_2^{(i)}, \Lambda_3^{(i)})$, $i = 1, 2, 3, 4$, are three-vectors.

REMARK 9.3.1 Let $\mathfrak{s}$ be a proxy for a 3×3 matrix. We use the notation $\mathfrak{s}_{\sharp,m}$ for the mth column and $\mathfrak{s}_{k,\sharp}$ for the kth row of the matrix $\mathfrak{s}$, respectively.

Substituting the ansatz (9.13) into (9.12) and separating terms that are proportional to $\widehat{E^{\bowtie}}(x)$ and to $\widehat{H^{\bowtie}}(x)$, we obtain that the components of $\Lambda^{(i)}$, $i = 1, 3$, solve the following system:

$$\begin{aligned} 0 &= \operatorname{div}_y((\varepsilon_{\mathfrak{L}})_{\sharp,j}) + \operatorname{div}_y(\varepsilon_{\mathfrak{L}} \operatorname{grad}_y \Lambda_j^{(1)}) + \operatorname{div}_y(\xi_{\mathfrak{L}} \operatorname{grad}_y \Lambda_j^{(3)}), \\ 0 &= \operatorname{div}_y((\zeta_{\mathfrak{L}})_{\sharp,j}) + \operatorname{div}_y(\zeta_{\mathfrak{L}} \operatorname{grad}_y \Lambda_j^{(1)}) + \operatorname{div}_y(\mu_{\mathfrak{L}} \operatorname{grad}_y \Lambda_j^{(3)}), \end{aligned} \tag{9.14}$$

for $j = 1, 2, 3$, and the components of $\Lambda^{(i)}$, $i = 2, 4$, solve the following system:

$$\begin{aligned} 0 &= \operatorname{div}_y((\xi_{\mathfrak{L}})_{\sharp,j}) + \operatorname{div}_y(\varepsilon_{\mathfrak{L}} \operatorname{grad}_y \Lambda_j^{(1)}) + \operatorname{div}_y(\xi_{\mathfrak{L}} \operatorname{grad}_y \Lambda_j^{(3)}), \\ 0 &= \operatorname{div}_y((\mu_{\mathfrak{L}})_{\sharp,j}) + \operatorname{div}_y(\zeta_{\mathfrak{L}} \operatorname{grad}_y \Lambda_j^{(1)}) + \operatorname{div}_y(\mu_{\mathfrak{L}} \operatorname{grad}_y \Lambda_j^{(3)}), \end{aligned} \tag{9.15}$$

for $j = 1, 2, 3$. The equations (9.14) and (9.15) are elliptic systems that by the Lax-Milgram lemma are well posed, yielding unique solutions for $\Lambda^{(i)}$, $i = 1, 2, 3, 4$, when considered periodic functions in y, with zero mean.

We now average (9.11) with respect to the y variable (the fast scale). In view of (9.9), we note that the terms containing the curls with respect to y give a zero contribution in the average because of periodicity. The remaining terms yield

$$\begin{aligned} \operatorname{curl}_x \widehat{H^{\bowtie}}(x) &= p\,(\varepsilon_{\mathfrak{L}}^h \, \widehat{E^{\bowtie}}(x) + \xi_{\mathfrak{L}}^h \, \widehat{H^{\bowtie}}(x)), \\ -\operatorname{curl}_x \widehat{E^{\bowtie}}(x) &= p\,(\zeta_{\mathfrak{L}}^h \, \widehat{E^{\bowtie}}(x) + \mu_{\mathfrak{L}}^h \, \widehat{H^{\bowtie}}(x)), \end{aligned}$$

for constant matrices $\varepsilon_{\mathfrak{L}}^{h}, \xi_{\mathfrak{L}}^{h}, \zeta_{\mathfrak{L}}^{h}, \mu_{\mathfrak{L}}^{h}$ given by

$$\begin{aligned}
\varepsilon_{\mathfrak{L}}^{h} &:= \left\langle \varepsilon_{\mathfrak{L}} + \varepsilon_{\mathfrak{L}} \operatorname{grad}_y \Lambda^{(1)} + \xi_{\mathfrak{L}} \operatorname{grad}_y \Lambda^{(3)} \right\rangle, \\
\xi_{\mathfrak{L}}^{h} &:= \left\langle \xi_{\mathfrak{L}} + \varepsilon_{\mathfrak{L}} \operatorname{grad}_y \Lambda^{(2)} + \xi_{\mathfrak{L}} \operatorname{grad}_y \Lambda^{(4)} \right\rangle, \\
\zeta_{\mathfrak{L}}^{h} &:= \left\langle \zeta_{\mathfrak{L}} + \zeta_{\mathfrak{L}} \operatorname{grad}_y \Lambda^{(1)} + \mu_{\mathfrak{L}} \operatorname{grad}_y \Lambda^{(3)} \right\rangle, \\
\mu_{\mathfrak{L}}^{h} &:= \left\langle \mu_{\mathfrak{L}} + \zeta_{\mathfrak{L}} \operatorname{grad}_y \Lambda^{(2)} + \mu_{\mathfrak{L}} \operatorname{grad}_y \Lambda^{(4)} \right\rangle,
\end{aligned} \tag{9.16}$$

where by $\operatorname{grad}_y \Lambda^{(k)}$, $k = 1, 2, 3, 4$, we denote the 3×3 matrix

$$\operatorname{grad}_y \Lambda^{(k)} = \begin{pmatrix} \partial_{y_1} \Lambda_1^{(k)} & \partial_{y_1} \Lambda_2^{(k)} & \partial_{y_1} \Lambda_3^{(k)} \\ \partial_{y_2} \Lambda_1^{(k)} & \partial_{y_2} \Lambda_2^{(k)} & \partial_{y_2} \Lambda_3^{(k)} \\ \partial_{y_3} \Lambda_1^{(k)} & \partial_{y_3} \Lambda_2^{(k)} & \partial_{y_3} \Lambda_3^{(k)} \end{pmatrix}.$$

Observe that (9.16) corresponds to the Laplace transform of the Maxwell equation for the average fields $\widehat{E^{\bowtie}}(x)$, $\widehat{H^{\bowtie}}(x)$. The Laplace transform of the corresponding constitutive relations is given as a multiplicative linear relation in terms of the constant matrices $\varepsilon^h, \xi^h, \zeta^h, \mu^h$ that when transformed back to the time domain, becomes a convolution-type linear operator with spatially independent coefficients. This is the form of the homogenised system whose coefficients are obtained as in (9.16) through averaging of the original coefficients multiplied by solutions of the cell equations (9.14)-(9.15).

REMARK 9.3.2 The above scheme for obtaining the homogenised system, useful and insightful as it may be, is *nothing but a formal argument* based on a simple power series expansion of the Maxwell equations and their solutions. It should be treated with caution, since there is no way for its validity to be checked unless a rigorous mathematical theory is formulated. There are many instances in which this formal argument may break down. First, the basic assumption that the solution of the Maxwell system (E^ϵ, H^ϵ) is expandable in a power series in ϵ has to be checked by a rigorous analysis of the full Maxwell system. Second, the argument is based on the series expansion, with only the first two terms in the series kept. Therefore, the homogenisation result obtained by this formal expansion is valid only if the remaining terms in the series that have been discarded are really negligible. This cannot be based simply on the observation that they are multiplied by terms of order ϵ^2, because there is no guarantee that $E^{(2)}$ and $H^{(2)}$ are (uniformly in x and t) of order 1. A rigorous proof that the remaining terms in the series do not change the results of the formal expansion needs to be completed before the result of this formal argument can be trusted.

The remaining sections of the chapter provide a rigorous justification of the above described homogenisation procedure.

9.4 THE OPTICAL RESPONSE REGION

We first consider the optical response region, assuming for the time being that $\mathsf{G}_{\mathrm{d}}(x) = 0$. The analysis extends the results of [51].

9.4.1 An auxiliary elliptic homogenisation problem

In this section we study the homogenisation theory for an auxilliary elliptic homogenisation problem, which will lead us to homogenisation of the Maxwell system. For ease of notation, and since the discussion below may be of more general interest than merely serving for the homogenisation of the Maxwell system, we state the elliptic system in more general form.

We first introduce the following notation:

DEFINITION 9.4.1 *Define*

$$\mathsf{A}_{e\ell} := \begin{pmatrix} a & b \\ c & d \end{pmatrix},$$

where the 3×3 *matrices* a, b, c, d *are periodic functions of* y. *The definition of* $\mathsf{A}^{\epsilon}_{e\ell}$ *is as above, replacing* $\mathfrak{s}$ *by* $\mathfrak{s}^{\epsilon}$, *where* $\mathfrak{s}$ *is a proxy for* a, b, c, d.

The following standing assumption holds throughout this chapter:

ASSUMPTION 9.4.2 The matrix $\mathsf{A}_{e\ell} \in L^{\infty}(\mathcal{O}, \mathbb{R}^{6\times 6})$ satisfies the following conditions:

(*i*) There exists a positive constant c_1 such that $|\mathsf{A}_{e\ell}(x)\,\mathsf{y}\cdot\mathsf{y}| \geq c_1\,|\mathsf{y}|$, for almost all $x \in \mathcal{O}$ and all $\mathsf{y} \in \mathbb{R}^6$.

(*ii*) There exists a positive constant c_2 such that $|\mathsf{A}^{-1}_{e\ell}(x)\,\mathsf{y}\cdot\mathsf{y}| \geq c_2\,|\mathsf{y}|$, for almost all $x \in \mathcal{O}$ and all $\mathsf{y} \in \mathbb{R}^6$.

REMARK 9.4.3 This assumption is technical (related to H-convergence) and strengthens Assumption 2.3.5 since we do not restrict ourselves to the homogenisation of symmetric elliptic systems only. Note that this assumption is equivalent to stating that $\mathsf{A}_{e\ell} \in \mathbb{M}(c_1, c_2, \mathcal{O})$ (see Definition C.3.1 in Appendix C and the discussion concerning elliptic equations on p. 81 in [407]).

In our subsequent analysis we will need the function space $H^1_{\mathrm{per}}(Y)$ (defined as the closure of the set of C^{∞}, Y-periodic functions in the H^1 norm; see Section C.1 in Appendix C), as well as the following auxiliary elliptic operators:

DEFINITION 9.4.4

(*i*) *The "cell" operator* $L_{per} : H^1_{per}(Y) \times H^1_{per}(Y) \to (H^1_{per}(Y) \times H^1_{per}(Y))'$:

$$L_{per} := -\mathrm{div}_y(\mathsf{A}_{e\ell}(y)\,\mathrm{grad}_y\,). \tag{9.17}$$

(*ii*) *The "microstructure" operator* $L^\epsilon : H^1_0(\mathcal{O}) \times H^1_0(\mathcal{O}) \to H^{-1}(\mathcal{O}) \times H^{-1}(\mathcal{O})$*:*

$$L^\epsilon := -\mathrm{div}_x(\mathsf{A}^\epsilon_{e\ell}\,(x)\,\mathrm{grad}_x\,). \tag{9.18}$$

REMARK 9.4.5 The elliptic operators defined above are *matrix* elliptic operators, used for abbreviating the notation for the corresponding systems. For instance,

$$L_{\mathrm{per}} = \begin{pmatrix} -\mathrm{div}_y(a\,\mathrm{grad}_y) & -\mathrm{div}_y(b\,\mathrm{grad}_y) \\ -\mathrm{div}_y(c\,\mathrm{grad}_y) & -\mathrm{div}_y(d\,\mathrm{grad}_y) \end{pmatrix},$$

where a, b, c, d are 3×3 matrices. This operator acts on two-vector $w = (w_1, w_2)^{tr}$ as follows:

$$L_{\mathrm{per}} \begin{pmatrix} w_1 \\ w_2 \end{pmatrix} = \begin{pmatrix} -\mathrm{div}_y(a\,\mathrm{grad}_y w_1) - \mathrm{div}_y(b\,\mathrm{grad}_y w_2) \\ -\mathrm{div}_y(c\,\mathrm{grad}_y w_1) - \mathrm{div}_y(d\,\mathrm{grad}_y w_2) \end{pmatrix}.$$

Further, the adjoint operator turns out to be

$$L^*_{\mathrm{per}} = \begin{pmatrix} -\mathrm{div}_y(a^{tr}\,\mathrm{grad}_y) & -\mathrm{div}_y(c^{tr}\,\mathrm{grad}_y) \\ -\mathrm{div}_y(b^{tr}\,\mathrm{grad}_y) & -\mathrm{div}_y(d^{tr}\,\mathrm{grad}_y) \end{pmatrix}.$$

Similarly, L^ϵ is given in an analogous block form where a, b, c, d are replaced by a^ϵ, b^ϵ, c^ϵ, d^ϵ.

The coercivity assumption 9.4.2 implies that L_{per} is *invertible modulo constants*, so that the elliptic system

$$L_{\mathrm{per}} u = f \tag{9.19}$$

has a unique solution in terms of the two-vector $u = (u_1, u_2)^{tr} \in H^1_{\mathrm{per}}(Y) \times H^1_{\mathrm{per}}(Y)$ for all $f \in (H^1_{\mathrm{per}}(Y) \times H^1_{\mathrm{per}}(Y))'$. Similarly, the coercivity assumption, in conjunction with the generalisation of the Lax-Milgram lemma for systems, guarantees the solvability in $H^1_0(\mathcal{O}) \times H^1_0(\mathcal{O})$ of the elliptic system

$$L^\epsilon w^\epsilon = g \tag{9.20}$$

for every two-vector $g = (g_1, g_2)^{tr} \in H^{-1}(\mathcal{O}) \times H^{-1}(\mathcal{O})$.

We consider the homogenisation theory for the system (9.20). To motivate the theory, we briefly describe a *formal two-scale expansion* based on the introduction of the independent variables x and $y = \frac{x}{\epsilon}$, and the subsequent expansion of the operator L^ϵ as

$$L^\epsilon = \epsilon^2\, L_{-2} + \epsilon^{-1}\, L_{-1} + L_0,$$

where

$$\begin{aligned} L_{-2} &:= -\mathrm{div}_y(\mathsf{A}_{e\ell}\ \mathrm{grad}_y\cdot\,), \\ L_{-1} &:= -\mathrm{div}_x(\mathsf{A}_{e\ell}\ \mathrm{grad}_y\cdot\,) - \mathrm{div}_y(\mathsf{A}_{e\ell}\ \mathrm{grad}_x\cdot\,), \\ L_0 &:= -\mathrm{div}_x(\mathsf{A}_{e\ell}\ \mathrm{grad}_x\cdot\,). \end{aligned}$$

Letting

$$w^\epsilon = w^{(0)} + \epsilon\, w^{(1)} + \epsilon^2\, w^{(2)} + \cdots$$

and substituting into the above operator expansion, we obtain[5] to various orders of powers of ϵ

$$\begin{aligned} L_{-2}\, w^{(0)} &= 0, \\ L_{-2}\, w^{(1)} &= -L_{-1}\, w^{(0)}, \\ L_{-2}\, w^{(2)} &= -L_{-1}\, w^{(1)} - L_0\, w^{(0)} + g \end{aligned} \tag{9.21}$$

Observe that $L_{-2} = L_{\text{per}}$.

A particular solution[6] of the first of equations (9.21) is $w^{(0)} = w^{(0)}(x) = (w_1^{(0)}(x), w_2^{(0)}(x))^{tr}$. We substitute this solution into the second of equations (9.21) to obtain

$$L_{\text{per}} w^{(1)} = \begin{pmatrix} \sum_j div_y a_{\sharp,j}\, \partial_{x_j} w_1^{(0)} + \sum_j div_y b_{\sharp,j}\, \partial_{x_j} w_2^{(0)} \\ \sum_j div_y c_{\sharp,j}\, \partial_{x_j} w_1^{(0)} + \sum_j div_y d_{\sharp,j}\, \partial_{x_j} w_2^{(0)} \end{pmatrix}. \tag{9.22}$$

We look for a solution of the above system of the form

$$w_\ell^{(1)} = \sum_j r_\ell^{(j)} \partial_{x_j} w_1^{(0)} + \sum_j v_\ell^{(j)} \partial_{x_j} w_2^{(0)}, \quad \ell = 1, 2, \tag{9.23}$$

where $r_\ell^{(j)}$, $v_\ell^{(j)}$, $j = 1, 2, 3$, $\ell = 1, 2$, are functions of y (12 in total), to be determined. Upon substitution of the ansatz (9.23) into (9.22), we observe that $r_\ell^{(j)}$, $v_\ell^{(j)}$, $j = 1, 2, 3$, $\ell = 1, 2$, can be chosen as the solutions of the elliptic systems

$$L_{\text{per}} \begin{pmatrix} r_1^{(j)} \\ r_2^{(j)} \end{pmatrix} = \begin{pmatrix} \text{div}_y a_{\sharp,j} \\ \text{div}_y c_{\sharp,j} \end{pmatrix}, \quad L_{\text{per}} \begin{pmatrix} v_1^{(j)} \\ v_2^{(j)} \end{pmatrix} = \begin{pmatrix} \text{div}_y b_{\sharp,j} \\ \text{div}_y d_{\sharp,j} \end{pmatrix}. \tag{9.24}$$

The system of equations (9.24) is called the *cell system* and has a unique periodic solution (modulo constants).

We now average the third equation of (9.21) over y. Observe first that

$$\begin{aligned} &\langle L_{-2}\, w^{(2)} \rangle = 0, \\ &\langle L_0 w^{(0)} \rangle = -\text{div}_x(\langle \mathsf{A}_{e\ell} \rangle\, \text{grad}_x w^{(0)}), \\ &\langle \text{div}_y(\mathsf{A}_{e\ell}\ \text{grad}_x w^{(0)}) \rangle = 0. \end{aligned}$$

Next we calculate the term $\text{div}_x(\mathsf{A}_{e\ell}\ \text{grad}_y w^{(0)})$, regrouping the terms proportional to $\text{grad}_x(w_1^{(0)})$ and $\text{grad}_x(w_2^{(0)})$. After some algebra we obtain

$$\langle \text{div}_x(\mathsf{A}_{e\ell}\ \text{grad}_y w^{(0)}) \rangle = \begin{pmatrix} \text{div}_x(\mathfrak{a}^h \text{grad}_x w_1^{(0)}) + \text{div}_x(\mathfrak{b}^h \text{grad}_x w_2^{(0)}) \\ \text{div}_x(\mathfrak{c}^h \text{grad}_x w_1^{(0)}) + \text{div}_x(\mathfrak{d}^h \text{grad}_x w_2^{(0)}) \end{pmatrix},$$

[5]In similar spirit as for the Maxwell equations in Section 9.3.

[6]The general solution contains a term related to the curl_y of a vector field; this is omitted here for simplicity but is fully taken into account in the subsequent rigorous analysis of the problem.

where $\mathfrak{a}^h$, $\mathfrak{b}^h$, $\mathfrak{c}^h$, $\mathfrak{d}^h$ are 3×3 matrices consisting of constant elements

$$(\mathfrak{a}^h)_{ij} = \langle \sum_{k=1}^{3} a_{ik} \partial_{y_k} r_1^{(j)} + \sum_{k=1}^{3} b_{ik} \partial_{y_k} r_2^{(j)} \rangle,$$

$$(\mathfrak{b}^h)_{ij} = \langle \sum_{k=1}^{3} a_{ik} \partial_{y_k} v_1^{(j)} + \sum_{k=1}^{3} b_{ik} \partial_{y_k} v_2^{(j)} \rangle,$$

$$(\mathfrak{c}^h)_{ij} = \langle \sum_{k=1}^{3} c_{ik} \partial_{y_k} r_1^{(j)} + \sum_{k=1}^{3} d_{ik} \partial_{y_k} r_2^{(j)} \rangle,$$

$$(\mathfrak{d}^h)_{ij} = \langle \sum_{k=1}^{3} c_{ik} \partial_{y_k} v_1^{(j)} + \sum_{k=1}^{3} d_{ik} \partial_{y_k} v_2^{(j)} \rangle.$$

Collecting all these terms together, we obtain that $w^{(0)}$ satisfies the elliptic system

$$L^h w^{(0)} = g\,,$$

where L^h is the matrix elliptic operator

$$L^h := -\mathrm{div}(\mathsf{A}^h_{e\ell}\ \mathrm{grad}_x)\,,$$

$\mathsf{A}^h_{e\ell}$ is the constant coefficients matrix, which in block form is expressed as

$$\mathsf{A}^h_{e\ell} = \begin{pmatrix} a^h & b^h \\ c^h & d^h \end{pmatrix}, \tag{9.25}$$

the 3×3 matrices a^h, b^h, c^h, d^h being defined as

$$\begin{aligned}
(a^h)_{ij} &= \langle a_{ij} + \sum_{k=1}^{3} a_{ik} \partial_{y_k} r_1^{(j)} + \sum_{k=1}^{3} b_{ik} \partial_{y_k} r_2^{(j)} \rangle, \\
(b^h)_{ij} &= \langle b_{ij} + \sum_{k=1}^{3} a_{ik} \partial_{y_k} v_1^{(j)} + \sum_{k=1}^{3} b_{ik} \partial_{y_k} v_2^{(j)} \rangle, \\
(c^h)_{ij} &= \langle c_{ij} + \sum_{k=1}^{3} c_{ik} \partial_{y_k} r_1^{(j)} + \sum_{k=1}^{3} d_{ik} \partial_{y_k} r_2^{(j)} \rangle, \\
(d^h)_{ij} &= \langle d_{ij} + \sum_{k=1}^{3} c_{ik} \partial_{y_k} v_1^{(j)} + \sum_{k=1}^{3} d_{ik} \partial_{y_k} v_2^{(j)} \rangle,
\end{aligned} \tag{9.26}$$

$r_\ell^{(j)}$, $v_\ell^{(j)}$, $j = 1, 2, 3$, $\ell = 1, 2$, being the solutions of the cell systems (9.24).

The above discussion motivates the following definitions:

DEFINITION 9.4.6 (HOMOGENISED DIFFUSION MATRIX) *The constant coefficient matrix* (9.25) *with coefficients defined as in* (9.26) *is called the homogenised diffusion matrix.*

With the aid of the homogenised matrix $\mathsf{A}^h_{e\ell}$ we define:

DEFINITION 9.4.7 (HOMOGENISED OPERATOR AND SYSTEM) *The homogenised (matrix) elliptic operator* $L^h : H_0^1(\mathcal{O}) \times H_0^1(\mathcal{O}) \to H^{-1}(\mathcal{O}) \times H^{-1}(\mathcal{O})$, *is defined by*

$$L^h = -\mathrm{div}_x(\mathsf{A}^h_{e\ell}\ \mathrm{grad}_x\ ). \tag{9.27}$$

The elliptic system

$$L^h w = g \tag{9.28}$$

is called the homogenised system.

It is not obvious, but it is easy to prove that the block matrix $\mathsf{A}^h_{e\ell}$ is positive definite. The operator L^h is a constant coefficient matrix elliptic operator.

The formal two-scale expansion for the elliptic system provided above suffers from the same problems concerning rigour as the relevant expansion for the Maxwell system (see Remark 9.3.2). The following theorem, which is a slight modification of Theorem 9.1 in [59], guarantees that the result obtained by this formal expansion indeed provides the right answer.

THEOREM 9.4.8 *For* $f = (f_1, f_2)^{tr}$, *consider the solution* $\mathsf{u}^\epsilon = (\mathsf{u}_1^\epsilon, \mathsf{u}_2^\epsilon)^{tr}$ *of the elliptic system* $L^\epsilon u^\epsilon = f$. *As* $\epsilon \to 0$, *we have that* $\mathsf{u}^\epsilon \rightharpoonup \mathsf{u}^*$, *in* $H_0^1(\mathcal{O}) \times H_0^1(\mathcal{O})$, *where* $\mathsf{u}^* = (\mathsf{u}_1^*, \mathsf{u}_2^*)^{tr}$ *is the solution of the elliptic system* $L^h \mathsf{u}^* = f$. *Furthermore,* $\mathsf{A}^\epsilon_{e\ell}\ \mathsf{u}^\epsilon \rightharpoonup \mathsf{A}^h_{e\ell}\ \mathsf{u}^*$, *in* $L^2(\mathcal{O})$.

REMARK 9.4.9 In the proof of Theorem 9.4.8, to ease notation, we denote grad_x, div_x, curl_x simply by grad, div, curl, respectively, but to avoid confusion, we always retain the explicit notation grad_y, div_y, curl_y when referring to the rescaled variable y.

Proof. We rewrite the cell equations in a more general form as

$$\begin{aligned} &-\mathrm{div}_y(a^{tr}\ (q_1 + \mathrm{grad}_y \mathsf{w}_1)) - \mathrm{div}_y(c^{tr}\ (q_2 + \mathrm{grad}_y \mathsf{w}_2) = 0, \\ &-\mathrm{div}_y(b^{tr}\ (q_1 + \mathrm{grad}_y \mathsf{w}_1)) - \mathrm{div}_y(d^{tr}\ (q_2 + \mathrm{grad}_y \mathsf{w}_2) = 0, \end{aligned} \tag{9.29}$$

where q_1, q_2 are arbitrary vectors in $\mathbb{R}^3$. Naturally, w depends on the choice of q_1, q_2, a fact that is suppressed in the subsequent notation. Let e_j be the unit vector in the j direction. Note that for the choice $q_1 = e_j$, $j = 1, 2, 3$, $q_2 = 0$, we recover the transposed cell equations for $(r_1^{(j)}, r_2^{(j)})$, whereas for the choice $q_1 = 0$, $q_2 = e_j$, $j = 1, 2, 3$, we recover the transposed cell equations for $(v_1^{(j)}, v_2^{(j)})$.

In what follows, by $\mathfrak{s}^\epsilon$ we denote $\mathfrak{s}\left(\frac{x}{\epsilon}\right)$ for $\mathfrak{s} = a, b, c, d$. We also define $w_1^\epsilon = \epsilon \mathsf{w}_1\left(\frac{x}{\epsilon}\right)$, $w_2^\epsilon = \epsilon \mathsf{w}_2\left(\frac{x}{\epsilon}\right)$. By the a priori estimates of the elliptic system and the properties of the matrix $\mathsf{A}_{e\ell}$ we have that $\mathrm{grad}\mathsf{u}^\epsilon$ is bounded in $H^1(\mathcal{O}) \times H^1(\mathcal{O})$, so that by the weak compactness results (see A.2 in Appendix A), we see that (at least up to a subsequence) there exists a function $\mathsf{u}^* = (\mathsf{u}_1^*, \mathsf{u}_2^*)$ such that $(\mathrm{grad}\mathsf{u}_1^\epsilon, \mathrm{grad}\mathsf{u}_2^\epsilon) \rightharpoonup (\mathrm{grad}\mathsf{u}_1^*, \mathrm{grad}\mathsf{u}_2^*)$ in $L^2(\mathcal{O}) \times L^2(\mathcal{O})$. Furthermore, by the a priori estimates of the elliptic system, and once more invoking weak compactness arguments, we have that there exists a function $r^* = (r_1^*, r_2^*)^{tr}$ such that

$$a^\epsilon \,\mathrm{grad}\mathsf{u}_1^\epsilon + b^\epsilon \,\mathrm{grad}\mathsf{u}_2^\epsilon \rightharpoonup r_1^*, \quad c^\epsilon \,\mathrm{grad}\mathsf{u}_1^\epsilon + d^\epsilon \,\mathrm{grad}\mathsf{u}_2^\epsilon \rightharpoonup r_2^*, \text{ in } L^2(\mathcal{O}).$$

By similar estimates for the cell equation and using periodicity and Theorem C.1.4 in Appendix C we have

$$\begin{aligned}
&\mathrm{grad} w_1^\epsilon = q_1 + \mathrm{grad}_y \mathsf{w}_1 \overset{*}{\rightharpoonup} q_1,\\
&\mathrm{grad} w_2^\epsilon = q_2 + \mathrm{grad}_y \mathsf{w}_2 \overset{*}{\rightharpoonup} q_2,\\
&a^{tr,\epsilon}\mathrm{grad} w_1^\epsilon + c^{tr,\epsilon}\mathrm{grad} w_2^\epsilon \overset{*}{\rightharpoonup} a^{tr,h} q_1 + c^{tr,h} q_2,\\
&b^{tr,\epsilon}\mathrm{grad} w_1^\epsilon + d^{tr,\epsilon}\mathrm{grad} w_2^\epsilon \overset{*}{\rightharpoonup} b^{tr,h} q_1 + d^{tr,h} q_2,
\end{aligned}$$

in $L^\infty(\mathcal{O})$, where the matrices a^h, b^h, c^h, d^h are such that

$$\begin{aligned}
a^{tr,h} q_1 + c^{tr,h} q_2 &= \langle a^{tr}(q_1 + \mathrm{grad}_y \mathsf{w}_1)\rangle + \langle c^{tr}(q_2 + \mathrm{grad}_y \mathsf{w}_2)\rangle,\\
b^{tr,h} q_1 + d^{tr,h} q_2 &= \langle b^{tr}(q_1 + \mathrm{grad}_y \mathsf{w}_1)\rangle + \langle d^{tr}(q_2 + \mathrm{grad}_y \mathsf{w}_2)\rangle,
\end{aligned} \tag{9.30}$$

for all $q_1, q_2 \in \mathbb{R}^3$.

We now observe that

$$\begin{aligned}
(a^{tr,\epsilon}\mathrm{grad} w_1^\epsilon + c^{tr,\epsilon}\mathrm{grad} w_2^\epsilon)^{tr}\mathrm{grad}\mathsf{u}_1^\epsilon &= \left((\mathrm{grad} w_1^\epsilon)^{tr} a^\epsilon + (\mathrm{grad} w_2^\epsilon)^{tr} c^\epsilon\right)\mathrm{grad}\mathsf{u}_1^\epsilon,\\
(b^{tr,\epsilon}\mathrm{grad} w_1^\epsilon + d^{tr,\epsilon}\mathrm{grad} w_2^\epsilon)^{tr}\mathrm{grad}\mathsf{u}_2^\epsilon &= \left((\mathrm{grad} w_1^\epsilon)^{tr} b^\epsilon + (\mathrm{grad} w_2^\epsilon)^{tr} d^\epsilon\right)\mathrm{grad}\mathsf{u}_2^\epsilon.
\end{aligned}$$

We add these two equations and rearrange terms to obtain

$$\begin{aligned}
&(a^{tr,\epsilon}\mathrm{grad} w_1^\epsilon + c^{tr,\epsilon}\mathrm{grad} w_2^\epsilon)^{tr}\mathrm{grad}\mathsf{u}_1^\epsilon + (b^{tr,\epsilon}\mathrm{grad} w_1^\epsilon + d^{tr,\epsilon}\mathrm{grad} w_2^\epsilon)^{tr}\mathrm{grad}\mathsf{u}_2^\epsilon =\\
&(\mathrm{grad} w_1^\epsilon)^{tr}(a^\epsilon \mathrm{grad}\mathsf{u}_1^\epsilon + b^\epsilon \mathrm{grad}\mathsf{u}_2^\epsilon) + (\mathrm{grad} w_2^\epsilon)^{tr}(c^\epsilon \mathrm{grad}\mathsf{u}_1^\epsilon + d^\epsilon \mathrm{grad}\mathsf{u}_2^\epsilon),
\end{aligned} \tag{9.31}$$

and apply the div-curl lemma (see Theorem C.2.2 in Appendix C) on the products of both sides. This is possible since $\mathrm{grad} w_1^\epsilon$, $\mathrm{grad} w_2^\epsilon$ are curl free, whereas

$$a^{tr,\epsilon}\mathrm{grad} w_1^\epsilon + c^{tr,\epsilon}\mathrm{grad} w_2^\epsilon = a^{tr}(y)(q_1 + \mathrm{grad}_y \mathsf{w}_1) + c^{tr}(y)(q_2 + \mathrm{grad}_y \mathsf{w}_2),$$

$$b^{tr,\epsilon}\mathrm{grad} w_1^\epsilon + d^{tr,\epsilon}\mathrm{grad} w_2^\epsilon = b^{tr}(y)(q_1 + \mathrm{grad}_y \mathsf{w}_1) + d^{tr}(y)(q_2 + \mathrm{grad}_y \mathsf{w}_2),$$

are divergence free, since $\mathsf{w} = (\mathsf{w}_1, \mathsf{w}_2)^{tr}$ satisfies the cell equations (9.29). Further, if $h = (h_1, h_2)^{tr}$ is such that $\mathrm{div} h_i = g_i$, $i = 1, 2$, then by the elliptic system $a^\epsilon \mathrm{grad}\mathsf{u}_1^\epsilon + b^\epsilon \mathrm{grad}\mathsf{u}_2^\epsilon - h_1$, $c^\epsilon \mathrm{grad}\mathsf{u}_1^\epsilon + d^\epsilon \mathrm{grad}\mathsf{u}_2^\epsilon - h_2$ are divergence free, and h is independent of ϵ. Using the above, we identify the limit of the products in (9.31) as

$$(a^{tr,h} q_1 + c^{tr,h} q_2)^{tr}\mathrm{grad}\mathsf{u}_1^* + (b^{tr,h} q_1 + d^{tr,h} q_2)^{tr}\mathrm{grad}\mathsf{u}_2^* = q_1^{tr} r_1^* + q_2^{tr} r_2^*.$$

Since this is valid for all $q_1, q_2 \in \mathbb{R}^3$, we finally obtain that

$$\begin{aligned}
r_1^* &= a^h \mathrm{grad}\mathsf{u}_1^* + b^h \mathrm{grad}\mathsf{u}_2^*,\\
r_2^* &= c^h \mathrm{grad}\mathsf{u}_1^* + d^h \mathrm{grad}\mathsf{u}_2^*,
\end{aligned}$$

which is the stated result. The identification of the homogenised diffusion matrix comes from (9.30) by proper choices of the vectors q_1, q_2. Noting that

$\mathsf{w} = (\mathsf{w}_1, \mathsf{w}_2)^{tr}$ are the solutions of the transposed cell equations (9.24) and using an integration by parts argument[7] similar to the one employed in [97] for the scalar case, we see that the homogenised diffusion matrix is given by (9.26). □

REMARK 9.4.10 The weak convergence in $H_0^1(\mathcal{O}) \times H_0^1(\mathcal{O})$ can be turned into a strong one as long as an appropriate term, called the *corrector*, is added to the solution (see, e.g., [59], [97], [407]).

9.4.2 The homogenisation of the Maxwell system

We now turn to the homogenisation of the Maxwell system. As mentioned earlier, we will treat this problem using a suitably selected auxiliary elliptic system and applying the results for the elliptic homogenisation theory provided in Section 9.4.1 and Section C.3 in Appendix C. For reasons that will become apparent in the proof of the subsequent Theorem 9.4.14, consider the following.

DEFINITION 9.4.11 *The auxiliary "microstructure" elliptic operator associated with the Maxwell system is*

$$L_M^\epsilon = -\mathrm{div}_x(\mathsf{A}_{or}^{\epsilon,tr}\mathrm{grad}_x\cdot)\,, \tag{9.32}$$

and the auxiliary "cell" elliptic operator associated with the Maxwell system is

$$L_{per,M} = -\mathrm{div}_y((\mathsf{A}_{or}^{per})^{tr}\mathrm{grad}_y\cdot)\,. \tag{9.33}$$

DEFINITION 9.4.12 *Set* $\mathsf{A}_{e\ell}^\epsilon = \mathsf{A}_{or}^{\epsilon,tr}$ *and let* $\mathsf{A}_{e\ell}^h$ *be the homogenised diffusion matrix defined as in* (9.26), *for* $a = \varepsilon^{tr}$, $b = \zeta^{tr}$, $c = \xi^{tr}$, $d = \mu^{tr}$. *The matrix* $\mathsf{A}_{or}^h := \mathsf{A}_{e\ell}^{h,tr}$ *is the homogenised optical response matrix of the medium.*

REMARK 9.4.13 By the symmetry of the optical response matrix (Assumption 2.3.5), we have $\mathsf{A}_{\mathrm{or}}^{tr} = \mathsf{A}_{\mathrm{or}}$. Nevertheless, since the arguments in the following proof will also be used for the dispersion part G_{d}, which is nonsymmetric, we perform the proof without assuming the symmetry of A_{or}.

The above procedure leads to explicit relations for the homogenised optical response matrix of the medium as

$$\begin{aligned}
\varepsilon^h &= \langle \varepsilon + \mathrm{grad}_y R_1\, \varepsilon + \mathrm{grad}_y R_2\, \zeta \rangle, \\
\zeta^h &= \langle \zeta + \mathrm{grad}_y V_1\, \varepsilon + \mathrm{grad}_y V_2\, \zeta \rangle, \\
\xi^h &= \langle \xi + \mathrm{grad}_y R_1\, \xi + \mathrm{grad}_y R_2\, \mu \rangle, \\
\mu^h &= \langle \mu + \mathrm{grad}_y V_1\, \xi + \mathrm{grad}_y V_2\, \mu \rangle,
\end{aligned} \tag{9.34}$$

where

$$\mathrm{grad}_y \mathfrak{S}_\ell = \begin{pmatrix} \partial_{y_1}\mathfrak{s}_\ell^{(1)} & \partial_{y_2}\mathfrak{s}_\ell^{(1)} & \partial_{y_3}\mathfrak{s}_\ell^{(1)} \\ \partial_{y_1}\mathfrak{s}_\ell^{(2)} & \partial_{y_2}\mathfrak{s}_\ell^{(2)} & \partial_{y_3}\mathfrak{s}_\ell^{(2)} \\ \partial_{y_1}\mathfrak{s}_\ell^{(3)} & \partial_{y_2}\mathfrak{s}_\ell^{(3)} & \partial_{y_3}\mathfrak{s}_\ell^{(3)} \end{pmatrix}, \quad \ell = 1, 2,$$

[7]The argument is similar to the one used in the proof of Proposition 9.4.16.

and $\mathfrak{S}$ and $\mathfrak{s}$ are proxies for R, V and r, v, respectively.

We shall prove the following.

THEOREM 9.4.14 *Let $j = (-J, 0)^{tr}$ satisfy the assumptions of Theorem 9.2.5 and let $u^\epsilon = (E^\epsilon, H^\epsilon)^{tr}$ be the solution of the Maxwell system (9.6)-(9.7). Then*

$$u^\epsilon \overset{*}{\rightharpoonup} u^*, \quad in\ L^\infty([0,T], \mathbb{X}),$$

where $u^* = (E^*, H^*)^{tr}$ is the unique solution of the Maxwell system

$$(\mathsf{d}^*)' = \mathsf{M}u^* + j, \quad in\ (0,T] \times \mathcal{O} \tag{9.35}$$

with the perfect conductor boundary condition and subject to the homogeneous constitutive relations

$$\mathsf{d}^* = \mathsf{A}^h_{or} u^*. \tag{9.36}$$

The matrices ε^h, ξ^h, ζ^h, μ^h that constitute the homogenised optical response A^h_{or} are given by (9.34).

Proof. We break the proof into three parts:

(*i*) *$u^* = (E^*, H^*)^{tr}$ satisfies the Maxwell system:*

By Theorem 9.2.5 we have the uniform bounds

$$\|E^\epsilon\|_{\mathsf{X}} \le C_1, \qquad \|H^\epsilon\|_{\mathsf{X}} \le C_2, \quad \epsilon > 0,\ t > 0. \tag{9.37}$$

The boundedness of A^ϵ_{or}, together with (9.37), implies that D^ϵ and B^ϵ are also bounded in X uniformly in $\epsilon, t > 0$. It is then standard [97, Theorem 1.26] that there exist $E^*, H^*, D^*, B^* \in L^\infty([0,T], \mathsf{X})$ such that, up to taking a subsequence $\epsilon \to 0$, there holds

$$\begin{matrix} E^\epsilon \overset{*}{\rightharpoonup} E^*, & H^\epsilon \overset{*}{\rightharpoonup} H^* \\ D^\epsilon \overset{*}{\rightharpoonup} D^*, & B^\epsilon \overset{*}{\rightharpoonup} B^* \end{matrix} \quad \text{in } L^\infty([0,T], \mathsf{X}). \tag{9.38}$$

The ensuing arguments will identify $(E^*, H^*)^{tr}$ and will establish that *any* $*$-weakly convergent subsequence of $(E^\epsilon, H^\epsilon)^{tr}$ has $(E^*, H^*)^{tr}$ as its limit. This implies the convergence of the full sequence $(E^\epsilon, H^\epsilon)^{tr}$ (see Theorem 1.26 in [97]).

Taking the Laplace transform of the Maxwell equations, we obtain for $p \in \mathbb{C}^+$, $x \in \mathcal{O}$,

$$\begin{aligned} p\widehat{D}^\epsilon &= \text{curl}\widehat{H}^\epsilon - \widehat{J}, \\ p\widehat{B}^\epsilon &= -\text{curl}\widehat{E}^\epsilon. \end{aligned} \tag{9.39}$$

Moreover, (9.38) implies that

$$\begin{matrix} \widehat{E}^\epsilon \rightharpoonup \widehat{E}^* & \widehat{H}^\epsilon \rightharpoonup \widehat{H}^* \\ \widehat{D}^\epsilon \rightharpoonup \widehat{D}^* & \widehat{B}^\epsilon \rightharpoonup \widehat{B}^* \end{matrix}, \text{ in } \mathsf{X} \quad (\text{fixed } p \in \mathbb{C}^+). \tag{9.40}$$

Combining (9.39) and (9.40) implies that for fixed $p \in \mathbb{C}^+$, the vector fields $\text{curl}\widehat{E}^\epsilon$ and $\text{curl}\widehat{H}^\epsilon$ remain bounded in X as $\epsilon \to 0$. Hence they have weak limits in X. It then follows from (9.40) that $\widehat{E}^*$ and $\widehat{H}^*$ belong to $H(\text{curl}, \mathcal{O})$, and moreover,

$$\widehat{E}^\epsilon \rightharpoonup \widehat{E}^*, \qquad \widehat{H}^\epsilon \rightharpoonup \widehat{H}^* \quad \text{in } H(\text{curl}, \mathcal{O}). \tag{9.41}$$

Letting $\epsilon \to 0$ in (9.39), then, for $p \in \mathbb{C}^+$, $x \in \mathcal{O}$, yields

$$\begin{aligned} p\widehat{D}^* &= \text{curl}\widehat{H}^* - \widehat{J}, \\ p\widehat{B}^* &= -\text{curl}\widehat{E}^*, \end{aligned} \tag{9.42}$$

which implies that E^*, H^*, D^*, B^* satisfy the Maxwell system

$$\begin{aligned} \partial_t D^* &= \text{curl}H^* - J, \text{ in } (0,T] \times \mathcal{O}, \\ \partial_t B^* &= -\text{curl}E^*, \text{ in } (0,T] \times \mathcal{O}, \\ E^*(x,0) &= 0, \quad H^*(x,0) = 0, \text{ in } \mathcal{O}. \end{aligned} \tag{9.43}$$

Hence it remains to establish that the boundary condition $n \times E^* = 0$ is also satisfied and that the vector fields E^*, H^*, D^*, B^* are related by the constitutive relations (9.36).

(ii) Validity of the boundary condition:

We first note that the boundary condition is understood in the sense of the trace operator $\gamma_\tau : H(\text{curl}, \mathcal{O}) \to H^{-1/2}(\partial\mathcal{O})$ (see Section 3.5). Let us fix a function $\phi \in H^{1/2}(\partial\mathcal{O})$. There exists (see [126] p. 341]) $\psi \in H^1(\mathcal{O})$ such that $\psi|_{\partial\mathcal{O}} = \phi$. Now, for $\epsilon > 0$ there holds

$$\begin{aligned} \int_{\mathcal{O}} \text{curl}\psi \cdot u^\epsilon \, dx &= \int_{\mathcal{O}} \text{curl}E^\epsilon \cdot \psi \, dx + \int_{\partial\mathcal{O}} \psi(n \times E^\epsilon) \, ds, \\ \int_{\mathcal{O}} \text{curl}\psi \cdot E^* &= \int_{\mathcal{O}} \text{curl}E^* \cdot \psi \, dx + \int_{\partial\mathcal{O}} \psi(n \times E^*) \, ds. \end{aligned}$$

In the limit as $\epsilon \to 0$ and using the fact that $n \times E^\epsilon|_{\partial\mathcal{O}} = 0$ and the relations

$$\begin{aligned} \int_{\mathcal{O}} \text{curl}\psi \cdot E^\epsilon \, dx &\to \int_{\mathcal{O}} \text{curl}\psi \cdot E^* \, dx, \\ \int_{\mathcal{O}} \text{curl}E^\epsilon \cdot \psi \, dx &\to \int_{\mathcal{O}} \text{curl}E^* \cdot \psi \, dx, \end{aligned}$$

we obtain

$$\int_{\partial\mathcal{O}} \phi(n \times E^*) \, ds = \int_{\partial\mathcal{O}} \psi(n \times E^*) \, ds = 0.$$

Since $\phi \in H^{1/2}(\partial\mathcal{O})$ is arbitrary, we conclude that $n \times E^* = 0$ on $\partial\mathcal{O}$.

(iii) Validity of the constitutive relations:

Let us fix a bounded domain $\mathcal{O}_0$ with $\overline{\mathcal{O}}_0 \subset \mathcal{O}$. Since $\text{div}\,\text{curl}\,u = 0$, (9.39) and (9.42) imply that $\text{div}\widehat{D}^\epsilon = \text{div}\widehat{D}^*$ and $\text{div}\widehat{B}^\epsilon = \text{div}\widehat{B}^*$, and (9.40) then yields

$$\widehat{D}^\epsilon \rightharpoonup \widehat{D}^*, \quad \widehat{B}^\epsilon \rightharpoonup \widehat{B}^* \quad \text{in } H(\text{div}, \mathcal{O}_0). \tag{9.44}$$

The elliptic operator L^ϵ (Definition 9.4.4) is invertible for all $\epsilon > 0$. For $g = (g_1, g_2)^{tr} \in H^{-1}(\mathcal{O}_0) \times H^{-1}(\mathcal{O}_0)$ fixed, let $w^\epsilon = (w_1^\epsilon, w_2^\epsilon)^{tr} \in H_0^1(\mathcal{O}_0) \times H_0^1(\mathcal{O}_0)$ solve the elliptic system

$$L^\epsilon w^\epsilon = g. \tag{9.45}$$

Moreover, let L^h be the constant coefficient operator elliptic operator of Definition 9.4.7, and consider the solution $w = (w_1, w_2)^{tr}$ of the problem

$$L^h w = g. \tag{9.46}$$

By Theorem 9.4.8,

$$\mathrm{grad} w_1^\epsilon \rightharpoonup \mathrm{grad} w_1, \quad \mathrm{grad} w_2^\epsilon \rightharpoonup \mathrm{grad} w_2 \quad \text{in } \mathsf{X}, \tag{9.47}$$

which, together with the fact that curl grad $= 0$, implies that

$$\mathrm{grad} w_1^\epsilon \rightharpoonup \mathrm{grad} w_1, \quad \mathrm{grad} w_2^\epsilon \rightharpoonup \mathrm{grad} w_2 \quad \text{in } H(\mathrm{curl}, \mathcal{O}_0). \tag{9.48}$$

Since $D^\epsilon \rightharpoonup D^*$ in $H(\mathrm{div}, \mathcal{O}_0)$ (by (9.44)) and $\mathrm{grad} w_1^\epsilon \rightharpoonup \mathrm{grad} w_1$ in $H(\mathrm{curl}, \mathcal{O}_0)$ (by (9.48)), the product of the two sequences will converge strongly to the product of the limits in $\mathcal{D}'(\mathcal{O}_0)$ by compensated compactness (see Theorem C.2.1 in Appendix C), and similarly for the sequences B^ϵ and w_2^ϵ. Therefore,

$$\left.\begin{array}{l} \widehat{D}^\epsilon \cdot \mathrm{grad} w_1^\epsilon \to \widehat{D}^* \cdot \mathrm{grad} w_1 \\ \widehat{B}^\epsilon \cdot \mathrm{grad} w_2^\epsilon \to \widehat{B}^* \cdot \mathrm{grad} w_2 \end{array}\right., \quad \text{in } \mathcal{D}'(\mathcal{O}_0). \tag{9.49}$$

Moreover, by Theorem 9.4.8,

$$\left.\begin{array}{l} a^\epsilon \mathrm{grad} w_1^\epsilon + b^\epsilon \mathrm{grad} w_2^\epsilon \rightharpoonup a^h \mathrm{grad} w_1 + b^h \mathrm{grad} w_2, \\ c^\epsilon \mathrm{grad} w_1^\epsilon + d^\epsilon \mathrm{grad} w_2^\epsilon \rightharpoonup c^h \mathrm{grad} w_1 + d^h \mathrm{grad} w_2 \end{array}\right., \quad \text{in } \mathsf{X}. \tag{9.50}$$

By the definition of w^ϵ and w, as solutions of (9.45) and (9.46), respectively,

$$\begin{array}{l} -\mathrm{div}(a^\epsilon \mathrm{grad} w_1^\epsilon + b^\epsilon \mathrm{grad} w_2^\epsilon) = g_1 = -\mathrm{div}(a^h \mathrm{grad} w_1 + b^h \mathrm{grad} w_2), \\ -\mathrm{div}(c^\epsilon \mathrm{grad} w_1^\epsilon + d^\epsilon \mathrm{grad} w_2^\epsilon) = g_2 = -\mathrm{div}(c^h \mathrm{grad} w_1 + d^h \mathrm{grad} w_2), \end{array}$$

and these, together with (9.50), imply

$$\left.\begin{array}{l} a^\epsilon \mathrm{grad} w_1^\epsilon + b^\epsilon \mathrm{grad} w_2^\epsilon \rightharpoonup a^h \mathrm{grad} w_1 + b^h \mathrm{grad} w_2 \\ c^\epsilon \mathrm{grad} w_1^\epsilon + d^\epsilon \mathrm{grad} w_2^\epsilon \rightharpoonup c^h \mathrm{grad} w_1 + d^h \mathrm{grad} w_2 \end{array}\right., \quad \text{in } H(\mathrm{div}, \mathcal{O}_0). \tag{9.51}$$

The convergence results (9.41) and (9.51) allow us to apply again the compensated compactness result (Theorem C.2.1 in Appendix C) twice, once to the pair of sequences $a^\epsilon \mathrm{grad} w_1^\epsilon + b^\epsilon \mathrm{grad} w_2^\epsilon$ and $\widehat{E}^\epsilon$ and once to the pair of sequences $c^\epsilon \mathrm{grad} w_1^\epsilon + d^\epsilon \mathrm{grad} w_2^\epsilon) \cdot \widehat{H}^\epsilon$ and $\widehat{H}^\epsilon$, to obtain

$$\left.\begin{array}{l} (a^\epsilon \mathrm{grad} w_1^\epsilon + b^\epsilon \mathrm{grad} w_2^\epsilon) \cdot \widehat{E}^\epsilon \to (a^h \mathrm{grad} w_1 + b^h \mathrm{grad} w_2) \cdot \widehat{E}^* \\ (c^\epsilon \mathrm{grad} w_1^\epsilon + d^\epsilon \mathrm{grad} w_2^\epsilon) \cdot \widehat{H}^\epsilon \to (c^h \mathrm{grad} w_1 + d^h \mathrm{grad} w_2) \cdot \widehat{H}^* \end{array}\right., \tag{9.52}$$

in $\mathcal{D}'(\mathcal{O}_0)$.

Now we observe that the left-hand side of the sum of the two relations in (9.49),

$$\widehat{D}^\epsilon \cdot \mathrm{grad} w_1^\epsilon + \widehat{B}^\epsilon \cdot \mathrm{grad} w_2^\epsilon$$

and the left-hand side of the sum of the two relations in (9.52),

$$(a^\epsilon \mathrm{grad} w_1^\epsilon + b^\epsilon \mathrm{grad} w_2^\epsilon) \cdot \widehat{E}^\epsilon + (c^\epsilon \mathrm{grad} w_1^\epsilon + d^\epsilon \mathrm{grad} w_2^\epsilon) \cdot \widehat{H}^\epsilon =$$
$$(a^{\epsilon,tr} \widehat{E}^\epsilon + c^{\epsilon,tr} \widehat{H}^\epsilon) \cdot \mathrm{grad} w_1^\epsilon + (b^{\epsilon,tr} \widehat{E}^\epsilon + d^{\epsilon,tr} \widehat{H}^\epsilon) \cdot \mathrm{grad} w_2^\epsilon,$$

coincide, as long as

$$a^{\epsilon,tr} = \varepsilon^\epsilon,\ c^{\epsilon,tr} = \xi^\epsilon, b^{\epsilon,tr} = \zeta^\epsilon, d^{\epsilon,tr} = \mu^\epsilon,$$

i.e., $\mathsf{A}_{e\ell} = \mathsf{A}_{\mathrm{or}}^{tr}$. For this choice, the corresponding right-hand sides

$$\widehat{D}^* \cdot \mathrm{grad} w_1 + \widehat{B}^* \cdot \mathrm{grad} w_2$$

and

$$(a^h \mathrm{grad} w_1 + b^h \mathrm{grad} w_2) \cdot \widehat{E}^* + (c^h \mathrm{grad} w_1 + d^h \mathrm{grad} w_2) \cdot \widehat{H}^*,$$

respectively, are also equal, which on rearrangement yields

$$\widehat{D}^* \cdot \mathrm{grad} w_1 + \widehat{B}^* \cdot \mathrm{grad} w_2 =$$
$$(a^{h,tr} \widehat{E}^* + c^{h,tr} \widehat{H}^*) \cdot \mathrm{grad} w_1 + (b^{h,tr} \widehat{E}^* + d^{h,tr} \widehat{H}^*) \cdot \mathrm{grad} w_2.$$

The fact that g_1 and g_2 were arbitrary, so that $\mathrm{grad} w_1, \mathrm{grad} w_2$ are also arbitrary, implies that

$$\widehat{D}^* = a^{h,tr} \widehat{E}^* + c^{h,tr} \widehat{H}^*,$$
$$\widehat{B}^* = b^{h,tr} \widehat{E}^* + d^{h,tr} \widehat{H}^*,$$

and this can be interpreted as a homogenised constitutive relation of the form

$$\widehat{D}^* = \varepsilon^h \widehat{E}^* + \xi^h \widehat{H}^*,$$
$$\widehat{B}^* = \zeta^h \widehat{E}^* + \mu^h \widehat{H}^*,$$

by setting

$$\varepsilon^h := a^{h,tr},\ \xi^h := c^{h,tr},\ \zeta^h := b^{h,tr}\ \mu^h := d^{h,tr},$$

or in compact notation, $\mathsf{A}_{\mathrm{or}}^h := (\mathsf{A}_{e\ell}^{\mathrm{M}\ ,h})^{tr}$.

Since $\mathcal{O}_0$ is arbitrary, we obtain the Laplace transforms of the stated constitutive relations. This completes the proof. □

REMARK 9.4.15 The result may be extended for $T = \infty$ using energy considerations for properly selected data.

We complete this section by showing the equivalence between the two representations for the homogenised coefficients.

PROPOSITION 9.4.16 *The representations* (9.16) *and* (9.34) *for the homogenised optical response matrix* A_{or}^h *are equivalent.*

Proof. We perform the calculation for ε^h only; the calculation for the other elements proceeds analogously. Our claim is that

$$\langle \mathrm{grad}_y R_1 \, \varepsilon + \mathrm{grad}_y R_2 \, \zeta \rangle = \langle \varepsilon \mathrm{grad}_y \Lambda^{(1)} + \xi \mathrm{grad}_y \Lambda^{(3)} \rangle.$$

Consider the (i,j) element of the matrix on the left-hand side:

$$\begin{aligned}(LHS)_{ij} :=& \langle \sum_{k=1}^{3} \varepsilon_{kj}\partial_{y_k} r_1^{(i)} + \sum_{k=1}^{3} \zeta_{kj}\partial_{y_k} r_2^{(i)} \rangle \\ =& - \langle (\sum_{k=1}^{3} \partial_{y_k}\varepsilon_{kj}) r_1^{(i)} + (\sum_{k=1}^{3} \partial_{y_k}\zeta_{kj}) r_2^{(i)} \rangle \\ =& - \langle (\mathrm{div}_y \varepsilon_{\sharp,j})\, r_1^{(i)} + (\mathrm{div}_y \zeta_{\sharp,j})\, r_2^{(i)} \rangle \,.\end{aligned}$$

We now use (9.14) to write

$$\begin{aligned}(LHS)_{ij} =& \langle (\mathrm{div}_y(\varepsilon\, \mathrm{grad}_y \Lambda_j^{(1)}) + \mathrm{div}_y(\xi\, \mathrm{grad}_y \Lambda_j^{(3)}))\, r_1^{(i)} + \\ & (\mathrm{div}_y(\zeta\, \mathrm{grad}_y \Lambda_j^{(1)}) + \mathrm{div}_y(\mu\, \mathrm{grad}_y \Lambda_j^{(3)}))\, r_2^{(i)} \rangle \\ =& - \langle \varepsilon\, \mathrm{grad}_y \Lambda_j^{(1)} \cdot \mathrm{grad}_y r_1^{(i)} + \xi\, \mathrm{grad}_y \Lambda_j^{(3)} \cdot \mathrm{grad}_y r_1^{(i)} + \\ & \zeta\, \mathrm{grad}_y \Lambda_j^{(1)} \cdot \mathrm{grad}_y r_2^{(i)} + \mu\, \mathrm{grad}_y \Lambda_j^{(3)} \cdot \mathrm{grad}_y r_2^{(i)} \rangle.\end{aligned}$$

Consider now the (i,j) element of the matrix on the right-hand side:

$$\begin{aligned}(RHS)_{ij} :=& \langle \sum_{k} \varepsilon_{ik}\partial_{y_k} \Lambda_j^{(1)} + \sum_{k} \xi_{ik}\partial_{y_k} \Lambda_j^{(3)} \rangle \\ =& - \langle (\mathrm{div}_y \varepsilon_{i,\sharp})\, \Lambda_j^{(1)} + (\mathrm{div}_y \xi_{i,\sharp})\, \Lambda_j^{(3)} \rangle \,.\end{aligned}$$

We now recall (9.24). Since $a = \varepsilon^{tr}$, $\mathrm{div}_y a_{\sharp,i} = \mathrm{div}_y \varepsilon_{i,\sharp}$, for $i = 1,2,3$. Similarly, since $c = \xi^{tr}$, $\mathrm{div}_y c_{\sharp,i} = \mathrm{div}_y \xi_{i,\sharp}$, for $i = 1,2,3$. Therefore,

$$\begin{aligned}(RHS)_{ij} =& \langle (\mathrm{div}_y(\varepsilon^{tr} \mathrm{grad}_y r_1^{(i)}) + \mathrm{div}_y(\zeta^{tr} \mathrm{grad}_y r_2^{(i)})\, \Lambda_j^{(1)} \\ & + (\mathrm{div}_y(\xi^{tr} \mathrm{grad}_y r_1^{(i)}) + \mathrm{div}_y(\mu^{tr} \mathrm{grad}_y r_2^{(i)})\, \Lambda_j^{(3)} \rangle \\ =& -\langle \varepsilon^{tr} \mathrm{grad}_y r_1^{(i)} \cdot \mathrm{grad}_y \Lambda_j^{(1)} + \zeta^{tr} \mathrm{grad}_y r_2^{(i)} \cdot \mathrm{grad}_y \Lambda_j^{(1)} \\ & + \xi^{tr} \mathrm{grad}_y r_1^{(i)} \cdot \mathrm{grad}_y \Lambda_j^{(3)} + \mu^{tr} \mathrm{grad}_y r_2^{(i)} \cdot \mathrm{grad}_y \Lambda_j^{(3)} \rangle \\ =& -\langle \mathrm{grad}_y r_1^{(i)} \cdot \varepsilon\, \mathrm{grad}_y \Lambda_j^{(1)} + \mathrm{grad}_y r_2^{(i)} \cdot \zeta\, \mathrm{grad}_y \Lambda_j^{(1)} \\ & + \mathrm{grad}_y r_1^{(i)} \cdot \xi\, \mathrm{grad}_y \Lambda_j^{(3)} + \mathrm{grad}_y r_2^{(i)} \cdot \mu\, \mathrm{grad}_y \Lambda_j^{(3)} \rangle \,.\end{aligned}$$

Clearly, $(LHS)_{ij} = (RHS)_{ij}$ for all $i,j = 1,2,3$. □

9.5 GENERAL BIANISOTROPIC MEDIA

We now include the effects of dispersion, modelled in the constitutive relation through $\mathsf{G}_{\mathrm{d}}^{\epsilon}(x)$. There are several ways to address this problem, each having its own advantages and disadvantages. We choose here to sketch two alternative ways based, respectively, either on the reduction to a properly selected elliptic homogenisation problem, which in essence is a generalisation of the approach adopted for the optical response model, or on a rigorisation of the double-scale expansion method, the *periodic unfolding method.*

In the first approach, we apply the Laplace transform on the Maxwell system, reducing the problem to a static one similar in form to that addressed for the optical response. In principle, the whole argument we have presented for the optical response may be translated almost verbatim for the Laplace transform of the full model, and we show that the Laplace transforms of the fields $\widehat{u}^\epsilon = (\widehat{E}^\epsilon, \widehat{H}^\epsilon)^{tr}$, $\widehat{\mathsf{d}}^\epsilon = (\widehat{D}^\epsilon, \widehat{B}^\epsilon)^{tr}$ have well-defined weak limits $\widehat{u}^* = (\widehat{E}_1^*, \widehat{H}_2^*)^{tr}$ and $\widehat{\mathsf{d}}^\epsilon = (\widehat{D}^\epsilon, \widehat{B}^\epsilon)^{tr}$, respectively, as $\epsilon \to 0$, which solves a static equation similar to (9.42). This homogenisation procedure, based on the connections with relevant problems for elliptic systems, provides a linear relationship between $\widehat{\mathsf{d}}^*$ and $\widehat{u}^*$ in terms of a constant coefficients matrix $\widehat{\mathsf{A}_{\mathrm{or}}}^h$, such that $\widehat{\mathsf{d}}^* = \widehat{\mathsf{A}_{\mathrm{or}}}^h \widehat{u}^*$. However, although this approach yields nice expressions for $\widehat{\mathsf{A}_{\mathrm{or}}}^h$, this is not exactly what we are interested in, and we need to invert the Laplace transform before we obtain a useful answer in the time domain. The inversion of Laplace transforms is not an easy task; however, various analytic and numerical techniques are available.

There are technical intricacies involved in this approach. Two of these consist (a) in proving that $\widehat{u}^*$, $\widehat{\mathsf{d}}^*$ are vector valued functions that correspond to the Laplace transform of suitable vector fields u^*, d^* in the time domain and (b) that the homogenised equation in a form similar to (9.42) can indeed be transformed by the inverse Laplace transform to a Maxwell-type system in the time domain. These problems are related to the fact that now the equation contains a complex parameter p; the solution of the equation must be a holomorphic function of this parameter so that it corresponds to a well-defined Laplace transform of a vector field. Furthermore, the corresponding auxiliary elliptic system will now be an elliptic system with complex coefficients, the homogenisation of which requires some extra care. These technical matters can be dealt with in a satisfactory manner, but rather advanced techniques from complex analysis are needed. However, leaving the above complications aside, homogenisation analysis for the Laplace transformed problem remains useful when treated for fixed p, as it leads to fixed frequency homogenisation results, which may often lead to interesting conclusions regarding the behaviour of composite materials (see, e.g., [428]). For this reason, in Section 9.5.1 we sketch the homogenisation problem for dispersive media using the Laplace transform and provide expressions for the coefficients of the homogenised medium.

The second approach to treating the problem with dispersion is to work directly in the time domain, introducing the concepts of two-scale convergence and the periodic unfolding method. This approach bypasses the need to use the Laplace transform and provides results for the homogenised constitutive relations directly in the time domain, rather than their Laplace transformed versions. In Section 9.5.2 we present a detailed discussion of the homogenisation problem using the periodic unfolding method.

9.5.1 Homogenisation using Laplace transforms

DEFINITION 9.5.1 *Let*

$$\mathsf{A}_{\mathfrak{L}}(y,p) := \begin{pmatrix} \varepsilon + \widehat{\varepsilon}_d & \xi + \widehat{\xi}_d \\ \zeta + \widehat{\zeta}_d & \mu + \widehat{\mu}_d \end{pmatrix} =: \begin{pmatrix} \varepsilon_{\mathfrak{L}} & \xi_{\mathfrak{L}} \\ \zeta_{\mathfrak{L}} & \mu_{\mathfrak{L}} \end{pmatrix},$$

where $\widehat{\mathfrak{s}}$ *denotes the Laplace transform of* $\mathfrak{s}$.

We impose the following assumption on the matrix $\mathsf{A}_{\mathfrak{L}}(y,p)$:

ASSUMPTION 9.5.2

(*i*) *The matrix* $\mathsf{A}_{\mathfrak{L}}(y,p)$ *satisfies*

$$(\mathcal{R}e\, \mathsf{A}_{\mathfrak{L}}(y,p))u \cdot u \geq c\|u\|^2, \quad y \in Y,\ p \in \mathbb{C}_+,\ u \in \mathbb{R}^6. \tag{9.53}$$

(*ii*) *The matrix* $\mathsf{A}_{\mathfrak{L}}(y,p)^{-1}$ *satisfies*

$$(\mathcal{R}e\, \mathsf{A}_{\mathfrak{L}}(y,p)^{-1})u \cdot u \geq c\|u\|^2, \quad y \in Y,\ p \in \mathbb{C}_+,\ u \in \mathbb{R}^6. \tag{9.54}$$

This is, e.g., satisfied if $\mathsf{G}_{\mathrm{d}}(y,t)$ is small compared to $\mathsf{A}_{\mathrm{or}}(y)$, $y \in Y$.

Let $r^{(j)} = (r_1^{(j)}, r_2^{(j)})^{tr}$, $v^{(j)} = (v_1^{(j)}, v_2^{(j)})^{tr}$, $j = 1,2,3$, be the solutions of the elliptic systems

$$L_{\mathfrak{L},\mathrm{per}}\begin{pmatrix} r_1^{(j)} \\ r_2^{(j)} \end{pmatrix} = \begin{pmatrix} \mathrm{div}_y(\varepsilon_{\mathfrak{L}})_{j,\sharp} \\ \mathrm{div}_y(\xi_{\mathfrak{L}})_{j,\sharp} \end{pmatrix},\ L_{\mathfrak{L},\mathrm{per}}\begin{pmatrix} v_1^{(j)} \\ v_2^{(j)} \end{pmatrix} = \begin{pmatrix} \mathrm{div}_y(\zeta_{\mathfrak{L}})_{j,\sharp} \\ \mathrm{div}_y(\mu_{\mathfrak{L}})_{j,\sharp} \end{pmatrix}, \tag{9.55}$$

where

$$L_{\mathfrak{L},\mathrm{per}} = \mathrm{div}_y(\mathsf{A}_{\mathfrak{L}}^{tr}(y,p)\,\mathrm{grad}_y).$$

Define

$$\begin{aligned} \varepsilon_{\mathfrak{L}}^h &:= \langle \varepsilon_{\mathfrak{L}} + \varepsilon_{\mathfrak{L}}\,\mathrm{grad}_y R_1 + \zeta_{\mathfrak{L}}\,\mathrm{grad}_y R_2 \rangle, \\ \zeta_{\mathfrak{L}}^h &:= \langle \zeta_{\mathfrak{L}} + \varepsilon_{\mathfrak{L}}\,\mathrm{grad}_y V_1 + \zeta_{\mathfrak{L}}\,\mathrm{grad}_y V_2 \rangle, \\ \xi_{\mathfrak{L}}^h &:= \langle \xi_{\mathfrak{L}} + \xi_{\mathfrak{L}}\,\mathrm{grad}_y R_1 + \mu_{\mathfrak{L}}\,\mathrm{grad}_y R_2 \rangle, \\ \mu_{\mathfrak{L}}^h &:= \langle \mu_{\mathfrak{L}} + \xi_{\mathfrak{L}}\,\mathrm{grad}_y V_1 + \mu_{\mathfrak{L}}\,\mathrm{grad}_y V_2 \rangle, \end{aligned} \tag{9.56}$$

where

$$\mathrm{grad}_y \mathfrak{S}_\ell = \begin{pmatrix} \partial_{y_1}\mathfrak{s}_\ell^{(1)} & \partial_{y_2}\mathfrak{s}_\ell^{(1)} & \partial_{y_3}\mathfrak{s}_\ell^{(1)} \\ \partial_{y_1}\mathfrak{s}_\ell^{(2)} & \partial_{y_2}\mathfrak{s}_\ell^{(2)} & \partial_{y_3}\mathfrak{s}_\ell^{(2)} \\ \partial_{y_1}\mathfrak{s}_\ell^{(3)} & \partial_{y_2}\mathfrak{s}_\ell^{(3)} & \partial_{y_3}\mathfrak{s}_\ell^{(3)} \end{pmatrix},\quad \ell = 1,2,$$

and $\mathfrak{S}$ and $\mathfrak{s}$ are proxies for R, V and r, v, respectively.

The matrices defined in (9.56) are constant matrices, depending on $p \in \mathbb{C}^+$ as a parameter. Assuming for the time being (see the comments in the proof of Theorem 9.5.4) that the dependence in p is such that the result corresponds to the Laplace transform of a function in the time domain, we have the following.

DEFINITION 9.5.3 *The Laplace transform of the homogenised constitutive relation is given by*

$$\widehat{\mathsf{d}}^h = \mathsf{A}^h_{\mathfrak{L}} \widehat{u},$$

where

$$\mathsf{A}^h_{\mathfrak{L}} = \begin{pmatrix} \varepsilon^h_{\mathfrak{L}} & \xi^h_{\mathfrak{L}} \\ \zeta^h_{\mathfrak{L}} & \mu^h_{\mathfrak{L}} \end{pmatrix}.$$

Inversion of the Laplace transform will then lead to a homogenised constitutive relation in the time domain in the form of a convolution operator

$$\mathsf{d} = \mathsf{A}^h_{\mathrm{or}} u + \mathsf{G}^h_{\mathrm{d}} \star u,$$

where the homogenised optical response matrix $\mathsf{A}^h_{\mathrm{or}}$ and the homogenised dispersion matrix $\mathsf{G}^h_{\mathrm{d}}$ are obtained from $\mathsf{A}^h_{\mathfrak{L}}$ by the inversion of the Laplace transform.

The following theorem holds.

THEOREM 9.5.4 *The solution $u^\epsilon = (E^\epsilon, H^\epsilon)^{tr}$ of system (9.6-9.7) satisfies*

$$u^\epsilon \overset{*}{\rightharpoonup} u^*, \quad in\ L^\infty([0,T], \mathbb{X}),$$

where $u^* = (E^*, H^*)^{tr}$ is the unique solution of the homogeneous Maxwell system

$$(\mathsf{d}^*)' = \mathsf{M}u^* + j, \quad in\ (0,T] \times \mathcal{O}, \tag{9.57}$$

with zero initial conditions and the perfect conductor boundary condition, and subject to the constitutive relations

$$\mathsf{d}^* = \mathsf{A}^h_{or} u^* + \mathsf{G}^h_d \star u^* \tag{9.58}$$

such that $\mathsf{A}^h_{or} + \widehat{\mathsf{G}^h_d} = \mathsf{A}^h_{\mathfrak{L}}$, where $\mathsf{A}^h_{\mathfrak{L}}$ is defined as in Definition 9.5.3.

Proof. The greatest part of the proof is analogous with that of the proof of the optical response case (see Theorem 9.4.14) and is therefore omitted. The elliptic homogenisation result is complicated by the observation that now the components of the elliptic system depend on $p \in \mathbb{C}^+$. Assumption 9.5.2 allows us to homogenise the auxiliary elliptic problems for all p and, following similar arguments as for the optical response problem, identify the Laplace transform of the homogenised constitutive relation as the one given in Definition 9.5.3. This relation depends on p as a parameter. For fixed p, these results can be used as fixed frequency homogenisation results. To be able to invert the Laplace transform and find homogenised constitutive relations in the time domain, we must show that the homogenised constitutive relation corresponds to the Laplace transform of a vector valued function. This requires detailed estimates on the dependence of the Laplace transform on p. These estimates can be obtained by extending the techniques of [206] or [374]. □

REMARK 9.5.5 Theorem 9.5.4 can be extended under additional assumptions about the data of the problem (e.g., under the assumption of positivity of the kernels G_{d}) to include the case $T = \infty$.

REMARK 9.5.6 It is clear that J can also depend on $\epsilon > 0$, provided one makes suitable assumptions about its behaviour as $\epsilon \to 0$.

9.5.2 The periodic unfolding method

In 1990, Arbogast, Douglas and Hornung defined a "dilation" operator in [17] to study homogenisation for a periodic medium with double porosity. In 2002, Cioranescu, Damlamian and Griso expanded this idea and presented, in [95], a general and simple approach for classical or multiscale periodic homogenisation under the name *periodic unfolding method*, a complete presentation of which can be found in [96]. The periodic unfolding method is essentially based on two ingredients: the "unfolding operator" (which is similar to the dilation operator and whose effect is to "zoom" the microscopic structure in a periodic manner), and the separation of the characteristic scales by decomposing every function $\phi \in W^{1,p}(\mathcal{O})$ into two parts; this scale splitting can be achieved either by using the local average or by a procedure inspired by the finite element method. Let us also mention that the periodic unfolding method simplifies many of the two-scale convergence proofs.

Let $Y = [0, \ell_1] \times [0, \ell_2] \times [0, \ell_3]$ be the reference periodic cell and for each $x \in \mathbb{R}^3$, define $[x]_Y := \sum_{i=1}^{3} k_i \ell_i$ as the unique integer combination of periods such that $\{x\}_Y := x - [x]_Y \in Y$. This definition implies that $x = \epsilon \left(\left[\frac{x}{\epsilon}\right]_Y + \left\{\frac{x}{\epsilon}\right\}_Y \right)$ a.e. for all $x \in \mathbb{R}^3$. Define $\widehat{\mathcal{O}}_\epsilon$ as the largest union of translated and rescaled $\epsilon\,(k+Y)$ cells that are included in $\mathcal{O}$, $\Lambda_\epsilon = \mathcal{O} \setminus \widehat{\mathcal{O}}_\epsilon$ is the subset of $\mathcal{O}$ containing the translated and rescaled cells that intersect $\partial\mathcal{O}$.

DEFINITION 9.5.7 (THE PERIODIC UNFOLDING OPERATOR) *The periodic unfolding operator $\mathcal{T}^\epsilon : L^2(\mathcal{O}) \to L^2(\mathcal{O} \times Y)$ is defined by*

$$\mathcal{T}^\epsilon(\mathfrak{u})(x, y) = \begin{cases} \mathfrak{u}\left(\epsilon \left[\frac{x}{\epsilon}\right]_Y + \epsilon y\right) & \text{for } x \in \widehat{\mathcal{O}}_\epsilon,\ y \in Y, \\ 0 & \text{for } x \in \Lambda_\epsilon,\ y \in Y. \end{cases}$$

If $\mathfrak{u} = a^\epsilon(x) = a_{\text{per}}\left(\frac{x}{\epsilon}\right)$, where a_{per} is a periodic function of period Y, then $\mathcal{T}^\epsilon(a^\epsilon)(x, y) = a_{\text{per}}(y)$. This shows that the action of the operator $\mathcal{T}^\epsilon$ is to "magnify" the periodic microstructure. Clearly, for functions of the special type considered above, $\mathcal{T}^\epsilon(a)(x, y) \to a_{\text{per}}(y)$, *a.e.* in $\mathcal{O} \times Y$. This result extends trivially for matrix valued functions of this special type.

Furthermore, the following properties of $\mathcal{T}^\epsilon$ are very important:

THEOREM 9.5.8 (PROPERTIES OF $\mathcal{T}^\epsilon$ [70])

(i) *$\mathcal{T}^\epsilon$ is a linear and continuous operator.*

(ii) *For all $\mathfrak{u}, \mathfrak{v} \in L^2(\mathcal{O})$, $\mathcal{T}^\epsilon(\mathfrak{u}\,\mathfrak{v}) = \mathcal{T}^\epsilon(\mathfrak{u})\,\mathcal{T}^\epsilon(\mathfrak{v})$.*

(iii) *For all $\mathfrak{u} \in L^2(\mathcal{O})$,*

$$\int_{\mathcal{O}} \mathfrak{u}(x)\, dx = \frac{1}{|Y|} \int_{\mathcal{O} \times Y} \mathcal{T}^\epsilon(\mathfrak{u})(x, y)\, dx\, dy + \mathscr{C}(\epsilon),$$

where $\mathscr{C}(\epsilon)$ is a correction term that may be shown to be negligible in the limit as $\epsilon \to 0$.

The following convergence results hold for $\mathcal{T}^\epsilon$:

THEOREM 9.5.9 (WEAK COMPACTNESS RESULTS [70])

(i) *If $\{\mathfrak{u}^\epsilon\}$ is uniformly bounded in $L^2(\mathcal{O})$, then there exists $\mathfrak{u} \in L^2(\mathcal{O}\times Y)$ such that $\mathcal{T}^\epsilon(\mathfrak{u}^\epsilon) \rightharpoonup \mathfrak{u}$ in $L^2(\mathcal{O} \times Y)$ (up to subsequences).*

(ii) *If $\{\mathfrak{u}^\epsilon\}$ is uniformly bounded in $H(\mathrm{curl}, \mathcal{O})$, then there exists a triplet $(\mathfrak{u}, \mathfrak{v}, \mathfrak{w}) \in H(curl, \mathcal{O}) \times L^2(\mathcal{O}, H^1_{per}(Y;\mathbb{R})) \times L^2(\mathcal{O}, H^1_{per}(Y;\mathbb{R}^3))$, with $\mathrm{div}_y \mathfrak{w} = 0$ so that*

$$\begin{aligned} &\mathfrak{u}^\epsilon \rightharpoonup \mathfrak{u}, \quad \textit{in } H(\mathrm{curl}, \mathcal{O}),\\ &\mathcal{T}^\epsilon(\mathfrak{u}^\epsilon) \rightharpoonup \mathfrak{u} + \mathrm{grad}_y \mathfrak{v} \quad \textit{in } L^2(\mathcal{O} \times Y; \mathbb{R}^3),\\ &\mathcal{T}^\epsilon(\mathrm{curl}\mathfrak{u}^\epsilon) \rightharpoonup \ \mathrm{curl}_x \mathfrak{u} + \mathrm{curl}_y \mathfrak{w} \quad \textit{in } L^2(\mathcal{O} \times Y; \mathbb{R}^3). \end{aligned}$$

(iii) *If $\{\mathfrak{u}^\epsilon\}$ is bounded in $L^2(\mathcal{O})$ and such that $\mathcal{T}^\epsilon(\mathfrak{u}^\epsilon) \rightharpoonup \widehat{\mathfrak{u}}$ in $L^2(\mathcal{O} \times Y)$, then*

$$\mathfrak{u}^\epsilon \rightharpoonup \mathfrak{u} := \frac{1}{|Y|}\int_Y \widehat{\mathfrak{u}}\, dy.$$

REMARK 9.5.10 The functions $\mathfrak{v}$, $\mathfrak{w}$ are to be understood as "correctors". $\mathfrak{v}(x, y)$ is a scalar and $\mathfrak{v}$ can be understood as a function $\mathfrak{v} : \mathcal{O} \to H^1_{\mathrm{per}}(Y)$, such that $\int_{\mathcal{O}} ||\mathfrak{v}||^2_{H^1_{\mathrm{per}}(Y)} dx < \infty$, whereas $\mathfrak{w}(x, y)$ is a three-vector and $\mathfrak{w}$ can be understood as a function $\mathfrak{w} : \mathcal{O} \to H^1_{\mathrm{per}}(Y; \mathbb{R}^3) \simeq (H^1_{\mathrm{per}}(Y))^3$, such that $\int_{\mathcal{O}} ||\mathfrak{w}||^2_{H^1_{\mathrm{per}}(Y;\mathbb{R}^3)} dx < \infty$.

REMARK 9.5.11 Some of the weak compactness results stated in Theorem 9.5.9 hold in the L^p setting for $1 \le p < \infty$.

The above weak compactness results allow us to derive a homogenised Maxwell equation. We first define the following auxiliary system:

DEFINITION 9.5.12 (CELL EQUATIONS) *Let $r_k \in H^1_{per}(Y) \times H^1_{per}(Y)$, $m_k \in W^{2,1}([0,T]; H^1_{per}(Y) \times H^1_{per}(Y))$, $h_k \in W^{1,1}([0,T]; H^1_{per}(Y) \times H^1_{per}(Y))$, $k = 1, \ldots, 6$ be the solutions of the following systems*

$$\begin{aligned} &-\mathrm{div}_y(\mathsf{A}_{or}(y)\mathrm{grad}_y r_k) = \mathrm{div}_y(\mathsf{A}_{or}(y)e_k),\\ &-\mathrm{div}_y\left(\mathsf{A}_{or}(y)\mathrm{grad}_y m_k(t,y) + (\mathscr{G} m_k)(t,y)\right) = -\mathrm{div}_y(\mathsf{A}_{or}(y)e_k),\\ &-\mathrm{div}_y\left(\mathsf{A}_{or}(y)\mathrm{grad}_y h_k(t,y) + (\mathscr{G} h_k)(t,y)\right) = -\mathrm{div}_y(\mathsf{G}_d(t,y)(e_k + \mathrm{grad}_y r_k(y))), \end{aligned}$$

where $(\mathscr{G}\mathfrak{s})(t,y) := \int_0^t \mathsf{G}_d(t-s,y)\mathrm{grad}_y \mathfrak{s}(s,y)ds$ and e_k is the canonical basis in $\mathbb{R}^6$.

REMARK 9.5.13 Note that if $k = 1, 2, 3$, then

$$\mathrm{div}_y(\mathsf{A}_{\mathrm{or}}(y)e_k) = (\mathrm{div}_y \varepsilon_{\sharp,k}, \mathrm{div}_y \zeta_{\sharp,k})^{tr}.$$

On the other hand, if $k = 4, 5, 6$, then

$$\mathrm{div}_y(\mathsf{A}_{\mathrm{or}}(y)e_k) = (\mathrm{div}_y \xi_{\sharp,k-3}, \mathrm{div}_y \mu_{\sharp,k-3})^{tr}.$$

By the symmetry properties of $\mathsf{A}_{\rm or}$, this shows that the equations for r_k, $k = 1, 2, \cdots, 6$ are related to the equations for $r^{(j)}$, $v^{(j)}$, $j = 1, 2, 3$ (see (9.24), where r_k, $k = 1, 2, 3$ corresponds to $r^{(k)}$ and r_k, $k = 4, 5, 6$ corresponds to $v^{(k-3)}$. The equations related to m_k were not encountered before since they correspond to correctors related to the initial conditions, and in the previous sections we were considering the initial conditions to be homogeneous. The integrodifferential equations related to h_k in their Laplace transformed version are similar to those encountered before in Section 9.5.1.

DEFINITION 9.5.14 (HOMOGENISED COEFFICIENTS) *The homogenised optical response matrix* A^h_{or} *and the family of homogenised dispersion matrices* $\{\mathsf{G}^h_d(t)\}$, $t \in [0, T]$ *consist of the columns*

$$(\mathsf{A}^h_{or})_{\sharp,k} = \int_Y \mathsf{A}_{or}(y)\mathfrak{r}_k(y)dy,$$

$$(\mathsf{G}^h_d(t))_{\sharp,k} = \int_Y \mathsf{G}_d(t,y)\mathfrak{r}_k(y)dy + \int_Y \mathsf{A}_{or}(y)\mathrm{grad}_y h_k(t,y)dy + \int_Y (\mathscr{G}h_k)(t,y)dy$$

for $k = 1, \ldots, 6$, *where* $\mathfrak{r}_k := e_k + \mathrm{grad}_y r_k(y))$ *and* $(\mathscr{G}h_k)(t, y)$ *is as in Definition 9.5.12.*

REMARK 9.5.15 In view of Remark 9.5.13, we observe that the results for the homogenised optical response $\mathsf{A}^h_{\rm or}$ provided by the periodic unfolding method in Definition 9.5.14 (see Theorem 9.5.16) coincide with the results for the homogenised optical response provided by the Laplace transform method (see Section 9.4.2, equation (9.34)). Furthermore, the results for the homogenised dispersive part, when Laplace transformed, coincide with those obtained by the Laplace transform method in Section 9.5.1.

The following theorem [70] provides a rigorous homogenisation result for the Maxwell equations, including the effects of dispersion in the time domain. Note that for the sake of completeness and to show the effects of nonhomogeneous initial conditions in the homogenisation procedure, for the rest of this section we relax the assumption that the initial condition of system (9.6) - (9.7) is the homogeneous one, and we substitute for it $u^\epsilon(0) = u_0$.

THEOREM 9.5.16 *The solution* u^ϵ *of (9.6-9.7) is such that*

$$u^\epsilon \overset{*}{\rightharpoonup} u \ \ in \ L^\infty([0,T], \mathbb{X}_{\mathsf{M}}),$$

where u *is the solution of the homogenised Maxwell system*

$$(\mathsf{A}^h_{or} u + \mathsf{G}^h_d \star u)' = \mathsf{M}u + j^h,$$

with A^h_{or}, G^h_d *given as in Definition 9.5.14 and* $j^h = j + (\mathscr{J}_0 u_0)'$ *where* $\mathscr{J}_0$ *is a matrix valued function of* t *with columns defined by*

$$(\mathscr{J}_0(t))_{\sharp,k} = \int_Y \left(\mathsf{A}_{or}(y)\,\mathrm{grad}_y m_k(t,y) + \int_0^t \mathsf{G}_d(t-s,y)\,\mathrm{grad}_y m_k(s,y)\right)dy.$$

Proof. The solvability results, which yield uniform bounds in ϵ and $t > 0$ for the sequences u^ϵ and $(u^\epsilon)'$ in $\mathbb{X}_{\mathsf{M}}$ and $\mathbb{X}$, respectively, combined with the weak compactness results of Theorem 9.5.9 guarantee the existence of a triplet of functions (u, v, w) such that

$$\begin{aligned}
&u^\epsilon \overset{*}{\rightharpoonup} u \text{ in } L^\infty([0,T], \mathbb{X}_{\mathsf{M}}),\\
&\mathcal{T}^\epsilon(u^\epsilon) \overset{*}{\rightharpoonup} u + \operatorname{grad}_y v \text{ in } L^\infty([0,T], L^2(\mathcal{O}, H(\operatorname{curl}, Y))),\\
&\mathcal{T}^\epsilon((u^\epsilon)') \overset{*}{\rightharpoonup} u' + \operatorname{grad}_y v' \text{ in } L^\infty([0,T], L^2(\mathcal{O} \times Y; \mathbb{R}^6)),\\
&\mathcal{T}^\epsilon(\operatorname{curl}_x u^\epsilon) \overset{*}{\rightharpoonup} \operatorname{curl}_x u + \operatorname{curl}_y w \text{ in } L^\infty([0,T], L^2(\mathcal{O} \times Y; \mathbb{R}^6)).
\end{aligned} \tag{9.59}$$

We now use Theorem 9.5.8 to deduce that

$$\mathcal{T}^\epsilon(\mathsf{A}^\epsilon_{\text{or}} u^\epsilon + \mathsf{G}^\epsilon_{\text{d}} \star u^\epsilon) \overset{*}{\rightharpoonup} \mathsf{A}^h_{\text{or}}(u + \operatorname{grad}_y v) + \mathsf{G}^h_{\text{d}} \star (u + \operatorname{grad}_y v)$$

in $L^\infty([0,T], L^2(\mathcal{O} \times Y))$, for some matrices A^h_{or}, G^h_{d} to be specified later on.

Consider the weak formulation of the original Maxwell system using a test function of the form $\theta^\epsilon = \phi(x)\psi\left(\left\{\frac{x}{\epsilon}\right\}\right)$ for $\phi \in \mathcal{D}(\mathcal{O})$, $\psi \in \mathcal{D}(Y)$. Applying the essential identity of the periodic unfolding method (Theorem 9.5.8, statement (*iii*)) three times, (i) for $\mathfrak{u} = (\mathsf{A}^\epsilon_{\text{or}} u^\epsilon + \mathsf{G}^\epsilon_{\text{d}} \star u^\epsilon) \cdot \theta^\epsilon$, (ii) for $\mathfrak{u} = \mathsf{M} u^\epsilon \cdot \theta^\epsilon$, and (iii) for $\mathfrak{u}$ being the product of the source terms and the initial conditions with the test function, and using the results concerning the action of the operator $\mathcal{T}^\epsilon$ on the test function, as well as the results on the weak star convergences presented in (9.59), we obtain that in the limit as $\epsilon \to 0$, the weak form of the full Maxwell equation converges to

$$\int_{\mathcal{O}\times Y} \mathcal{L}(u+\operatorname{grad}_y v)\cdot\theta \, dx\, dy = \int_{\mathcal{O}\times Y} (\mathsf{M}_x u + \mathsf{M}_y w + j + \mathsf{A}^h_{\text{or}} u_0)\cdot\theta \, dx\, dy, \tag{9.60}$$

for all $v \in L^2(\mathcal{O} \times Y)$, where we have also used the density of the tensor product $\mathcal{D}(\mathcal{O}) \otimes \mathcal{D}(Y)$ in $L^2(\mathcal{O}\times Y)$, and $\mathcal{L}$ denotes the constitutive operator.

We now wish to identify the matrices A^h_{or} and G^h_{d} in (9.60). This is done by considering properly selected test functions. We first choose in (9.60) test functions of the form

$$\theta(x,y) = \phi(x) \operatorname{grad}_y \psi = (\phi(x) \operatorname{grad}_y \psi_1(y), \phi(x) \operatorname{grad}_y \phi_2(y))^{tr},$$

where $\phi \in L^2(\mathcal{O})$, $\psi_1, \psi_2 \in H^1_{\text{per}}(Y)$ are scalar functions. For this choice of test functions, the only nonvanishing terms in the weak formulation (9.60) are

$$\int_Y \mathcal{L}(u + \operatorname{grad}_y v) \cdot \operatorname{grad}_y \psi dy = \int_Y \mathsf{A}^h_{\text{or}} u_0 \cdot \operatorname{grad}_y \psi dy, \tag{9.61}$$

for all $\psi \in H^1_{\text{per}}(Y) \times H^1_{\text{per}}(Y)$. In the above equation, $u = u(t,x)$, $u_0 = u_0(x)$ are known fields (in principle), whereas $v = v(t,x,y)$ is a corrector field to be determined. Thus, equation (9.61) can be considered an integrodifferential equation for v, driven by $\mathcal{L}u$, so that we may look for a solution in the form

$$v(t,x,y) = R^\ddagger(y)\cdot u(t,x) + M^\ddagger(y)\cdot u_0(x) + \int_0^t H^\ddagger(t-s,y)\cdot u(s,x)\, ds \tag{9.62}$$

where $R^\ddagger, M^\ddagger, H^\ddagger$ are six-vectors. Rather than splitting the six-vectors in two three-vectors, as, e.g., in $u = (u_1, u_2)^{tr}$, it is convenient to re-express u as $u = (u^1, \ldots, u^6)^{tr}$, and similarly for the initial condition, to avoid any confusion. On the other hand, for the components of $R^\ddagger$, $M^\ddagger$ and $H^\ddagger$, we use the standard notation for the components $\mathfrak{S} = (\mathfrak{s}_1, \ldots, \mathfrak{s}_6)^{tr}$, to avoid awkward notation in the homogenisation formulae. With this notation, (9.62) becomes

$$v = \sum_{k=1}^{6} r_k u^k + \sum_{k=1}^{6} m_k u_0^k + \sum_{k=1}^{6} h_k \star u^k, \tag{9.63}$$

where $v = v(t,x,y)$, $r_k = r_k(y)$, $m_k = m_k(t,y)$, $h_k = h_k(t,y)$. Substituting this ansatz to (9.61), we observe that r_k, m_k and h_k, $k = 1, \ldots, 6$, satisfy the weak form of the systems in Definition 9.5.12. Observe that the first equation in the system of Definition 9.5.12 is equivalent to the elliptic system obtained in Section 9.4.2.

We now try to obtain the homogenised Maxwell system. We choose test functions $\theta \in \mathbb{X}$ depending only on x and insert them into (9.60). Note that this is equivalent to averaging the system over the fast scale y. As a result of that, we have

$$\int_{\mathcal{O}\times Y} (\mathsf{M}_x u + \mathsf{M}_y w) \cdot \theta \, dx\, dy = \int_{\mathcal{O}\times Y} \mathsf{M}_x u \cdot \theta \, dx\, dy$$

for all t, which is equivalent to the Maxwell operator acting on the averaged field over the fast scale. We consider next the term $\int_{\mathcal{O}\times Y} \mathcal{L}(u + \mathrm{grad}_y v) \cdot \theta$, which will give us the homogenised constitutive relation. Substituting the expression (9.63) into this integral and integrating by parts over Y, we obtain the required result. This concludes the proof. □

REMARK 9.5.17 As seen by Theorem 9.5.16 the effect of nonzero initial conditions is to introduce a correction term in the homogenised source term j^h which is related to the initial condition u_0, as well as the composition of the material through A_{or}, G_{d}. For zero initial conditions we have $j^h = j$. The effect of nonzero initial conditions on homogenisation is an important issue and can cause interesting phenomena depending on whether the initial condition u_0 presents small-scale variability or not. This phenomenon appears in many important differential equations; to the best of our knowledge it was first observed for the wave equation in [77].

9.6 MISCELLANEA

9.6.1 Two-scale convergence

Two-scale convergence is a special type of convergence in L^p spaces. Let $\mathcal{O}$ be an open bounded subset in $\mathbb{R}^N$, Y the unit cube in $\mathbb{R}^N$ and $\{\epsilon\}$ a sequence of positive numbers converging to 0. In 1989 Nguetseng (see [337]) proved that for each bounded sequence $\{u^\epsilon\}$ in $L^2(\mathcal{O})$ there exists a subsequence, still indexed by ϵ, and a $u \in L^2(\mathcal{O} \times Y)$ such that

$$\int_{\mathcal{O}} u^\epsilon(x)\, \phi\left(x, \frac{x}{\epsilon}\right) dx \to \int_{\mathcal{O}\times Y} u(x,y)\, \phi(x,y)\, \mathtt{dv}\,, \tag{9.64}$$

where $\mathtt{dv} = dx\, dy$, for every sufficiently smooth $\phi(x, y)$ that is Y–periodic in y. Nguetseng also proved that for a bounded sequence $\{u^\epsilon\} \in H^2(\mathcal{O})$ there exist functions $u \in L^2(\mathcal{O} \times Y)$ and $u_1 \in L^2(\mathcal{O}, H^2_{\text{per}}(Y))$ such that, up to a subsequence,

$$u^\epsilon \rightharpoonup u\,, \text{ in } H^2(\mathcal{O})\,,$$

$$\int_{\mathcal{O}} \operatorname{grad} u^\epsilon(x) \cdot \psi\left(x, \frac{x}{\epsilon}\right) dx \to \int_{\mathcal{O}\times Y} \left(\operatorname{grad} u(x) + \operatorname{grad}_y u_1(x,y)\right) \cdot \psi(x,y)\mathtt{dv},$$

for every sufficiently smooth $\psi(x, y)$. The latter result made it possible for Nguetseng to make a new proof of the homogenisation result corresponding to the linear elliptic equations of the general form studied in Section 9.4.1. Later on Allaire (see [2]) started to call the type of convergence defined by (9.64) *two-scale convergence*. Allaire also developed the theory further by studying some general properties of two-scale convergence. Moreover, he used two-scale convergence to analyse several homogenisation problems, both linear and nonlinear. Two-scale convergence has also been generalised to n-scale convergence (or multiscale convergence) in the obvious way. Two-scale convergence is now a well-known concept within the (rigorous) homogenisation community. A lot of different homogenisation problems have been analysed using this tool. There is also a relation of two-scale convergence to strong and weak convergence in $L^p(\mathcal{O})$, [96]. Let us also mention that the periodic unfolding method simplifies many of the two-scale convergence proofs.

9.6.2 On the term "structure" in homogenisation

Homogenisation theory requires the passage to a suitable limit in a family of media. This question is a meaningful one if the medium displays certain properties that allow this transition to the limit; loosely speaking the medium must display some sort of "self-repeating" (or "self-averaging") structure. In the classical case, the medium is assumed to have a periodic structure. However, for the majority of inhomogeneous media the right structure hypothesis is far from being the periodicity hypothesis. In such cases periodicity has to be replaced by quasi-periodicity (the medium coefficients are linear combinations of periodic functions of incommensurate periods) or almost-periodicity (closure of quasi-periodic). Homogenisation problems in structures beyond the above mentioned remain in general unsolved because of the lack of the necessary mathematical framework.

In [338], Nguetseng assigned a self-contained mathematical meaning to the word structure in the context of homogenisation. New tools were required like the notion of a homogenisation algebra and the underlying concept of mean value; homogenisation algebras were introduced in [216], under the names of algebras with mean values. In [338] a systematic utilisation of such algebras leads, by means of the Gelfand representation theory and two-scale convergence, to a general mathematical framework that includes the model of classical periodic homogenisation theory.

9.6.3 Memory effects in homogenisation

In the 1970s E. Sanchez-Palencia made the observation that the homogenisation procedure of the Maxwell equations for periodic materials even in the optical response region may lead to a homogenised system with memory terms of the convolution type, as long as the effects of conductivity are taken into account. This remarkable observation was taken up by Tartar, who studied the problem extensively in its general form (for general equations of the hyperbolic type) and showed, using rigorous homogenisation theory, that this phenomenon is common in hyperbolic equations (see [407] for a thorough exposition). In the early 1990s Antonić studied memory effects in the homogenisation of the Maxwell equations [15] and reconfirmed the observation of Sanchez-Palencia. The theme of memory terms introduced through homogenisation to the Maxwell equations was taken up by other authors as well (see, e.g., [215] or [216]). These works consider dielectric electromagnetic media, but their arguments may be generalised to complex media. To clarify the effect of conductivity in generating the memory term we summarise the relevant arguments (see, e.g., [215]) in a very simple model where A_{or} consists of $\varepsilon^\epsilon(x) = \varepsilon(\frac{x}{\epsilon})$, $\xi = \zeta = 0$, $\mu = I$, but with a spatially dependent conductivity term $\sigma^\epsilon(x) = \sigma(\frac{x}{\epsilon})$. This contributes to the Maxwell equations an effective source term $j^\epsilon = (\sigma^\epsilon(x)u_1, 0)^{tr}$, where as usual $u = (u_1, u_2)^{tr} = (E, H)^{tr}$. Repeating the formal two-scale expansion for this simplified version of the Maxwell equations (in the time domain) we obtain from the $O(\epsilon^{-1})$ terms that the average of the fields over the periodic cell Y satisfy $E^{(0)}(x, y) - \widehat{E^{\bowtie}}(x) = \text{grad}_y \Psi_1(x, y)$, $H^{(0)}(x, y) - \widehat{H^{\bowtie}}(x) = \text{grad}_y \Psi_2(x, y)$, where Ψ_i, $i = 1, 2$ are scalar functions (we use the notation of Section 9.3, dropping the hats, since we no longer work in terms of the Laplace transforms of the fields). After insertion of this expression into the expansions and subsequent application of the div_y operator, we obtain

$$\begin{aligned} &\text{div}_y \left[(\partial_t + \sigma(y))(\text{grad}_y \Psi_1 + \widehat{E^{\bowtie}})\right] = 0, \\ &\text{div}_y \text{grad}_y \Psi_2 = 0. \end{aligned} \tag{9.65}$$

Application of the mean operator to $O(\epsilon^0)$ terms yields the homogenised Maxwell equations. Repeating the arguments of Section 9.3 and using the second of equations (9.65), we obtain

$$\partial_t B^{\bowtie} = -\text{curl} E^{\bowtie},$$

where $B^{\bowtie} = \langle B^\epsilon \rangle$. We now consider the first of equations (9.65). Following [15] (and restricting our arguments to the simplified version of the model employed here[8]), we rewrite this equation in variational form (including the periodic boundary conditions in y), and then treat it as an ordinary differential equation in $H^1_{\text{per}}(Y)$,

$$(\Psi_1 + hE^{\bowtie})' + \mathsf{S}\Psi_1 + g \cdot E^{\bowtie} = 0, \tag{9.66}$$

with initial condition Ψ_1, where S is the bounded linear operator generated in variational form by the coercive bilinear form $a : H^1_{\text{per}}(Y) \times H^1_{\text{per}}(Y) \to \mathbb{R}$,

$$a(\phi, \theta) := \int_Y \sigma(y)\text{grad}_y\phi \cdot \text{grad}_y\theta, \quad \forall \phi, \theta \in H^1_{\text{per}}(Y),$$

$$(\mathsf{S}\phi, \theta)_{H^1_{\text{per}}(Y)} := a(\phi, \theta),$$

and $h, g \in (H^1_{\text{per}}(Y))^3$ are such that

$$(h, v)_{(H^1_{\text{per}}(Y))^3} = \int_Y \epsilon(y) \cdot \text{grad}_y v\, dy, \quad \forall v \in (H^1_{\text{per}}(Y))^3,$$

$$(g, v)_{(H^1_{\text{per}}(Y))^3} = \int_Y \sigma(y) \cdot \text{grad}_y v\, dy, \quad \forall v \in (H^1_{\text{per}}(Y))^3.$$

The existence and the properties of the operator S, as well as the existence of h, g, are guaranteed by the Lax-Milgram lemma and the Riesz representation theorem, respectively. The differential equation (9.66) can be solved in terms of the semigroup generated by S as

$$\Psi_1 = -h \cdot E^{\bowtie} + \int_0^t \exp(-(t-s)\mathsf{S})(\mathsf{S}h - g)\, ds\,,$$

and this specifies completely $E^{(0)}$. Inserting this expression for $E^{(0)}$ into the expansions and averaging over y, we obtain the homogenised equation

$$\partial_t(\varepsilon^h E^{\bowtie} + K_1 \star E^{\bowtie}) = \text{curl}_x H^{\bowtie} + \sigma^h E^{\bowtie} + K_2 \star E^{\bowtie}\,,$$

where ε^h is a constant obtained by an averaging formula equivalent to those provided in Section 9.4.2 and

$$K_1 = \langle \varepsilon \cdot (\text{grad}_y(\exp(-(t-s)\mathsf{S})\,(\mathsf{S}h - g))^{tr}\rangle\,,$$

and similarly for σ^h and K_2. The important observation here is that the homogenised constitutive relation now includes a convolution term, which is generated by the periodic conductivity coefficient. In the absense of conductivity $\sigma = 0$, this term vanishes. The above results can be made rigorous (see, e.g., [15], [407]).

The work in [215] is devoted to the memory effect induced by homogenisation of the Maxwell system for conducting media. The memory kernel is described by a Volterra integral equation. Furthermore, it can be characterised explicitly in terms of Young measures, and the kinetic formulation of the homogenised equation is also obtained. The kinetic formulation allows obtaining the homogenisation of the energy density and the associated conservation law with the Poynting vector. The interesting interaction phenomenon of the microscopic and macroscopic scales is also discussed, and the memory effect qualitatively explains something about irreversibility.

---

[8] Antonić [15] deals with the dielectric case where A_{or} is a block diagonal matrix with spatially dependent (periodic) coefficients, while dispersive terms are absent.

9.6.4 The Bloch-wave homogenisation method:

The Floquet-Bloch expansion and the corresponding *Bloch-wave homogenisation method* (see [59], [110] and the references in the latter) is a high-frequency method that provides dispersion relations for wave propagation in periodic structures.

9.6.5 Further references for the Maxwell equations:

Regarding the homogenisation of the Maxwell equations, an indicative but incomplete list of important studies conducted within the framework of rigorous mathematical analysis would comprise [18], [52], [59], [86], [215], [302], [374], [387], [389], and the important contributions of Wellander ([193], [390], [425], [426], [427], [428]).

9.6.6 Numerics

The numerical treatment of the homogenisation problems for the Maxwell equations in complex media is a very interesting topic in its own right. Such problems require the use of the analytic results of homogenisation theory, such as multiscale expansions, correctors, etc. For example, in [86] the multiscale analysis of the Maxwell equations in composite materials with a periodic microstructure is discussed. The new contributions in this paper are the determination of higher-order correctors and the explicit convergence rate for the approximate solutions. Further, the multiscale finite element method is presented and the convergence result is derived. The numerical results demonstrate that higher-order correctors are essential for solving the Maxwell equations in composite materials. See also [449].

Chapter Ten

Towards a Scattering Theory

10.1 INTRODUCTION

Scattering theories can provide methods for developing robust approximation methods for solving wave problems. In this chapter we indicate how such theories can be developed when wave motions in chiral media are studied; we show, for the sake of illustration, how a relatively simple scattering theory involving achiral materials can be modified to accommodate problems involving a class of chiral materials (see also [368]).

We begin by remarking that a scattering process describes the effects of a perturbation on a system about which everything is known in the absence of the perturbation. Such a process can be conveniently characterised in terms of three main features; generation, interaction and measurement. In the generation stage an incident wave, a signal, is generated, far away in both space and time (to ensure complete independence of the two systems) from any perturbation that might have to be considered, e.g., a target body or some potential. At this stage the interaction between an incident wave and the perturbation is negligible and the system evolves as though it were a free system, that is, a system in which there are no perturbations. Eventually, however, the incident wave and the perturbation interact and exert considerable influence on each other. The resulting effects, that is, the scattered waves, often have a very complicated structure. After the interaction during which the scattering occurred, the now scattered wave and the perturbation can once more become quite distant from each other and the interaction effects again become negligible. Consequently, any measurement of the scattered wave at this stage would indicate that the system is once again evolving as a free system, but not necessarily the same free system as that considered originally.

In practical situations, measurements of a wave far away from any perturbation are really the only data available. Consequently, the asymptotic behaviour of solutions to wave equations, and especially the asymptotic equality of solutions of the associated free and perturbed systems becomes of particular interest.

The structure of the chapter is as follows. In Section 10.2 we present a general formulation of the class of problems described above in terms of evolution operators, while in Section 10.3 we outline an approach to scattering theory in the time domain. In this section we introduce the basic concepts and tools of scattering theory, such as the wave operators and the scattering operator. In Section 10.4 we show how the use of spectral theory allows the explicit construction of solutions to abstract initial boundary value problems in terms of generalised integral transforms. In Section 10.5 we show how these generalised integral transforms can be used for the construction of the wave operators and the scattering operator. Finally, in Section 10.6 we explore the extension of these ideas to the study of electromagnetics of complex media.

10.2 FORMULATION

The above remarks can be conveniently expressed in symbolic form as follows.

Consider first a system that has no inhomogeneities. Let $f_{\mathsf{f}}(s,x)$ be a quantity that characterises the state of the system at some initial time $t = s$ and let $u_{\mathsf{f}}(t,x)$ be a quantity that characterises the state of the system at some later time $t > s$. We shall be concerned with systems for which states can be related by means of an "evolution rule", denoted by $U_{\mathsf{f}}(t-s)$, which determines the *evolution*, in time, of the system from its initial state $f_{\mathsf{f}}(s,x)$ to a state $u_{\mathsf{f}}(t,x)$ at a later time $t > s$. This being the case, we write

$$u_{\mathsf{f}}(t,x) = U_{\mathsf{f}}(t-s)f_{\mathsf{f}}(s,x),$$

where it is understood that $U_{\mathsf{f}}(0) = I$, the identity. The evolution rule can be understood as a semigroup or as an evolution family.

In a similar manner, when inhomogeneities are present in the system, then we will assume we can express the evolution of the system from an initial state $f_{\mathsf{p}}(s,x)$ to a state $u_{\mathsf{p}}(t,x)$ at a later time $t > s$ in the form

$$u_{\mathsf{p}}(t,x) = U_{\mathsf{p}}(t-s)f_{\mathsf{p}}(s,x), \quad U_{\mathsf{p}}(0) = I,$$

where $U_{\mathsf{p}}(t-s)$ denotes an appropriate "evolution rule". Thus, we see that we are concerned with two classes of problem. When there are no inhomogeneities present in the system, we shall say that we have a *free problem*. When inhomogeneities are present in a system, we shall say that we have a *perturbed problem*. We express this situation symbolically in the form

$$u_{\kappa}(t,x) = U_{\kappa}(t-s)f_{\kappa}(s,x), \quad U_{\kappa}(0) = I, \quad \kappa = \mathsf{f}, \mathsf{p},$$

where, when $\kappa = \mathsf{f}$, we shall assume that we have a free problem, while when $\kappa = \mathsf{p}$ we will have a perturbed problem.

With the construction of approximation methods in mind, we shall always asssume that the free problem is one that can be readily solved. Consequently, it would seem natural to investigate under what conditions solutions of the free problem and perturbed problem could be considered equal. This is essentially what a scattering theory does.

In connection with the possible equality of the $u_{\kappa}(t,x)$, $\kappa = \mathsf{f}, \mathsf{p}$, we first recognise and make use of the fact that in most experimental procedures, measurements in a system are made far away from any inhomogeneities that might exist in the system. Consequently, we are mainly concerned here with the nature of the solutions $u_\kappa(t,x)$, $\kappa = \mathsf{f},\mathsf{p}$ and of their difference in the so-called *far-field* of any inhomogeneity, that is, with the behaviour of the $u_\kappa(t,x)$, $\kappa = \mathsf{f},\mathsf{p}$ as $t \to \infty$. Once this asymptotic behaviour is known, then we can clarify what we mean by the equality of the $u_\kappa(t,x)$, $\kappa = \mathsf{f},\mathsf{p}$ and turn attention to determining the conditions that will actually ensure when, in the far field at least, the $u_\kappa(t,x)$, $\kappa = \mathsf{f},\mathsf{p}$ can be considered equal. A detailed mathematical analysis of such questions can be technically very demanding. However, this chapter is simply meant to provide a guide through the various technical areas, with the intention of highlighting their uses in practical problems of interest. Consequently, the presentation in this chapter is frequently quite formal, and we rely very much on the frequently expressed view that "any formal manipulations which are not obviously wrong are assumed to be correct" ([162], p. 173). Nevertheless, references will always be given in the text to where more precise and often quite general details can be found. Furthermore, we emphasise that in this monograph we are not interested in investigating the evolutionary processes mentioned above in full generality but rather confine our attention to those systems involving waves.

One of the major differences from the problems that have been addressed so far is that we are interested both in the case of unbounded domains, or even the whole of $\mathbb{R}^3$, and in asymptotic behaviour with respect to time; i.e., we allow $t \in \mathbb{R}^+$ or even $t \in \mathbb{R}$. This introduces several technical differences that call for treatment using different techniques. One such candidate is spectral theory. Furthermore, when perturbations are taken into account, we may resort to the rich and extensive literature of perturbation theory for linear operators (see, e.g., [229]) to draw useful and interesting results on the qualitative or even quantitative behaviour of the inhomogeneous (perturbed) system.

10.3 SOME BASIC STRATEGIES

Before dealing with wave problems in chiral materials, we first illustrate some of the strategies that will have to be adopted when developing scattering theories by considering somewhat simpler systems that are governed by initial value problems of the form

$$\{\partial_t^2 + L_\mathsf{f}\}u_\mathsf{f}(t,x) = 0\,,\ u_\mathsf{f}(0,x) = \varphi_\mathsf{f}(x)\,,\ \partial_t u_\mathsf{f}(0,x) = \psi_\mathsf{f}(x), \tag{10.1}$$

$$\{\partial_t^2 + L_\mathsf{p}\}u_\mathsf{p}(t,x) = 0\,,\ u_\mathsf{p}(0,x) = \varphi_\mathsf{p}(x)\,,\ \partial_t u_\mathsf{p}(0,x) = \psi_\mathsf{p}(x), \tag{10.2}$$

where $(t,x) \in \mathbb{R} \times \mathbb{R}^3$ and we have set, for ease of presentation and without any loss of generality, the initial time $t = 0$.

In (10.2) the differential operator L_p is assumed to be some perturbation of the differential operator $L_\mathsf{f} = -\Delta$, the three-dimensional Laplacian. Therefore, (10.1) is one component of a vector Helmholtz equation, which often arises in the modelling of wave phenomena in electromagnetics in the absense of chirality. Equation (10.2) will be considered a perturbation of the vector Helmholtz equation when taking into account the effects of chirality. We shall refer to (10.1) as a *free problem* and to (10.2) as a *perturbed problem.*

We shall consider the initial value problems (10.1), (10.2) in appropriate function spaces. In terms of the spatial behaviour of the solutions, a suitable choice is the function space $L^2(\mathbb{R}^3)$, of (possibly complex-valued) functions that are square integrable with respect to the spatial variables. Such a selection is compatible with energy considerations for the wave equation. We then introduce spatial operators, $A_\kappa : D(A_\kappa) \to L^2(\mathbb{R}^3)$, $\kappa = \mathsf{f}, \mathsf{p}$, defined by $A_\kappa u_\kappa = L_\kappa u_\kappa, u_\kappa \in D(A_\kappa)$, with

$$D(A_\kappa) = \{u_\kappa \in L^2(\mathbb{R}^3) : L_\kappa u_\kappa \in L^2(\mathbb{R}^3),\ u_\kappa \in (bc)\},$$

where the qualifier (bc) denotes that u_κ satisfies an appropriate boundary condition; in the event that we are dealing only with potential scattering problems, this qualifier is omitted from the definition of $D(A_\kappa)$. The definition of $D(A_\kappa)$ ensures that throughout any mathematical manipulations involving A_κ we always "stay in $L^2(\mathbb{R}^3)$".

REMARK 10.3.1 We have assumed, for ease of presentation, that (10.1) and (10.2) can be posed on the same function space. This will not always be the case.

If we now define the functions $u_\kappa : \mathbb{R} \to L^2(\mathbb{R}^3)$, such that for each $t \in \mathbb{R}$, $u_\kappa(t) \in L^2(\mathbb{R}^3)$ and $(u_\kappa(t))(x) = u_\kappa(t, x)$, $\kappa = \mathsf{f}, \mathsf{p}$, then the *classical initial value problems* (10.1) and (10.2) can now be replaced by the abstract differential equations on $L^2(\mathbb{R}^3)$:

$$\begin{aligned} u_\kappa''(t) + A_\kappa u_\kappa(t) &= 0, \\ u_\kappa(0) &= \varphi_\kappa, \\ u_\kappa'(0) &= \psi_\kappa, \end{aligned} \tag{10.3}$$

for $\kappa = \mathsf{f}, \mathsf{p}$. Consequently, with this understanding, the problems (10.3) have solutions (in $L^2(\mathbb{R} \times \mathbb{R}^3)$), which can (formally) be expressed as

$$u_\kappa(t) = (\cos tA_\kappa^{1/2})\varphi_\kappa + A_\kappa^{-1/2}(\sin tA_\kappa^{1/2})\psi_\kappa, \qquad \kappa = \mathsf{f}, \mathsf{p}, \tag{10.4}$$

where $\{\cos(tA_\kappa)\}_{t\in\mathbb{R}}$, $\{\sin(tA_\kappa)\}_{t\in\mathbb{R}}$ are families of linear operators generated by the operators A_κ. These families (the cosine and the sine family, respectively) are similar to the exponential family through which the semigroups were defined. In fact, this is a generalisation of semigroup theory that allows the treatment of second-order abstract ordinary differential equations in function spaces, in a fashion similar to the treatment of first-order abstract ODEs in function spaces, by semigroup theory.

The solutions of (10.3) can be expressed more compactly using the complex-valued functions

$$\begin{aligned} v_\kappa(t) &= \exp(-itA_\kappa^{1/2})h_\kappa =: U_\kappa(t)h_\kappa, \\ h_\kappa &= \varphi_\kappa + iA_\kappa^{-1/2}\psi_\kappa, \end{aligned} \tag{10.5}$$

so that

$$u_\kappa(t) = \mathcal{R}e\,\{v_\kappa(t)\}. \tag{10.6}$$

REMARK 10.3.2 The above problems are written in second-order form so as to resemble the wave equation that has been a standard paradigm for the development of scattering theory. By a standard transformation of variables it may be brought into the form of a first-order system $v' = Gv$, where $v = (u, u')^{tr}$ and G is a matrix operator of the form

$$G = \begin{pmatrix} 0 & I \\ -A & 0 \end{pmatrix}.$$

The system is then subject to the usual semigroup theory treatment.

With all this in mind, we shall compare the solutions of the free problem and the perturbed problem by considering an expression of the form $\|v_{\mathsf{p}}(t) - v_{\mathsf{f}}(t)\|$, where $\|\cdot\|$ denotes the norm in $L^2(\mathbb{R}^3)$. Assume that $U_{\mathsf{f}}(t)$ is a unitary operator whose formal adjoint $U_{\mathsf{f}}^*(t)$ is equal to its inverse $U^{-1}(t)$ for all $t \in \mathbb{R}$. We then find that

$$\begin{aligned} &\|v_{\mathsf{p}}(t) - v_{\mathsf{f}}(t)\| = \|U_{\mathsf{p}}(t)h_{\mathsf{p}} - U_{\mathsf{f}}(t)h_{\mathsf{f}}\| = \\ &\|U_{\mathsf{f}}^*(t)U_{\mathsf{p}}(t)h_{\mathsf{p}} - h_{\mathsf{f}}\| = \|W(t)h_{\mathsf{p}} - h_{\mathsf{f}}\|, \end{aligned} \tag{10.7}$$

where $W(t) := U_{\mathsf{f}}^*(t)U_{\mathsf{p}}(t)$. Assuming that limits as $t \to \pm\infty$ exist we obtain

$$\lim_{t\to\pm\infty} \|v_{\mathsf{p}}(t) - v_{\mathsf{f}}(t)\| = \|W_\pm h_{\mathsf{p}} - h_{\mathsf{f}}\|, \tag{10.8}$$

where

$$W_\pm := \lim_{t\to\pm\infty} W(t) = \lim_{t\to\pm\infty} U_{\mathsf{f}}^*(t)U_{\mathsf{p}}(t) = \lim_{t\to\pm\infty} \exp(itA_{\mathsf{f}}^{1/2})\exp(-itA_{\mathsf{p}}^{1/2}) \tag{10.9}$$

are the so-called *wave operators.* The rigorous proof of existence of the wave operators is an important problem.

When all of the above has been achieved, we see that if the initial data for the free problem and the perturbed problem are related according to

$$h_{\mathsf{f}} = W_\pm h_{\mathsf{p}}, \tag{10.10}$$

then the limit in (10.8) is zero, thus indicating that the perturbed problem is *asymptotically free* as $t \to \pm\infty$. That is, solutions of the perturbed problem with initial data h_{p} are *asymptotically equal* in time to solutions of a free problem with initial data h_{f} which is given by (10.10).

Consequently, if solutions of the two systems are known to exist, then we would expect the existence of elements $h_\pm \in L^2(\mathbb{R}^3)$ such that

$$v_{\mathsf{p}}(t) \sim U_{\mathsf{f}}(t)h_\pm \quad \text{as} \quad t \to \pm\infty, \tag{10.11}$$

where $\sim$ denotes *asymptotic equality* and the $\pm$ signs are used to indicate (the possibly different) limits as $t \to \pm\infty$. We emphasise that it is not automatic that both limits implied by (10.11) should exist. Indeed, a solution such as v_{p} could be asymptotically free as $t \to +\infty$ but not as $t \to -\infty$. Thus, the wave operators $W_{\pm}$ can be considered as mapping h_1 to the elements $h_{\pm}$. Indeed, if we use (10.11) in conjunction with the definition $v_{\kappa} = U_{\kappa} h_{\kappa}$, $\kappa = \mathsf{f}, \mathsf{p}$, then we have

$$U_{\mathsf{f}}^*(t)U_{\mathsf{p}}(t)h_{\mathsf{p}} = W(t)h_{\mathsf{p}} \sim h_{\pm},$$

which in the limit as $t \to \pm\infty$ yields the stated result.

Furthermore, the two initial conditions $h_{\pm}$ for the free problem are related by

$$h_+ = W_+ h_{\mathsf{p}} = W_+ W_-^* h_- =: \mathsf{S} h_-.$$

The operator $\mathsf{S} := W_+ W_-^*$ that connects h_- with h_+ is called the *scattering operator* for the problem. The scattering operator provides important information on the effect of the perturbation on the system. The proof of the existence of the scattering operator, through the construction of the wave operators, the study of its properties, and its calculation, is one of the fundamental problems of scattering theory.

10.4 ON THE CONSTRUCTION OF SOLUTIONS

Questions regarding the existence and uniqueness of solutions of the initial value problems (10.3) can be treated using techniques from abstract differential equations, e.g., semigroup methods or variational methods such as the Faedo-Galerkin method (see Chapter 7 and Section A.8 in Appendix A). Once questions of existence and uniqueness of solution have been settled, we can turn our attention to methods for actually determining such solutions. This can be effected using the spectral decomposition of the operator A_{f}.

We first notice that whenever the spectral theorem (see Section A.4 in Appendix A) is applicable, it provides a useful tool for interpreting the solution forms (10.4) and (10.6). For example, according to this theorem, if $\{E_{\kappa}(\lambda)\}_{\lambda \in \sigma(A_{\kappa})}$ denotes the spectral family of A_{κ} (see Section A.4 in Appendix A), then we have the spectral representation

$$A_{\kappa} = \int_0^{\infty} \lambda dE_{\kappa}(\lambda). \tag{10.12}$$

Furthermore, using the spectral theorem we may define functions of the operators A_{κ}. For $\Phi : \mathbb{C} \to \mathbb{C}$, bounded and Lebesgue measurable, we may define the operators $\Phi(A_{\kappa})$ by the spectral representation

$$\Phi(A_{\kappa}) = \int_0^{\infty} \Phi(\lambda) dE_{\kappa}(\lambda). \tag{10.13}$$

However, a difficulty associated with the results (10.12) and (10.13) concerns the practical determination of the spectral family $\{E_{\kappa}(\lambda)\}_{\lambda \in \sigma(A_{\kappa})}$.

For the case of the free problem ($\kappa = \mathsf{f}$) that we are concerned with, the situation can be eased by introducing the results of Fourier transforms in

$L^2(\mathbb{R}^3)$. The Plancherel theory indicates that the operator $F_{\mathrm{f}} : L^2(\mathbb{R}^3) \to L^2(\mathbb{R}^3)$, such that

$$(F_{\mathrm{f}} f)(q) = \widetilde{f}(q) := \frac{1}{(2\pi)^{\frac{3}{2}}} \int_{\mathbb{R}^3} \exp(-ix \cdot q) f(x) dx, \tag{10.14}$$

is well defined (where the improper integral is considered a limit in the $L^2(\mathbb{R}^3)$ sense). This operator is the Fourier transform and according to the Plancherel theory is a unitary operator defining an isometry in $L^2(\mathbb{R}^3)$ so that $F_{\mathrm{f}}^* = F_{\mathrm{f}}^{-1}$; therefore, if $\widetilde{f} = F_{\mathrm{f}} f$ is the Fourier transform of $f \in L^2(\mathbb{R}^3)$, then

$$f(x) = (F_{\mathrm{f}}^* \widetilde{f})(x) := \frac{1}{(2\pi)^{\frac{3}{2}}} \int_{\mathbb{R}^3} \exp(ix \cdot q) \widetilde{f}(q) dq, \quad x, \in \mathbb{R}^3. \tag{10.15}$$

It can also be shown that, if $A_{\mathrm{f}} = -\Delta$, for any bounded, Lebesgue measurable function Φ we have

$$(\Phi(A_{\mathrm{f}}) f)(x) = \frac{1}{(2\pi)^{\frac{3}{2}}} \int_{\mathbb{R}^3} \exp(ix \cdot q) \Phi(|q|^2) \widetilde{f}(q) dq, \tag{10.16}$$

i.e., $F_{\mathrm{f}}(\Phi(A_{\mathrm{f}} f)) = \Phi(|q|^2)\widetilde{f}(q)$.

The above limits are very close in form to the spectral expansions needed for the operator A_{f} in fact, as we shall see, they can be identified exactly as such. Note that for $q, x \in \mathbb{R}^3$, $w_{\mathrm{f}}(q, x) = (2\pi)^{-3/2} e^{ix \cdot q}$ satisfies the Helmholtz equation $(\Delta + |q|^2) w_{\mathrm{f}}(q, x) = 0$, so that w_{f} might be thought of as an eigenfunction of $A_{\mathrm{f}} = -\Delta$ with associated eigenvalue $|q|^2$. However, a direct calculation shows that $w_{\mathrm{f}} \notin L^2(\mathbb{R}^3)$, and so w_{f} must be a generalised eigenfunction of A_{f}. Nevertheless, the Fourier-Plancherel theory[1] indicates that (10.15) can be perceived as an expansion of the function $f \in L^2(\mathbb{R}^3)$ in terms of the generalised eigenfunctions w_{f} of the Laplace operator, with a similar interpretation for (10.16). This implies that the spectral decomposition of A_{f} can be written as a generalised eigenfunction expansion in the form

$$(\Phi(A_{\mathrm{f}}) f)(x) = \int_{q \in \mathbb{R}^3} w_{\mathrm{f}}(q, x) \Phi(|q|^2) \widetilde{f}(q) dq. \tag{10.17}$$

It will be useful later on to bear in mind the decomposition (10.17) can also be written in the form

$$F_{\mathrm{f}}(\Phi(A_{\mathrm{f}}) f)(q) = \Phi(|q|^2)(F_{\mathrm{f}} f)(q).$$

These various results imply that the wave function v_{f} introduced in (10.5) can be interpreted in the form

$$v_{\mathrm{f}}(t, x) = \int_{\mathbb{R}^3} w_{\mathrm{f}}(q, x) \exp(-it\,|q|) \widetilde{h}_{\mathrm{f}}(q) dq, \tag{10.18}$$

where $\widetilde{h}_{\mathrm{f}} = F_{\mathrm{f}} h_{\mathrm{f}}$ is the Fourier transform of the initial condition. Therefore, once the $\widetilde{h}_{\mathrm{f}}$ is known, the solution is reconstructed in terms of the improper

[1] Which has been developed independently of any scattering aspects.

integral (10.18), which involves only the Fourier basis w_{f}. This representation has an interesting physical meaning, which becomes clear as long as we notice that

$$w_{\mathsf{f}}(q,x)\exp(-it\,|q|) = \frac{1}{(2\pi)^{\frac{3}{2}}}\exp(i(x\cdot q - t\,|q|)) \tag{10.19}$$

are solutions of (10.3) with $\kappa = \mathsf{f}$ and as such represent plane waves propagating in the direction of the vector q. Therefore, the wave function given by (10.18) is a representation of a wave (acoustic) in terms of elementary plane waves (10.19).

The above result concerning expansion of the solution in terms of eigenfunctions holds more generally than in the special case where $A_{\mathsf{f}} = -\Delta$. Of course, in the more general case the eigenfunctions will no longer coincide with the Fourier basis and our line of reasoning will no longer be able to stand on the shoulder of Fourier-Placherel theory. However, this theory can be generalised in terms of generalised integral transforms, in which the generalised eigenfunctions of the operators A_{κ} play the rôle of the Fourier basis. As a consequence, we could interpret (10.5), for $\kappa = \mathsf{p}$, in a form similar to (10.18), where of course now w_{f} will have to be replaced by a more suitable set of functions. Specifically, associated with A_{p} we want a generalised eigenfunction expansion theorem, now in terms of the set of generalised eigenfunctions $\{w_{\mathsf{p}}\}$ of the operator A_{p}. This set consists of the kernels $w_{\mathsf{p}}(q,x)$, which are solutions of

$$(A_{\mathsf{p}} - |q|^2)w_{\mathsf{p}}(q,x) = 0, x, q \in \mathbb{R}^3\,.$$

Using the kernels $w_{\mathsf{p}}(q,x)$ we may define the operator $F_{\mathsf{p}} : L^2(\mathbb{R}^3) \to L^2(\mathbb{R}^3)$ by

$$(F_{\mathsf{p}}f)(q) := \check{f}(q) := \int_{\mathbb{R}^3} \overline{w_{\mathsf{p}}(q,x)}f(x)dx. \tag{10.20}$$

The operator F_{p} can be considered as defining an integral transform that is a generalisation of the Fourier transform (10.14) in terms of the new set of functions $\{w_{\mathsf{p}}\}$. As long as $\{w_{\mathsf{p}}\}$ enjoy certain properties, it can be shown that $F_{\mathsf{p}}^* = F_{\mathsf{p}}^{-1}$ has an integral representation of the form

$$f(x) = (F_{\mathsf{p}}^*\check{f})(x) := \int_{\mathbb{R}^3} w_{\mathsf{p}}(q,x)\check{f}(q)dq. \tag{10.21}$$

Furthermore, the following spectral representation holds:

$$(\Phi(A_{\mathsf{p}})f)(x) = \int_{\mathbb{R}^3} w_{\mathsf{p}}(q,x)\Phi(|q|^2)\check{f}(q)dq, \tag{10.22}$$

so that in terms of the new integral transform, we have $F_{\mathsf{p}}(\Phi(A_{\mathsf{p}})f) = \Phi(|q|^2)\check{f}$. As in the case of the Fourier transform, the above improper integrals have to be interpreted as limits in the $L^2(\mathbb{R}^3)$ sense.

We emphasise that for any specific perturbed problem, it has to be proved that a generalised eigenfunction expansion (spectral decomposition) such as (10.20) to (10.22) is indeed available for use. For specific physical problems this can often involve a great deal of work. A full spectral analysis of A_{p} is required, and functions such as w_{p}, that are intimately connected with the particular problem being considered have to be determined. For the remainder of this chapter, we assume that such generalised eigenfunction expansions are available. Consequently, we are able to write (10.5) in the following form:

$$v_{\mathsf{p}}(t,x) = \int_{\mathbb{R}^3} w_{\mathsf{p}}(q,x) \exp(-it\,|q|) \check{f}_{\mathsf{p}}(q) dq, \tag{10.23}$$

which is interpreted in the same way as (10.18).

We remark that in (10.18) and (10.23), the q need not be the same for both. It is associated with eigenvalues of A_{f} in (10.18) and with eigenvalues of A_{p} in (10.23).

From (10.18) and (10.23), it is a straightforward matter to obtain the representations

$$u_{\mathsf{f}}(t,x) = \int_{\mathbb{R}^3} w_{\mathsf{f}}(q,x) \left\{ \widetilde{\varphi}_{\mathsf{f}}(q) \cos t\,|q| + \widetilde{\psi}_{\mathsf{f}}(q) \frac{\sin t\,|q|}{|q|} \right\} dq, \tag{10.24}$$

$$u_{\mathsf{p}}(t,x) = \int_{\mathbb{R}^3} w_{\mathsf{p}}(q,x) \left\{ \check{\varphi}_{\mathsf{p}}(q) \cos t\,|q| + \check{\psi}_{\mathsf{p}}(q) \frac{\sin t\,|q|}{|q|} \right\} dq. \tag{10.25}$$

Hence, provided we can establish an eigenfunction expansion theorem of the form (10.20) - (10.22), then, since all the terms in (10.24) and (10.25) are computable, we have available, in (10.24) and (10.25), a practical means of constructing solutions to the free problem and perturbed problem, respectively.

For the purpose of developing a scattering theory, it remains to investigate whether or not these solutions can be considered asymptotically equal, in some sense, as $t \to \pm\infty$. We begin to investigate this aspect in the next section.

10.5 WAVE OPERATORS AND THEIR CONSTRUCTION

The discussion in Section 10.3 can be restated more abstractly as follows: let $\mathbb{H}$ be the Hilbert space that serves as the state space of the system[2] and consider the existence of two orthogonal subspaces $D_\pm \subset \mathbb{H}$ such that

(*i*) $U(t)D_\pm \subset D_\pm,\ t \in \mathbb{R}^\pm$.

(*ii*) $\bigcap_{t\in\mathbb{R}^\pm} U(t)D_\pm = \{0\}$.

(*iii*) $\bigcup_{t\in\mathbb{R}^\mp} U(t)D_\pm$ is dense in $\mathbb{H}$.

[2] For example, $L^2(\mathbb{R}^3)$ in the case of the Helmholtz equation or $(L^2(\mathbb{R}^3))^3 \times (L^2(\mathbb{R}^3))^3$ in the case of the Maxwell system.

These subspaces are called the *outgoing* and the *incoming subspace*, respectively. With the use of these two subspaces, every element of $\mathbb{H}$ can be represented as the sum of an outgoing and an incoming element. The scattering operator S is an operator that maps the incoming component w_- to the outgoing component w_+. In this respect, it quantifies the effect that the physical system has to whatever is incoming from the distant past (and from $x \to -\infty$) to the state of the system in the distant future (and to $x \to \infty$). If we consider a wave propagation phenomenon with finite speed of propagation, it is physically right to connect space with time in this way. The usual context of scattering theory is the comparison of an unperturbed with a perturbed system. The concept of perturbation is rather general; for instance, we may consider a complex medium that is situated in a bounded domain $\mathcal{O} \subset \mathbb{R}^3$ while the rest of $\mathbb{R}^3$ is the vacuum. Then the unperturbed system is the Maxwell equations in the vacuum, whereas the perturbed system is the Maxwell equations in the chiral medium in $\mathcal{O}$, complemented with the Maxwell equations in $\mathbb{R}^3 \setminus \mathcal{O}$. This is a typical problem of scattering by a chiral obstacle. We may consider the incoming element to be an electromagnetic wave in the vacuum, whereas the outgoing element (wave) conveys the information of the interaction of the incoming wave with the complex medium. Therefore, the scattering operator provides information concerning the effects of the medium on the electromagnetic wave. Other similar examples can be sketched.

The abstract formulation calls for two evolution operators, $U_{\mathsf{f}}(t)$, $U_{\mathsf{p}}(t)$, where the first one corresponds to the unperturbed system while the second corresponds to the perturbed system. The scattering operator is expressed in terms of the wave operators $W_\pm$, defined as

$$W_\pm = \lim_{t\to\pm\infty} U_{\mathsf{p}}(-t)U_{\mathsf{f}}(t),$$

by the relation $\mathsf{S} = W_+^{-1}W_-$. The unperturbed problem is frequently used to motivate the definitions of the incoming and the outgoing elements $D_\pm$. This requires a detailed study of the unperturbed system. In the case of interest here, this corresponds to the Maxwell equations in the vacuum. This to some extent can be effected using the relevant theory for the vector Helmholtz equation, via the reduction of the Maxwell equation to the wave equation, but it can also be done using the original form of the Maxwell system.

We now turn to the construction of the wave operators $W_\pm$ and the scattering operator S, and show how the spectral expansions of the solutions introduced above may provide us with explicit expressions for these quantities. To do this, it is important to identify the subspaces $D_\pm$. This problem depends on the particular nature of the application under consideration. We consider here the relevant problem for the Helmholtz equation before turning to the Maxwell system.

It is easy to check that $u(t,x) = \exp(-i\varpi t)(u_+(x)+u_-(x))$, where $u_\pm(x) = e^{\pm i\varpi x}$, is a solution of the one-dimensional wave equation. Recalling the d'Alembert representation of solutions of the wave equation, we see that u_+

characterises a wave moving from left to right and u_- a wave moving from right to left, both having the same time dependence $\exp(-i\varpi t)$. Equivalently, we can say that u_+ is an *outgoing wave* since it is moving away from the origin while u_- is an *incoming wave* as it is moving towards the origin. Motivated by the above discussion in one dimension, we generalise this approach to higher spatial dimensions ($n = 2, 3$), where now the shapes of the incoming and the outgoing waves will be the solutions $u_\pm$ of the equation

$$(\Delta + \varpi^2)u_\pm(x) = f(x), \quad x \in \mathbb{R}^n,$$

which are assumed to satisfy (see Remark 2.4.3) the *Sommerfeld radiation conditions*

$$\{\partial_r \mp i\varpi\}\, u_\pm(x) = o\left(r^{\frac{1-n}{2}}\right), \quad u_\pm(x) = O\left(r^{\frac{1-n}{2}}\right) \text{ as } r = |x| \to \infty, \quad (10.26)$$

uniformly with respect to the direction $x/|x|$. The estimate (10.26) taken with a minus (plus) sign is called the *Sommerfeld outgoing (incoming)* radiation condition. These outgoing and incoming solutions provide the basis for the construction of a scattering theory.

We now consider the problem of expressing w_{p}, the kernel function in the generalised eigenfunction expansion theorem (10.20) - (10.22), in a form convenient for perturbation analysis. Considering the nature of the problem, it is reasonable to assume that w_{p} is a perturbation of w_{f} and since w_{f} characterises a plane wave, we shall refer to w_{p} as a *distorted plane wave.*

DEFINITION 10.5.1 *An outgoing (resp. incoming) distorted plane wave $w_+(q,x)$ (resp. $w_-(q,x)$) satisfies*

(i) $(\Delta + \varpi^2)w_\pm(x) = 0$, *for* $x, q \in \mathbb{R}^n$, $|q|^2 = \varpi^2$.

(ii) $w_+(q,x) - w_{\mathsf{f}}(q,x)$ *satisfies the outgoing radiation condition (resp.* $w_-(q,x) - w_{\mathsf{f}}(q,x)$ *satisfies the incoming radiation condition).*

Consequently, we assume here that the kernel $w_1\ (q,x)$ is either an outgoing or an incoming distorted plane wave, and we write

$$w_1\ (q,x) \equiv w_\pm\ (q,x) = w_0\ (q,x) + \mathsf{w}_\pm(q,x), \quad (10.27)$$

where $\mathsf{w}_+ (\mathsf{w}_-)$ behaves like an outgoing (incoming) wave.

REMARK 10.5.2 Of course, when dealing with specific physical problems, we must establish the existence and structure of the distorted plane waves. One way of achieving this is by means of the *limiting absorption principle*[3] ([136], [137], [367]), which is based on noticing that if A is a self-adjoint, linear operator in a Hilbert space $\mathbb{H}$ and if $\lambda = \tau + i\nu \in \mathbb{C}$, with $\nu \neq 0$, then the equation $(A - \lambda I)u(\lambda, x) = f(x)$ has a solution $u(\cdot, \lambda) \in \mathbb{H}$ for each $f \in \mathbb{H}$ because $\lambda \notin \sigma(A)$. In the limiting absorption principle method we look for solutions in the form $u_\pm(\tau, x) = \lim_{\nu \to 0^\pm} u(\lambda, x)$. The difficulty with this approach is centred on the interpretation of this limit, which can be

[3] See Section 5.7.4.

understood as the limit only in appropriate subspaces of $\mathbb{H}$. Physically, the quantity $u(\lambda, x)$, $\nu \neq 0$ describes a steady-state wave in an energy-absorbing medium with an absorption coefficient proportional to ν ([432]).

If we assume the existence of the $w_{\pm}$ and, moreover, that they form two complete sets of generalised eigenfunctions for A_{p}, then we may define the generalised integral transforms $F_{\pm}$, using $w_{\pm}$ as kernels in (10.20). In particular,

$$\check{f}_{\pm}(q) = (F_{\pm}f)(q) = \lim_{r\to\infty} \int_{|x|\le r} \overline{w_{\pm}\,(q,x)} f(x)dx, \tag{10.28}$$

with similar expressions for $F_{\pm}^{*}$ and the spectral representation of A_{p}, provided these limits exist. We refer to F_{+} as an *outgoing generalised Fourier transform* and F_{-} as an *incoming generalised Fourier transform.*

On the basis of these various assumptions, we try to express v_{p} in two different manners, as

$$v_{\mathsf{p}}(t,x) = v_{\mathsf{f}}^{\pm}(t,x) + v^{\pm}(t,x).$$

In the above expressions, $v_{\mathsf{f}}^{\pm}$ are solutions of the unperturbed system to which the solution of the perturbed system converges as $t \to \pm\infty$, respectively. The terms $v^{\pm}$ are terms such that $\lim_{t\to\pm\infty} v^{\pm} = 0$, and that model the transient effects of the perturbation on the system. On the other hand, the terms $v_{\mathsf{f}}^{\pm}$ model the "permanent" effects of the perturbation. As stated above it is of interest to specify the initial conditions $h_{\pm}$ of the unperturbed system, such that $U_{\mathsf{f}}(t)h_{\pm} = v_{\mathsf{f}}^{\pm}(t,x)$. Furthermore, the comparison of h_{+} and h_{-}, through the scattering operator S will provide information on the effect of the perturbation.

The solution of the perturbed system $v_{\mathsf{p}}(t,x)$ given in (10.23) has two spectral representations, depending on whether w_{+} or w_{-} is used in the expansion theorem (10.28). Specifically, we have ([369])

$$v_{\mathsf{p}}(t,x) = \lim_{r\to\infty} \int_{|q|\le r} w_{+}(q,x) \exp(-it\,|q|)\check{h}_{+}(q)dq \tag{10.29}$$

and

$$v_{\mathsf{p}}(t,x) = \lim_{r\to\infty} \int_{|q|\le r} w_{-}(q,x) \exp(-it\,|q|)\check{h}_{-}(q)dq, \tag{10.30}$$

where

$$\check{h}_{\pm}(q) = \int_{\mathbb{R}^3} \overline{w_{\pm}(q,x)} h(x)dx.$$

Since w_{+}(resp. w_{-}) is an outgoing (resp. incoming) distorted plane wave, we refer to (10.29) (resp. (10.30)) as the outgoing (resp. incoming) spectral representations of v_{p}.

We are now in a position to construct a useful form for the wave operators $W_{\pm}$. If we substitute the decomposition (10.27) for w_{-} into (10.30), we obtain

$$v_{\mathsf{p}}(t,x) = v_{\mathsf{f}}^{-}(t,x) + v^{-}(t,x), \tag{10.31}$$

where

$$v_{\mathsf{f}}^{-}(t,x) = \lim_{r\to\infty}\int_{|q|\le r} w_{\mathsf{f}}(q,x)\exp(-it\,|q|)\widetilde{h}_{-}(q)dq, \tag{10.32}$$

$$v^{-}(t,x) = \lim_{r\to\infty}\int_{|q|\le r} \mathsf{w}_{-}(q,x)\exp(-it\,|q|)\widetilde{h}_{-}(q)dq. \tag{10.33}$$

We now notice that since the kernel function in the integral (10.32) is w_{f}, then it follows that v_{f}^{-} represents a free wave. Therefore we can write

$$v_{\mathsf{f}}^{-}(t,x) = U_{\mathsf{f}}(t)h_{\mathsf{f}}^{-}(x) = \exp(-itA_{\mathsf{f}}^{1/2})h_{\mathsf{f}}^{-}(x).$$

The above discussion motivates the relation

$$h_{\mathsf{f}}^{-}(x) = v_{\mathsf{f}}^{-}(0,x),$$

which leads to

$$h_{\mathsf{f}}^{-}(x) = v_{\mathsf{f}}^{-}(0,x) = (F_{\mathsf{f}}^{*}\widetilde{h}_{-})(x) = (F_{\mathsf{f}}^{*}F_{-}h_{\mathsf{p}})(x). \tag{10.34}$$

Now, (10.34) relates the initial data for a free problem and the initial data for an associated perturbed problem. Therefore, we conclude that as $t\to-\infty$, we might expect that

$$h_{\mathsf{f}}^{-}(x) = (F_{\mathsf{f}}^{*}F_{-}h)(x) = W_{-}h(x),$$

that is, we might expect that

$$W_{-} = F_{\mathsf{f}}^{*}F_{-}.$$

It turns out that this is indeed the case, provided we have local energy decay of the form $\lim_{t\to-\infty} v^{-}(t,\cdot) = 0$, which by (10.31) is equivalent to $\lim_{t\to-\infty}\|v_{\mathsf{p}}(t,\cdot) - v_{\mathsf{f}}^{-}(t,\cdot)\| = 0$. It now follows that

$$\begin{aligned}\left\|v_{\mathsf{p}}(t,\cdot) - v_{\mathsf{f}}^{-}(t,\cdot)\right\| &= \left\|\exp(-itA_{\mathsf{p}}^{1/2})h_{\mathsf{p}} - \exp(-itA_{\mathsf{f}}^{1/2})h_{\mathsf{f}}^{-}\right\| \\ &= \left\|\{\exp(itA_{\mathsf{f}}^{1/2})\exp(-itA_{\mathsf{p}}^{1/2}) - F_{\mathsf{f}}^{*}F_{-}\}h_{\mathsf{p}}\right\|,\end{aligned} \tag{10.35}$$

so that by the definition of the wave operators given in (10.9) W_{-} exists and is given by

$$W_{-} = F_{\mathsf{f}}^{*}F_{-}.$$

Working in similar fashion in terms of the decomposition $v_{\mathsf{p}} = v_{\mathsf{f}}^{+} + v^{+}$, we find that

$$W_{+} = F_{\mathsf{f}}^{*}F_{+}.$$

Once we have determined the existence and the form of the wave operators $W_{\pm}$, then a scattering operator, S, that links the initial conditions $h_{\mathsf{f}}^{\pm}$ can be introduced as follows.

The above results indicate that

$$h_{\mathsf{f}}^{\pm} = W_{\pm}h_{\mathsf{p}} = F_{\mathsf{f}}^{*}F_{\pm}h_{\mathsf{p}}.$$

This in turn implies

$$F_{\mathsf{f}} h_{\mathsf{f}}^{\pm} = \check{h}_{\mathsf{f}}^{\pm} = F_{\pm} h_{\mathsf{p}}.$$

Hence,

$$\check{h}_{\mathsf{f}}^{+} = F_{+} h_{\mathsf{p}} = F_{+} F_{-}^{*} \check{h}_{\mathsf{f}}^{-} =: \mathsf{S}\, \check{h}_{\mathsf{f}}^{-},$$

and we see that

$$\mathsf{S} := F_{+} F_{-}^{*} : \check{h}_{\mathsf{f}}^{-} \mapsto \check{h}_{\mathsf{f}}^{+}.$$

This operator and the unitarily equivalent operator

$$F_{\mathsf{f}}^{*}\, \mathsf{S}\, F_{\mathsf{f}} := F_{\mathsf{f}}^{*} F_{+} F_{-}^{*} F_{\mathsf{f}} : h_{\mathsf{f}}^{-} \mapsto h_{\mathsf{f}}^{+}$$

are particularly useful when discussing the theoretical and practical details of the asymptotic condition and the associated asymptotic equality results.

10.6 COMPLEX MEDIA ELECTROMAGNETICS

In this section we present the first steps in an attempt towards a Lax-Phillip - like scattering theory, for general linear complex media in the time domain. Such a theory involves complicated technical issues, not analogous to the ones presented in detail for bounded domains in Chapter 7. It is intended to give a sketch of the approach required and as such is only a descriptive account of work in progress. For noncomplex linear media this problem is studied in detail in [376]. For a particular class of constitutive relations for complex media and under specific assumptions on the fields, this problem has been treated in detail in [37].

In this section we consider electromagnetic waves propagating in a homogeneous, three-dimensional chiral medium. Then the Maxwell equations assume, in terms of the six-vector notation $u = (u_1, u_2)^{tr} = (E, H)^{tr}$, the following form:

$$u' = \mathsf{M}_{\mathrm{A}}\, u + \mathsf{G}_{\mathrm{A}} \star u + \mathsf{J}_{\mathrm{A}}, \tag{10.36}$$

and the fields are assumed to be divergence free. We assume without loss of generality that $\mathsf{J}_{\mathrm{A}} = 0$ and that the material parameters are spatially homogeneous and isotropic.

To apply the general theory sketched in the previous sections, it is convenient to reduce the Maxwell equations (10.36) to an equation for the electric field component resembling the wave equation. We need to define the following intermediate operators and matrices:

$$C = \begin{pmatrix} \mathrm{curl} & 0 \\ 0 & \mathrm{curl} \end{pmatrix}, \quad L = \begin{pmatrix} 0 & -\varepsilon^{-1}\Delta \\ \mu^{-1}\Delta & 0 \end{pmatrix}, \quad \Upsilon = \begin{pmatrix} 0 & \varepsilon^{-1} \\ -\mu^{-1} & 0 \end{pmatrix}.$$

In what follows, for simplicity of presentation and without loss of generality, we assume $\epsilon = \mu = 1$. We now take the curl of the Maxwell equation and use the vector identity $\mathrm{curl}\,\mathrm{curl}\, v = \mathrm{grad}\,\mathrm{div}\, v - \Delta\, v$, where Δ denotes the vector Laplacian, to obtain

$$(Cu)' = Lu + \mathsf{G}_{\mathrm{A}} \star Cu. \tag{10.37}$$

Furthermore, we differentiate the Maxwell equation with respect to time to obtain

$$u'' = \Upsilon(Cu)' + \mathsf{G}'_{\mathsf{A}} \star u, \tag{10.38}$$

where we have assumed appropriate regularity for G_{A} and (without loss of generality) that $\mathsf{G}_{\mathsf{A}}(0) = 0$. After some algebraic manipulations, this Maxwell system reduces to the following system of coupled vector Helmholtz equations:

$$u'' + \mathbb{A}_{\mathsf{f}} = \mathsf{G}'_{\mathsf{A}} \star u + \mathsf{K} \star Cu, \tag{10.39}$$

where

$$\mathbb{A}_{\mathsf{f}} = diag(-\Delta, -\Delta), \;\; \mathsf{K} = \Upsilon \mathsf{G}_{\mathsf{A}}.$$

In the absence of chirality and dispersion effects, the right-hand side of (10.39) vanishes and this system reduces to the vector Helmholz equation, whereas for complex media we obtain a perturbed vector Helmholtz equation. This may be rewritten in abstract form as

$$u'' + \mathbb{A}_{\mathsf{p}} u = 0, \tag{10.40}$$

where $\mathbb{A}_{\mathsf{p}} u = \mathbb{A}_{\mathsf{f}} u - \mathsf{G}'_{\mathsf{A}} \star u - \mathsf{K} \star Cu$. We are thus in the familiar situation described in Section 10.3, where now $A_{\mathsf{f}} = \mathbb{A}_{\mathsf{f}}$ and $A_{\mathsf{p}} = \mathbb{A}_{\mathsf{p}}$.

Concerning the unperturbed problem (10.40), this is simply equivalent to the Maxwell equation in $\mathbb{R}^3$. This system is well studied (see, e.g., [278], [376]). All the various assumptions introduced in the abstract formulation above hold for M_{A} for instance, the existence of the group U_{f} for the unperturbed problem in $\mathbb{R}^3$ is given by the following result.

THEOREM 10.6.1 *The Maxwell operator* M_A *is skew adjoint on* $H(\mathrm{div}0, \mathbb{R}^3) \times H(\mathrm{div}0, \mathbb{R}^3)$*, and generates a group.*

Furthermore, there is a one-to-one correspondence between the solutions of the Helmholtz equation (10.40) and the Maxwell equations ([376]). This correspondence has several useful implications. For instance, as $\mathbb{A}_{\mathsf{f}}$ is nothing but multiple copies of the Laplacian, the spectral decomposition of $\mathbb{A}_{\mathsf{f}}$ can be effected through a standard extension of the Fourier-Plancherel theory in $(L^2(\mathbb{R}^3))^3 \times (L^2(\mathbb{R}^3))^3$. Also, this correspondence allows us to obtain some information concerning the propagation properties of the wave solutions to the unperturbed Maxwell problem. These are summarised in the next theorem.

THEOREM 10.6.2 ([376]) *Suppose that the initial data are in* $H(\mathrm{div}0, \mathbb{R}^3) \times H(\mathrm{div}0, \mathbb{R}^3)$.

(*i*) *If* $u_{\mathsf{f},0}$ *has support in* $|x - x_0| \geq \rho$*, then for* $|t| < \rho$, $U_{\mathsf{f}}(t) u_{\mathsf{f},0}$ *has support in* $|x - x_0| \geq \rho - |t|$.

(*ii*) *If* $u_{\mathsf{f},0}$ *has support in* $|x - x_0| \leq \rho$*, then for* $|t| < \rho$, $U_{\mathsf{f}}(t) u_{\mathsf{f},0}$ *has support in* $|x - x_0| \geq |t| - \rho$.

(*iii*) *If* $u_{\mathsf{f},0}$, $v_{\mathsf{f},0}$ *are two initial conditions such that* $U_{\mathsf{f}}(t)u_{\mathsf{f},0}$ *vanishes for* $|x| < t$, *for all* $t > 0$, *and* $U_{\mathsf{f}}(t)v_{\mathsf{f},0}$ *vanishes for* $|x| < -t$, *for all* $t < 0$, *then* $u_{\mathsf{f},0}$ *and* $v_{\mathsf{f},0}$ *are orthogonal.*

(*iv*) *Let* v *be an element of* $D(\mathsf{M}_A)$. *If* $(\mathsf{M}_A - i\lambda)v$ *vanishes for* $|x| > \rho$ *for some real but nonzero value of* λ, *then* v *vanishes for* $|x| > \rho$.

The above propagation properties help in the construction of the outgoing and incoming subspaces for the Maxwell system as

$$D_+^\rho = \{v \in (L^2(\mathbb{R}^3))^3 \times (L^2(\mathbb{R}^3))^3 \ : \ U_{\mathsf{f}}v = 0 \text{ for } |x| < \rho + t, \ t > 0\},$$
$$D_-^\rho = \{v \in (L^2(\mathbb{R}^3))^3 \times (L^2(\mathbb{R}^3))^3 \ : \ U_{\mathsf{f}}v = 0 \text{ for } |x| < \rho - t, \ t < 0,$$

which can be shown to be orthogonal ([376]). Finally, the spectral theory of the Maxwell operator guarantees that the spectrum is purely absolutely continuous ([278]). The above facts allow us to validate for the unperturbed Maxwell system all the assumptions that were introduced in the formal development of the scattering theory above.

We now turn to the perturbed problem. Our first concern is to show global existence for the complex Maxwell equations in the whole of $\mathbb{R}^3$. This can be guaranteed if the Fourier transform with respect to time $\widetilde{\mathsf{G}_\mathrm{A}}$ of G_A is antisymmetric. Then the Fourier transform of the Maxwell system yields $iq\widetilde{u} = \mathsf{M}_\mathrm{A}\widetilde{\mathsf{G}_\mathrm{A}}\widetilde{u}$, and by the antisymmetry of $\widetilde{\mathsf{G}_\mathrm{A}}$ the operator $\mathsf{M}_\mathrm{A} + \widetilde{\mathsf{G}_\mathrm{A}}$, acting on Fourier transformed fields, is skew adjoint. This provides information on the spectrum of $\mathsf{M}_\mathrm{A} + \widetilde{\mathsf{G}_\mathrm{A}}$, which guarantees that the solution $\widetilde{u}$ of $iq\widetilde{u} = \mathsf{M}_\mathrm{A}\widetilde{\mathsf{G}_\mathrm{A}}\widetilde{u}$ is such that its inverse Fourier transform with respect to time is well defined over the whole of $\mathbb{R} \times \mathbb{R}^3$. This yields global solutions.

The scattering theory for the perturbed operator $A_{\mathsf{p}} = \mathbb{A}_{\mathsf{p}}$ is a more delicate issue, especially since the perturbation is no longer autonomous as in the case described in Section 10.3 but rather nonautonomous. However, a perturbation theory approach (of the general type presented in [368]) may be employed in order to develop an appropriate scattering theory.

The key to this approach is the generalisation of the one-parameter family of operators $U_{\mathsf{p}}(t)$, to a two-parameter family, $U_{\mathsf{p}}(t, s)$, called the *propagator*. This family plays the rôle of the evolution operator in nonautonomous systems. Of course, the unperturbed system, being an autonomous system can still be studied in terms of the evolution operator $U_{\mathsf{f}}(t)$. For the special case in which the perturbation is due to a convolution-type integral operator, the evolution operator can be given in terms of a one-parameter family of linear operators called the *resolvent operator*.

Definition 10.6.3 *Consider the problem*

$$u' = \mathsf{M}_A u + \mathsf{G}_A \star u + \mathsf{J}_A. \tag{10.41}$$

A family of continuous linear operators $\{R(t)\}_{t \geq 0}$ *is called a resolvent to* (10.41) *if*

(*i*) $R(0) = I$.

(*ii*) *For all* $x \in \mathbb{H}$, *the map* $t \mapsto R(t)x$ *is a continuous function* $\mathbb{R}_+ \to \mathbb{H}$.

(*iii*) *For all* $t \geq s$, $U(s,t)$ *is a continuous linear operator on* $D(\mathsf{M}_A)$, *endowed with the graph norm, and for all* $y \in D(\mathsf{M}_A)$ *the map* $t \mapsto R(t)y$ *belongs to* $C^1(\mathbb{R}^+, D(\mathsf{M}_A)) \cap C^1(\mathbb{R}_+, \mathbb{H})$, *and satisfies*

$$(R(t)y)' = \mathsf{M}_A R(t)y + \int_0^t \mathsf{G}_A(t-\tau)R(\tau)y\,d\tau,$$

$$(R(t)y)' = R(t)\mathsf{M}_A y + \int_0^t R(t-\tau)\mathsf{G}_A(\tau)y\,d\tau$$

For properties of the resolvent, see [133]. For the resolvent, existence implies uniqueness. The solution in $C(\mathbb{R}^+, D(\mathsf{M}_\mathrm{A})) \cap C^1(\mathbb{R}^+, \mathbb{H})$ of the problem (10.41) can be expressed in terms of the resolvent as

$$u(t) = R(t)u_0 + \int_0^t R(t-s)\mathsf{J}_\mathrm{A}(s)\,ds$$

as long as $\mathsf{J}_\mathrm{A} \in C(\mathbb{R}^+, \mathbb{H})$. It can be seen that for $\mathsf{J}_\mathrm{A} = 0$, the resolvent family plays the rôle of the propagator. In the special case where $\mathsf{G}_\mathrm{A} \equiv 0$, the resolvent becomes a C_0 semigroup with infinitesimal generator A.

THEOREM 10.6.4 *Under the assumptions on* G_A *of Theorem 7.4.12, with* $[0,T]$ *replaced by* $\mathbb{R}^+$, *a resolvent family exists for* (10.41).

This follows from the general results of [133] on perturbations of resolvent families, in conjunction with Theorem 10.6.1.

PROPOSITION 10.6.5 *The propagator can be obtained through the solution of a Volterra integral equation of the form*

$$R(t) = \mathsf{T}_{\mathsf{M}_A}(t) + \int_0^t \mathsf{T}_{\mathsf{M}_A}(t-\tau)\left(\int_0^\tau \mathsf{G}_A(\tau-r)R(r)\,dr\right)d\tau, \qquad (10.42)$$

where $\mathsf{T}_{\mathsf{M}_A}$ *is the semigroup generated by* M_A.

The proof follows easily by noting that the solution of (10.41) for $\mathsf{J}_\mathrm{A} = 0$ can be expressed as $u(t) = R(t)u_0$, while at the same time, employing a fixed point scheme, it can be expressed in terms of $\mathsf{T}_{\mathsf{M}_\mathrm{A}}$ as $u(t) = \mathsf{T}_{\mathsf{M}_\mathrm{A}}(t)u_0 + \int_0^t \mathsf{T}_{\mathsf{M}_\mathrm{A}}(t-\tau)(\int_0^\tau \mathsf{G}_\mathrm{A}(\tau-r)u(r)\,dr)\,d\tau$.

The expression of the resolvent through the solution of the Volterra integral equation (10.42) is useful because it allows the perturbative construction of the resolvent. Assuming that G_A is small, in an appropriate norm, a Neumann series expansion for $R(t)$ can be constructed. The assumption of smallness of G_A is physically a reasonable one, as in general, the chirality effects are considered to be small. Assume that $\mathsf{G}_\mathrm{A} = \vartheta \mathsf{G}_\mathrm{A}^\sharp$, where ϑ is a small parameter denoting the order of magnitude of G_A in the relevant norm, the resolvent can be expanded in the Neumann series

$$R(t) = \sum_{j=0}^\infty \vartheta^j R^{(j)}(t),$$

where the $R^{(j)}$ are given by the iterative procedure

$$R^{(0)}(t) = \mathsf{T}_{\mathsf{M}_\mathsf{A}}(t),$$

$$R^{(j)}(t) = \int_0^t \mathsf{T}_{\mathsf{M}_\mathsf{A}}(t-\tau)\left(\int_0^\tau \mathsf{G}^\sharp_\mathsf{A}(\tau - r)R^{(j-1)}(r)\,dr\right)d\tau, \quad j = 1, 2, \cdots .$$

The terms in the series are easy to calculate, thus yielding approximate expressions for $u(t, x)$.

The resolvent family $R(t)$ coincides with $U(t, 0)$, where U is the propagator. Clearly, by a simple modification, all the above results can be stated for $s \geq 0$. Furthermore, for the needs of scattering theory, we have to work in $\mathbb{R}$ instead of $\mathbb{R}^+$; the above results may be extended to this setting under proper technical modifications, thus allowing for $s < 0$. In this task we need to consider the two-parameter family generalisation of the resolvent family as introduced in Definition 10.6.3, $R(t, s)$, $s \leq t$.

The wave operators in the present case have to be modified, and understood as the operators that connect the condition of the perturbed and the unperturbed system, respectively, at time s. Note that since the system is nonautonomous, they can be defined as the strong limits

$$W_{\pm s} = \lim_{t \pm \infty} U_\mathsf{f}(s, t)U_\mathsf{p}(t, s),$$

where $U_\mathsf{f}(s, t) = U_\mathsf{f}(s - t)$ is the propagator for the unperturbed system, which can be expressed in terms of the Fourier-Plancherel theory. As stated above, $U_\mathsf{p}(t, s)$ may be expressed through a convergent Neumann series involving U_f and the perturbation operator $\mathbb{A}_\mathsf{p} - \mathbb{A}_\mathsf{f}$, which in turn may be approximated using the Fourier-Plancherel theory. Therefore, in principle, the wave operators, if they exist, can be approximated constructively. The existence of the wave operators is related once more to the spectral theory of the perturbed Maxwell operator and can be treated by the extension of methods used in the perturbation theory of linear operators, similar to those presented in [229]; see also [277], [362].

10.7 MISCELLANEA

10.7.1 Scattering problems with boundary conditions

Extension to the analysis of electromagnetic scattering problems involving boundary conditions is in principle a reasonably straightforward but lengthy matter; see [432] and [369] for more details. Specific applications for special types of complex media can be found in [37]. Consideration of these problems when, e.g., the chiral parameters are time dependent[4] can be found in [368].

10.7.2 Perturbed stratified media

Another class of very interesting problems of scattering theory refers to stratified media. Stratified media are media whose physical properties depend

[4]For a general introduction to time-dependent scattering theory, see [377].

on a single coordinate, and they can produce guided waves propagating in directions orthogonal to that of the stratification, in addition to the free waves propagating in a homogeneous medium. The perturbation of such a stratified medium may produce waves whose propagation properties are very important for a number of applications. In the case of noncomplex media, this problem has been studied in [423]; see also [32]. The extension of these ideas to complex media is an interesting problem that to the best of our knowledge has not been studied.

10.7.3 Inverse scattering problems

In this chapter we have introduced a few ideas concerning *direct scattering problems*, i.e., determining the scattering operator S and the initial data $h^{=}$ from knowledge of U_{f} and U_{p} and $u_{\mathsf{p}}(0)$. Of at least equal importance are *inverse scattering problems*, which consist in determining U_{p} from the knowledge of S and U_{f}. Inverse scattering problems are nonlinear and ill posed.

Inverse problems are particularly useful in applications since S can be determined from experimental data; therefore, the solution of the inverse problem allows us to obtain information on the structure of the perturbed system such as the composition of the medium. For example, one could formulate an inverse scattering problem so as to reconstruct the susceptibility kernel matrix, G_{A}, from measurements of the incoming and outgoing waves. To the best of our knowledge, this problem is open.

Chapter Eleven

Nonlinear Problems

11.1 INTRODUCTION

Nonlinearity is inherent in nature and accounts for a number of interesting phenomena. To keep within the context of electromagnetic media, nonlinearity appears in a number of cases in which the dispersion relations or equivalently the coefficients of the constitutive relations change as a function of the field amplitudes. This behaviour is very common in dielectrics, where it has been experimentally verified, theoretically studied, and widely studied mathematically. Furthermore, the interplay between dispersion and nonlinearity in dielectrics has led to the observation of solitary waves, which has important applications to optical communications. Regarding chiral media, although third-order nonlinear effects were predicted as early as 1967, nonlinear optical rotation experiments were not undertaken before 1993. Out of a long list, the papers [313] and [314] can serve as representative of related important experimental work.

The structure of this chapter is as follows. In Section 11.2 we introduce a model for the study of nonlinear phenomena in complex media, and in Section 11.3 we provide some rigorous results concerning the solvability and well posedness of the nonlinear model using techniques from the theory of nonlinear PDEs.

11.2 FORMULATION

We consider a nonlinear complex electromagnetic medium modelled by constitutive relations of the form

$$\mathsf{d} = \mathcal{L}u = \mathsf{A}_0 u + \mathsf{G}_0 \star u + \mathsf{G}_{0,\mathrm{n}\ell} \star \mathsf{N}(u)u$$

(see Section 2.3.5, equation (2.24)). The spatiotemporal evolution of the fields is given by the Maxwell equation

$$(\mathsf{A}_0 u + \mathsf{G}_0 \star u + \mathsf{G}_{0,\mathrm{n}\ell} \star \mathsf{N}(u)u)' = \mathsf{M}u + j,$$

in a domain $\mathcal{O}$, supplemented with appropriate boundary conditions, e.g., the perfect conductor boundary condition $n \times u_1 = 0$, on $\partial\mathcal{O}$.

Assuming that the convolution kernels G_0 and $\mathsf{G}_{0,\mathrm{n}\ell}$ are smooth enough that we may differentiate (weakly) $\mathcal{L}u$ with respect to time, we obtain

$$(\mathcal{L}u)' = \mathsf{A}_0 u' + \mathsf{G}_0\,(0)\, u + \mathsf{G}_{0,\mathrm{n}\ell}\,(0)\, \mathsf{N}(u)u + \mathsf{G}_0' \star u + \mathsf{G}_{0,\mathrm{n}\ell}' \star \mathcal{N}(u)u.$$

Assuming further, as usual without loss of generality, that $\mathsf{G}_0(0) = 0$ and defining $\mathsf{B}_{\mathrm{A}} := \mathsf{A}_{\mathrm{or}}^{-1}\mathsf{G}_{0,\mathrm{n}\ell}(0)$ and $\mathsf{G}_{\mathrm{A},\mathrm{n}\ell} := \mathsf{A}_{\mathrm{or}}^{-1}\mathsf{G}'_{0,\mathrm{n}\ell}$, the Maxwell system assumes in $\mathcal{O} \times [0,T]$ the explicit form

$$u' + \mathsf{B}_{\mathrm{A}}\mathsf{N}(u)u + \mathsf{G}_{\mathrm{A},\mathrm{n}\ell} \star \mathsf{N}(u)u = \mathsf{M}_{\mathrm{A}}u + \mathsf{G}_{\mathrm{A}} \star u + \mathsf{J}_{\mathrm{A}} \tag{11.1}$$

with initial condition

$$u(0,x) = u_0(x), \quad x \in \mathcal{O}, \tag{11.2}$$

and the perfect conductor boundary condition

$$n \times u_1 = 0, \quad (t,x) \in [0,T] \times \partial\mathcal{O}. \tag{11.3}$$

REMARK 11.2.1 An alternative problem would be to add Ohmic effects, possibly nonlinear, to the above system. The inclusion of such a term does not introduce any significant complications into the mathematical treatment of the problem.

11.3 WELL POSEDNESS OF THE MODEL

In this section we study the well posedness of the nonlinear evolution equation (11.1). To make the presentation clearer we make a simplifying assumption according to which the nonlinear convolution effects in (11.1) are weaker than all the other effects, so that they may be neglected. This assumption models the fact that dispersion and nonlinearity are both weak effects, so that in general, $\mathsf{G}_{\mathrm{A},\mathrm{n}\ell} \star \mathsf{N}(u)u \ll \mathsf{B}_{\mathrm{A}}\ \mathsf{N}(u)u$. This assumption is a temporary one and will soon be relaxed to take into account the effect of both terms (see Section 11.3.4).

The above discussion motivates the following:

ASSUMPTION 11.3.1 *$\mathsf{G}_{0,n\ell}$ is weakly differentiable and the order of magnitude of $\mathsf{G}'_{0,n\ell}$ is negligible compared to the order of magnitude of $\mathsf{G}_{0,n\ell}(0)$.*

Under this assumption, the governing equation (11.1) becomes

$$u' + \mathsf{B}_{\mathrm{A}}\mathsf{N}(u)u = \mathsf{M}_{\mathrm{A}}u + \mathsf{G}_{\mathrm{A}} \star u + \mathsf{J}_{\mathrm{A}}. \tag{11.4}$$

As far as the linear dispersive effects modelled by the convolution kernels G_{A} are concerned, we keep the standing assumptions of Chapter 7. However, we need some extra assumptions about the nonlinear terms.

We will adopt the following assumptions concerning the nonlinearity.

ASSUMPTION 11.3.2 The nonlinearity $\mathsf{B}_{\mathrm{A}}\mathsf{N}(u)u$ is *monotone*, i.e., there exists a $p \in \mathbb{N}$ such that for all $\mathsf{v} \in \mathbb{R}^6$,

$$\mathsf{N}(\mathsf{v})\mathsf{v} \cdot \mathsf{v} \geq \mathsf{c}\,|\mathsf{v}|^p, \quad \mathsf{c} > 0.$$

ASSUMPTION 11.3.3 There exists a constant α such that for all $\mathsf{v} \in \mathbb{R}^6$,

$$|\mathsf{N}(\mathsf{v})\mathsf{v}| \leq \alpha\,(1 + |\mathsf{v}|^{p-1}).$$

ASSUMPTION 11.3.4 The nonlinearity $\mathsf{B}_{\mathrm{A}}\mathsf{N}(u)u$ satisfies

$$D_{\mathsf{v}}\Big(\mathsf{B}_{\mathrm{A}}\mathsf{N}(\mathsf{v})\mathsf{v}\Big)\mathsf{w}\cdot\mathsf{w}\geq 0 \text{ for all } \mathsf{v},\mathsf{w}\in\mathbb{R}^6,$$

where D_{v} denotes the derivative with respect to the vector v.

REMARK 11.3.5 It is easily seen that for nonlinearities of the form given by equations (2.24) and (2.25) assumed here, Assumptions 11.3.2 (for $p = q+2$) and 11.3.4 hold provided that the matrices N_1, N_2 out of which N is composed, are positive definite.

REMARK 11.3.6 The above assumptions on the nonlinearity are common in the mathematical modelling and study of nonlinear electromagnetic media. For instance, they are used by Wellander [426] in the study of nonlinear dielectric materials. On the mathematical side, these assumptions allow us to employ the powerful techniques of monotone operators to study the well posedness of these models. The second assumption corresponds to the convexity of the energy functional of the medium.

We now turn to the problem of existence. We will consider weak solutions, which are defined in a similar manner to that of Chapter 7, with the important difference that because of the nonlinearity the test functions will no longer belong to the same space as the solution, hence the corresponding integrals are to be understood as duality pairings between the involved spaces. The strategy of the proof is to construct finite-dimensional approximations of the system, show the existence of these approximations, and then use the a priori bounds to show that a limit exists that is a solution of the original (infinite-dimensional) problem.

11.3.1 A priori bounds

To prove the local existence of a weak solution of the initial boundary value problem (11.2)-(11.4), we will apply the Faedo-Galerkin method. At the first step of the method we need some a priori estimates.

We prove the following.

PROPOSITION 11.3.7 *Assume that the nonlinearity satisfies the monotonicity Assumption 11.3.2, the kernel* $\mathsf{G}_A \in W^{1,1}([0,T],(L^\infty(\mathcal{O}))^{36})$, *the initial condition* $u_0 \in (L^{q+2}(\mathcal{O}))^3 \times (L^{q+2}(\mathcal{O}))^3$, *and the forcing term* $\mathsf{J}_A \in L^1([0,T],\mathbb{X})$. *Then any sufficiently regular solution* u *of (11.3)-(11.4) satisfies the following energy estimates:*

$$\sup_{0\leq t\leq T} \| u(t) \|_{(L^{q+2}(\mathcal{O}))^6} \leq C, \quad \sup_{0\leq t\leq T} \|u(t)\|_{\mathbb{X}} \leq C, \tag{11.5}$$

where the constant C *depends on* $\mathcal{O}, q$ *and* T.

Proof. To obtain the energy estimates (11.5), we use a standard technique introduced first for parabolic equations ([263]). Taking the $L^2(\mathcal{O})$ inner product of (11.4) with u, we obtain

$$\int_{\mathcal{O}} u' \cdot u\, dx + \int_{\mathcal{O}} (\mathsf{B}_{\mathrm{A}} \mathsf{N}(u)u) \cdot u\, dx$$
$$= \int_{\mathcal{O}} \mathsf{M}_{\mathrm{A}} u \cdot u dx + \int_{\mathcal{O}} (\mathsf{G}_{\mathrm{A}} \star u) \cdot u dx + \int_{\mathcal{O}} \mathsf{J}_{\mathrm{A}} \cdot u\, dx\,,$$

or, equivalently,

$$\frac{1}{2}\frac{d}{dt}||u||_{\mathbb{X}}^2 + \int_{\mathcal{O}} \mathsf{B}_{\mathrm{A}} \mathsf{N}(u)u\, dx \leq C_1(T) \sup_{s\in[0,T]} ||u(s)||_{\mathbb{X}}^2 + \frac{\epsilon}{2}||u||_{\mathbb{X}}^2 + \frac{1}{2\epsilon}||\mathsf{J}_{\mathrm{A}}||_{\mathbb{X}}^2\,, \tag{11.6}$$

where we have used the property that $\int_{\mathcal{O}} \mathsf{M}_{\mathrm{A}} u \cdot u\, dx = 0$ (recall the properties of the matrix A_0) and the Cauchy-Schwarz inequality to estimate $\int_{\mathcal{O}} \mathsf{J}_{\mathrm{A}} \cdot u\, dx$, for an arbitrary $\epsilon > 0$. On the right-hand side we have estimates of the convolution term $\mathsf{G}_{\mathrm{A}} \star u$, and this explains the explicit dependence of C_1 and C_2 on the length of the time interval $[0, T]$.

Using Assumption 11.3.2 (monotonicity of the nonlinearity N), the estimate (11.6) yields

$$\frac{1}{2}\frac{d}{dt}||u||_{\mathbb{X}}^2 \leq C_1(T) \sup_{s\in[0,T]} ||u(s)||_{\mathbb{X}}^2 + \frac{1}{2\epsilon}||\mathsf{J}_{\mathrm{A}}||_{\mathbb{X}}^2, \tag{11.7}$$

which, by Gronwall's inequality, for any bounded $T > 0$ gives

$$\sup_{t\in[0,T]} ||u(t)||_{\mathbb{X}}^2 < C(T) < \infty\,,$$

for a constant C in general dependent on the time horizon T. Therefore, this a priori bound guarantees that $u \in L^\infty([0,T]; \mathbb{X})$.

We now return to estimate (11.6), reinserting the nonlinear term into the left-hand side, which, combined with the monotonicity assumption for the nonlinearity, yields

$$\frac{1}{2}\frac{d}{dt}||u||_{\mathbb{X}}^2 + C\lambda||u||_{(L^{q+2}(\mathcal{O}))^6}^{q+2} \leq C_1(T) \sup_{s\in[0,T]} ||u(s)||_{\mathbb{X}}^2 + ||\mathsf{J}_{\mathrm{A}}||_{\mathbb{X}}^2\,. \tag{11.8}$$

Integrating with respect to time over the interval $[0, T]$ and using the previous estimate, we obtain

$$\int_0^T ||u||_{(L^{q+2}(\mathcal{O}))^6}^{q+2}(s)\, ds < C_3(T) < \infty$$

for some constant C_3; hence $u \in L^{q+2}([0,T]; (L^{q+2}(\mathcal{O}))^3 \times (L^{q+2}(\mathcal{O}))^3)$. By (11.7) and (11.8) we obtain the $L^\infty([0,T]; (L^{q+2}(\mathcal{O}))^3 \times (L^{q+2}(\mathcal{O}))^3)$ bound. The assumption $T < \infty$ is crucial in the above estimates. □

REMARK 11.3.8 The above result can be obtained under a weaker assumption on the forcing term: $\mathsf{J}_{\mathrm{A}} \in L^1([0,T], (L^{1+\frac{1}{p}}(\mathcal{O}))^3 \times (L^{1+\frac{1}{p}}(\mathcal{O}))^3)$.

We now obtain a priori bounds for the temporal derivative of the fields.

PROPOSITION 11.3.9 *Under the assumptions of Proposition 11.3.7, and additionally if Assumptions 11.3.3, 11.3.4 are true and* $\mathsf{J}_A \in W^{1,1}([0,T];\mathbb{X})$, *the following a priori bounds hold:*

$$||u'||_{L^2([0,\,T];\,\mathbb{X})} \leq C_1\,, \quad ||\mathsf{M}_A u||_{L^{p'}([0,\,T];\,(L^{p'}(\mathcal{O}))^6} \leq C_2\,,$$

where $p = q+2$, *and* $\frac{1}{p'} + \frac{1}{p} = 1$. *Furthermore,*

$$\sup_{t\in[0,T]} ||u'(t)||_{\mathbb{X}} < C_3, \quad \sup_{t\in[0,T]} ||\mathsf{M}_A u(t)||_{(L^{p'}(\mathcal{O}))^6} < C_4.$$

Proof. We differentiate the equation (11.1) formally with respect to t to obtain

$$u'' + D_u(\mathsf{B}_\mathrm{A}\mathsf{N}(u)u)u' = \mathsf{M}_\mathrm{A}u' + \mathcal{C} + \mathsf{J}_\mathrm{A}'\,, \tag{11.9}$$

where $\mathcal{C}(t) = (\mathsf{G}_\mathrm{A} \star u)' = \mathsf{G}_\mathrm{A}(0)u(t) + \mathsf{G}_\mathrm{A}' \star u$, the second equality being true provided appropriate temporal regularity of the convolution kernels is assumed.

We now take the inner product of (11.9) with u' in $\mathbb{X}$. Since $\int_{\mathcal{O}} \mathsf{M}_\mathrm{A}u' \cdot u'\,dx = 0$, this gives

$$\int_{\mathcal{O}} u'' \cdot u'\,dx + \int_{\mathcal{O}} D_u(\mathsf{B}_\mathrm{A}\mathsf{N}(u)u)u' \cdot u'\,dx = \int_{\mathcal{O}} \mathcal{C}(t)\cdot u'\,dx + \int_{\mathcal{O}} \mathsf{J}_\mathrm{A}' \cdot u'\,dx.$$

The term $\int_{\mathcal{O}} \mathcal{C}(t)\cdot u'\,dx$ consists of convolution integrals which can be estimated as

$$\left|\int_{\mathcal{O}} (\mathsf{G}_\mathrm{A}' \star u)(t)\,u'(t)\,dx\right| \leq \epsilon||u'(t)||_{\mathbb{X}} + \frac{1}{2\epsilon}C(t) \sup_{s\in[0,T]} ||u(s)||_{\mathbb{X}}.$$

Therefore, for every $\epsilon > 0$,

$$\frac{1}{2}\frac{d}{dt}||u'||_{\mathbb{X}}^2 + \int_{\mathcal{O}} \mathsf{B}_\mathrm{A} D_u(\mathsf{N}(u)u)u' \cdot u' dx \leq \epsilon||u'||_{\mathbb{X}}^2 + \frac{1}{4\epsilon}||\mathsf{J}_\mathrm{A}'||_{\mathbb{X}}^2 + C,$$

where C is an appropriate constant. By Assumption 11.3.4, the second term in the left-hand side of the above inequality is always non-negative. Hence, by choosing ϵ as small as necessary and applying Gronwall's inequality, we obtain that $u' \in L^2([0,T];\mathbb{X})$. Then, integration of the above inequality yields that u' is bounded in $L^\infty([0,T];\mathbb{X})$.

We now use the above obtained bounds for u' to obtain a bound for $\mathsf{M}_\mathrm{A}u$. By equation (11.4) we have

$$\mathsf{M}_\mathrm{A}u = u' + \mathsf{B}_\mathrm{A}\mathsf{N}(u)u - \mathsf{G}_\mathrm{A} \star u - \mathsf{J}_\mathrm{A}.$$

Since $u \in L^p([0,T],(L^p(\mathcal{O}))^3 \times (L^p(\mathcal{O}))^3)$ (by Proposition 11.3.7), we observe by Assumption 11.3.3 that $\mathsf{N}(u)u \in L^{p'}([0,T],(L^{p'}(\mathcal{O}))^3 \times (L^{p'}(\mathcal{O}))^3)$, where $\frac{1}{p} + \frac{1}{p'} = 1$. Furthermore, since $u' \in L^2([0,T];\mathbb{X})$, we have that $\mathsf{M}_\mathrm{A}u \in L^{p'}([0,T],(L^{p'}(\mathcal{O}))^3 \times (L^{p'}(\mathcal{O}))^3)$ by the standard Lebesgue embeddings (for $p > 2$). By further applying the $L^\infty([0,T];\mathbb{X})$ bound for u', we obtain the required $L^\infty([0,T];(L^{p'}(\mathcal{O}))^6)$ bound for $\mathsf{M}_\mathrm{A}u$. In all the above estimates the assumption that $T < \infty$ is crucial. □

REMARK 11.3.10 Analogous a priori estimates can also be established under the assumption of Remark 11.3.8. Clearly, in this case u' will no longer be in L^2.

11.3.2 Finite-dimensional approximation

To construct an approximation sequence to a (weak) solution to (11.4), (11.2), (11.3), we first consider an orthonormal base $\{e_k\}_{k=1}^{\infty}$ of $\mathbb{X}_{\mathsf{M}}$. Then we define for any $n > 1$ the approximating sequence $\{u_n\}$, where $u_n(t,x) = \sum_{k=1}^{n} \vartheta_k(t) e_k(x)$ and the (scalar) coefficients $\vartheta_k(t)$ are to be determined. This expansion simplifies the finite-dimensional projection of the nonlinear terms. Since $\{e_k\}$ is also a basis of $\mathbb{X}$, we can expand the initial datum and source terms similarly. In what follows we identify $u_n = (\vartheta_1, \ldots, \vartheta_n)^{tr} \in \mathbb{R}^n$ with an element (also denoted by u_n) $u_n \in V_n := span(\{e_k\}_{k=1}^{n}) \subset \mathbb{X}$.

The approximating problem satisfied by u_n is the following:

$$\begin{aligned} u_n(t) = \phi(u_n(0)) - \int_0^t \mathsf{B}_{\mathsf{A}}\, \mathsf{N}(u_n(s)) u_n(s)\, ds - \int_0^t (\mathsf{G}_{\mathsf{A}} \star u_n)(s)\, ds \\ + \int_0^t \mathsf{M}_{\mathsf{A}} u_n(s)\, ds + \int_0^t (\mathsf{J}_{\mathsf{A}})_n(s)\, ds\,, \end{aligned} \tag{11.10}$$

where the subscripts n denote the projections of the relevant quantities onto the space spanned by the first n elements of $\{e_k\}$, and in $\phi(u_n(0))$ we collect all contributions of the initial conditions. In accordance with the previous comment on the nature of u_n, this integral equation can be treated either as an equation in $V_n \subset \mathbb{X}$ or as an equation in $\mathbb{R}^n$. In fact $V_n \simeq \mathbb{R}^n$.

REMARK 11.3.11 To arrive at (11.10), we need to take the weak form of the equation using appropriate test functions (e.g., $\Phi(x,t) \in H^1([0,T], \mathbb{X}_{\mathsf{M}})$, with $\Phi(x,T) = 0$). We also consider as initial condition $u_{n,0}$ the projection of $u_0(x)$ on $\mathbb{X}_{\mathsf{M}}$.

The finite-dimensional approximation (11.10) is a nonlinear integral equation of Volterra type, the well posedness of which is shown in the next proposition.

PROPOSITION 11.3.12 *Assume that the nonlinearity has the following properties*[1]*:*

(*i*) *For all* $\mathsf{v} \in \mathbb{R}^6$ *such that* $|\mathsf{v}| \leq M$, *there exist* $p \in \mathbb{N}$ *and* $\alpha > 0$ *such*[2] *that* $|\mathsf{B}_A \mathsf{N}(\mathsf{v})\mathsf{v}| \leq \alpha M^p$.

(*ii*) *For all* $\mathsf{v}_i \in \mathbb{R}^6$ *such that* $|\mathsf{v}_i| \leq M$, $i = 1, 2$, *the nonlinearity is locally Lipschitz, i.e.,* $|\mathsf{B}_A \mathsf{N}(\mathsf{v}_1)\mathsf{v}_1 - \mathsf{B}_A \mathsf{N}(\mathsf{v}_2)\mathsf{v}_2| \leq \Lambda(\alpha, M)|\mathsf{v}_1 - \mathsf{v}_2|$.

Then the finite-dimensional approximation (11.10) *admits a unique solution in* $C([0,T], \mathbb{R}^n)$, *for all* $n \in \mathbb{N}$.

---

[1] These assumptions on the nonlinearity $\mathsf{B}_{\mathsf{A}}\, \mathsf{N}(\mathsf{v})\mathsf{v}$ are very reasonable and hold for, e.g., polynomial-type nonlinearities.

[2] The value of α may depend on M and by proper choice of M, α may be chosen as small as possible.

Proof. The proof uses a standard fixed point argument (see, e.g., [82]), which we sketch briefly. We choose an interval $[t_0, t_0 + \theta]$ and consider (11.10) in this interval. This yields the following nonlinear Volterra equation in $\mathbb{R}^n$:

$$u_n(t) = \phi(u_n(t_0)) - \int_{t_0}^{t} \mathsf{B}_{\mathrm{A}}\, \mathsf{N}(u_n(s))u_n(s)\, ds - \int_{t_0}^{t} (\mathsf{G}_{\mathrm{A}} \star u_n)(s)\, ds \\ + \int_{t_0}^{t} \mathsf{M}_{\mathrm{A}} u_n(s)\, ds + \int_{t_0}^{t} (\mathsf{J}_{\mathrm{A}})_n(s)\, ds\,, \tag{11.11}$$

where in $\phi(u_n(t_0))$ we collect all the "initial" value terms, i.e., the terms that contain the functions u_n calculated at t_0.

We now define the operator $\mathsf{F} : C([0,T], \mathbb{R}^n) \to C([0,T], \mathbb{R}^n)$ by

$$(\mathsf{F}\mathfrak{s})(t) := \phi(\mathfrak{s}(t_0)) - \int_{t_0}^{t} \mathsf{B}_{\mathrm{A}}\, \mathsf{N}(\mathfrak{s}(s))\mathfrak{s}(s)\, ds - \int_{t_0}^{t} (\mathsf{G}_{\mathrm{A}} \star \mathfrak{s})(s)\, ds \\ + \int_{t_0}^{t} \mathsf{M}_{\mathrm{A}} \mathfrak{s}(s)\, ds + \int_{t_0}^{t} (\mathsf{J}_{\mathrm{A}})_n(s)\, ds\,,$$

or, in compact notation,

$$(\mathsf{F}\mathfrak{s})(t) = \phi(\mathfrak{s}(t_0)) + \mathsf{F}_{\ell}(\mathfrak{s})(t) + \mathsf{F}_{\mathrm{n}\ell}(\mathfrak{s})(t)\,,$$

where F_{ℓ} contains all the linear terms and $\mathsf{F}_{\mathrm{n}\ell}$ contains all the nonlinear terms of the operator F.

Then the nonlinear integral equation (11.11) is written in operator form as

$$u_n = \mathsf{F} u_n\,,$$

and the solution u_n being the fixed point of the operator F. We will employ a fixed point theorem to show that the operator F has a fixed point.

Consider:

$$\mathbb{S}_{\theta,\nu,\lambda} := \{\mathfrak{s} \in C([t_0, t_0+\theta]; \mathbb{R}^n \;:\; |\mathfrak{s}(t) - \phi| \le \nu\lambda,\;\; |\mathfrak{s}(t_1) - \mathfrak{s}(t_2)| \le \lambda\, |t_1 - t_2|\},$$

where $\phi := \phi(u_n(t_0))$ and θ, ν, λ are to be determined later on. The set $\mathbb{S}_{\theta,\nu,\lambda}$ is convex. Further, by the Arzelà-Ascoli theorem (see Theorem A.5.1 in Appendix A), it is compact.

We will show first that F leaves $\mathbb{S}_{\theta,\nu,\lambda}$, invariant for a suitable choice of θ, ν, λ.

As a first step we can show that if $|\mathfrak{s}(t) - \phi| \le \nu\lambda$ then $|\mathsf{F}\mathfrak{s}(t) - \phi| \le 1$. Indeed,

$$|(\mathsf{F}\mathfrak{s})(t) - \phi| \le |(\mathsf{F}_{\ell}\mathfrak{s})(t) + (\mathsf{F}_{\mathrm{n}\ell}\mathfrak{s})(t)| \le C(\alpha, M, \theta, \nu, \lambda)\,,$$

where $C(\alpha, M, \theta, \nu, \lambda)$ is a finite constant that can be made as small as possible for small enough α, θ, ν and λ, as can be easily seen by the properties of the nonlinearity. We choose the constants so that $C(\alpha, \theta, \nu, \lambda) < \nu\lambda$.

We now address the question of whether $(\mathsf{F}(\mathfrak{s})(t)$ is a Lipschitz continuous function of t. To this end we calculate $|(\mathsf{F}\mathfrak{s})(t_1) - (\mathsf{F}\mathfrak{s})(t_2)|$, which consists of integrals of linear and nonlinear terms, as follows:

$$|(\mathsf{F}\mathfrak{s})(t_1) - (\mathsf{F}\mathfrak{s})(t_2)| \le |(\mathsf{F}_{\ell}\mathfrak{s})(t_1) - (\mathsf{F}_{\ell}\mathfrak{s})(t_2)| + |(\mathsf{F}_{\mathrm{n}\ell}\mathfrak{s})(t_1) - (\mathsf{F}_{\mathrm{n}\ell}\mathfrak{s})(t_2)|.$$

The contribution of F_ℓ consists of terms of the form

$$\int_{t_0}^{t_1} \mathsf{K}_{\mathsf{G}}(t_1,s)u(s)ds - \int_{t_0}^{t_2} \mathsf{K}_{\mathsf{G}}(t_2,s)u(s)ds = \int_{t_0}^{t_2} \mathsf{K}_{\mathsf{G}}^{\diamond}(s)ds + \int_{t_1}^{t_2} \mathsf{K}_{\mathsf{G}}(t_1,s)u(s)ds,$$

where the kernel $\mathsf{K}_{\mathsf{G}}(t,s)$ contains the convolution integral $\mathsf{G}'_{\mathsf{A}} \star u$ (i.e., the above term consists of iterated integrals) and $\mathsf{K}_{\mathsf{G}}^{\diamond}(s) := \mathsf{K}_{\mathsf{G}}(t_1,s) - \mathsf{K}_{\mathsf{G}}(t_2,s)$. The fact that K_{G} is defined in terms of convolution integrals guarantees that $\mathsf{K}_{\mathsf{G}}(t,s)$ is a Lipschitz function with Lipschitz constant L_K with respect to its first argument. Combining the above, we see that

$$|(\mathsf{F}_\ell \mathfrak{s})(t_1) - (\mathsf{F}_\ell \mathfrak{s})(t_2)| \leq C_1(L_K, M)\,|t_1 - t_2|.$$

A similar argument in which we also use the fact that the nonlinearity is locally Lipschitz shows that a Lipschitz estimate holds for the nonlinear term:

$$|(\mathsf{F}_{\mathrm{n}\ell} \mathfrak{s})(t_1) - (\mathsf{F}_{\mathrm{n}\ell} \mathfrak{s})(t_2)| \leq C_2(L_K, \lambda, M, \alpha)\,|t_1 - t_2|\,.$$

Therefore,

$$|(\mathsf{F}\mathfrak{s})(t_1) - (\mathsf{F}\mathfrak{s})(t_2)| \leq (C_1(L_K, M) + C_2(L_K, \lambda, M, \alpha))\,|t_1 - t_2|\,,$$

so that as long as λ is chosen so that $(C_1(L_K, M) + C_2(L_K, \lambda, M, \alpha)) \leq \lambda$ (which is always feasible as long as α is small enough), F leaves $\mathbb{S}_{\theta,\nu,\lambda}$ invariant. Therefore, by the Schauder fixed point theorem (see Theorem A.9.3 in Appendix A), the operator F has a fixed point that corresponds to a local solution of the nonlinear integral equation in the interval $[t_0, t_0 + \theta]$. Using a standard continuation argument (see, e.g., [82]), we extend this solution in the whole interval $[0, T]$. Uniqueness follows by the Lipschitz assumption. □

11.3.3 Solvability

We now pass to the limit as $n \to \infty$ in the finite-dimensional approximations $\{u_n\}$. It is immediately evident that the finite-dimensional approximations of the solution satisfy the a priori bounds obtained in the previous section (see Proposition 11.3.7), and in particular with the right-hand side independent of n.

We will need the following lemma.

LEMMA 11.3.13 *Let $b \geq 0$. Then for any $y_1, y_2 \in \mathbb{R}^m$, $m \in \mathbb{N}$, there holds*

$$(|y_1|^b y_1 - |y_2|^b y_2) \cdot (y_1 - y_2) \geq C(m,b)\,|y_1 - y_2|^{b+2}.$$

Here "$\cdot$" stands for the usual inner product in $\mathbb{R}^m$.

For a proof, see [134], p. 13.

Now we are ready to prove our main result.

THEOREM 11.3.14 *Under the assumptions of Propositions 11.3.7 and 11.3.9 and further assuming that $\mathsf{G}_d \in W^{1,\infty}([0,T]; L^\infty(\mathcal{O})^{36})$[3], problem (11.2)-(11.4) has a unique weak solution,*

$$u \in L^\infty([0,T], \mathbb{X}_{\mathsf{M}}) \cap L^\infty([0,T], (L^{q+2}(\mathcal{O}))^6)\,.$$

[3]This assumption is used for simplicity and can be relaxed.

Proof. The basic argument of the proof goes as follows: let $p = q + 2$. By the a priori estimates and the general weak compactness results (see, e.g., Section A.2 in Appendix A) it follows that:

(*i*) Since u_n is uniformly bounded in $L^p([0,T];(L^p(\mathcal{O}))^3 \times (L^p(\mathcal{O}))^3)$, there exists a subsequence $\{u_{n_k}\}$ and an element u such that
$$u_{n_k} \rightharpoonup u \ \text{ in } L^p([0,T],(L^p(\mathcal{O}))^3 \times (L^p(\mathcal{O}))^3).$$

(*ii*) Since $\{v_n\} := \{\mathsf{M}_{\mathrm{A}} u_n\}$ is uniformly bounded in $L^{p'}([0,T],(L^{p'}(\mathcal{O}))^3 \times (L^{p'}(\mathcal{O}))^3) \simeq L^{p'}([0,T],(L^{p'}(\mathcal{O}))^6)$, there exists a subsequence $\{v_{n_m}\}$ and an element v such that
$$v_{n_m} \rightharpoonup v \text{ in } L^{p'}([0,T];(L^{p'}(\mathcal{O}))^3 \times (L^{p'}(\mathcal{O}))^3).$$

(*iii*) Since $\{z_n\} := \{\mathsf{N}(u_n)u_n\}$ is uniformly bounded in $L^{p'}([0,T],(L^{p'}(\mathcal{O}))^3 \times (L^{p'}(\mathcal{O}))^3)$, there exists a subsequence $\{z_{n_r}\}$ and an element z such that
$$z_{n_r} \rightharpoonup z \text{ in } L^{p'}([0,T],(L^{p'}(\mathcal{O}))^3 \times (L^{p'}(\mathcal{O}))^3).$$

As usual, for simplicity and to ease notation in what follows, all weakly convergent subsequences are denoted by the same notation as their original sequences.

By standard arguments the weak limit v can be identified as $v = \mathsf{M}_{\mathrm{A}} u$. Indeed, $\langle \mathsf{M}_{\mathrm{A}} u_n, \phi\rangle = -\langle u_n, \mathsf{M}_{\mathrm{A}}\phi\rangle$ for every test function ϕ, and using the weak convergence properties of $\{u_n\}$ leads us to $\langle \mathsf{M}_{\mathrm{A}} u_n, \phi\rangle \to -\langle u, \mathsf{M}_{\mathrm{A}}\phi\rangle = \langle \mathsf{M}_{\mathrm{A}} u, \phi\rangle$ for every test function ϕ; therefore this identification is valid.

Furthermore, by the properties of the convolution kernel G_{A}, it holds that
$$\mathsf{G}'_{\mathrm{A}} \star u_n \rightharpoonup \mathsf{G}'_{\mathrm{A}} \star u \text{ in } L^p([0,T],(L^p(\mathcal{O}))^3 \times (L^p(\mathcal{O}))^3).$$

The troublesome term is the nonlinear one. Showing the convergence of this term requires special attention to take into account the monotonicity properties of the nonlinearity. We take the weak form of the equation, replace the test function by u_n and integrate with respect to time to obtain
$$\begin{aligned}&\int_0^T \langle \mathsf{B}_{\mathrm{A}}\mathsf{N}(u_n)u_n, u_n\rangle dt = \frac{1}{2}(u_n(0),u_n(0)) - \frac{1}{2}(u_n(T),u_n(T))\\ &- \int_0^T \langle \mathsf{G}'_{\mathrm{A}} \star u_n, u_n\rangle dt + \int_0^T \langle (\mathsf{J}_{\mathrm{A}})_n, u_n\rangle\, dt\,,\end{aligned} \tag{11.12}$$
where we have used the properties of the Maxwell operator.

We assume further that the initial condition is such that $u_n(0) \to u(0)$ in $(L^p(\mathcal{O}))^3 \times (L^p(\mathcal{O}))^3$. Then, by the lower semicontinuity of the norm for weak convergence, we have
$$\liminf_n ||u_n(T)||_{(L^p(\mathcal{O}))^6} \geq ||u(T)||_{(L^p(\mathcal{O}))^6}.$$

Applying the above to (11.12) leads to
$$\begin{aligned}&\limsup_{n\to\infty} \int_0^T \langle \mathsf{B}_{\mathrm{A}}\mathsf{N}(u_n)u_n, u_n\rangle dt \leq \int_0^T \langle \mathsf{J}_{\mathrm{A}}, u\rangle dt +\\ &\frac{1}{2}(u(0),u(0)) - \frac{1}{2}(u(T),u(T)) - \int_0^T \langle (\mathsf{G}'_{\mathrm{A}} \star u)(t), u(t)\rangle\, dt\,,\end{aligned}$$

and the right-hand side is identified as $\int_0^T \langle z, u(t)\rangle dt$. The last equality is obtained from the observation that $u' = \mathsf{M}_{\mathrm{A}} u - \mathsf{G}'_{\mathrm{A}} \star u - z + \mathsf{J}_{\mathrm{A}}$, which is then multiplied by u and integrated over $\mathcal{O}$.

Thus,

$$\limsup_{n\to\infty} \int_0^T \langle \mathsf{B}_{\mathrm{A}} \mathsf{N}(u_n)u_n, u_n\rangle dt \le \int_0^T \langle z, u(t)\rangle\, dt\,. \tag{11.13}$$

We now show that the monotonicity and continuity of the nonlinearity allow us to conclude from (11.13) that $z = \mathsf{B}_{\mathrm{A}}\mathsf{N}(u)u$. In the terminology of Showalter (see [380], p. 38), we need to show that the nonlinear operator $\mathsf{N}(u)u$ is an *M-operator*. According to Lemma 2.1 of [380] (see also Section A.12, in Appendix A) it suffices to show that $\mathsf{B}_{\mathrm{A}}\mathsf{N}(u)u$ is hemicontinuous and monotone. Monotonicity follows directly from Lemma 11.3.13. To prove hemicontinuity, we need to prove that the real-valued function $t \mapsto \mathsf{B}_{\mathrm{A}}\mathsf{N}(u+tv)(u+tv)v$ is continuous. This is immediately evident from the polynomial nature of the nonlinearity and the fact that u, v are in $L^{p'}(\mathcal{O})$. To show that the solution is actually in $L^\infty([0,T], \mathbb{X}_{\mathrm{M}}) \cap L^\infty([0,T], (L^{q+2}(\mathcal{O}))^6)$, we use the relevant a priori bounds and the following observation, in the spirit of Lemma I.4.4.1 in [181], is used: if $v_n \in L^\infty([0,T];X)$, where X is a reflexive Banach space, uniformly bounded in n, then[4] there exists a subsequence v_{n_k} such that $v_{n_k} \rightharpoonup v$ in $L^r([0,T];X)$ for every $r \in (1,\infty)$, where in addition $v \in L^\infty([0,T];X)$. Working with this particular subsequence yields the required result.

Uniqueness follows by considering the evolution equation for the difference between two solutions $u^\diamond = u_1 - u_2$. Using the monotonicity of the nonlinear operator and the a priori estimates above, we readily conclude that $u^\diamond \equiv 0$, from which the uniqueness follows. □

11.3.4 The effect of dispersion

We may now include the effect of dispersion in the constitutive relations and reinsert the term $\mathsf{G}_{\mathrm{A},n\ell} \star \mathsf{N}(u)u$ into the Maxwell equations. The inclusion of this term, which according to our model is a weak term, does not change considerably the treatment of the evolution equations as long as the convolution kernels are such that the monotonicity and the convexity of the nonlinearity are conserved. For instance, consider the following assumption of steep kernels for the nonlinear convolutions.

ASSUMPTION 11.3.15 We assume that the convolution kernel $\mathsf{G}_{\mathrm{A},n\ell}$, is steep enough that there exists a constant $C > 0$ such that

$$(\mathsf{G}_{\mathrm{A},n\ell} \star \mathsf{N}(u)u)(t) \cdot u(t) \ge C\,(\mathsf{B}_{\mathrm{A}}\mathsf{N}(u)u)(t)\cdot u(t)\,,$$

for every function u, $t \in [0,T]$.

REMARK 11.3.16 This notion can be considered a generalisation in the framework of nonlinear equations of the condition that the kernels are functions of positive type (see the corresponding footnote on page 17), which is consistent with energy considerations; for the linear case, see [227].

---

[4] This result follows from the weak compactness of bounded sets in reflexive spaces (see Section A.2 in Appendix A).

Under this assumption, the a priori estimates derived in Propositions 11.3.7, 11.3.9 can be easily obtained with minor modifications and then the existence result can be established since the nonlinear operator with action $\mathsf{B}_{\mathrm{A}}\mathsf{N}(u)u + \mathsf{G}_{\mathrm{A},n\ell} \star \mathsf{N}(u)u$ is monotone. However, this assumption can be overly restrictive and may be relaxed (see [230]) using more advanced techniques from the perturbation theory of monotone operators (see, e.g., [451]).

11.4 MISCELLANEA

11.4.1 On global existence

For global existence we need to impose extra conditions that will allow us to obtain a priori bounds that hold globally in time. Such conditions can be, e.g., $\mathsf{B}_{\mathrm{A}}\mathsf{N}(\mathsf{y})\mathsf{y} \cdot \mathsf{y} \geq 0$, for all $\mathsf{y} \in \mathbb{R}^6$ and when $-\mathsf{G}_0(0)$ is a positive definite matrix. These two conditions help us obtain some global bounds as follows. Taking the inner product in $\mathbb{X}$ by u and using the monotonicity of the nonlinear term and the Cauchy-Schwarz inequality on the convolution term gives

$$\frac{d}{dt}||u||_{\mathbb{X}}^2 - \mathsf{G}_0(0)\,||u||_{\mathbb{X}}^2 \leq \epsilon||u||_{\mathbb{X}}^2 + \frac{1}{2\epsilon}||\mathsf{J}_{\mathrm{A}}||_{\mathbb{X}}^2, \tag{11.14}$$

and by Gronwall's inequality (using the positivity of $-\mathsf{G}_0(0)$) we obtain

$$||u||_{\mathbb{X}}^2(t) \leq C(||u||_{\mathbb{X}}^2(0)), \quad \forall t \in [0, \infty).$$

Furthermore by the positivity of $-\mathsf{G}_0(0)$, (11.14) gives

$$\frac{d}{dt}||u||_{\mathbb{X}}^2 \leq \frac{1}{2\epsilon}||\mathsf{J}_{\mathrm{A}}||_{\mathbb{X}}^2.$$

Hence, repeating the arguments of Proposition 11.3.7, and using the above inequality, $u \in L^\infty([0,\infty), L^{q+2}(\mathcal{O}))$.

11.4.2 Other approaches for solvability

In this chapter we have only touched the surface of the very important problem of nonlinear complex media, using an approach based on the Faedo-Galerkin approximation and monotonicity arguments. Other approaches are possible that may yield well-posedness results under possibly weaker conditions on the data of the problem. To the best of our knowledge, the problem of nonlinear complex media has not been subjected to rigorous mathematical analysis in its full generality, even though there exist a number of interesting mathematical works for certain classes of nonlinear electromagnetic media. We present a very partial list of approaches: An interesting approach to nonlinear ferromagnetic media has been made by Ammari and Hamdache [7], while Sjöberg has published interesting works ([386], [388]) on modelling issues as well as on well-posedness results for nonlinear media. It is worth mentioning both the abstract approach of Milani and Picard [320], which is based on operator theoretic considerations, and the approach of Visintin [421], which relies on convex analysis techniques. The latter approach, which is based on the assumption that the constitutive relations may be considered the subdifferential of a convex functional, apart from its interesting mathematical aspects, provides some interesting physical intuition. Interpreting this freely in terms of physics, one may state that Visintin's approach is inspired by (generalised) energy considerations, which makes it very attractive as an approach for handling such problems. Therefore, it is very interesting to see how this approach may be generalised for the study of general complex electromagnetic media. In closing this very brief tour through the literature on the mathematical analysis of nonlinear problems in electromagnetics it is worth mentioning the work of Wei and Yin [424] and [440] dealing with singular nonlinear problems related to the Maxwell equations, which have a parabolic limit, and the important work by Babin and Figotin [48]: assuming the classical nonlinear optics representation for the nonlinear polarisation as a power series, they show that - provided the excitation current is not too large - the Maxwell equations are uniquely solvable for nonlinear inhomogeneous media; this solution is represented in a certain time interval as a power series in terms of the solution to the corresponding linear Maxwell equations.

11.4.3 Homogenisation of nonlinear media

Homogenisation theory can be extended, with necessary modifications, to study composite nonlinear complex media. The basic compactness arguments used for the existence of solutions can be extended to study the behaviour of families of solutions depending on a small parameter characterising the periodic structure of the media. Wellander [426] was one of the first to address such issues for nonlinear dielectrics. Visintin's convex analysis approach [421] was also employed to obtain some very interesting homogenisation results for specific class of media. However, the rigorous theory for general nonlinear complex media still remains open. For certain results from the engineering viewpoint see [297].

The strong property fluctuation theory (SPFT) is an alternative approach to homogenisation wherein a Feynman diagram method is employed in an iterative scheme to calculate successive refinements to a *comparison medium.* Iterates are expressed in terms of statistical correlation functions of the spatial distribution of the component phases. The zeroth-order SPFT coincides with the Bruggeman homogenisation formalism [316]. The second-order SPFT (known as BA-SPFT; or bilocally approximated SPFT) is most commonly implemented for practical applications, although in principle, correlation functions of arbitrary order may be considered. In [298] the SPFT is formulated for cubically nonlinear isotropic chiral composite media. The BA-SPFT and the third-order SPFT are developed from the corresponding linear theories using Maclaurin expansions to accommodate nonlinear behaviour. By means of a numerical example, convergence is established at the level of BA-SPFT, with respect to both linear and nonlinear properties. The phenomenon of nonlinearity enhancement is also explored.

11.4.4 Controllability of nonlinear media

The study of controllability may be extended to the case of nonlinear media. The fixed point technique introduced in Chapter 8 can be modified accordingly to prove the controllability of nonlinear media. The controllability of nonlinear problems may require detailed and delicate estimates for the solutions of either the forward or the backward problems and is beyond the scope of the present work. Similar comments hold for boundary controllability issues as well as for optimal control problems. For an excellent account of control for nonlinear systems, see the monograph of Coron [112].

11.4.5 Soliton-type solutions

Nonlinear systems present interesting types of solutions in the form of travelling waves that propagate with unchanged shape through the medium as an effect of the interplay between dispersion and nonlinearity. This type of behaviour is typical and is very well studied in integrable systems (see, e.g., [148]); however, solutions of similar type are often present in nonintegrable systems and find important applications in various branches of science (see, e.g., [200]). A *formal* approach to the evolution of nonlinear waves in chiral media with weak dispersion and weak nonlinearity of the Kerr type in the low chirality case is presented in [152], [413]. A set of modulation equations is obtained for the evolution of the slowly varying field envelopes that is in the form of four coupled nonlinear Schrödinger equations. This set of equations is nonintegrable; however, with the use of reductive perturbation theory, under certain conditions these equations may be reduced to an integrable system, the Melnikov system. This system is known to possess vector soliton solutions. Thus, by the above reduction, in certain (limiting) cases the existence of vector solitons in chiral media may be shown; these appear in pairs of dark and bright solitons . Depending on the boundary conditions at infinity chosen, the dark component can be along the right-handed component of the field and the bright component along the left-handed component of the field, or vice versa. We refer to [191] for an account of the reductive perturbation theory, to [243] for the properties of bright and dark solitons, and to [310], [311], [312] for the properties of the Melnikov system. Soliton-type solutions for the nonlinear wave equation in Kerr-type nonlinear materials with uniaxial chirality have also been studied in [100], [55] and [186].

PART 4
Stochastic Problems

Chapter Twelve

Well Posedness

12.1 INTRODUCTION

The aim of this part of the book is to examine the effects of randomness on the evolution and behaviour of electromagnetic fields in chiral media. Since our basic interest is the development of a mathematically rigorous framework for the study of this problem, our approach is to model random effects through the introduction of a general class of random fields[1] into the system of equations that governs the evolution of the electromagnetic fields. We do not take up the question of self-consistent mathematical modelling of the random effects, starting from the starting point of first principles physical theory. Such a task, although extremely interesting and important, is well beyond the scope of this work. Instead, we assume that the randomness follows some rather general and reasonable qualitative assumptions, e.g., Gaussian behaviour, etc., and use a well-founded mathematical model for it, in terms of the Wiener process. This model is a very reasonable first approximation to the problem: because of arguments based on the central limit theorem, the effect of independent and identically distributed perturbations, under minimal assumptions on the existence of second moments, may be approximated as an infinite-dimensional generalisation of the Wiener process (see Appendix B for the relevant definitions). Then we introduce this type of random perturbation into our model as part of the externally imposed sources or as part of the medium parameters, or both. The former models the effect of uncertainty of the externally imposed sources in the medium, e.g., experimental errors when designing source terms for the media. The latter models more carefully the situation in which the chiral medium itself presents random properties, and in some sense reflects our uncertainty about the actual properties of the medium. There are several reasons for such effects, ranging from inability to experimentally specify the parameters of the medium to the possible feedback of the various uncertainties and of the fields themselves on the medium's properties.

The introduction of these random perturbations has important effects on the electromagnetic fields. The evolution equations for the fields now assume the form of stochastic integrodifferential equations. It is very important and interesting to study the effects of the randomness on well posedness and

[1] We use the term random field here in the standard fashion employed in the probability literature, i.e., to denote a stochastic process with a multidimensional index set.

the properties of the solutions of the stochastic integrodifferential equations for the field evolution. We shall see, for instance, that the introduction of randomness into the model has important consequences on the temporal regularity of the electromagnetic fields, as predicted by the solution of the equations, and this may affect the spatial regularity as well.

The structure of this chapter is as follows. In Section 12.2 we present a model for the Maxwell equations for complex random media in terms of stochastic integrodifferential equations. In Section 12.3 we present the appropriate functional framework that allows us to treat the problem as an abstract stochastic integrodifferential equation in Hilbert space, and, exploiting this framework in Section 12.4, we provide some solvability and well-posedness results for this model, using a semigroup-based approach. In Section 12.5 we propose alternative approaches to solvability and well posedness for the stochastic integrodifferential equations that arise in the modelling of random complex electromagnetic media, using either the finite-dimensional approximation (Faedo-Galerkin method) or the Wiener chaos approach.

12.2 MAXWELL EQUATIONS FOR RANDOM MEDIA

The starting point for the analysis of this chapter is the Maxwell equations for random chiral media in a bounded domain $\mathcal{O}$, as expressed in Section 2.4.11 and in particular equation (2.56). We refer to that section for details on the model and its assumptions. Here we merely state again the stochastic integrodifferential equation for convenience. The integral form of the equation is

$$\begin{aligned} u(t) = u(0) + \int_0^t \left(\mathsf{M}_{\mathrm{A}} u(s) + \int_0^s \mathsf{G}_{\mathrm{A}}(s-r)u(r)\, dr + \mathsf{J}_{\mathrm{A}}(s) \right) ds \\ + \int_0^t \mathsf{Q}_{\mathrm{A}}(s,\omega)\ dW(s), \end{aligned} \tag{12.1}$$

where the first integral is considered a Riemann-Stieltjes integral, whereas the second integral is considered an Itō integral over the infinite-dimensional Wiener process $W(t)$. In the above equation Q_{A} is an operator valued stochastic process that models the effect of spatial correlations of the fluctuating terms. In the next section we provide the details of the exact conditions this operator process has to satisfy. We just note here that Q_{A} may be either independent of or dependent[2] on the electromagnetic field. The first case, especially if G_{A} is not a random process, is called the *additive noise* case; the second case is called the *multiplicative noise* case. The boundary conditions are considered to be those of the perfect conductor. To simplify the notation, occasionally we write the integral equation (12.1) in differential form as

$$du(t) = (\mathsf{M}_{\mathrm{A}} u(t) + (\mathsf{G}_{\mathrm{A}} \star u)(t) + \mathsf{J}_{\mathrm{A}})\, dt + \mathsf{Q}_{\mathrm{A}}\, dW(t)\,, \tag{12.2}$$

[2] Linearly or not.

where we use the notation of Chapter 7. In principle, the matrices A_{or}, G_{A} may be random matrices, i.e., functionals of the Wiener process. To simplify the exposition of our results, we assume here that they are deterministic matrices and that the stochastic effects are introduced into the system only through the operator term $\mathsf{Q}_{\mathrm{A}}\, dW(t)$. This assumption is relaxed in Chapter 14, where we consider the medium to be random and discuss the problem of homogenisation.

The evolution equation for the electromagnetic fields is in the form of a stochastic integrodifferential equation of Volterra type. While the literature on deterministic integrodifferential equations is extensive, relatively little work has been done on stochastic integrodifferential equations of type (12.1) (see, e.g., [167], [221], [231], [232], [233]). These studies concern only adapted processes. We follow the approach of [120], [168] and [199] to stochastic differential equations in Hilbert space concerning predictable processes, which we modify accordingly for the case of stochastic integrodifferential equations.

Similarly, as for the deterministic case, equation (12.2) will be considered either without reference to the divergence or with divergence-free conditions on the electromagnetic field. The properties of the electromagnetic field with respect to the divergence are included in the choice of functional setting for the problem.

12.3 FUNCTIONAL SETTING

We will assume that the noise terms take values on a Hilbert space U, and that the stochastic process $\{W(t)\}_{t\,\in\,[0,\,T]}$ is a $Q-$Wiener process on a probability space $(\Omega, \mathscr{F}, P)$, where Q is a trace class operator from U into another Hilbert space $\mathbb{H}$, the state space of the system. We consider as $\mathbb{F} = \{\mathscr{F}_t\}_t$ the filtration generated by the Wiener process. We will further need the space $L_2^0 = L_2(Q^{1/2}(U), \mathbb{H})$, where by $L_2(X, Y)$ we denote the space of all Hilbert Schmidt operators for the Hilbert space X to the Hilbert space Y. For definitions of the above concepts, see Appendix B.

As in the deterministic case, we consider a pair of Hilbert spaces $(\mathbb{H}, \mathbb{H}_{\mathsf{M}})$ to be either $(\mathbb{X}, \mathbb{X}_{\mathsf{M}})$ or $(\mathbb{X}, \mathfrak{X}_{\mathsf{M}})$, where (see Section 3.9.1):

$$
\begin{aligned}
&\mathsf{X} := (L^2(\mathcal{O}))^3,\\
&\mathbb{X} := \mathsf{X} \times \mathsf{X},\\
&\mathbb{X}_{\mathsf{M}} := H_0(\mathrm{curl}, \mathcal{O}) \times H(\mathrm{curl}, \mathcal{O}),\\
&\mathfrak{X}_1 := H_0(\mathrm{curl}, \mathcal{O}) \cap H(\mathrm{div}0, \mathcal{O}), \;\; \mathfrak{X}_2 := H(\mathrm{curl}, \mathcal{O}) \cap H_0(\mathrm{div}0, \mathcal{O}),\\
&\mathfrak{X}_{\mathsf{M}} := \mathfrak{X}_1 \times \mathfrak{X}_2 .
\end{aligned}
$$

As in the deterministic case, we use the pair of function spaces $(\mathbb{X}, \mathbb{X}_{\mathsf{M}})$ when considering the evolution equation without special reference to the divergence properties of the electromagnetic field, whereas the pair $(\mathbb{X}, \mathfrak{X}_{\mathsf{M}})$ is used when we require divergence-free properties for the electromagnetic field.

The important difference in the stochastic case is that the electromagnetic field is no longer considered a function from $[0,T] \to \mathbb{H}$, as in the deterministic case, but rather a stochastic process on the probability space $(\Omega, \mathscr{F}, P)$, taking values on $\mathbb{H}$. We will thus employ the space of all continuous (in mean square) and square integrable predictable processes:

$$\mathcal{C}([0,T];\mathbb{H}) = \{Y \in C([0,T]; L^2(\Omega, \mathbb{H}) : Y \text{ is predictable}\}.$$

This space equipped with the norm

$$\|Y\|_{\mathcal{C}} = \sup_{t\in[0,T]} \left(\mathbb{E}\left[\|Y_t\|^2_{\mathbb{H}}\right]\right)^{1/2}$$

is a Banach space (see Section 3.2.5). Furthermore, we will use the notation $L^p([0,T]; L^2(\Omega, \mathfrak{F}, P; \mathbb{H}))$ for the random fields $u : [0,T] \to \mathbb{H}$ which constitute a square integrable (with respect to the probability measure P) stochastic process, adapted to the filtration $\mathfrak{F} = \{\mathscr{F}_t\}$ such that $\int_0^T \mathbb{E}[||u(t)||^p_{\mathbb{H}}]dt < \infty$. For $1 \le p < \infty$ this is a Banach space with the obvious norm. The case $p = \infty$ requires the usual modification of replacing the integral over $[0,T]$ by the essential supremum.

REMARK 12.3.1 The use of predictable processes in the definition of solutions of stochastic Volterra equations is in line with other works (see, e.g., the monograph [225]). An alternative approach would be to work in terms of adapted processes ([120]). The main arguments of this chapter remain valid under minor modifications since an adapted process that is stochastically continuous on an interval $[0,T]$ has a predictable version (see, e.g., Proposition 3.6 of [120]; see also Section B.2 in Appendix B).

Let the operator M_{A} be defined from $\mathbb{H}_{\mathsf{M}} := D(\mathsf{M}_{\mathrm{A}})$ to $\mathbb{H}$, where $\mathbb{H}$ and $\mathbb{H}_{\mathsf{M}}$ are two Hilbert spaces (in the present context the pair $(\mathbb{H}, \mathbb{H}_{\mathsf{M}})$ will be used as a proxy either for the pair $(\mathbb{X}, \mathbb{X}_{\mathsf{M}})$ or for $(\mathbb{X}, \mathfrak{X}_{\mathsf{M}})$, where the second constituent of the pairs is considered to be endowed with the graph norm).

The following assumption on the operator M_{A} is crucial for our approach:

ASSUMPTION 12.3.2 *The operator* $\mathsf{M}_A : \mathbb{H}_{\mathsf{M}} \to \mathbb{H}$ *is the generator of a unitary group* $\{\mathsf{T}_{\mathsf{M}_A}(t)\}_{t\in\mathbb{R}}$ *on a Hilbert space* $\mathbb{H}$.

REMARK 12.3.3 As in the deterministic case, we will choose the functional environment for the problem so as to guarantee that the generalised Maxwell operator M_{A} satisfies this assumption.

12.4 WELL POSEDNESS

We now consider the problem of well posedness of the mathematical model governing the evolution of electromagnetic fields in complex media under the effects of noise in the time domain, as expressed by the stochastic integro-differential equation (12.2).

As in the deterministic case, several alternative approaches to treating the system (12.2) can be used. For the reasons presented in Chapter 7, our primary approach to this problem is to adopt a semigroup formulation of the problem, based on the semigroup generated by the Maxwell operator, and to treat the convolution terms, which model the chirality and the dispersive effects, as perturbations of this semigroup. The results of this section are based on [285].

12.4.1 Different notions of solutions

We now consider the abstract stochastic integrodifferential equation (12.2) on a Hilbert space $\mathbb{H}$.

A variety of different types of solutions for (12.2) of varying spatial regularity may be defined. The temporal regularity is not expected to be as good as the spatial regularity because of the pathological properties of the Wiener process with respect to temporal regularity that are inherited by the solution of (12.2). We proceed with the relevant definitions.

DEFINITION 12.4.1 *A stochastic process* $u \in \mathcal{C}([0,T];\mathbb{H})$ *is called a mild solution of* (12.2) *if:*

$$u(t) = \mathsf{T}_{\mathsf{M}_A}(t)u_0 + \int_0^t \mathsf{T}_{\mathsf{M}_A}(t-s)\int_0^s \mathsf{G}_A(s-r)u(r)\,dr\,ds +$$
$$+ \int_0^t \mathsf{T}_{\mathsf{M}_A}(t-s)\mathsf{J}_A(s)\,ds + \int_0^t \mathsf{T}_{\mathsf{M}_A}(t-s)\mathsf{Q}_A(u(s))\,dW(s),$$

for every $t \in [0,T]$, *P-a.s.*

REMARK 12.4.2 Note that we include the condition of continuity of the solution in the definition of a mild solution. An alternative would be to require that $u \in L^2([0,T],\mathbb{H})$ and then show using, e.g., the Lévy continuity criterion (see, e.g., [120]) that this process admits a continuous version.

DEFINITION 12.4.3 *An* $\mathbb{H}$*-valued predictable process* $\{u(t)\}_{t\,\in\,[0,\,T]}$, *is called a weak solution of* (12.2) *if:*

(i) $\int_0^T \|u(s)\|_{\mathbb{H}}\ \ ds < \infty$, *P-a.s.*

(ii) For every $\phi \in D(\mathsf{M}_A^*) = D(\mathsf{M}_A)$ *it holds that*

$$(u(t),\phi) = (u_0,\phi) + \int_0^t [-(u(s),\mathsf{M}_A\phi) + (\mathsf{J}_A(s),\phi)]\,ds$$
$$+ \int_0^t (\int_0^s \mathsf{G}_A(s-r)u(r)dr,\phi)ds + \int_0^t (\mathsf{Q}_A(u(s))dW(s),\phi)\,,$$

where $(\cdot,\cdot)$ *denotes the inner product of the involved spaces, for every* $t \in [0,T]$, *P-a.s.*

REMARK 12.4.4 In similar manner as for deterministic equations, the notions of mild and weak solutions for stochastic equations coincide.

Definition 12.4.5 *An* $\mathbb{H}$*-valued predictable process* $\{u(t)\}_{t\,\in\,[0,\,T]}$ *is called a strong solution of* (12.2) *if:*

(*i*) $u(t) \in D(\mathsf{M}_A)$, *P-a.s., a.e. on* $[0, T]$.

(*ii*) $\int_0^T [\ \|\mathsf{M}_A u(s)\|_{\mathbb{H}} + \|\int_0^s \mathsf{G}_A(s-r)u(r)\ dr\|_{\mathbb{H}}\ ]\ ds < \infty$, *P-a.s.*

(*iii*) *The equality*

$$u(t) = u_0 + \int_0^t \mathsf{M}_A u(s)\ ds\ + \int_0^t \mathsf{J}_A(s)\ ds$$
$$+ \int_0^t \int_0^s \mathsf{G}_A(s-r)u(r)dr\ ds + \int_0^t \mathsf{Q}_A(u(s))\ dW(s),$$

holds for all $t \in [0, T]$, *P-a.s.*

Remark 12.4.6 It is important to emphasise that the notions of weak and strong solutions employed here are considered in the sense used by the PDE community rather than in the sense used by the probabilist community. In particular, we consider the Wiener process $W(t)$ to be given and not to be determined, as is the case in the notion of weak solution used by the probabilist community (see, e.g., [224] for stochastic ODEs and [159] for stochastic PDEs).

We have the following result:

Theorem 12.4.7 *Assume that* $u_0 \in L^2(\Omega, \mathscr{F}_0, \mathbb{H})$, *and that*

(*i*) $\mathsf{J}_A \in L^1([0, T]; L^2(\Omega, \mathfrak{F}, P; \mathbb{H}))$.

(*ii*) *The operator* $\mathsf{Q}_A(u) = \mathsf{Q}_{A,1} + \mathsf{Q}_{A,2}u : \mathbb{H} \to L_2^0$ *satisfies the bound* $||\mathsf{Q}_A(u)||_{L_2^0} \leq C_0 + C_1||u||_{\mathbb{H}}$ *for all* $u \in \mathbb{H}$.

(*iii*) $\mathsf{G}_A \in L^2([0, T]; (L^\infty(\mathcal{O}))^{36})$.

Then (12.2) *is weakly well posed in* $\mathbb{H}$.

Proof. Let $b > 0$. Consider the Banach space $\mathcal{C}([0, T]; \mathbb{H})$, endowed with the norm $\|u\|_b := \sup_{t\in[0,T]} e^{-bt} \left(\mathbb{E}\left[\|u(t)\|_{\mathbb{H}}^2\right]\right)^{1/2}$, which is clearly equivalent to the usual norm of $\mathcal{C}([0, T]; \mathbb{H})$ for $T < \infty$. Define the map $\mathsf{F} : \mathcal{C}([0, T]; \mathbb{H}) \to \mathcal{C}([0, T]; \mathbb{H})$ as follows:

$$\mathsf{F}(u(t)) = \mathsf{T}_{\mathsf{M_A}}(t)u_0 + \mathsf{F}_1(u(t)) + \mathsf{F}_2(u(t)) + \mathsf{F}_3(u(t)), \qquad (12.3)$$

where

$$\mathsf{F}_1(u(t)) := \int_0^t \mathsf{T}_{\mathsf{M_A}}(t-s) \int_0^s \mathsf{G_A}(s-r)u(r)\ dr\ ds,$$
$$\mathsf{F}_2(u(t)) := \int_0^t \mathsf{T}_{\mathsf{M_A}}(t-s)\mathsf{J_A}(s)\ ds,$$
$$\mathsf{F}_3(u(t)) := \int_0^t \mathsf{T}_{\mathsf{M_A}}(t-s)\mathsf{Q_A}(u(s))\ dW(s).$$

Note that F_1, F_2 are expressed as Riemann-Stieltjes integrals while F_3 is expressed as an Itō integral. We have already encountered F_1, F_2 in the treatment of the deterministic problem (see Chapter 7), so we focus our attention on F_3, which is a stochastic convolution.

A straightforward application of Itō's isometry (see Section B.5 in Appendix B), in conjunction with the properties of Q_{A}, shows that F_3 leaves $\mathcal{C}([0,T];\mathbb{H})$ invariant. It remains to show that F_3 is a contraction in $\mathcal{C}([0,T];\mathbb{H})$.

To this end, take any pair of processes $u^{(1)}, u^{(2)} \in \mathcal{C}([0,T];\mathbb{H})$ and set

$$u^{\diamond}(t) := u^{(1)}(t) - u^{(2)}(t),$$
$$\mathsf{Q}_{\mathrm{A}}^{\diamond}(t) := \mathsf{Q}_{\mathrm{A}}(u^{(1)}(t)) - \mathsf{Q}_{\mathrm{A}}(u^{(2)}(t)) = \mathsf{Q}_{\mathrm{A},1} u^{\diamond}(t).$$

Following the proof of Theorem 1 in [199], we first use the Itō isometry (see Section B.5 in Appendix B) to obtain

$$\mathbb{E}[\| \int_0^t \mathsf{T}_{\mathsf{M}_{\mathrm{A}}}(t-s)\mathsf{Q}_{\mathrm{A}}^{\diamond}(s)dW(s)\|_{\mathbb{H}}^2] = \mathbb{E}[\int_0^t \|\mathsf{T}_{\mathsf{M}_{\mathrm{A}}}(t-s)\mathsf{Q}_{\mathrm{A}}^{\diamond}(s)\|_{L_2^0}^2 ds],$$

and then, using that in conjunction with the Cauchy-Schwarz inequality, we estimate

$$e^{-2bt}\mathbb{E}[\| \int_0^t \mathsf{T}_{\mathsf{M}_{\mathrm{A}}}(t-s)\mathsf{Q}_{\mathrm{A}}^{\diamond}(s)dW(s)\|_{\mathbb{H}}^2] \leq \int_0^t e^{-2bt}\mathbb{E}[\|\mathsf{Q}_{\mathrm{A}}^{\diamond}(s)\|_{L_2^0}^2]ds$$
$$\leq C_1\{\sup_{s\in[0,t]} e^{-2bs}\mathbb{E}[\|u^{\diamond}(s)\|_{\mathbb{H}}^2]\}\left(\int_0^t e^{-2b(t-s)}ds\right).$$

A straightforward calculation shows that

$$\|\mathsf{F}_3(u^{\diamond}(t))\|_b \leq C_1(2b)^{-1/2}\|u^{\diamond}\|_b. \tag{12.4}$$

The components of F, i.e., $\mathsf{F}_1, \mathsf{F}_2$, present no major problems. For instance, to show that F_1 is a contraction, we estimate

$$e^{-2bt}\mathbb{E}\left[\| \int_0^t \mathsf{T}_{\mathsf{M}_{\mathrm{A}}}(t-s)\int_0^s \mathsf{G}_{\mathrm{A}}(s-r)u^{\diamond}(r)drds\|_{\mathbb{H}}^2\right]$$
$$\leq Te^{-2bt}\mathbb{E}\left[\int_0^t \| \int_0^s \mathsf{G}_{\mathrm{A}}(s-r)u^{\diamond}(r)dr\|_{\mathbb{H}}^2 ds\right]$$
$$\leq T^2 e^{-2bt}\mathbb{E}\left[\sup_{s\in[0,t]} \| \int_0^s \mathsf{G}_{\mathrm{A}}(s-r)u^{\diamond}(r)dr\|_{\mathbb{H}}^2\right]$$
$$\leq T^3 e^{-2bt}\mathbb{E}\left[\sup_{s\in[0,t]} \int_0^s \|\mathsf{G}_{\mathrm{A}}(s-r)u^{\diamond}(r)\|_{\mathbb{H}}^2\, dr\right]$$
$$\leq T^3 e^{-2bt}\mathbb{E}\left[\int_0^t \|\mathsf{G}_{\mathrm{A}}(t-r)u^{\diamond}(r)\|_{\mathbb{H}}^2\, dr\right]$$
$$\leq C_3 T^3 \left(\int_0^t e^{-2b(t-s)}ds\right) \sup_{r\in[0,t]} e^{-2br}\mathbb{E}\left[\|u^{\diamond}(r)\|_{\mathbb{H}}^2\right],$$

where C_3 is a bound for G_{A}, whereby

$$\|\mathsf{F}_1(u^{\diamond}(t))\|_b \leq T^{3/2}C_3\ (2b)^{-1/2}\ \|u^{\diamond}\|_b. \tag{12.5}$$

Assuming for simplicity that J_{A} is independent of u, and using the estimates (12.4), (12.5), we obtain

$$\|\mathsf{F}(u^{\diamond}(t))\|_b \leq (2b)^{-1/2}\,(T^{3/2}C_3 + C_1)\|u^{\diamond}\|_b.$$

Hence, for $b > 0$ sufficiently large, the map F is a contraction on $\mathcal{C}([0,T];\mathbb{H})$, and thus has a unique fixed point in $\mathcal{C}([0,T];\mathbb{H})$. Therefore, (12.2) has a unique mild solution that is b-exponentially bounded. By a straightforward modification of Theorem 6.5 in [120], this mild solution is also a weak solution. By the form of the solution we can check that (12.2) is weakly well posed. □

Under more restrictive conditions on the data of the problem we may prove the existence of strong solutions.

THEOREM 12.4.8 *Assume that*

(*i*) $u_0 \in D(\mathsf{M}_A)$, J_A *is adapted in* $\mathfrak{F}$ *with* $\mathsf{J}_A(t) \in D(\mathsf{M}_A)$ *and* $\mathsf{Q}_A(\mathsf{y})Q^{1/2}h \in D(\mathsf{M}_A)$ *P-a.s. for all* $t \in [0,T]$, $\mathsf{y} \in \mathbb{H}$, $h \in U$.

(*ii*) $\|\mathsf{M}_A\mathsf{J}_A\|_{\mathbb{H}} \leq g_1(t)$, $g_1 \in L^1([0,T];\mathbb{R})$.

(*iii*) $\|\mathsf{M}_A\mathsf{Q}_A(\mathsf{y})\|_{L_2^0} \leq g_2(t)\|\mathsf{y}\|_{\mathbb{H}}$, $g_2 \in L^2([0,T];\mathbb{R})$, $\mathsf{y} \in \mathbb{H}$.

(*iv*) $\exists\, C > 0$: $||\mathsf{G}_A(t)\mathsf{y}||_{\mathbb{H}_{\mathsf{M}}} < C\,||\mathsf{y}||_{\mathbb{H}_{\mathsf{M}}}$ $\forall\, \mathsf{y} \in \mathbb{H}$ *a.e. in* $[0,T]$.

Then (12.2) *is strongly well posed in* $\mathbb{H}$.

Proof. For $b > 0$, consider the Banach space $L^1([0,T];\ L^2(\Omega,\mathfrak{F},P;\mathbb{H}_{\mathsf{M}}))$, endowed with the norm $\|u\|_b = \int_0^T e^{-bt}\left(\mathbb{E}\left[\|u(t)\|^2_{\mathbb{H}_{\mathsf{M}}}\right]\right)^{1/2} dt$, which, for $T < \infty$, is clearly equivalent to the usual norm of $L^1([0,T];\ L^2(\Omega;\mathbb{H}_{\mathsf{M}}))$. Define the map $\mathsf{F} : L^1([0,T];\ L^2(\Omega,\mathfrak{F},P;\mathbb{H}_{\mathsf{M}})) \to L^1([0,T];\ L^2(\Omega,\mathfrak{F},P;\mathbb{H}_{\mathsf{M}}))$ as in (12.3). By a combination of Theorem 2 and Proposition 2.3 of [199] (or Theorem 2.1 in [168]), one can show that F has a unique fixed point in the space $(L^1([0,T];\ L^2(\Omega,\mathfrak{F},P;\mathbb{H}_{\mathsf{M}})), \|\cdot\|_b)$ that satisfies properties (*i*) and (*ii*) of Definition 12.4.5. Following [199], p. 26, by the Fubini theorem, and the fact that in $D(\mathsf{M}_{\mathrm{A}}) = \mathbb{H}_{\mathsf{M}}$, it holds[3] that $\mathsf{M}_{\mathrm{A}}\mathsf{T}_{\mathsf{M}_{\mathrm{A}}}(r)u = (\mathsf{T}_{\mathsf{M}_{\mathrm{A}}}(r)u)'$, we can see that, upon defining $u_{\mathsf{G}}(t) := \int_0^r \mathsf{G}_{\mathrm{A}}(t-\ell)u(\ell)d\ell$,

$$\int_0^t\int_0^s \mathsf{M}_{\mathrm{A}}\mathsf{T}_{\mathsf{M}_{\mathrm{A}}}(s-r)u_{\mathsf{G}}(r)drds = \int_0^t\int_r^t \mathsf{M}_{\mathrm{A}}\mathsf{T}_{\mathsf{M}_{\mathrm{A}}}(s-r)u_{\mathsf{G}}(r)dsdr =$$
$$\int_0^t \mathsf{T}_{\mathsf{M}_{\mathrm{A}}}(t-r)u_{\mathsf{G}}(r)dr - \int_0^t u_{\mathsf{G}}(r)dr =: I_1$$

and that

$$\int_0^t\int_0^s \mathsf{M}_{\mathrm{A}}\mathsf{T}_{\mathsf{M}_{\mathrm{A}}}(s-r)\mathsf{J}_{\mathrm{A}}(r)drds = \int_0^t \mathsf{T}_{\mathsf{M}_{\mathrm{A}}}(t-r)\mathsf{J}_{\mathrm{A}}(r)dr - \int_0^t \mathsf{J}_{\mathrm{A}}(r)dr =: I_2.$$

[3] See Theorem A.8.12 in Appendix A. Recall also that $\mathsf{T}_{\mathsf{M}_{\mathrm{A}}}(t)\mathsf{M}_{\mathrm{A}}u = \mathsf{M}_{\mathrm{A}}\mathsf{T}_{\mathsf{M}_{\mathrm{A}}}(t)$ for all $u \in D(\mathsf{M}_{\mathrm{A}})$.

Let $\mathsf{Q}_{\mathrm{A}}(r) := \mathsf{Q}_{\mathrm{A}}(u(r))$. By the stochastic Fubini theorem, we also have

$$\int_0^t \int_0^s \mathsf{M}_{\mathrm{A}} \mathsf{T}_{\mathsf{M}_{\mathrm{A}}}(s-r)\mathsf{Q}_{\mathrm{A}}(r)dW(r)ds = \int_0^t \int_r^t \mathsf{M}_{\mathrm{A}} \mathsf{T}_{\mathsf{M}_{\mathrm{A}}}(s-r)\mathsf{Q}_{\mathrm{A}}(r)dsdW(r)$$
$$= \int_0^t \mathsf{T}_{\mathsf{M}_{\mathrm{A}}}(t-r)\mathsf{Q}_{\mathrm{A}}(r)\, dW(r) - \int_0^t \mathsf{Q}_{\mathrm{A}}(r)\, dW(r) =: I_3.$$

Hence, applying M_{A} to the fixed point equation $u = \mathsf{F}u$ and using the closedness of M_{A} and the above results, we have

$$\int_0^t \mathsf{M}_{\mathrm{A}} u(s)\, ds = \int_0^t \mathsf{M}_{\mathrm{A}} \mathsf{T}_{\mathsf{M}_{\mathrm{A}}}(s) u_0\, ds + \int_0^t \int_0^s \mathsf{M}_{\mathrm{A}} \mathsf{T}_{\mathsf{M}_{\mathrm{A}}}(s-r) u_{\mathsf{G}}(r) dr ds$$
$$+ \int_0^t \int_0^s \mathsf{M}_{\mathrm{A}} \mathsf{T}_{\mathsf{M}_{\mathrm{A}}}(s-r)\mathsf{J}_{\mathrm{A}}(r)\, dr\, ds + \int_0^t \int_0^s \mathsf{M}_{\mathrm{A}} \mathsf{T}_{\mathsf{M}_{\mathrm{A}}}(s-r)\mathsf{Q}_{\mathrm{A}}(r)\, dW(r)\, ds$$
$$= \mathsf{T}_{\mathsf{M}_{\mathrm{A}}}(t)u_0 - u_0 + I_1 + I_2 + I_3$$
$$= u(t) - u_0 - \int_0^t u_{\mathsf{G}}(r)\, dr - \int_0^t \mathsf{J}_{\mathrm{A}}(r)\, dr - \int_0^t \mathsf{Q}_{\mathrm{A}}(r)\, dW(r).$$

Therefore, the unique fixed point satisfies property *(iii)* of Definition 12.4.5. Thus, we conclude that (12.2) is strongly well posed. □

REMARK 12.4.9 (FINITE ENERGY SOLUTIONS) As in the deterministic case, depending on the choice of the Hilbert space $\mathbb{H}$, one may obtain a variety of different solutions for the abstract system (12.2). The choice $(\mathbb{H}, \mathbb{H}_{\mathsf{M}}) = (\mathbb{X}, \mathbb{X}_{\mathsf{M}})$ corresponds to finite energy solutions, without any special reference to the divergence-free property. Working in similar manner as in the deterministic case, one sees that the abstract setting covers this case, providing mild, weak and strong well posedness of the model, depending on the regularity of the data of the problem.

REMARK 12.4.10 (DIVERGENCE-FREE SOLUTIONS) The choice $(\mathbb{H}, \mathbb{H}_{\mathsf{M}}) = (\mathbb{X}, \mathfrak{X}_{\mathsf{M}})$ corresponds to finite-energy, divergence-free solutions. Working in similar manner as in the deterministic case, one sees that the abstract setting covers this case, providing mild, weak and strong well posedness of the model, depending on the regularity of the data of the problem. Other choices can be found (e.g., in [286]).

12.5 OTHER POSSIBLE APPROACHES TO SOLVABILITY

12.5.1 Evolution families approach

A possible alternative approach to the solvability of the stochastic problem in the case where G_{A} is nonrandom is to use the theory of evolution families instead of the theory of semigroups (see Section 7.5.1). The solution of system (12.2) can be expressed as

$$u(t) = R(t)u_0 + \int_0^t R(t-s)\mathsf{J}_{\mathrm{A}}(s)\, ds + \int_0^t R(t-s)\, dW(s)\,, \qquad (12.6)$$

where we now use the stochastic convolution with respect to the evolution family $\{R(t)\}$, generated by the system

$$v' = \mathsf{M}_{\mathrm{A}}\, v + \mathsf{G}_{\mathrm{A}} \star v,$$

instead of the convolution with respect to the group $\{\mathsf{T}_{\mathsf{M}_{\mathrm{A}}}(t)\}$, generated by the Maxwell operator. By using fixed point arguments, similar to the one employed in the previous section, one may show the existence of mild and strong solutions of (12.2) satisfying the representation formula (12.6). Of course, the usual drawbacks of using the evolution family $\{R(t)\}$ instead of the well-studied Maxwell group $\{\mathsf{T}_{\mathsf{M}_{\mathrm{A}}}(t)\}$ apply here as well. Furthermore, in the case where G_{A} is a random kernel, this representation is rather complicated, since then $\{R(t)\}$ is a random evolution family. This makes the representation formula (12.6) more difficult to handle, and some care should be taken when applied in conjunction with a fixed point scheme to show well-posedness results.

12.5.2 The Faedo-Galerkin approach

A different approach is to use finite-dimensional approximations of the solution of the stochastic integrodifferential equation (12.2) similar to the Faedo-Galerkin approach used for the deterministic problem. The idea of the method is similar to that followed in Section 7.5.2. Using a basis $\{e_n\}$ of $\mathbb{H}$ in $\mathbb{H}_{\mathsf{M}}$, where as before, $(\mathbb{H}, \mathbb{H}_{\mathsf{M}})$ is used as a proxy for $(\mathbb{X}, \mathbb{X}_{\mathsf{M}})$ or $(\mathbb{X}, \mathfrak{X}_{\mathsf{M}})$ (in the notation of Section 7.3), we project the stochastic integrodifferential equation (12.2) onto $V_m := span(e_1, \cdots, e_m)$. The projection is now a system of stochastic integrodiferential equations on the finite-dimensional space $\mathbb{R}^m$. Using standard theorems for finite-dimensional stochastic integrodifferential equations, one can show that the finite-dimensional approximation is well posed for any finite m. We must now pass to the limit as $m \to \infty$. This requires the existence of a priori estimates on the sequence u_m of finite-dimensional appoximations to (12.2) that are independent of m. To obtain such a priori bounds, one must use properly Itō's lemma and Gronwall's inequality. The existence proof then follows from weak convergence arguments, properly modified so that they are applicable to the function spaces needed for the study of the stochastic case. We briefly sketch the main differences as compared with the deterministic case, keeping the notation of Section 7.5.2.

The finite-dimensional approximation is performed with the aid of the sequence of functions $u_m = \sum_{i=1}^{m} u^{(i)}\, e_i$, where $u^{(i)}$ are scalar functions. The projection of (12.2) onto V_m results in a system of stochastic integral equations in $\mathbb{R}^m$ of the general form

$$Au_m + K \star u_m - \int_0^t M\, u_m(s)\, ds = \int_0^t j_m(s)\, ds + \int_0^t B(s, u_m(s))\, dw(s)\,,$$

where now $w(s)$ is a finite-dimensional approximation of the infinite-dimensional Wiener process $W(s)$ (of dimension m) and A, K, M, B are the finite-dimensional projections of A_{or}, G_{d}, M, Q_{A}, respectively. The solvability of the above equation may be obtained through the use of a Picard-type iteration scheme (see, e.g., [301]).

We now proceed to the necessary a priori bounds. To this end, we first apply the finite-dimensional version of Itō's lemma to the quantity $\sum_{i=1}^{m}(u^{(i)})^2$, which, if $\{e_i\}$ are chosen to be orthonormal - this is always feasible using the Gram-Schmidt procedure - coincides with the quantity $||u_m||_{\mathbb{X}}$. After straightforward algebra, we arive at an equality like (7.7), with an extra term expressing the *Itō drift*[4]. The Itō drift term results in a quadratic term in U_m; therefore, the equality is exactly of the form (7.7) with a modified Θ, which now contains terms related to the characteristics of the noise. Application of the Gronwall inequality yields that $u_m(x, t; \omega)$ is bounded in $L^\infty([0, T]; \mathbb{X})$ a.s. The bound is independent of m, and assuming that $j \in L^1([0, T];\ L^2(\Omega, \mathfrak{F}, P; \mathbb{H}))$, we readily obtain that $\sup_{t\in[0,T]} \mathbb{E}[\,||u_m||_{\mathbb{X}}^2] < C$, where C depends only on T but not on m. We further need a similar bound for the sequence $\mathsf{M}_{\mathrm{A}} u_m$. The corresponding result in the deterministic case was obtained through bounds on the sequence $(u_m)'$. However, here the stochastic term introduces problems related to the temporal differentiability of the solution; therefore, an alternative approach must be taken. To get around this difficulty we have to assume higher spatial regularity of J_{A} and the initial data.

12.5.3 The Wiener chaos approach

An interesting alternative approach is the Wiener chaos approach. This important approach has been proposed recently in a number of interesting publications by Rozovskii, Lototsky and co-workers ([292], [293], [294]) for the study of parabolic stochastic PDEs and has allowed an increased understanding of the structure and the properties of solutions of SPDEs. This approach has been extended to hyperbolic equations in [220] as well as to backward stochastic evolution equations in [439]. The numerical treatment of SPDEs using the Wiener chaos approach is an active area of research.

In this section, we sketch briefly how the Wiener chaos expansion can be used as a theoretical tool to construct the solutions of evolution problems, in terms of stochastic integrodifferential systems of the form (12.2).

The theoretical basis of the Wiener chaos expansion is the following fundamental result concerning square integrable real-valued stochastic processes, adapted with respect to the filtration generated by the Wiener process.

Let V be a separable Hilbert space with inner product $(\cdot,\cdot)_V$ and orthonormal basis $\{e_n\}_{n\,\in\,\mathbb{N}}$, and $\check{W}$ a cylindrical Brownian motion defined on a complete, filtered probability space $(\Omega, \mathscr{F}, P, \{\mathscr{F}_t\}_{0\le t\le T})$ (see Section B.4 in Appendix B). Recall that $\check{W}$ can be considered as a collection of real-valued and independent Brownian motions, in the sense that it can be represented in terms of standard, real-valued and independent Wiener processes $\{w_n(t)\}$, $n \in \mathbb{N}$, $t \in [0, T]$, as $\check{W}(t, \omega)(v) = \sum_{n\in\mathbb{N}} \left(v, e_n\right)_V w_n(t)$,

[4]See Theorem B.6.1 in Appendix B.

$(t, v) \in [0,T] \times V$.

For each $t \in [0,T]$, we shall denote by $L^2(\Omega, \mathscr{F}_t, P)$ the Hilbert space consisting of all $\mathscr{F}_t$−measurable and $\mathbb{P}$−square integrable real-valued random variables. Further, let $H_\ell(x)$, $\ell \in \mathbb{N}$, $x \in \mathbb{R}$ be the Hermite polynomial[5] of order ℓ, and $\mathcal{M} = \{m_i(\cdot)\}_{i\in\mathbb{N}}$ be an orthonormal basis of $L^2([0,T])$ such that $m_i(\cdot) \in L^\infty([0,T])$, $i \in \mathbb{N}$. A possible choice for $\mathcal{M}$ is the Fourier basis. Let α be a multi-index, and $\alpha! := \prod_{i,n} \alpha_i^n!$. The Hilbert space $\mathbb{L}^2(\mathbb{W}) := L^2(\Omega, \mathscr{F}_T, P)$, called the *Wiener chaos space,* can be constructed using the (orthonormal) *Cameron-Martin basis* [85]:

$$\xi_\alpha := \frac{1}{\sqrt{\alpha!}} \prod_{(i,n)\in\mathbb{N}^2} H_{\alpha_i^n}(\xi_{in}), \text{ where } \xi_{in} := \int_0^T m_i(t) dw_n(t) ,$$

and the multi-index α is an element of

$$\mathcal{J} := \left\{ \left(\alpha_i^n\right)_{(i,n)\in\mathbb{N}^2} : \alpha_i^n \in \mathbb{N}_0 , \ |\alpha| := \sum_{i,n} \alpha_i^n < \infty \right\},$$

which is countable and - for each $\alpha \in \mathcal{J}$ - only finitely many α_i^n are nonzero.

Therefore, every element v of $\mathbb{L}^2(\mathbb{W})$ can be expressed as

$$\mathrm{v} = \sum_{\alpha\in\mathcal{J}} \mathrm{v}_\alpha \xi_\alpha , \qquad \text{where} \quad \mathrm{v}_\alpha := \mathbb{E}[\mathrm{v}\xi_\alpha] , \tag{12.7}$$

which is called the *Wiener chaos expansion* of v. Using the Wiener chaos expansion of (12.7), we may construct measurable functions taking values in some other Hilbert space $\mathbb{H}$, using formal Fourier series expansions of the form $u = \sum_{\alpha\in\mathcal{J}} u_\alpha \, \xi_\alpha$. Here the elements $u_\alpha \in \mathbb{H}$ can be thought of as “generalised Fourier coefficients” for the measurable function $u : \Omega \to \mathbb{H}$. Such functions are generalisations of the concept of random variable and will be called “random elements” in $\mathbb{H}$. The space of random elements in $\mathbb{H}$ will be denoted by $\mathbb{L}^2(\mathbb{W}; \mathbb{H})$.

To use the Wiener chaos expansion to solve equation (12.2) we assume that for almost all (a.e.) $t \in [0,T]$, the solution $u(t)$ has an expansion of the form

$$u(t) = \sum_{\alpha\in\mathcal{J}} u_\alpha(t) \, \xi_\alpha, \tag{12.8}$$

where $u_\alpha(\cdot)$ are $\mathbb{H}$-valued deterministic functions. This is guaranteed by the extension of the Cameron Martin theorem for square integrable random elements. The above idea allows us to construct stochastic processes taking values in $\mathbb{H}$; the space of stochastic processes, adapted to the filtration $\mathfrak{F} = \{\mathscr{F}_t\}_{t\in[0,T]}$, for which the expansion (12.8) converges in the $L^2(\Omega \times [0,T])$ sense will be denoted by $L^2([0,T], \mathbb{W}; \mathbb{H})$ and is a Hilbert space[6] when equipped with the norm $(\mathbb{E}[\int_0^T ||u(s)||_{\mathbb{H}}^2 ds])^{1/2}$. Our aim is to

---

[5]Recall that $H_\ell(x) := (-1)^\ell \, e^{x^2/2} \, \frac{d^\ell}{dx^\ell} e^{-x^2/2}$.

[6]This space coincides with $L^2([0,T]; \, L^2(\Omega, \mathfrak{F}, P; \mathbb{H}))$ under certain circumstances (see [439]).

show that the deterministic functions $u_\alpha(\cdot)$ satisfy a hierarchy of deterministic integrodifferential equations very similar to those we have studied in Chapter 7.

In what follows we assume that $\mathsf{Q}_{A,2}$ is deterministic.

THEOREM 12.5.1 *Let $u \in L^2([0,T],\mathbb{W};\mathbb{H})$ be a square integrable solution of (12.2). Then, u has a Wiener chaos expansion of the form* (12.8) *and $\{u_\alpha\}$, $\alpha \in \mathcal{J}$, satisfy the infinite hierarchy of deterministic integrodifferential equations*

$$u'_\alpha(t) = \mathsf{M}_A u_\alpha + \mathsf{G}_A \star u_\alpha + (\mathsf{J}_A)_\alpha + \sum_{i,n} \sqrt{\alpha_i^n}(\mathsf{Q}_A(u))_\alpha m_i(t), \qquad (12.9)$$

where $(\mathsf{J}_A)_\alpha$ are the Wiener chaos expansion coefficients of the random field J_A, $(\mathsf{Q}_A(u))_\alpha = (\mathsf{Q}_{A,1})_{\alpha^-(i,n)} + \mathsf{Q}_{A,2}\, u_{\alpha^-(i,n)}$, and

$$(\alpha^-(i,n))_j^\ell = \begin{cases} max(a_i^n - 1) & \text{if } i = n,\ j = \ell \\ a_i^n & \text{otherwise.} \end{cases}$$

Proof. The existence of the representation (12.8) for each $t \in [0,T]$ is guaranteed by the generalisation of the Cameron Martin theorem for square integrable random elements (see [293]).

To specify the deterministic evolution equations (12.9) that the expansion coefficients u_α have to satisfy, we need to use the equalities $u_\alpha(t) = \mathbb{E}[u(t)\,\xi_\alpha]$, which are valid a.e. in t, along with the stochastic evolution equation (12.2). To reach this result formally, for a fixed $\alpha \in \mathcal{J}$, multiply (12.2) by ξ_α and take expectations. By the closedness of the operator M_A we see that $\mathbb{E}[(\mathsf{M}_A u)\,\xi_\alpha] = \mathsf{M}_A u_\alpha$. We also have that $\mathbb{E}[(\mathsf{G}_A \star u)\,\xi_\alpha] = \mathsf{G}_A \star u_\alpha$. Finally, by the properties of the Wiener basis (see, e.g., [293]) we have that $\mathbb{E}[(\int_0^t \mathsf{Q}_A\, dW(s))\,\xi_\alpha] = \sum_{i,n} \sqrt{a_i^n}\,(\mathsf{Q}_A)_\alpha\, m_i(t)$. This formal approach can be converted to a fully rigorous treatment by repeating the steps in [292]. □

Based on Theorem 12.5.1, we may construct solutions of (12.2) using the Wiener chaos expansion and the infinite hierarchy of deterministic integrodifferential equations (12.9). Observe that this is a hierarchy of uncoupled deterministic integrodifferential equations of the general form studied in Chapter 7, and therefore the general results obtained there may be used for the well posedness of (12.9), for each $\alpha \in \mathcal{J}$. Then the solution to the stochastic problem may be reconstructed using the expansion (12.8).

We may thus use Theorem 12.5.1 in conjunction with the existence results of Chapter 7 to provide an alternative existence proof for the solutions of (12.2) via the Wiener chaos expansion.

THEOREM 12.5.2 *Suppose that $\mathsf{J}_A \in L^2([0,T],\mathbb{W};\mathbb{H})$ and that G_A satisfies appropriate boundedness conditions. Then (12.2) has a unique solution $u \in L^2([0,T],\mathbb{W};\mathbb{H})$ admitting a Wiener chaos expansion of the form* (12.8)*, where u_α solve the deterministic hierarchy (12.9) for each $\alpha \in \mathcal{J}$.*

Proof. Assume first that $\mathsf{Q}_{A,2} = 0$, so that (12.9) is an infinite system of decoupled integrodifferential equations. Applying one of the theorems in

Chapter 7, e.g., Theorem 7.4.8, we have for each $\alpha \in \mathcal{J}$ there exists a unique u_α solving (12.9), such that $||u_\alpha||^2_{\mathbb{H}} \leq C\,||(\mathsf{J}_{\mathrm{A}})_\alpha||^2$, where $||\cdot||$ is a relevant norm for the Wiener chaos coefficient $(\mathsf{J}_{\mathrm{A}})_\alpha$, e.g., the $L^2([0,T],\mathbb{H})$ norm. Adding over all $\alpha \in \mathcal{J}$, we get that

$$\sum_{\alpha\in\mathcal{J}} ||u_\alpha||^2_{\mathbb{H}} \leq C \sum_{\alpha\in\mathcal{J}} ||(\mathsf{J}_{\mathrm{A}})_\alpha||^2.$$

If $\mathsf{J}_{\mathrm{A}} \in L^2([0,T],\mathbb{W};\mathbb{H})$, we recognise the right-hand side of the above inequality as the norm of J_{A} in the space $L^2([0,T],\mathbb{W};\mathbb{H})$; therefore, the right-hand side is bounded. Therefore the numerical series $\sum_{\alpha\in\mathcal{J}} ||u_\alpha||^2_{\mathbb{H}}$ converges and this implies that the random element $\sum_{\alpha\in\mathcal{J}} u_\alpha\xi_\alpha$ converges in $L^2([0,T],\mathbb{W};\mathbb{H})$. This can be easily seen with the use of the Parseval identity, which holds for the orthonormal basis $\{\xi_\alpha\}$. Therefore, a straightforward application of Theorem 12.5.1 yields the stated result. In the case of $\mathsf{Q}_{\mathrm{A},2} \neq 0$, we have that (12.9) is an infinite system of coupled deterministic integro-differential equations, of lower triangular form. Therefore, using an iterative scheme along with an induction argument (see [292], [220]), we obtain the required result. □

REMARK 12.5.3 The Wiener chaos approach to (12.2) is particularly interesting since:

(*i*) It reduces the original stochastic problem to an infinite hierarchy of deterministic problems. This hierarchy is decoupled in the case of additive noise and has a lower triangular structure in the case where the system is subject to a multiplicative noise introduced into the model through the term $\mathsf{Q}_{\mathrm{A}}(u) = \mathsf{Q}_{\mathrm{A},1} + \mathsf{Q}_{\mathrm{A},2}u$. In both cases the hierarchy is easy to solve using any of the approaches we have presented in Chapter 7. The first case is a straightforward application of the results of Chapter 7, whereas the second case calls for an additional simple induction argument.

(*ii*) It can be seen as a Galerkin-type approach, which separates the effects of randomness from the effects of the spatiotemporal dynamics, through the proposed expansion. It is important to note that the stochastic basis $\{\xi_\alpha\}$ is a "universal basis" that is *independent* of the problem under consideration. Therefore, this approach is well suited for numerical analysis and simulation purposes, especially when statistical moments of the solutions are needed.

(*iii*) The spatial regularity of the solutions can be treated very easily simply by considering more regular solutions of (12.9).

REMARK 12.5.4 The Wiener chaos approach may be generalised to allow the construction of a wide variety of solutions ranging from very regular to very irregular as far as their behaviour as random variables is concerned. This can be achieved through the generalisation of the Wiener chaos expansion in terms of weighted Wiener chaos spaces (see, e.g., [292], [293], [294] for definitions and applications to parabolic SPDEs and [220] for applications to hyperbolic SPDEs). Finally, we should mention that the Wiener chaos approach in principle works for the case in which the coefficients of the problem (12.2) are random as well. However, in such cases the hierarchy of deterministic evolution equations for u_α does not have such a simple structure as the one depicted in (12.9), thus making the analysis more complicated.

12.6 MISCELLANEA

12.6.1 Alternative forms of randomness

In this chapter we have introduced the Wiener process as a possible source of the randomness in the medium or the external sources. However, the analysis presented here can be extended to alternative forms of randomness that may be more general. Such a generalisation, which falls within the framework of Itō's theory of stochastic integration, is the use of Lévy processes for the modelling of the noise terms. In modelling the noise by Lévy processes, we keep the basic assumption of independent increments of the random effects, relaxing, however, the assumption that it is possible to obtain a scaling of space and time so that these effects can be modelled as continuous. Therefore, we allow random jumps and discontinuities in the noise terms, which of course introduce similar effects to the electromagnetic field. There is still relatively little work on the solutions of stochastic PDEs driven by Lévy noise (see, e.g., [347] or [353]) and the extension of the aforementioned general results to models for complex media seems like a promising plan for future research, from both the mathematical and the modelling point of view.

Another interesting generalisation is to relax the assumption of independence between the random increments of the noise term. This is reasonable, especially in physical models, where temporal (as well as spatial) correlations are often present. In such cases the general framework of Lévy processes can no longer be used and has to be abandoned, and with it, the convenient framework of the generalisation of Itō's theory of stochastic integration, as presented in Appendix B, for integrators that can be decomposed into a martingale part and a bounded variation part. One possibility is to use *coloured* [7] noise to model such correlations, another, even more challenging from the stochastic analysis point of view, is to use generalisations of the fractional Brownian motion in infinite-dimensional spaces. This is certainly a very interesting direction for further research.

[7] Coloured noise is a Gaussian process that does not have independent increments. The "colour" of the noise characterises this dependence and is related to the spectral properties of the noise signal. White noise is considered to be the limit where independence of increments is recovered.

12.6.2 Nonlinear stochastic problems

In this chapter we have treated the stochastic Maxwell equations for linear complex media. Our analysis can be extended for the study of nonlinear problems. In fact, under the assumption that the nonlinearities are globally Lipschitz, most of the results in this chapter may be extended almost with minor changes, employing the fixed point schemes that are used for the proofs of the well-posedness results in the linear case[8]. In the case where the nonlinearities are not globally Lipschitz, we need to apply more sophisticated techniques to complete the fixed point arguments, such as approximation techniques, etc. Another class of techniques may be an approach based on monotonicity arguments, which is close to the one employed for the deterministic case in Chapter 11. Monotonicity techniques have been used successfully for nonlinear stochastic problems (see, e.g., [352]).

[8]This was one of the reasons we chose to employ fixed point arguments for the study of linear media.

Chapter Thirteen

Controllability

13.1 INTRODUCTION

This chapter addresses the problem of controllability for stochastic complex electromagnetic media. Our starting point is the stochastic integrodifferential equations that model the evolution of the fields, with a control procedure to be selected so that the system is driven to a desired final state. The controllability problem for stochastic media is more complicated than that for deterministic media and includes subtleties that must be addressed to reach a satisfactory answer.

The stucture of this chapter is as follows: In Section 13.2 we set the model, while in Section 13.3 we discuss the subtle issues introduced by the stochasticity in the controllability problem. We then propose two different approaches towards controllability: in Section 13.4 an approach using PDEs with random coefficients and in Section 13.5 an approach using backward stochastic evolution equations (BSEEs). In Section 13.6 we list several comments concerning boundary controllability and optimal control.

13.2 FORMULATION

We consider the problem of controllability of the stochastic complex medium with an additive noise. The problem can be modelled through the use of the stochastic integrodifferential equation

$$du = (\mathsf{M}_{\rm A} u + \mathsf{G}_{\rm A} \star u + \mathsf{J}_{\rm A} + \mathcal{B}v)\, dt + \sum_{j=1}^{N} g_j \, dw_j(t), \tag{13.1}$$

where for the sake of simplicity we assume the noise to be finite dimensional. In particular, by w_j we denote the jth component of the N-dimensional[1] Wiener process defined on a filtered probability space $(\Omega, \mathscr{F}, \{\mathscr{F}_t\}_{[0,T]}, P)$. We will use the notation $\mathfrak{F} = \{\mathscr{F}_t\}$. Further, $g_j = (g_j^{(1)}, g_j^{(2)})^{tr} \in \mathbb{R}^6$ is a vector function of the spatial variables multiplying the noise terms such that

$$\text{div} g_j^{(k)} = 0 \text{ for } k = 1, 2,\ j = 1 \ldots N.$$

We use the same notation as for the deterministic case in Chapter 8; in particular, by v we denote the internal control. Consider first the stochastic

[1] The case where $N = \infty$ may be considered with minor technical modifications.

achiral case. Assume that the corresponding linear system

$$du = (\mathsf{M}_{\mathrm{A}} u + \mathsf{J}_{\mathrm{A}} + \mathcal{B} v)\, dt + \sum_{j=1}^{N} g_j \, d w_j(t)$$

is controllable. Our approach to treating the chiral case is based on a fixed point scheme: assume that the state space of the system is a Hilbert space $\mathbb{H}$; at this point, we do not specify this space. Fix a function $z(\cdot) \in \mathbb{H}$, and consider the linear system

$$du = (\mathsf{M}_{\mathrm{A}} u + \mathsf{J}^{\square}_{\mathrm{A}} + \mathcal{B} v)\, dt + \sum_{j=1}^{N} g_j \, d w_j(t)\,, \tag{13.2}$$

where $\mathsf{J}^{\square}_{\mathrm{A}} = \mathsf{J}_{\mathrm{A}} + \mathsf{G}_{\mathrm{A}} \star z$ (recall that the chirality effects are introduced via the part $\mathsf{G}_{\mathrm{A}} \star \cdot$ of $\mathsf{J}^{\square}_{\mathrm{A}}$). We denote by $v_z(\cdot)$ the control procedure that drives the system (13.2) from $u(0) = U_0$ to $u(T) = U_T$. It is important to note here that U_T can in principle be a random variable as long as it is measurable with respect to the σ-algebra $\mathscr{F}_T$. Let $u_z(\cdot)$ be the solution of

$$du = (\mathsf{M}_{\mathrm{A}} u + \mathsf{J}^{\square}_{\mathrm{A}} + \mathcal{B}\, v_z)\, dt + \sum_{j=1}^{N} g_j \, d w_j(t)\,,$$

with initial condition $u(0) = U_0$. By the definition of v_z we know that $u(T) = U_T$. Let $\mathbb{X}_{st}$ denote either $\mathcal{C}([0,T];\mathbb{H})$ or $L^2([0,T];\ L^2(\Omega, \mathfrak{F}, P; \mathbb{H}))$. Define the map $\mathsf{F} : \mathbb{X}_{st} \to \mathbb{X}_{st}$ by

$$\mathsf{F}(z(\cdot)) = u(\cdot)\,.$$

If this map has a fixed point, i.e., if there exists a function $\mathfrak{u} \in \mathbb{X}_{st}$, such that $\mathsf{F}(\mathfrak{u}) = \mathfrak{u}$, then this $\mathfrak{u}$ is the solution of (13.2) that connects the states U_0 and U_T in time T, and furthermore $v_{\mathfrak{u}}$ is the required control.

13.3 SUBTLETIES OF STOCHASTIC CONTROLLABILITY

We now introduce the problem of approximate controllability:

> *Given an initial condition $\xi_0 \in L^2(P, \mathscr{F}_0, \mathbb{H})$, where $\mathbb{H}$ is an appropriately chosen Hilbert space, and a given final state $\xi_T \in L^2(P, \mathscr{F}_T, \mathbb{H})$, is it possible to find an adapted control procedure $v \in L^2([0,T];\ L^2(\Omega, \mathfrak{F}, P; \mathbb{V}))$, where $\mathbb{V}$ is the control space, such that the system (13.1) is driven ϵ-close to the final condition ξ_T in the chosen time period?*

The choice of functional setting for this problem is the same as the one used for the deterministic problem (see Chapter 8 for details and notation).

The question of mild well posedness of the problem (13.1) was settled in Chapter 12. Using generalisations of these results, we may show the well posedness of the problem for a wide range of possible control functions $v(\cdot)$.

Further regularity conditions on the solution may be obtained by straightforward modifications of the Faedo-Galerkin method proposed for the wave equation in [240] (see also Remark 2.4 in [240]). In analogy with the deterministic controllability problem, we choose $\mathbb{H} = \mathbb{V} = \mathfrak{X}_{\mathsf{M}}$ (see Chapter 8).

LEMMA 13.3.1 *The stochastic control system (13.1) is mildly well posed in* $\mathbb{X}$ *for every* $v \in L^2([0,T];\ L^2(\Omega, \mathfrak{F}, P; \mathfrak{X}_{\mathsf{M}}))$.

We are now ready to propose a procedure for approximate controllability. The procedure is based on the fixed point procedure used for the deterministic control problem in combination with the approximation result in Lemma 13.4.1.

So far there has been no reference to approximate controllability. Therefore, some comments are due concerning the need to consider the problem of approximate controllability at this point. This need arises from the controllability of problem (13.2). Recall the way that we treated the problem of construction of the control procedure v_z. The first auxiliary problem we need to consider is the uncontrolled system

$$\begin{aligned} dV &= (\mathsf{M}_{\mathrm{A}} V + \mathsf{J}_{\mathrm{A}}^{\square})\, dt + \sum_{j=1}^{N} g_j\, d\, w_j(t)\,, \\ V(0) &= U_0\,. \end{aligned} \tag{13.3}$$

The solution of this problem will give us the final state of the system $V(T)$, which is an $\mathscr{F}_T$-measurable random variable.

We then consider the backward adjoint problem

$$\begin{aligned} \Phi' &= \mathsf{M}_{\mathrm{A}} \Phi\,, \\ \Phi(T) &= \Phi_T\,, \end{aligned} \tag{13.4}$$

for a final condition that is now in principle a random variable, and the forward problem

$$\begin{aligned} \Psi' &= \mathsf{M}_{\mathrm{A}} \Psi + \mathcal{B}\, \Phi\,, \\ \Psi(0) &= 0\,, \end{aligned} \tag{13.5}$$

where Ψ is the solution of problem (13.4). Note that even though there is no stochasticity in the equation, the source term is a random function, and as a result, Ψ is a random function as well. In complete analogy with the deterministic problem in Chapter 8, we define the map Λ that connects the random variable Ψ_T with the final value of the solution of the forward problem (13.5) (which is itself a random variable).

Assume that the random operator equation

$$\Lambda(\Phi_T^{\flat}) = U_T - V(T) \tag{13.6}$$

is satisfied for some random variable $\Phi_T^{\flat} \in \mathfrak{X}_{\mathsf{M}}$. Now, using the same procedure as we did for the deterministic problem in Chapter 8, we see that setting $u(t) = V(t) + \Psi(t)$, where $\Psi(t)$ is now the solution of (13.5), with $\Phi(t)$ as obtained by the solution of (13.4) and the final condition $\Phi_T^\flat$ as provided by the solution of the operator equation (13.6), we have obtained a solution of the original control system such that $u(0) = U_0$ and $u(T) = U_T$ in time T. Therefore, the desired control procedure is the random process $\Phi(t)$ as defined above.

Even though the stochastic case gives the impression of formally proceeding in the same manner as the deterministic case, there is an important qualitative difference that does not allow us such an approach. The reason is very simple and goes as follows: Since $V(T)$ is an $\mathscr{F}_T$-measurable random variable, then $U_T - V(T)$ is also an $\mathscr{F}_T$-measurable random variable. Therefore, in principle, the solution of the operator equation (13.6) $\Phi_T^\flat$ is an $\mathscr{F}_T$-measurable random variable. Consider now the backward equation (13.4). This is a differential equation, with no stochasticity but with a final condition that is a random variable. Therefore, the solution is a random process that inherits the measurability properties of the final condition, i.e., $\Phi(t)$ will be an $\mathscr{F}_T$-measurable random variable for all $t \in [0, T]$. Now consider the forward equation (13.5). This is a random differential equation, the randomness being inserted by the random process $\Phi(t)$, which acts as a source term in the right-hand side of (13.5). By the same arguments, the solution of this equation $\Psi(t)$ inherits the measurability properties of the source term; therefore the solution $\Psi(t)$ is measurable with respect to the σ-algebra $\mathscr{F}_T$ for all $t \in [0, T]$. Since the control procedure obtained is $v = \Psi(t)$, we see that neither the control procedure nor the state of the system $u(t)$ has the desired property of being adapted to the filtration generated by the noise process, since that would require that $v(t)$ and $u(t)$ be measurable with respect to $\mathscr{F}_t$ for all $t \in [0, T]$ and not with respect to $\mathscr{F}_T$, as it actually happens. Put differently, the control procedure we have proposed above gives us a control scheme that requires knowledge of future states of the system, and not simply the history of the system. This property renders the control proposed above nonphysical and not practical to implement.

In this chapter we propose two possible ways to bypass this difficulty introduced by the noise. One is through an approach to the problem using PDEs with properly selected random initial or final conditions. The other is through the use of backward stochastic differential equations. Both approaches are interesting, for reasons that will be analysed in the relevant sections.

13.4 APPROXIMATE CONTROLLABILITY I: RANDOM PDES

Assume that we wish to solve the system (13.4) - (13.5) in the interval $[T - \tau, T]$ for some $\tau \in [0, T]$. The system is modified to the two problems

$$\begin{aligned} \Phi' &= \mathsf{M}_{\mathsf{A}} \Phi \,, \\ \Phi(T) &= \Phi_T \,, \end{aligned} \tag{13.7}$$

$$\begin{aligned} \Psi' &= \mathsf{M}_{\mathrm{A}} \Psi + \mathcal{B}\, \Phi\,, \\ \Psi(T - \tau) &= 0\,. \end{aligned} \tag{13.8}$$

As we commented in the section above, as long as Φ_T is measurable with respect to $\mathscr{F}_T$, the solution to system (13.7) is measurable with respect to $\mathscr{F}_T$, and this renders the solution to (13.8) measurable with respect to $\mathscr{F}_T$ as well for all times $t \in [T - \tau, T]$. This is unphysical, as both the control (related to $\Phi(\cdot)$) and the state of the system (related to $\Psi(\cdot)$) are forward looking (looking into the future) in the sense that they can only be fully determined as long as the final state of the uncertainty is known. One way to remedy this is to assume that Φ_T is measurable with respect to $\mathscr{F}_{T-\tau}$ instead of being measurable with respect to $\mathscr{F}_T$. Then, both $\Phi(\cdot)$ and $\Psi(\cdot)$ are measurable with respect to $\mathscr{F}_{T-\tau}$ for all $t \in [T - \tau, T]$, and this immediately makes both processes dependent only on the past states of the uncertainty. This makes the solution acceptable from a physical point of view and thus useful for the construction of the control procedure.

However, an important issue remains unsettled. One does not in principle have the freedom to pick the measurability of Φ_T; this is fixed and determined. One may rather modify the final condition Φ_T by choosing another final condition, $\Phi_T^\flat$, as close as possible to the original final condition Φ_T but satisfying the condition that it is measurable with respect to $\mathscr{F}_{T-\tau}$.

The major technical result, essential in our work, is a generalisation of the martingale representation theorem in finite-dimensional spaces (see, e.g., [224] and Theorem B.9.1 in Appendix B), which is summarised in the following result, [240]:

LEMMA 13.4.1 *Let $\mathbb{H}$ be a Hilbert space and $\epsilon > 0$. Given an $\mathscr{F}_T$-measurable $\mathbb{H}$-valued random variable ξ, there exist $\tau > 0$ and an $\mathscr{F}_{T-\tau}$-measurable random variable ξ_ϵ such that for some $\epsilon = \epsilon(\tau) > 0$, $\mathbb{E}[\|\ \xi - \xi_\epsilon\ \|_{\mathbb{H}}^2] < \epsilon$ and $\mathbb{E}[\|\ \xi_\epsilon\ \|_{\mathbb{H}}^2] \leq \mathbb{E}[\|\ \xi\ \|_{\mathbb{H}}^2]$.*

REMARK 13.4.2 The choice of τ can be made explicit using the Clark-Ocone form for the martingale representation theorem (see, e.g., [88]) in the case where ξ is Malliavin differentiable; estimates for τ can be established in terms of the Malliavin derivative of ξ.

After this preliminary discussion, we may present a first approach towards approximate controllability. To make the argument more transparent, we first omit the chiral terms (integral terms), which will be treated by the fixed point argument shortly.

- ▷ Choose a τ.
- ▷ Start the uncontrolled system at U_0 at time $t = 0$ and run until time $t = T - \tau$. The state of the uncontrolled system at $t = T - \tau$ is a random variable $u_0(T - \tau)$, which is measurable with respect to $\mathscr{F}_{T-\tau}$.

- ▷ We now wish to find a control v, that acting from $t = T - \tau$ to $t = T$, is able to drive the controlled system from $u_0(T - \tau)$ to the desired final state U_T.

- ▷ However, U_T is in principle an $\mathscr{F}_T$-measurable random variable. In reference to the discussion above, using the auxiliary systems (13.7)-(13.8) we cannot construct an adapted control procedure that will achieve this task. Therefore, we may approximate U_T as closely as possible by a random variable $\mathring{U}_T$, measurable with respect to $\mathscr{F}_{T-\tau}$ (a step that is guaranteed by Lemma 13.4.1).

- ▷ Then find the control procedure that drives the controlled system from $u_0(T - \tau)$ to $\mathring{U}_T$ in the time interval $[T - \tau, T]$. This control procedure can be constructed consistently with the use of the auxiliary systems (13.7) - (13.8) following a procedure close to that adopted for the deterministic problem in Chapter 8.

The remaining crucial step in this procedure is to show that the map Λ, defined by (13.6), is invertible. This operator is defined from a suitable space of square integrable $\mathscr{F}_{T-\tau}$-measurable random variables to itself. It is straightforward to repeat the arguments of Section 8.4.2 replacing the initial and final conditions by $\mathscr{F}_{T-\tau}$-measurable random variables, working pointwise and then passing to the $L^2(\Omega, \mathscr{F}, P; \mathfrak{X}_{\mathsf{M}})$ setting using Lebesgue dominated convergence type arguments, based on a priori estimates. We leave the details to the reader, and here we simply state the following generalisation of Lemma 8.4.5.

LEMMA 13.4.3 *There exists an $\alpha^\flat$ such that $\Lambda(\alpha^\flat) = \xi$, for every $\xi \in L^2(\Omega, \mathscr{F}_{T-\tau}, P; \mathfrak{X}_{\mathsf{M}})$.*

Finally, we have to address the chirality terms. These terms have to be treated using a fixed point scheme similar to the one proposed in Section 8.2 for the deterministic problem. The existence of a fixed point for the map F is equivalent to the proof of approximate controllability. The existence of a fixed point may be shown through the use of a fixed point theorem. Contrary to the deterministic case, where we resorted to Schauder's fixed point theorem in order to establish the existence of a fixed point of F, now we shall employ Banach's contraction mapping theorem; the reasons are that (a) the chirality terms are typically assumed to be small and (b) it allows us to bypass issues related to the characterisation of compact sets in $L^2(\Omega, \mathscr{F}, P; \mathbb{H})$, where $\mathbb{H}$ is a Hilbert space.

This leads to the following controllability result.

THEOREM 13.4.4 *For small enough chirality, the stochastic control system (13.1) is approximately controllable.*

13.5 APPROXIMATE CONTROLLABILITY II: BSPDES

An alternative to the above approach that does not require the approximation of the final condition for the stochastic auxiliary equation (13.4) by one that is measurable with respect to a diferent σ-algebra in order to achieve the adaptivity of the solution with respect to the filtration generated by the Wiener process is to use the theory of backward stochastic partial differential equations (BSPDEs).

As we commented above, the solution of (13.4) is not in general adapted to the filtration $\mathscr{F}_t$ as required, a fact that is rather annoying from the modelling point of view. Is there any way of guaranteeing that (13.4) achieves a solution that satisfies the adaptivity property? This can be done as long as we decide to modify the equation slightly, to

$$\begin{aligned} d\Phi &= \mathsf{M}_{\mathrm{A}}\Phi\, dt + Z\, dW(t)\,, \\ \Phi(T) &= \Phi_T\,, \end{aligned} \tag{13.9}$$

where Z is an (operator-valued) unknown stochastic process that will be selected so as to render the solution of (13.9) adapted and be sure that it satisfies the right final condition. For example if Φ takes values in a Hilbert space $\mathbb{H}$ and W takes values in a Hilbert space $\mathbb{V}$ then Z should take values in $L_2(\mathbb{V}, \mathbb{H})$ the space of Hilbert-Schmidt operators from $\mathbb{V}$ to $\mathbb{H}$. It is worth noting that equation (13.9) is a stochastic differential equation in two unknowns (Φ, Z) that is solved with a final condition. This type of equation is called a *backward stochastic evolution equation* (BSEE) or a *backward stochastic partial differential equation* (BSPDE). Even though we have a single equation in two unknowns, under certain conditions the solution is unique, as the processes Φ and Z are related to each other.

Assume for the time being that (13.9) has a unique solution (Φ, Z) for any final condition $\Phi(T) = \Xi$ measurable with respect to $\mathscr{F}_T$, and let us show how this pair of processes may be used to construct the desired control.

To this end, let (Φ, Z) be the solution of the BSEE (13.9) with final condition Ξ, and consider the solution of the forward random equation (13.5) with initial condition $\Psi(0) = 0$. Since by construction Φ is an adapted stochastic process, then Ψ is also an adapted process. Define the map Λ that connects the final condition Ξ of the BSEE (13.9) with the final condition $\Psi(T)$ of the forward equation (13.5). Furthermore, solve the uncontrolled stochastic system (13.3) with initial condition $V(0) = U_0$, and obtain the $\mathscr{F}_T$-measurable random variable $V(T)$. We now add equations (13.5) and (13.3) and by linearity observe that the process $U = V + \Psi$ solves equation

$$dU = (\mathsf{M}_{\mathrm{A}} I + \mathsf{J}_{\mathrm{A}}^{\square} + \mathcal{B}\Psi)\, dt + \sum_{j=1}^{N} g_j\, d w_j(t)\,, \tag{13.10}$$

with initial condition $U(0) = U_0$ and final condition $U(T) = \Psi(T) + V(T)$. If we can choose Ξ so that $\Psi(T) = U(T) - V(T)$, then the desired control procedure is $v = \Psi$. Therefore, the control procedure may be constructed through the use of the BSEE.

There are two missing steps in this analysis. The first is to show that the BSEE (13.9) indeed has an adapted solution for every square integrable final condition Ξ. This result is provided by the general theory of BSEEs, which is by now a well-established subject of stochastic analysis. Characteristic references on the subject are, e.g., [198], [340], [409]; see also references therein. The conditions for solvability of this problem are particularly simple, and in fact the square integrability of the final condition Ξ and the property that M_{A} is the generator of a C_0 group are enough to guarantee the well posedness of (13.9). The following proposition summarises these comments.

PROPOSITION 13.5.1 *For every* $\Xi \in L^2(\Omega, \mathscr{F}_T, P; \mathbb{H})$, *there exists a unique pair*[2] $(\Phi, Z) \in L^\infty([0,T]; L^2(\Omega, \mathfrak{F}, \mathbb{H}) \times L^2([0,T]; L^2(\Omega, \mathfrak{F}; L_2(\mathbb{V}, \mathbb{H}))$ *such that*

$$\Xi - \Phi(t) = \int_t^T \mathsf{M}_A \, \Phi(s)\, ds + \int_t^T Z(s)\, dW(s) \tag{13.11}$$

for all $t \in [0,T]$.

Proof. The proof is simple and hinges on the martingale representation theorem for Hilbert space-valued random variables. It is essentially an adaptation of similar proofs in, e.g., [409], etc.

Since M_{A} is the infinitesimal generator of a C_0 group of linear operators, $\{\mathsf{T}_{\mathsf{M}_{\mathrm{A}}}(t)\}_{t\in\mathbb{R}}$, $\mathsf{T}_{\mathsf{M}_{\mathrm{A}}}(t) = e^{t\,\mathsf{M}_{\mathrm{A}}}$, one may easily see that $\Phi^\flat(t) = \mathsf{T}_{\mathsf{M}_{\mathrm{A}}}(t)\Phi(t)$ satisfies the BSEE

$$\begin{aligned} d\Phi^\flat(t) &= \mathsf{T}_{\mathsf{M}_{\mathrm{A}}}(t)\, Z(t)\, dW(t)\,, \\ \Phi(T) &= \mathsf{T}_{\mathsf{M}_{\mathrm{A}}}(T)\, \Xi\,. \end{aligned} \tag{13.12}$$

Consider now the stochastic process $M(t) := \mathbb{E}[\mathsf{T}_{\mathsf{M}_{\mathrm{A}}}(T)\,\Xi \mid \mathscr{F}_t]$, which is a square integrable martingale. To this martingale we apply the martingale representation theorem (Theorem B.9.1 in Appendix B), to guarantee the existence of an operator-valued stochastic process $\mathring{Z}$ such that

$$M(t) = M(0) + \int_0^t \mathring{Z}(s)\, dW(s) \tag{13.13}$$

for every $t \in [0,T]$. Set $t = T$ in this expression to obtain

$$\Phi^\flat(T) = M(0) + \int_0^T \mathring{Z}(s)\, dW(s)\,, \tag{13.14}$$

where we have used the fact that

$$M(T) := \mathbb{E}[\mathsf{T}_{\mathsf{M}_{\mathrm{A}}}(T)\,\Xi \mid \mathscr{F}_t] = \mathsf{T}_{\mathsf{M}_{\mathrm{A}}}(T)\,\Xi = \Phi^\flat(T).$$

We now subtract (13.13) and (13.14) to obtain that

$$\Phi^\flat(T) - \Phi^\flat(t) = \int_t^T \mathring{Z}(s)\, dW(s)\,. \tag{13.15}$$

A direct comparison of (13.11) with (13.12) shows that $M(t)$ as defined above is a candidate for the unknown process $\Phi^\flat(t)$, whereas a candidate for

---

[2] These processes are adapted to the filtration $\mathfrak{F} = \{\mathscr{F}_t\}_{t\in[0,T]}$.

the process Z is the process $\mathring{Z}$ provided by the martingale representation theorem. By the group property of $\{\mathsf{T}_{\mathsf{M_A}}(t)\}_{t\in\mathbb{R}}$, we return to the original processes $\Phi(t)$ and Z. The uniqueness follows by an application of Itō's lemma. □

The second remaining step is to show that we can always find a final condition Ξ for (13.9) so that the solution of (13.5) satisfies $\Psi(T) = U_T - V(T)$. This relies on the invertibility of the map Λ. For approximate controllability we would like to show that for every ξ in $L^2(\Omega, \mathscr{F}_T, P; \mathfrak{X}_{\mathsf{M}})$ we may find a Ξ such that the solution of the system (13.9) and (13.5) satisfies the property that for every $\epsilon > 0$, $||\Psi(T) - \xi|| < \epsilon$. An alternative way of looking at this is to treat the approximate solvability of the forward-backward stochastic differential problem

$$
\begin{aligned}
d\Psi &= (\mathsf{M_A}\Psi + \mathcal{B}\,\Phi)\,dt\,,\\
d\Phi &= \mathsf{M_A}\Phi\,dt + Z\,dW(t)\,,\\
\Psi(0) &= \xi\,,\\
\Phi(T) &= \Xi\,.
\end{aligned}
$$

This is in some sense a stochastic two-point boundary value problem. The solvability of this problem is equivalent to the approximate controllability of the original problem at hand.

Consider now the linear manifold

$$
\mathcal{F} = \{\Psi(T) : \text{there exists a } \Xi \text{ so that } (\Psi, \Phi) \text{ solves } (13.9) \text{ and} (13.5)\}\,.
$$

We will show that $\mathcal{F}$ is dense in $L^2(\Omega, \mathscr{F}_T, P; \mathfrak{X}_{\mathsf{M}})$.

PROPOSITION 13.5.2 *The linear manifold $\mathcal{F}$ is dense in $L^2(\Omega, \mathscr{F}_T, P; \mathfrak{X}_{\mathsf{M}})$ if $\mathcal{B}$ is invertible.*

Proof. Take $X \in L^2(\Omega, \mathscr{F}_T, P; \mathfrak{X}_{\mathsf{M}})$ and assume that $(X, \Psi(T)) = 0$ for any $\Psi(T) \in \mathcal{F}$. We will show that this implies that $X = 0$.

For the given X, consider the BSEE

$$
X - p(t) = \int_t^T \mathsf{M_A} p(s)\,ds + \int_t^T q(s)\,dW(s)\,. \tag{13.16}
$$

According to Proposition 13.5.1, this problem is well posed, so there exists a unique pair of stochastic processes (p, q) satisfying (13.16). Now, let us apply Itō's lemma on the inner product $(p(t), \Psi(t))$. After taking into account the fact that the operator $\mathsf{M_A}$ is skew adjoint, we see that

$$
(X, \Psi(T)) = \int_0^T (p(t), \mathcal{B}\,\Phi(t))\,dt\,.
$$

This is equal to 0 if $\int_0^T (p(t), \mathcal{B}\,\Phi(t))\,dt = 0$, or if $\int_0^T (\mathcal{B}^*\,p(t), \Phi(t))\,dt = 0$. Since we want $(X, \Psi(T)) = 0$ to hold true for all $\Psi(T) \in \mathcal{F}$, the above implies that $\mathcal{B}^*\,p(t) = 0$ for all t. If the operator $\mathcal{B}^*$ is such that $\mathcal{B}^* p(t) = 0$ implies that $p(t) = 0$ for all t, then $X = p(T) = 0$, and that concludes the proof of density of $\mathcal{F}$ in $L^2(\Omega, \mathscr{F}_T, P; \mathfrak{X}_\mathsf{M})$. If $\mathcal{B}^*$ is an invertible operator, then one can immediately see that the above condition holds, so the system satisfies the approximate controllability property. □

If $\mathcal{B}^*$ is not an invertible operator, then Carleman-type[3] estimates for the backward problem are needed to show this property. Carleman-type inequalities for backward PDEs constitute an interesting research field that is relatively unexplored. Because of the technical nature of the subject we do not address it here; let us just mention that a Carleman inequality for the stochastic wave equation has been proved recently by Zhang in [447]. Similar results for forward and backward stochastic parabolic equations have been reported in [54] and [401]. See also the interesting unified approach by the same author in [448].

13.6 MISCELLANEA

There are several possible extensions of the approach presented in this chapter on the problem of controllability of stochastic integrodifferential equations modelling complex media.

13.6.1 Boundary controllability

The problem of boundary controllability is still open. In principle either approach employed in this chapter to study approximate controllability can be used, along with a fixed point procedure to prove approximate controllability. Extra care must be taken here to define properly the map $\mathcal{B}$.

13.6.2 Multiplicative noise

Some of the controllability results and techniques in this chapter can be generalised to the case where the model is subject to multiplicative noise. This will result in a model of the general form

$$du = (\mathsf{M}_\mathrm{A} u + \mathsf{G}_\mathrm{A} \star u + \mathcal{B}v)\, dt + (\mathsf{Q}_{\mathrm{A},1} + \mathsf{Q}_{\mathrm{A},2} u) dW\,, \tag{13.17}$$

where W can either be a finite- or infinite-dimensional Wiener process.

If one decides to use the backward stochastic differential approach, then it may be shown that the approximate controllability of the control system

[3]In 1939, Torsten Carleman introduced some energy estimates with exponential weights to prove a uniqueness result for a class of PDEs with smooth coefficients in dimension two. This type of estimates, now referred to as "Carleman estimates", or "Carleman inequalities", were generalised and systematised by Lars Hörmander in 1963 (and others later) for a large class of differential operators in arbitrary dimensions. More recently, the field of applications of Carleman estimates has exceeded by far the original domain they had been introduced for, i.e., a quantitative result for unique continuation. They are also extensively used in the study of inverse problems and control theory for PDEs, and more recently in stochastic differential equations. The existing bibliography on Carleman estimates is very extended; the main ideas can be found, e.g., in [280].

(13.17) is related to the properties of the solution of the backward stochastic evolution equation

$$\begin{aligned} dp &= (-\mathsf{M}_{\mathrm{A}}^{*}p + \mathsf{Q}_{\mathrm{A},1}^{*}Z)\,dt + Z\,dW(t)\,, \\ u(T) &= \xi\,. \end{aligned} \tag{13.18}$$

This is similar to the BSEE obtained for the additive noise case, with the main difference that now the auxiliary process Z is contained in the drift of the equation. The well posedness of this equation can be shown by extending the simple argument we have used in Lemma 13.5.1 along with a fixed point scheme of the form[4]

$$du^{(n+1)} = (\mathsf{M}_{\mathrm{A}}u^{(n+1)} + \mathsf{Q}_{\mathrm{A},1}^{*}Z^{(n)})\,dt + Z^{(n)}\,dW(t)\,.$$

A criterion for approximate controllability of (13.17), similar to that proposed in the case of additive noise, is the following:

The system (13.17) is approximately controllable if and only if for any finite horizon T, the solution of (13.18) that satisfies $\mathcal{B}^{*}p(t) = 0$ for all $t \in [0,T]$ necessarily satisfies $p(t) = 0$ for all $t \in [0,T]$. This result has recently been proved for the multiplicative case in [166].

Of course, proving that this property holds brings us back to the issue of deriving Carleman inequalities for stochastic evolution equations; see the relevant discussion in Section 13.5.

Furthermore, the methods presented in this chapter can be extended for the case of nonlinear problems using fixed point techniques.

13.6.3 Stochastic optimal control problems

The optimal control problems sketched briefly in Section 8.6.3 can be generalised for the case of stochastic problems. The relevant cost functional now involves expectations, e.g., stochastic linear quadratic control problems involve functionals of the form

$$\mathcal{J}(u,v) = \mathbb{E}[\int_0^T (||\mathscr{C}_1 u(t)||_{\mathbb{H}}^2 + ||\mathscr{C}_2 v(r)||_{\mathbb{V}}^2)dt] + \mathbb{E}[||\mathscr{C}_3(u(T) - U_T)||_{\mathbb{H}}^2]\,,$$

where $\mathbb{H}$ and $\mathbb{V}$ are proxies for the state space and the control space of the system, respectively, and the expectation is with respect to the probability measure. The operators $\mathscr{C}_1, \mathscr{C}_2, \mathscr{C}_3$ are operators that quantify the relative importance of the intertemporal deviation from a desired state, the cost of the control procedure and the deviation from the desired final state, respectively, in the overall cost functional.

Problems of this form can be solved in terms of feedback laws of the form $v = \mathbb{F}u$, where $\mathbb{F}$ is the feedback operator that connects the state of the system with the control required. The feedback operator can be determined using the infinite-dimensional Riccati equation. An alternative is to use generalisations of the Pontryagin maximum principle using adjoint variables. The optimal control law can be determined through the adjoint variable. Of

[4] We have taken into account that $\mathsf{M}_{\mathrm{A}}^{*} = -\mathsf{M}_{\mathrm{A}}$.

importance, the adjoint variable solves a BSEE of the form introduced and studied in Section 13.5.

This theory can be extended in a perturbative fashion, via fixed point arguments, to include the effects of the convolution terms; however, this is beyond the scope of the present work.

Chapter Fourteen

Homogenisation

14.1 INTRODUCTION

In certain classes of materials the spatial structure is not sufficiently regular as to be modelled by periodic functions. Such materials can be modelled as random media having some sort of statistical periodicity. This statistical periodicity is expressed mathematically through the concept of *ergodicity*. This concept is powerful enough to generalise periodicity and allows us to build a homogenisation theory that bypasses the need for periodic structure. From the applications point of view, this generalisation leads to more realistic models. In nature there is actually no such thing as a deterministic periodic structure; materials are subject to random imperfections that may depart from standard periodicity but still leave a random structure reminiscent of periodicity. Note that, according to some authors (see, e.g., [8]), the very nature of chirality is due to such effects, i.e., to randomly positioned helices in the medium.

There exists a rich theory concerning random homogenisation problems. For linear elliptic and parabolic problems see, e.g., [59], [67], [216] and references therein. It is the aim of this chapter to present a theory of homogenisation for random bianisotropic media exibiting an ergodic structure. We show that for such a medium there exists a homogenised system of the Maxwell type (see [52], [395]). The homogenisation problem can be reduced to a random elliptic system, and the homogenised coefficients for the complex medium may be computed using proper averaging procedures over the solutions of this elliptic system. This leads us to a generalisation of the homogenisation theory for deterministic complex media (see Chapter 9).

The structure of the chapter is as follows. In Section 14.2 we present an introduction to the necessary notions from the theory of ergodicity that will be used in our treatment of homogenisation. In Section 14.3 we present a model for a random complex medium, on which we will base our analysis. In Section 14.4 we present a formal two-scale approach that allows us to set ideas and understand the basic mechanisms that will lead us to a homogenised system, as well as to identify the coefficients of the homogenised system. In Section 14.5 we present some rigorous results on the homogenisation of random complex electromagnetic media.

14.2 ERGODIC MEDIA

We will consider a special class of random media, namely, *ergodic media*, for which there has been a resurgence of interest in the context of homogenisation of first- and second-order PDEs[1]. The ergodic assumption allows for the generation of a "self-repeating" structure, in the statistical sense, for the medium which is a generalisation in the stochastic framework of the periodicity condition for deterministic media (see Section 9.6.2).

We will consider the following framework. Let $(\Omega, \mathscr{F}, P)$ be a probability space, and let $\mathfrak{G}$ be a group of transformations on Ω. Such a group is often called a *dynamical system* on Ω. We say that the probability measure P is preserved under the action τ of the group $\mathfrak{G}$, if $P(\tau A) = P(A)$ for every $A \in \mathscr{F}$.

As mentioned already in Section 2.4.13, a typical example of this setup is the case where $\Omega = \mathbb{R}^3$, i.e., each ω is identified with a point $x \in \mathbb{R}^3$. Then the group $\mathfrak{G}$ is the translation group on $\mathbb{R}^3$ with the action τ denoted by $\tau_y = x + y$, when the point $x = \omega \in \Omega$ is translated by $y \in \mathbb{R}^3$. The invariance property of the measure P under the action of the group in this case takes the form,

$$P(\tau_y A) = P(A), \quad \forall A \in \mathscr{F}, \ \forall y \in \mathbb{R}^3 .$$

The Lebesgue measure is an example of such an invariant measure.

The probability space $(\Omega, \mathscr{F}, P)$ is to be interpreted as follows: each realisation ω is to be understood as a particular configuration of the medium. In other words, each experiment we perform on a particular medium corresponds to a particular choice of $\omega \in \Omega$. However, it is neither known beforehand nor known with certainty which medium is to be realised at the time the experiment is performed. The probability that a particular medium is realised is given by the probability measure P. A random variable $F : \Omega \to X$, where X is an appropriate metric space, will serve as a mathematical model for a medium. For instance, when $X \in \mathbb{R}^{6\times 6}$, we may consider F as a particular outcome of the electromagnetic parameters of a random complex medium (e.g., a particular outcome of $\mathsf{A}_{\mathrm{or},\omega}$ or G_ω). More precisely, if F is a measurable function on Ω, we will call for each fixed $\omega \in \Omega$, $F(\tau_x\omega)$ (often denoted by $\check{F}(\tau_x\omega)$) a *realisation* of F. A measurable function F is called *invariant under the group action* (dynamical system) if $F(\tau_x\omega) = F(\omega)$ for every x and ω.

We shall assume the *ergodicity* and *stationarity* of the random coefficients.

Definition 14.2.1 (ergodicity and stationarity)

(*i*) *The action τ is called ergodic if for all $A \in \mathscr{F}$,*

$$\tau_x A = A \ \forall x \in \mathbb{R}^3 \implies P(A) = 0, \ \text{or } P(A) = 1.$$

[1]This is a very general setting that allows us to bypass (by resorting to the ergodic theorem and stationarity properties) certain technical difficulties arising in the study of spaces of random fields such as the lack of the standard compact Sobolev embedding theorems for $L^2(\Omega, P, \mu)$, where (Ω, P, μ) is a general probability space. Such embeddings are important, among others, for the well posedness of the cell equations (see, e.g., [279]).

(*ii*) *A random variable* F *is called stationary if*

$$\forall y \in \mathbb{R}^3, \;\; F(x+y,\omega) = F(x, \tau_y \omega), \;\; a.e. \; in \; x, \; a.s.$$

REMARK 14.2.2 An alternative definition of ergodicity is to say that an action is ergodic if every invariant function[2] under this action is the constant function almost surely in Ω (P-a.s.).

These properties guarantee that, in a statistical sense, parts of the material located at different positions will present the same properties, i.e., that the statistical properties of the medium are invariant under translations that are to be understood as the transformation τ_y. This fact allows us to look at average properties of the material at long scales and obtain nice expressions for these quantities. In fact, by the ergodic theorem (see, e.g., [66], [67] and Theorem C.4.2 in Appendix C) we may obtain that

$$F\left(\frac{x}{\epsilon}, \omega\right) \overset{*}{\rightharpoonup} \mathbb{E}[F], \;\; \text{as } \epsilon \to 0 \text{ a.s.} \quad \text{in } L^\infty(\mathbb{R}^3),$$

where $\mathbb{E}$ is the expectation over the measure P. The ergodic hypothesis implies that instead of looking at an ensemble average of media and averaging the properties of the medium on the ensemble average, we may consider a single realisation of the medium whose spatial dimensions are large and sample its properties by traversing this single realisation for large enough distances.

Below we provide some examples where this approach is relevant.

EXAMPLE 14.2.3 An example of a stationary random variable is a random field $u : \mathbb{R}^3 \times \Omega \to X$, where X is a metric space[3] that may be represented as $u(x,\omega) = F(\tau_x \omega)$ for a fixed random variable F. Indeed, by the group property, $u(x+y,\omega) = F(\tau_{x+y}\omega) = F(\tau_x \tau_y \omega) = u(x, \tau_y \omega)$. By the invariance of P under τ, the above equality means that for *any* $x, y \in \mathbb{R}^3$, the distribution of $u(x+y,\omega)$ does not depend on y.

EXAMPLE 14.2.4 [344] Periodic media naturally fall within this framework. It is easily seen that periodicity is a special case of the stationarity condition. Take Ω to be the unit torus $\mathbb{T}$ in $\mathbb{R}^d$ and each realisation ω to be a point in the unit torus. While in this case the medium is deterministic, the meaning of ω is that we may consider any point in the unit torus (unit cell of the medium) as the origin of the coordinate system, chosen randomly according to the Lebesgue measure on the unit torus. Let $\bar{a}(\omega)$ be any periodic function and define $a(x,\omega) = \bar{a}(\omega - x)$. Clearly, $a(x,\omega)$ satisfies the stationarity condition according to which for every $x_1, \ldots, x_n \in \mathbb{T}$, it holds that

$$(a(x_1,\omega), \cdots, a(x_n,\omega)) \overset{d}{=} (a(x_1+h,\omega), \cdots, a(x_n+h,\omega))$$

---

[2]As this would imply that the random variable F is measurable with respect to the null σ-algebra $\mathscr{F}_0 = \mathscr{O} = \{\emptyset, \Omega\}$, which implies that F is a constant function.

[3]That is, for each $x \in \mathbb{R}^3$, $\omega \in \Omega$ we get a random realisation $u(x,\omega) \in X$, e.g., $X \in \mathbb{R}^6$ is we think of the electromagnetic field, or $X \in \mathbb{R}^{6\times 6}$ if we think of the medium parameters.

for every vector $h \in \mathbb{R}^d$, where $\stackrel{d}{=}$ denotes equality in distribution of the relevant random variables. In this context, the transformation τ_x acts on ω by $\tau_x\omega = \omega + x \; mod(1) = \omega + x - [x]$, i.e., as a shift. The transformation is invertible and $(\tau_x)^{-1} = \tau_{-x}$. It obviously holds that $a(x+y,\omega) = a(x,\tau_y\omega)$.

EXAMPLE 14.2.5 [344] Consider a medium that may exhibit imperfections, with centres that are randomly distributed. To simplify matters, assume that the imperfections may be modelled as points. In a particular realisation of the medium ω, there will be n imperfections centred at the points $(r_1, \dots, r_n)$ where both the positions of points as well as their number n are random variables. So $\omega = (r_1, \dots, r_n)$, and Ω is the space of sequences in $\mathbb{R}^d$. We assume that if A is a Borel set in $\mathscr{B}(\mathbb{R}^d)$, then the number of imperfections in A, $\nu(A)$ is distributed according to the Poisson distribution, i.e., by

$$P(\nu(A) = n) = \exp(-\lambda \mid A \mid) \frac{(\lambda \mid A \mid)^n}{n!},$$

where $\mid A \mid$ is the Lebesgue measure of A and λ is a parameter. The exact medium configuration can be described by the random measure

$$\mu = \sum_{j=1}^{n} \delta_{r_j},$$

where δ_x is the Dirac point measure centred at x. Assuming that the positions of the imperfections are uniformly distributed, one may calculate the Laplace functional of the random measures as

$$\Phi := \mathbb{E}[\exp(-(\phi,\mu))] = \mathbb{E}[\exp(-\sum_{j=1}^{n} \phi(r_j))] = \exp\left(\lambda \int (e^{-\phi(r)} - 1)\, dr\right),$$

where ϕ is any smooth function of compact support. The Laplace functional has the stationarity property which is inherited by the stationarity property of the Poisson point process. Indeed, let h be any vector in $\mathbb{R}^d$ and define $\mu_h(dx) = \mu(dx+h)$. Then,

$$\mathbb{E}[\exp(-(\phi,\mu_h))] = \mathbb{E}[\exp(-(\phi,\mu))],$$

which shows the stationarity property for the Laplace functional. Any function $a(x,\omega)$ can be considered as a special case of a Laplace functional for the proper choice of ϕ. Hence, such a medium may be modelled by the framework described in this section. In this context the transformation τ_x acts on ω by

$$\tau_x\omega = \tau_x(r_1, \dots, r_n) = \omega + x = (r_1 + x, \dots, r_n + x),$$

i.e., as a shift. The transformation is invertible and $(\tau_x)^{-1} = \tau_{-x}$. The ergodicity of this medium relies on the ergodic properties of the Poisson process (see, e.g., [408] or [344]).

For more examples of media that fall within this description, we refer to [408].

14.3 FORMULATION

Let $\mathcal{O}$ be a domain in $\mathbb{R}^3$, filled with a random complex linear electromagnetic medium. For the time being, we consider the case where the randomness is assumed to be spatial only, so that the structure of the medium does not change in time. The constitutive relations for the medium are given as in Section 2.3.4, equation (2.23). The effect of the random structure of the medium is that now the electromagnetic fields are random fields whose evolution is given by the random Maxwell equations

$$\partial_t(\mathsf{A}_{\mathrm{or},\omega}u + \mathsf{G}_\omega \star u) = \mathsf{M}u + j, \text{ in } (0,T] \times \mathcal{O}, \tag{14.1}$$

subject to the perfect conductor boundary condition

$$n \times u_1 = 0, \text{ in } [0,T] \times \partial\mathcal{O}\,,$$

and for homogeneous initial conditions $u(x,0) = 0$, $x \in \mathcal{O}$. The coefficients of the medium are now random variables, with a spatial dependence[4]; the same can also hold for the source term j. The fields $u = u(t,x;\omega)$ are random fields (vector space-valued random variables). All these random variables are assumed to be defined on a suitable probability space $(\Omega, \mathscr{F}, P)$ (which we do not need to specify at this point) related to the nature of the random structure of the medium. The differential equation (14.1) is now an equation between random variables and is assumed to hold almost surely in P.

REMARK 14.3.1 Equation (14.1) is a *random differential equation*. In this model we assume that the source term j, even though it may be a random process, is of bounded variation (with respect to time), so that equation (14.1) may be considered a differential equation pointwise in ω. This is in contrast to the situation in, e.g., Chapter 12, where the randomness in the term j was considered as arising from the summation of a large number of independent error terms, thus leading - through central limit theorem type of arguments - to a model for j as a process of infinite variation, so that equation (14.1) would require a special treatment in terms of the theory of the Itō integral and stochastic differential equations. To mark the qualitative and technical differences between these two types of differential equations, we will retain the terminology *random* for the first and *stochastic* for the second. See also Remark 2.4.6.

The medium coefficients must be of such form as to allow us to model small-scale (fast-varying) random microstructure. From now on we assume medium coefficients of a specific form:

ASSUMPTION 14.3.2

(*i*) Consider a probability space $(\Omega, \mathscr{F}, P)$ and let $\Phi(\cdot,\omega) : \mathbb{R}^3 \to \mathbb{R}^3$ be a random diffeomorphism P-a.s. with the property[5] that $\mathrm{grad}\Phi$ is

---

[4]We use the subscript ω to emphasise the randomness.

[5]We also make the technical assumptions $\operatorname{ess\,inf}_{\omega\in\Omega,\ x\in\mathbb{R}^3}[\det(\mathrm{grad}\Phi(x,\omega)] > 0$, $\operatorname{ess\,sup}_{\omega\in\Omega,\ x\in\mathbb{R}^3} \mathrm{grad}|\Phi(x,\omega)| < \infty$.

stationary under an ergodic group action, i.e.,

$$\forall y \in \mathbb{R}^3, \quad \operatorname{grad}\Phi(x+y,\omega) = \operatorname{grad}\Phi(x,\tau_y,\omega).$$

(*ii*) The random medium can be modelled with coefficients of the form

$$\begin{aligned} \mathsf{A}_{\text{or},\omega} &= \mathsf{A}^{\epsilon}_{\text{or}}(x,\omega) = \mathsf{A}_{\text{or}}\left(\Phi^{-1}\left(\frac{x}{\epsilon},\omega\right)\right), \\ \mathsf{G}_{\omega} &= \mathsf{G}^{\epsilon}_{\text{d}}(t,x,\omega) = \mathsf{G}_{\text{d}}\left(t,\Phi^{-1}\left(\frac{x}{\epsilon},\omega\right)\right), \end{aligned} \tag{14.2}$$

where $\mathsf{A}_{\text{or}}(y)$ and $\mathsf{G}_{\text{d}}(y,t)$ are deterministic matrix-valued functions periodic in $y \in \mathbb{R}^3$ with common period Y, where Φ is P-a.s random diffeomorphism with stationary gradients under an ergodic action.

Remark 14.3.3 This assumption on the coefficients of the medium is inspired by recent very interesting work by X. Blanc, C. Le Bris and P.-L. Lions [67] on stochastic elliptic homogenisation, and models some kind of statistical periodicity of the medium and guarantees ergodicity.

To be able to model the small-scale periodic microstructure, we must let ϵ vary over a range of arbitrarily small values. We are therefore led to a sequence of random boundary value problems,

$$(\mathsf{A}^{\epsilon}_{\text{or}} u^{\epsilon} + \mathsf{G}^{\epsilon}_{\text{d}} \star u^{\epsilon})' = \mathsf{M} u^{\epsilon} + j, \tag{14.3}$$

with initial condition $u^{\epsilon} = 0$, and the perfect conductor boundary condition

$$n \times u^{\epsilon}_1 = 0, \quad \text{on } \mathcal{O}. \tag{14.4}$$

The explicit t, x and ω dependence is omitted for simplicity.

If the solution of the above sequence of random boundary value problems exists for all $\epsilon > 0$ (a hypothesis that will be verified by Theorem 14.3.4), then this will generate a sequence of random fields $\{u^{\epsilon}\} = \{u^{\epsilon}(t,x,\omega)\}$. To understand the effects of small-scale random microstructure, we must go to the limit as $\epsilon \to 0$. Then questions similar to those we addressed in Chapter 9 arise, i.e., whether the sequence of random fields $\{u^{\epsilon}\}$ converges in some weak sense to a limit random field u^*, and whether this random field is the solution of a differential equation similar in type to the original Maxwell system

$$(\mathsf{A}^{h}_{\text{or}} u^* + \mathsf{G}^{h}_{\text{d}} \star u^*)' = \mathsf{M} u^* + j\,,$$

but now with constant coefficients $\mathsf{A}^{h}_{\text{or}}$, $\mathsf{G}^{h}_{\text{d}}$. Can these coefficients be specified by the coefficients of the original medium, in a manner similar to what we have seen in Chapter 9 for deterministic media? Furthermore, can it be that under certain circumstances, the limiting field as well as the limiting differential equation are not random? The answer to these questions is complicated in the case of random media by the fact that now the quantities involved are random fields defined on a probability space and not only deterministic functions. This fact requires some special care as far as convergence properties of the sequences of random fields are involved. However, the powerful tools of ergodic theory will be of assistance, leading in the end to positive answers to most of our questions, as well as elegant formulae for the calculation of the homogenised coefficients.

We end this section by providing a theorem showing the well posedness of the random Maxwell system.

THEOREM 14.3.4 *The Maxwell system (14.1) is uniquely solvable for all* $\epsilon > 0$ *and* $\omega \in \Omega$ *and the solution satisfies*

$$\|u^\epsilon(t)\|_{\mathbb{X}} \leq C, \ \ \textit{for all } \epsilon, \ t > 0, \ P - a.s.$$

and

$$\|u^\epsilon(t)\|_{L^2(\Omega, \mathscr{F}, P; \mathbb{X})} \leq C, \ \ \textit{for all } \epsilon, \ t > 0.$$

Proof. The existence of a solution P-a.s. may be proved for any fixed $\epsilon > 0$ by the Faedo-Galerkin method, with minor modifications of Theorem 7.5.2 in Section 7.5.2, (see also Proposition 1 in [70]), and thus the details are omitted. We show only that the solution obtained pointwise in ω is square integrable with respect to the probability measure. To this end, fix any $\epsilon > 0$ and consider the finite-dimensional approximation of the solution $\{u_m^\epsilon\}$, $u_m^\epsilon = P_m u^\epsilon$, where P_m is the projection operator on the subspace spanned by the first m vectors of the chosen basis. The sequence $\{u_{m^\epsilon}\}_{m\in\mathbb{N}}$ is now a sequence of random fields. This sequence for all m satisfies a finite-dimensional equation of the form

$$(\mathsf{A}^\epsilon_{\text{or}} u^\epsilon_m + \mathsf{G}^\epsilon_{\text{d}} \star u^\epsilon_m)' = \mathsf{M} u^\epsilon_m + j_m, \tag{14.5}$$

where $\mathsf{A}^\epsilon_{\text{or}}, \mathsf{G}^\epsilon_{\text{d}}, \mathsf{M}, j_m$ are appropriate finite-dimensional projections of the relevant quantities (see the proof of Theorem 7.5.2 for their definitions; here for simplicity they are denoted the same as their infinite-dimensional counterparts).

As an intermediate step using the Faedo-Galerkin method, we find the following a priori estimates for the finite-dimensional approximations,

$$\| \, u^\epsilon_m \, \|_{L^\infty([0,\,T];\,\mathbb{X})} \leq c \, (\| \, j \, \|_{L^1([0,\,T];\,\mathbb{X})} + \| \, u^\epsilon(0) \, \|_{\mathbb{X}_\mathsf{M}}) \tag{14.6}$$

and

$$\| \, (u^\epsilon_m)'(t) \, \|_{\mathbb{X}} \leq c \, (\| \, j \, \|_{W^{1,1}([0,\,T];\,\mathbb{X})} + \| \, u^\epsilon(0) \, \|_{\mathbb{X}_\mathsf{M}}) \,, \tag{14.7}$$

which hold P-a.s. We start from the a priori estimate (14.6), which holds P-a.s. for any m; we square this inequality and take the expectation with respect to the probability measure P. As long as

$$\mathbb{E}[\| \, j \, \|^2_{W^{1,1}([0,\,T];\,\mathbb{X})}] < \infty, \ \ \mathbb{E}[\| \, u^\epsilon(0) \, \|^2_{\mathbb{X}_\mathsf{M}}] < \infty,$$

we obtain that the sequences u^ϵ_m and $(u^\epsilon_m)'$ are uniformly bounded in m, for all $t \in [0, T]$ in the norms of $L^2(\Omega, \mathscr{F}, P; \mathbb{X}_\mathsf{M})$ and $L^2(\Omega, \mathscr{F}, P; \mathbb{X})$, respectively. Therefore, u^ϵ_m, $(u^\epsilon_m)'$ converge to u^ϵ, $(u^\epsilon)'$, respectively, weakly in $L^2(\Omega, \mathscr{F}, P, \mathbb{X}_\mathsf{M})$ and $L^2(\Omega, \mathscr{F}, P; \mathbb{X})$, respectively. However, one may show that these sequences converge strongly in $L^2(\Omega, \mathscr{F}, P; \mathbb{X}_\mathsf{M})$ and $L^2(\Omega, \mathscr{F}, P, \mathbb{X})$, respectively. Consider (14.5) for n and m and subtract. The difference

$u^{\epsilon,\diamondsuit}(t) := u_n^\epsilon(t) - u_m^\epsilon(t)$ solves an equation of the same form as (14.5) with $P_n j - P_m j$ on the right-hand side and with $u^{\epsilon,\diamondsuit}(0) = u_n(0) - u_m(0)$ as the initial condition. By working in exactly the same manner as above we obtain the a.s. estimates

$$\begin{aligned}&\| \, u_n^\epsilon - u_m^\epsilon \, \|_{L^\infty([0,\,T];\,\mathbb{X})} \leq \\ &c\,(\| \, P_n j - P_m j \, \|_{L^1([0,\,T];\,\mathbb{X})} + \| \, u_n^\epsilon(0) - u_m^\epsilon(0) \, \|_{\mathbb{X}_{\mathsf{M}}})\,,\end{aligned} \tag{14.8}$$

and similar estimates for the temporal derivatives. Taking expectations with respect to the measure P,

$$\begin{aligned}&\mathbb{E}[\, \| \, u_n^\epsilon - u_m^\epsilon \, \|_{L^\infty([0,\,T];\,\mathbb{X})} \,] \leq \\ &c\,(\mathbb{E}[\, \| \, P_n j - P_m j \, \|_{L^1([0,\,T];\,\mathbb{X})} \,] + \mathbb{E}[\, \| \, u_n^\epsilon(0) - u_m^\epsilon(0) \, \|_{\mathbb{X}_{\mathsf{M}}} \,])\,.\end{aligned} \tag{14.9}$$

If we assume that $j \in L^2(\Omega, \mathscr{F}, P, L^1([0,T]; \mathbb{X}))$ and $u(0) \in L^2(\Omega, \mathscr{F}, P; \mathbb{X}_{\mathsf{M}})$, then the right-hand side of the inequalities tends to 0 as $n, m \to \infty$. Therefore, we obtain that u_m^ϵ, $(u_m^\epsilon)'$ are Cauchy sequences in $L^2(\Omega, \mathscr{F}, P; \mathbb{X}_{\mathsf{M}})$ and $L^2(\Omega, \mathscr{F}, P; \mathbb{X})$, respectively; therefore they converge strongly, to u^ϵ and $(u^\epsilon)'$, respectively. We may now take the limit as $m \to \infty$ in (14.6) and reach the stated result.

Finally, working similarly as for the a.s. solution, we obtain that the limit is the solution of

$$\mathbb{E}[\langle \mathsf{A}_{\text{or}} u^\epsilon + \mathsf{G}_{\text{d}} u^\epsilon, \, v \rangle'] = \mathbb{E}[\langle \mathsf{M} u^\epsilon, v \rangle] + \mathbb{E}[\langle j, v \rangle]\,,$$

for all $v \in L^2(\Omega, \mathscr{F}, P; \mathbb{X}_{\mathsf{M}})$. This concludes the proof. □

14.4 A FORMAL TWO-SCALE EXPANSION

To provide some insight concerning the structure of the homogenised system for equation (14.3) that will facilitate the mathematically rigorous treatment that follows, we provide some formal arguments using a two-scale expansion. As most of this formal expansion proceeds in parallel with the periodic case presented in Section 9.3, we shall be very brief, simply focusing on the differences arising in the random case. To save space, we will use six-vector notation.

In view of the comments in Remark 14.3.1, we may take the Laplace tranform of (14.3) with respect to time and use its properties to reduce this equation to a random partial differential equation in terms of the spatial variable only of the form

$$p(\mathsf{A}_{\text{or}} \widehat{u}^\epsilon + \widehat{\mathsf{G}_{\text{d}}} \widehat{u}^\epsilon) = \mathsf{M} \widehat{u}^\epsilon + \widehat{j} \tag{14.10}$$

where $\mathsf{A}_{\text{or}} = \mathsf{A}_{\text{or}}\left(\Phi^{-1}\left(\frac{x}{\epsilon}, \omega\right)\right)$ and $\widehat{\mathsf{G}_{\text{d}}} = \widehat{\mathsf{G}_{\text{d}}}\left(\Phi^{-1}\left(\frac{x}{\epsilon}, \omega; p\right)\right)$. In order to simplify the exposition we will use the notation $\mathfrak{s} + \widehat{\mathfrak{s}}_d =: \widehat{\mathfrak{s}}$, where $\mathfrak{s}$ is a proxy for ε, ξ, ζ, μ. As in the periodic case, we assume that the fields may be expanded in power series, in terms of $\epsilon \ll 1$, as $\widehat{u}^\epsilon(x,\omega) = \sum_{j=0}^\infty \epsilon^j \widehat{u}^{(j)}(x,\omega)$ where $\widehat{u}^{(j)}(x,\omega) = (\widehat{E}^{(j)}(x,\omega), \widehat{H}^{(j)}(x,\omega))^{tr}$ and we have dropped the explicit p dependence for simplicity of notation. In view of the special choice for the structure of the random coefficients (14.2), these functions assume the special form $\widehat{u}^{(j)}(x,\omega) = \widehat{u}^{(j)}\left(x, \Phi^{-1}(y,\omega)\right)$, where $y = \frac{x}{\epsilon}$ is considered an independent variable from x (in the spirit of the two-scale expansion). Owing to the stationarity of the gradients of the random diffeomorphism Φ, the functions $\mathrm{grad}_y \widehat{u}^{(j)}$ will be stationary in y.

Proceeding with the two-scale expansion in parallel with the periodic case (see Section 9.3), we observe that to order $O(\epsilon^{-1})$, we have that $\widehat{u}^{(0)}(x,y) = u^{\bowtie}(x) + \mathrm{grad}_y \Psi(x, \Phi^{-1}(y,\omega))$, where $\Psi = (\Psi_1, \Psi_2)^{tr}$ for Ψ_i, $i = 1, 2$, scalar functions, and $\mathrm{grad}_y \Psi = (\mathrm{grad}_y \Psi_1, \mathrm{grad}_y \Psi_2)^{tr}$. The function $\Psi(x,z)$ is deterministic and periodic with respect to the second variable with periodicity Y. The introduction of randomness requires a generalisation of the concept of averaging. This comes from the fact that $\Psi(x,z)$ is deterministic and Y-periodic in the variable z, but it is calculated at $\Phi^{-1}(y,\omega)$, which is a random variable. Then

$$\mathbb{E}\left[\int_{\Phi(Y,\omega)} \mathrm{grad}_y \Psi(x, \Phi^{-1}(y,\omega))\, dy\right] = 0\,,$$

and this shows that the averaging operation should be modified to

$$\langle \mathfrak{f} \rangle := \frac{1}{\mathbb{E}[|\Phi(Y,\omega)|]} \mathbb{E}\left[\int_{\Phi(Y,\omega)} \mathfrak{f}(x, \Phi^{-1}(y,\omega))\, dy\right]\,, \tag{14.11}$$

where $|\Phi(Y,\omega)|$ is the Lebesgue measure of the image of the cell Y under the (random) mapping $\Phi(\cdot,\omega)$. It is clear that $|\Phi(Y,\omega)|$ is a real-valued random variable, so that the averaging operator requires normalisation over the average volume of the image of the periodicity cell Y. The effect of applying this averaging operator on $\widehat{u}^{(0)}$ is $\langle \widehat{u}^{(0)} \rangle = u^{\bowtie}(x)$.

We now proceed to the next order $O(\epsilon^0)$, substitute the expression obtained above for $\widehat{u}^{(0)}$ and take the divergence with respect to y to obtain (in complete analogy with the periodic case) that Ψ must satisfy the random elliptic system

$$\mathrm{div}_y(\mathsf{A}_{\mathrm{or}} \mathrm{grad}_y \Psi) = \mathrm{div}_y(\mathsf{A}_{\mathrm{or}} u^{\bowtie})\,,$$

where all the terms are to be understood exactly as in (9.12) (see p. 186) with the difference that now Ψ is a random field. The analogue of periodicity is now to be understood in a random sense as above. By the same arguments as in Section 9.3, this elliptic system has a solution in the form $\Psi_i = R^{(i)} \cdot u_1^{\bowtie} + V^{(i)} u_2^{\bowtie}$, $i = 1, 2$ where $R^{(i)}, V^{(i)}$ are three-vectors ($R^{(1)} = \Lambda^{(1)}$, $R^{(2)} = \Lambda^{(3)}$, $V^{(1)} = \Lambda^{(2)}$, $V^{(2)} = \Lambda^{(4)}$ in the notation of (9.13) p. 186). Substitution of this ansatz into the elliptic system yields that the three-vectors $R^{(i)} = (R_1^{(i)}, R_2^{(i)}, R_3^{(i)})$, $V^{(i)} = (R_1^{(i)}, R_2^{(i)}, R_3^{(i)})$ should solve the elliptic systems

$$\begin{aligned} 0 &= \mathrm{div}_y(\mathsf{A}_{\mathrm{or}}\, \mathrm{grad}_y R_k) + \mathrm{div}_y(\mathsf{A}_{\mathrm{or}} e_k), \quad k = 1, 2, 3\,, \\ 0 &= \mathrm{div}_y(\mathsf{A}_{\mathrm{or}}\, \mathrm{grad}_y V_{k-3}) + \mathrm{div}_y(\mathsf{A}_{\mathrm{or}} e_k), \quad k = 4, 5, 6\,, \end{aligned} \tag{14.12}$$

where $R_k = (R_k^{(1)}, R_k^{(2)})^{tr}$ and $V_k = (V_k^{(1)}, V_k^{(2)})^{tr}$, $k = 1, 2, 3$ and e_k is the canonical basis of $\mathbb{R}^6$. These equations are of the exact form as (9.14) and

(9.15) (see Section 9.3 p. 186) but with random coefficients, and expressed now in more compact notation to save space.

We next apply the averaging operator on the equation obtained by the $O(\epsilon^0)$ expansion, and we see that the average fields $u^{\bowtie}$ satisfy an equation of the form

$$p\widehat{\mathsf{A}_{\mathrm{or}}}^h u^{\bowtie} = \mathsf{M}u^{\bowtie},$$

with $\widehat{\mathsf{A}_{\mathrm{or}}}^h$ given by averaging $\widehat{\mathsf{A}_{\mathrm{or}}}$ over the solutions of the cell equations (14.12) in terms of the averaging operator (14.11). The expressions are the same as in (9.16) in Section 9.3 (p. 187) but of course with the new interpretation of the solutions of the cell equations and the operator $\langle \cdot \rangle$.

14.5 HOMOGENISATION OF THE MAXWELL SYSTEM

We now turn to the mathematically rigorous study of the Maxwell equations for random complex media exhibiting ergodic properties. The treatment of the problem proceeds similarly as for the periodic problem treated in Sections 9.4 and 9.5, so that we will focus only on the main differences that arise in the random case.

14.5.1 An auxiliary random elliptic problem

As motivated by the formal two-scale expansion of Section 14.4 (see also Section 9.4.1 for the periodic case), the following random elliptic system is closely related to the Laplace transformed random Maxwell problem. Consider a 6×6 random matrix $\mathsf{A}_{e\ell}$ expressed in block form as

$$\mathsf{A}_{e\ell} = \begin{pmatrix} a & b \\ c & d \end{pmatrix}, \tag{14.13}$$

where a, b, c, d are random matrices of the form assumed in Assumption 14.3.2, and in particular in (14.2).

DEFINITION 14.5.1 *For a random matrix* $\mathsf{A}_{e\ell}\,(x,\omega)$ *as in equation* (14.13), *consider the random elliptic operator* $L^\epsilon : H_0^1(\mathcal{O}) \times H_0^1(\mathcal{O}) \to H^{-1}(\mathcal{O}) \times H^{-1}(\mathcal{O})$, *defined as*

$$L^\epsilon = \mathrm{div}_x(\mathsf{A}_{e\ell}\,(x,\omega)\,\mathrm{grad}_x \cdot).$$

REMARK 14.5.2 The operators defined above are random matrix elliptic operators acting on two-vectors $w = (w_1, w_2)^{tr}$, as follows:

$$L^\epsilon \begin{pmatrix} w_1 \\ w_2 \end{pmatrix} = \begin{pmatrix} -\mathrm{div}_x(a\,\mathrm{grad}_x w_1) - \mathrm{div}_x(b\,\mathrm{grad}_x w_2) \\ -\mathrm{div}_x(c\,\mathrm{grad}_x w_1) - \mathrm{div}_y(d\,\mathrm{grad}_y w_2) \end{pmatrix}$$

and $\mathfrak{s} = \mathfrak{s}\left(\Phi^{-1}\left(\frac{x}{\epsilon},\omega\right)\right)$, where $\mathfrak{s}$ is a proxy for a, b, c and d.

REMARK 14.5.3 Note that by Assumption 14.3.2, it is $\mathfrak{s}\left(\Phi^{-1}\left(\frac{x}{\epsilon},\omega\right)\right)$ that is random, i.e., the randomness is inserted into the model by taking the composition of $\mathfrak{s}$ with the random diffeomorphism $\Phi^{-1}(y,\omega)$. The matrix $\mathfrak{s}$ itself is not random; if calculated on a deterministic argument y, then $\mathfrak{s}(y)$ is a deterministic matrix-valued function of period Y.

ASSUMPTION 14.5.4 The matrix $\mathsf{A}_{e\ell} \in L^\infty(\mathcal{O},\mathbb{R}^{6\times 6})$ satisfies the following conditions:

(*i*) There exists a positive constant c_1 such that $|\mathsf{A}_{e\ell}\,(z)\,\mathsf{y}\cdot\mathsf{y}| \geq c_1\,|\mathsf{y}|$ for almost all $z \in \mathcal{O}$ and all (deterministic) $\mathsf{y} \in \mathbb{R}^6$.

(*ii*) There exists a positive constant c_2 such that $|\mathsf{A}_{e\ell}^{-1}(z)\,\mathsf{y}\cdot\mathsf{y}| \geq c_2\,|\mathsf{y}|$ for almost all $z \in \mathcal{O}$ and all $\mathsf{y} \in \mathbb{R}^6$.

REMARK 14.5.5 When choosing $z = \Phi^{-1}\left(\frac{x}{\epsilon},\omega\right)$, the above assumption holds P-a.s. for the family of random matrices $\{\mathsf{A}_{e\ell}^{\epsilon}\}$.

We now define the random averaging operator.

DEFINITION 14.5.6 (RANDOM AVERAGING OPERATOR) *Let $\mathfrak{s}$ be a random field of the form $\mathfrak{s}(x,y,\omega) = \mathfrak{s}\left(x,\Phi^{-1}(y,\omega)\right)$. The random averaging operator is defined as*

$$\langle\mathfrak{s}\rangle = \left(\mathbb{E}[\int_Y \det(\mathrm{grad}\Phi(y,\omega))\,dy]\right)^{-1}\mathbb{E}[\int_{\Phi(Y)}\mathfrak{s}(x,\Phi^{-1}(y,\omega))dy]\,.$$

REMARK 14.5.7 This is clearly the random generalisation of the periodic averaging operator used in Chapter 9. In fact if the medium is deterministic and periodic then $\langle\mathfrak{s}\rangle = \frac{1}{|Y|}\int_Y \mathfrak{s}(x,y)dy$. Furthermore, in the random case, this operator is essentially an averaging operator since by the ergodic theorem $\langle\mathfrak{s}\rangle = \lim_{L\to\infty}\int_{-L}^{L}\mathfrak{s}(x,y)dy$. This justifies the term *self-averaging* environment often used in homogenisation theory.

REMARK 14.5.8 By standard calculus arguments it follows that,

$$\mathbb{E}[\int_Y \det(\mathrm{grad}\Phi(y,\omega))\,dy] = \mathbb{E}[|\Phi(Y,\omega)|],$$

so that the random averaging operator coincides with the averaging operator employed in the formal approach in Section 14.4. Furthermore, using the ergodic theorem in this context, it can be shown (see Remark 1.9 in [67]) that

$$\mathbb{E}[\int_Y \det(\mathrm{grad}\Phi(y,\omega))\,dy] = \det\left(\mathbb{E}[\int_Y (\mathrm{grad}\Phi(y,\omega))\,dy]\right).$$

Consider now the following random cell elliptic problems and their solutions:

Definition 14.5.9 (Random cell systems) *For $j = 1, 2, 3$, $\ell = 1, 2$, the random cell systems are the random elliptic systems*

$$L_c \begin{pmatrix} r_1^{(j)} \\ r_2^{(j)} \end{pmatrix} = \begin{pmatrix} \mathrm{div}_y a_{\sharp,j} \\ \mathrm{div}_y c_{\sharp,j} \end{pmatrix}, \quad L_c \begin{pmatrix} v_1^{(j)} \\ v_2^{(j)} \end{pmatrix} = \begin{pmatrix} \mathrm{div}_y b_{\sharp,j} \\ \mathrm{div}_y d_{\sharp,j} \end{pmatrix}, \tag{14.14}$$

where L_c is the random matrix operator $L_c = -\mathrm{div}_y \mathsf{A}_{e\ell} \left(\Phi^{-1}(y, \omega) \; \mathrm{grad}_y \cdot \right)$ and $\mathfrak{s}(y) = \mathfrak{s}(\Phi^{-1}(y, \omega))$, where $\mathfrak{s}$ is a proxy for the matrices a, b, c, d. These equations are supplemented with the conditions

$$\mathrm{grad}_y \mathfrak{s} = \mathrm{grad}_y \check{\mathfrak{s}}(\Phi^{-1}(y, \omega)), \quad \mathrm{grad}_y \check{\mathfrak{s}} \text{ is stationary}, \quad \langle \mathrm{grad}_y \mathfrak{s} \rangle = 0, \tag{14.15}$$

where $\mathfrak{s}$ is a proxy for the random fields $r^{(j)}$, $v^{(j)}$, $j = 1, 2, 3$.

Remark 14.5.10 The conditions (14.15) constitute a generalisation of the periodic boundary conditions used in Chapter 9.

This system of equations (14.14) is called the *cell system* and, complemented with the boundary conditions (14.15), has a unique solution (modulo random constants). Note that in (14.14), $y \in \mathbb{R}^3$ rather than in Y; it is $\Phi^{-1}(y, \omega)$ that belongs in Y. The solvability follows by a proper application of the Lax-Milgram lemma, generalising the approach of [66] (see also [67]) for elliptic systems[6]. The solution is in $L^2(\Omega, \mathscr{F}, P; H^1_{loc}(\mathbb{R}^3) \times H^1_{loc}(\mathbb{R}^3))$.

Consider now the 3×3 matrices a^h, b^h, c^h, d^h, defined as

$$\begin{aligned}
(a^h)_{ij} &= \langle a_{ij} + \sum_{k=1}^{3} a_{ik} \partial_{y_k} r_1^{(j)} + \sum_{k=1}^{3} b_{ik} \partial_{y_k} r_2^{(j)} \rangle, \\
(b^h)_{ij} &= \langle b_{ij} + \sum_{k=1}^{3} a_{ik} \partial_{y_k} v_1^{(j)} + \sum_{k=1}^{3} b_{ik} \partial_{y_k} v_2^{(j)} \rangle, \\
(c^h)_{ij} &= \langle c_{ij} + \sum_{k=1}^{3} c_{ik} \partial_{y_k} r_1^{(j)} + \sum_{k=1}^{3} d_{ik} \partial_{y_k} r_2^{(j)} \rangle, \\
(d^h)_{ij} &= \langle d_{ij} + \sum_{k=1}^{3} c_{ik} \partial_{y_k} v_1^{(j)} + \sum_{k=1}^{3} d_{ik} \partial_{y_k} v_2^{(j)} \rangle,
\end{aligned} \tag{14.16}$$

where $r_\ell^{(j)}$, $v_\ell^{(j)}$, $j = 1, 2, 3$, $\ell = 1, 2$, are the solutions of the cell systems (14.14) and the averaging operation is to be understood in the sense of Definition 14.5.6.

Definition 14.5.11 (Homogenised diffusion matrix) *The constant coefficient matrix*

$$\mathsf{A}^h_{e\ell} = \begin{pmatrix} a^h & b^h \\ c^h & d^h \end{pmatrix}, \tag{14.17}$$

where a^h, b^h, c^h, d^h are defined as in (14.16), is called the homogenised diffusion matrix.

[6] Making explicit use of the stationarity hypothesis.

The following random homogenisation theorem holds for the elliptic problem.

THEOREM 14.5.12 *Consider the solution u^ϵ of the random elliptic problem $L^\epsilon u^\epsilon = f$. As $\epsilon \to 0$, we have that $u^\epsilon \rightharpoonup u^h$, in $H_0^1(\mathcal{O}) \times H_0^1(\mathcal{O})$, P-a.s., where u^h is the solution of the elliptic problem $L^h u^h = f$, where the homogenised matrix $\mathsf{A}^h_{e\ell}$ is given as in (14.17). Furthermore, $\mathsf{A}^\epsilon_{e\ell}\, u^\epsilon \rightharpoonup \mathsf{A}^h_{e\ell}\, u^h$, in $L^2(\mathcal{O})$.*

Proof. The proof follows closely the treatment of [67], [66] for the case of scalar elliptic equations and is sketched only briefly here. The key result is a convergence result for stationary essentially bounded (in $\mathbb{R}^3$) and integrable (with respect to the probability measure) functions $\mathfrak{s}$, according to which

$$\mathfrak{s}\left(\Phi^{-1}\left(\frac{x}{\epsilon},\omega\right),\omega\right) \overset{*}{\rightharpoonup} \left(\mathbb{E}[\int_Y \det(\mathrm{grad}\Phi(y,\omega))\,dy]\right)^{-1} \mathbb{E}[\int_{\Phi(Y)} \mathfrak{s}(\Phi^{-1}(y,\omega))dy],$$

in $L^\infty(\mathbb{R}^3)$. P-a.s. in the limit as $\epsilon \to 0$, where the right-hand side is nothing else but $\langle \mathfrak{s} \rangle$ (see Definition 14.5.6). This result follows from the ergodic theorem (see [67]). We work with the weak form of the random elliptic system and follow closely the same steps as for the periodic case (see Theorem 9.4.8), along with the random version of the div-curl lemma to obtain the desired result. □

14.5.2 Homogenisation of the random Maxwell system

We now turn to the homogenisation of the random Maxwell system. We work with the Laplace transform of the system and make the following assumptions and definitions:

ASSUMPTION 14.5.13 *The block matrix*

$$\mathsf{A}_{or,\omega}(y,p) := \mathsf{A}_{or,\omega}(y) + \widehat{\mathsf{G}_\omega}(y,p) = \begin{pmatrix} \varepsilon + \widehat{\varepsilon}_d & \xi + \widehat{\xi}_d \\ \xi^{tr} + \widehat{\xi}_d^{tr} & \mu + \widehat{\mu}_d \end{pmatrix} =: \begin{pmatrix} \varepsilon_{\mathfrak{L}} & \xi_{\mathfrak{L}} \\ \zeta_{\mathfrak{L}} & \mu_{\mathfrak{L}} \end{pmatrix}$$

satisfies the conditions of Assumption 14.5.4.

The following random elliptic operators will be needed.

DEFINITION 14.5.14 *The auxiliary "microstructure" random elliptic operator associated with the Maxwell system is*

$$L^\epsilon_M = -\mathrm{div}_x(\mathsf{A}^{\epsilon,tr}_{or}\mathrm{grad}_x\cdot), \tag{14.18}$$

and the auxiliary "cell" random elliptic operator associated with the Maxwell system is

$$L_{c,M} = -\mathrm{div}_y((\mathsf{A}^{per}_{or})^{tr}\mathrm{grad}_y\cdot\,). \tag{14.19}$$

DEFINITION 14.5.15 *Let $\mathsf{A}^h_{or,\omega}$ be the homogenised matrix for the random elliptic system of Definition 14.5.14, obtained as in Section 14.5.1. The Laplace transform of the homogenised constitutive relation is given by*

$$\widehat{\mathsf{d}}^h = \mathsf{A}^h_{or,\omega}\widehat{u}.$$

REMARK 14.5.16 Because of the symmetry condition on A_{or} it follows that $\mathsf{A}_{\mathrm{or}}^{tr} = \mathsf{A}_{\mathrm{or}}$.

We now perform random elliptic homogenisation for the auxiliary elliptic systems in 14.5.14 as in Section 14.5.1 and obtain the constant coefficient matrix $\mathsf{A}_{\mathrm{or}}^{h}$ using the expressions in equations (14.17) and (14.16) with $\mathsf{A}_{e\ell} = \mathsf{A}_{\mathrm{or}}$. Note that by the ergodicity of the medium, the homogenised coefficients are deterministic.

THEOREM 14.5.17 *The solution* $u^{\epsilon} = (E^{\epsilon}, H^{\epsilon})^{tr}$ *of the random Maxwell system* (14.3) *satisfies*

$$u^{\epsilon} \overset{*}{\rightharpoonup} u^{*}, \quad \text{in } L^{2}(\Omega, \mathscr{F}, P; L^{\infty}([0,T], \mathbb{X})),$$

where $u^{*} = (E^{*}, H^{*})^{tr}$ *is the unique solution of the Maxwell system*

$$(\mathsf{A}_{or}^{h} u + \mathsf{G}_{d}^{h} \star u)' = \mathsf{M}u + j \tag{14.20}$$

with homogeneous initial conditions and perfect conductor boundary conditions, where the homogenised coefficients A_{or}^{h} *and* G_{d}^{h} *are given as in Definition 14.5.15.*

Proof. The proof proceeds in parallel with the one for the periodic case, so we stress only the differences needed for the random case. The existence of the limit u^{*} follows from the results of Theorem 14.3.4 and weak compactness arguments. These guarantee the existence of a subsequence $\{u^{\epsilon_k}\}$ converging to u^{*} weak star in $L^{2}(\Omega, \mathscr{F}, P; L^{\infty}([0,T], \mathbb{X}))$ and, by standard properties of Lebesgue spaces, of a further subsequence $\{u^{\epsilon_{k_\ell}}\}$ converging to u^{*} P-a.s. Similar arguments, along with the properties of A_{or}, hold for the sequence $\{\mathsf{d}^{\epsilon}\}$. We will work with these subsequences, which will be denoted by $\{u^{\epsilon}\}$, $\{\mathsf{d}^{\epsilon}\}$ for simplicity, to identify the relation between the limits d^{*} and u^{*} that will lead us to the homogenised constitutive relation.

For these subsequences, we repeat the proof of Theorem 9.4.14 but replace the auxiliary periodic elliptic systems with their random counterparts. Using the results of the random elliptic homogenisation theory stated in Section 14.5.1 we obtain exactly the same results as in Theorem 9.4.14, with the difference that the relevant quantities are now random fields and the convergence holds P-a.s. Minor modification of the arguments provides the form of the homogenised system. □

14.6 MISCELLANEA

The form of random coefficients used in this chapter is only a special form, one that models random deviations from a periodic material. This form facilitates the calculation of the homogenised coefficients, especially in the case where the deviation from periodicity is small. This can be modelled in the present framework by assuming that $\Phi = I + \theta\Phi_1$, where I is the identity transformation, $\theta \ll 1$ is a small parameter and all the randomness is included in the random diffeomorphism Φ_1. Such an assumption allows a perturbative study of the homogenisation formulae and can lead to convenient numerical methods for their calculation; it has been introduced and provided very interesting results for random elliptic equations in [66], [67].

Furthermore, techniques like the two-scale convergence or the periodic unfolding method may be extended for the study of random homogenisation (see, e.g., [419]). The techniques for random homogenisation introduced here may be also be extended to nonlinear media. Finally, the theory of correctors can be generalised for random homogenisation, see, e.g., [217].

PART 5
Appendices

Appendix A

Some Facts from Functional Analysis

For a thorough introduction to functional analysis the reader may consult, e.g., [47], [60], [79], [276], [361], [364], [451].

A.1 DUALITY

Let X be a normed space. Its dual, denoted by X', is the space $\mathcal{L}(X, \mathbb{C})$ of bounded linear functionals $f : X \to \mathbb{C}$. The value $f(v)$ of the functional $f \in X'$ at the point $v \in X$ is denoted by $\langle f, v\rangle$. By the definition of the norm in $\mathcal{L}(\mathcal{X}, \mathbb{C})$ we have

$$|\langle f, v\rangle| \leq ||f||_{X'} ||v||_X \quad \text{and} \quad ||f||_{X'} = \sup_{0 \neq v \in X} \frac{|\langle f, v\rangle|}{||v||_X}.$$

Let us note that even if X is not complete, X' is a Banach space. This section closely follows [309], where a detailed exposition appears. See also [79].

An important result for the study of duality is the celebrated *Hahn-Banach theorem*; we consider the following version.

THEOREM A.1.1 *If W is a subspace of a normed space X, then every functional in W' can be extended to a functional in X' having the same norm.*

A simple corollary is that if $0 \neq v \in X$, there exists a $f \in X'$ such that $\langle f, v\rangle = ||v||_X$ and $||f||_{X'} = 1$.

Let $X'' = (X')'$ be the *second* (or *double*) *dual* of X. Let $\mathbb{I} : X \to X''$ be defined by $\langle \mathbb{I}v, f\rangle = \langle f, v\rangle$, $v \in X, f \in X'$. It easily follows that $\mathbb{I}$ is an isometric isomorphism from X onto a subspace $\mathbb{I}(X)$ of X''. So the identification of X with $\mathbb{I}(X)$ is permitted; we write $X \subseteq X''$. X is closed in the complete space X'' if and only if X is complete. If $X = X''$ the space X is called *reflexive*. Clearly, every reflexive space is complete.

Let X_1, X_2 be normed vector spaces and $A : X_1 \to X_2$ be a linear map. Its *transpose* $A^{tr} : X_2' \to X_1'$ is the linear map defined by $\langle A^{tr}v, w\rangle = \langle v, Aw\rangle$, $\forall v \in X_2', w \in X_1$. It is well known that A^{tr} is bounded if and only if A is bounded; further, $||A^{tr}||_{\mathcal{L}(X_2', X_1')} = ||A||_{\mathcal{L}(X_1, X_2)}$. In the case that $A : H \supset D(A) \to H$ is a densely defined linear operator, where H is a Hilbert space, the term *adjoint* and the notation A^* is used instead of transpose.

In many applications, it is convenient to employ a space Y that is isomorphic to X' as follows: suppose there exists a bounded bilinear form

$\langle\!\langle\cdot,\cdot\rangle\!\rangle : Y\times X \to \mathbb{C}$. We then can define a bounded linear operator $\mathbb{I} : Y \to X'$ by $\langle \mathbb{I}f, v\rangle = \langle\!\langle f, v\rangle\!\rangle$. If $\mathbb{I}$ is an isomorphism of Banach spaces (by assuming that it has a bounded inverse), then Y is called a *realisation* of X'. The pair $\langle\!\langle\cdot,\cdot\rangle\!\rangle$ is called the *duality pairing* for $Y \times X$ and is denoted by $\langle\!\langle\cdot,\cdot\rangle\!\rangle_{Y,X}$. In this case $||\mathbb{I}f||_{X'}$ is equivalent[1] to $||f||_Y$, which is equivalent to $\sup_{0\neq v\in X} \frac{|\langle\!\langle f,v\rangle\!\rangle|}{||v||_X}$. Y is usually identified with X', and the distinction between f and $\mathbb{I}f$, as well that between $\langle\cdot,\cdot\rangle$ and $\langle\!\langle\cdot,\cdot\rangle\!\rangle$, and between $\langle\cdot,\cdot\rangle_{Y,X}$ and $\langle\!\langle\cdot,\cdot\rangle\!\rangle_{Y,X}$, is suppressed.

Now let H be a Hilbert space with inner product $(\cdot,\cdot)_H$. Given $v \in H$, the inner product defines a bounded linear functional $\mathbb{I}_1 v \in H'$ by $\langle \mathbb{I}_1 v, w\rangle = (v, w)_H$. We have the celebrated *Riesz representation theorem.*

THEOREM A.1.2 *Let H be a Hilbert space. For each $f \in H'$ there exists a unique $v \in H$ such that $\langle f, w\rangle = (v, w)_H$, $\forall w \in H$. Additionally, $||f||_{H'} = ||v||_H$.*

Let us note that H' is a Hilbert space, a fact established in view of the above representation by defining $(f, g)_{H'} = (\mathbb{I}_1^{-1}g, \mathbb{I}_1^{-1}f)_H$ and then obtaining a conjugate-linear isometry $\mathbb{I}_2 : H' \to H''$ given by $\langle \mathbb{I}_2 f, g\rangle = (f, g)_{H'}$. It can then be seen that $\mathbb{I}_2 \circ \mathbb{I}_1 : H \to H''$ coincides with the natural embedding of H into H'', so that every Hilbert space is reflexive. A linear isometry $\mathbb{I} : H \to H'$ can be defined by $\langle \mathbb{I}v, w\rangle = (v, w)_H$. In view of $\mathbb{I}$, the spaces H and H' can be identified.

Consider two Hilbert spaces, V and H, where V is a dense subspace of H, with $||v||_H \le ||v||_V$ for all $v \in V$. Assume that H is equipped with a conjugation, which induces - by restriction - a conjugation on V. In particular, $||\overline{v}||_V$ is equivalent to $||v||_V$ and $||\overline{v}||_H$ is equivalent to $||v||_H$. We identify H with H', but *not* V with V'. Hence, the inclusion map $\mathfrak{i} : V \to H$ is bounded and one-to-one, with dense range. Therefore, the transposed map $\mathfrak{i}^{tr} : H = H' \to V'$ enjoys the same properties. Identifying H with a dense subspace of V' via the transposed map, we say that H acts as a *pivot space* for V' and use the notation $V \subseteq H \subseteq V'$. This inclusion relation is called the *Gelfand triple* of V, H, V'.

Typical examples are[2]

1. $V = H_0^1(\mathcal{O}) \subseteq H = L^2(\mathcal{O}) \subseteq V' = H^{-1}(\mathcal{O})$
2. $V = H^s(\mathbb{R}^n) \subseteq H = L^2(\mathbb{R}^n) \subseteq V' = H^{-s}(\mathbb{R}^n)$, $s \ge 0$.

Let us note that although the above setting is by far the most common, in certain cases it does not make sense to identify a space with its topological

---

[1] Recall that two norms $||\cdot||$ and $|||\cdot|||$ on a vector space X are equivalent if $c||v|| \le |||v||| \le C||v||$, $\forall x \in X$, where c, C are constants.

[2] Recall that for $s \ge 0$, $H^s(\mathbb{R}^n) := \{u \in L^2(\mathbb{R}^n) : (1 + |z|^2)^{s/2}\widetilde{u}(z) \in L^2(\mathbb{R}^n)\}$, with $(u, v)_{H^s(\mathbb{R}^n)} = \int_{\mathbb{R}^n} (1 + |z|^2)^s\, \widetilde{u}(z)\widetilde{v}(z)\, dz$, where $\widetilde{u}$ denotes the Fourier transform of u. By $S'(\mathbb{R}^n)$ we denote the *tempered distributions* in $\mathbb{R}^n$, i.e., the dual space of $S(\mathbb{R}^n)$, the *rapidly decreasing functions* in $\mathbb{R}^n$. Then $H^{-s}(\mathbb{R}^n) = \{f \in S'(\mathbb{R}^n) : (1 + |z|^2)^{-s/2}\widetilde{f}(z) \in L^2(\mathbb{R}^n)\}$.

dual: let $H = L^2(\mathbb{R})$ and consider the weighted (dense) subspace $V = L^2(\mathbb{R}; (1+|x|))$ of H. The space V is endowed with the (standard) inner product $(v,w)_V = \int_{\mathbb{R}}(1+|x|)v(x)w(x)dx$. Any element $\phi \in H'$ is also an element of V'. If we identify ϕ with an element $f \in H$, this function does not define a linear form on V and the expression $\phi(v) = \langle f, v\rangle_V$ is meaningless on V. In such a situation it is necessary to work in a nonpivot Hilbert space. For example, weighted variational inequalities in nonpivot Hilbert spaces constitute an interesting topic that also finds a variety of applications (see [50]).

A.2 STRONG, WEAK AND WEAK-$*$ CONVERGENCE

The notions of weak and weak-$*$ convergence are extremely important and useful in analysis. We introduce them briefly, paying attention to particular examples in Lebesgue spaces that are of interest in this work.

DEFINITION A.2.1 *Consider a Banach space X, its dual X', $\langle\cdot,\cdot\rangle$ the duality pairing between these spaces and a sequence $\{x_n\}$ in X.*

(*i*) *The sequence $\{x_n\}$ converges strongly to $x \in X$ if and only if $||x_n - x|| \to 0$.*

(*ii*) *The sequence $\{x_n\}$ converges weakly to $x \in X$ if and only if for every $x' \in X$ we have that $\langle x_n, x'\rangle \to \langle x, x'\rangle$. We denote this by $x_n \rightharpoonup x$.*

(*iii*) *Suppose that there exists a Banach space Y such that $X = Y'$. Then we say that the sequence $\{x_n\}$ converges weak-$*$ to $x \in X$ if and only if $\langle x_n, y\rangle_{Y',Y} \to \langle x, y\rangle_{Y',Y}$ for all $y \in Y$. We denote this by $x_n \overset{*}{\rightharpoonup} x$.*

One could say that the strong convergence corresponds to convergence in the norm topology. On the other hand, the weak convergence corresponds to convergence in the topology of the dual space, whereas the weak-$*$ convergence corresponds to convergence in the topology of the double dual space.

The following scheme may easily be shown:

$$\begin{array}{c}\text{STRONG}\\ \text{CONVERGENCE}\end{array} \to \begin{array}{c}\text{WEAK}\\ \text{CONVERGENCE}\end{array} \to \begin{array}{c}\text{WEAK-}*\\ \text{CONVERGENCE}\end{array}$$

but of course it does not work the other way around! It may also be seen directly from the definition that if X is a reflexive space, i.e., if $(X')' = X$, then weak-$*$ convergence coincides with weak convergence. Therefore, weak$-*$ convergence is important when we deal with nonreflexive spaces. Note that all Hilbert spaces are reflexive, since by the Riesz identity $H' \simeq H$. However, this is not true for Lebesgue spaces in general. If $\mathcal{O} \subset \mathbb{R}^N$, then $L^p(\mathcal{O})$, $1 < p < \infty$ is a reflexive and separable space whose dual is $L^{p'}(\mathcal{O})$, $(\frac{1}{p} + \frac{1}{p'} = 1)$; $L^1(\mathcal{O})$ is separable but nonreflexive, and its dual is $L^\infty(\mathcal{O})$; while $L^\infty(\mathcal{O})$ is neither reflexive nor separable, and its dual is the space of

Radon measures (i.e., signed measures on $\mathcal{O}$ of bounded total variation), which is strictly bigger than $L^1(\mathcal{O})$.

The above definitions assume the following specific forms in the context of Lebesgue spaces:

DEFINITION A.2.2 (WEAK AND WEAK-$*$ CONVERGENCE IN L^p SPACES)

(i) $u_n \rightharpoonup u$ *in* $L^p(\mathcal{O})$, $1 < p < \infty$, *if*

$$\int_{\mathcal{O}} u_n \, \phi \, dx \to \int_{\mathcal{O}} u \, \phi \, dx, \;\; \forall \, \phi \in L^{p'}(\mathcal{O}), \; \frac{1}{p} + \frac{1}{p'} = 1.$$

(ii) $u_n \rightharpoonup u$ *in* $L^1(\mathcal{O})$, *if*

$$\int_{\mathcal{O}} u_n \, \phi \, dx \to \int_{\mathcal{O}} u \, \phi \, dx, \;\; \forall \, \phi \in L^{\infty}(\mathcal{O}).$$

(iii) $u_n \overset{*}{\rightharpoonup} u$ *in* $L^\infty(\mathcal{O})$, *if*

$$\int_{\mathcal{O}} u_n \, \phi \, dx \to \int_{\mathcal{O}} u \, \phi \, dx, \;\; \forall \, \phi \in L^{1}(\mathcal{O}).$$

The following proposition is often useful (see, e.g., [97]).

PROPOSITION A.2.3 *Suppose that* EITHER $u_n \rightharpoonup u$ *in* $L^p(\mathcal{O})$ OR $u_n \overset{*}{\rightharpoonup} u$ *in* $L^\infty(\mathcal{O})$. *Then the sequence* $\{u_n\}$ *is uniformly bounded (in* n*) in* $L^p(\mathcal{O})$ *or* $L^\infty(\mathcal{O})$, *respectively, and*

$$\int_{\mathcal{O}_0} u_n \, \phi \, dx \to \int_{\mathcal{O}_0} u \, \phi \, dx, \quad \textit{for every open } \mathcal{O}_0 \subset \mathcal{O}\,.$$

The following result (Eberlein-Šmulian theorem) provides weak compactness results.

THEOREM A.2.4 *Let* X *be a reflexive space. Then any bounded sequence* $\{x_n\}$ *in* X *has at least a weakly convergent subsequence* $\{x_{n_k}\}$. *Furthermore, if each weakly convergent subsequence converges to the same limit* x, *then the whole sequence weakly converges to* x.

The following theorem provides weak-$*$ compactness results.

THEOREM A.2.5 *Let* Y *be a separable space, such that* $X = Y'$. *Then any bounded sequence* $\{x_n\}$ *in* X *has at least a weakly-$*$ convergent subsequence* $\{x_{n_k}\}$. *Furthermore, if each weakly-$*$ convergent subsequence converges to the same limit* x, *then the whole sequence weakly-$*$ converges to* x.

This theorem works in particular for the case, where $X = L^\infty(\mathcal{O})$ (in which case $Y = L^1(\mathcal{O})$). Furthermore, if $u_n \overset{*}{\rightharpoonup} u$ in $L^\infty(\mathcal{O})$, then $u_n \rightharpoonup u$ in $L^p(\mathcal{O})$, $1 \le p < \infty$. This can be easily seen since, if $\{u_n\}$ weakly-$*$ converges in $L^\infty(\mathcal{O})$, it is bounded in this space, but then the standard embeddings for Lebesgue spaces imply that $\{u_n\}$ is bounded in every L^p, $1 \le p < \infty$. In turn, the weak compactness result in Theorem A.2.4 guarantees the existence of a weakly convergent subsequence in $L^p(\mathcal{O})$.

REMARK A.2.6 The product of weakly convergent sequences need *not* necessarily converge weakly. This problem of weak convergence, which often arises in a number of applications, e.g., in homogenisation theory, is rectified with the use of *compensated compactness* (see Section C.2 in Appendix C), which assumes further properties of the sequences (e.g., properties of their derivatives) to guarantee convergence of the product.

A.3 CALCULUS IN BANACH SPACES

A.3.1 Derivatives of vector-valued functions

Let $F : X \to Y$ be a map between two Banach spaces X, Y.

DEFINITION A.3.1 *A linear map $T : X \to Y$ is the Fréchet derivative of F at $x_0 \in X$ if*

$$\lim_{||u||_X \to 0} \frac{1}{||u||_X} ||F(x_0 + u) - F(x_0) - Tu||_Y = 0.$$

We denote T by $F'(x_0)$.

The standard theorems of differential calculus can now be transferred in the Banach space setting using the Fréchet derivative. For instance, the chain rule has a Banach space version, as follows.

THEOREM A.3.2 *Let X, Y, Z be Banach spaces and $F : X \to Y$, $G : Y \to Z$ be continuous mappings such that the composition $G \circ F$ is defined. If F and G are Fréchet differentiable, then $G \circ F$ is also Fréchet differentiable, and*

$$(G \circ F)'(x) = G'(F(x)) \circ F'(x).$$

Furthermore:

THEOREM A.3.3 *Let $F : X \to Y$ be an invertible mapping, and denote the inverse by $G = F^{-1}$. If F is Fréchet differentiable at a point x_0 and $F'(x_0) : X \to Y$ is a linear homeomorphism, then the inverse mapping G is also Fréchet differentiable at $y_0 = F(x_0)$, and*

$$G'(y_0) = (F^{-1})'(y_0) = (F'(x_0))^{-1}.$$

With the use of the Fréchet derivative we may generalise Newton's method for the solution of nonlinear equations of the form $F(x) = 0$ in a Banach space setting. The Newton scheme for this case would be the iterative scheme

$$x_{n+1} = x_n - (F'(x_n))^{-1} F(x_n), \tag{A.1}$$

where now $(F'(x_n))^{-1}$ is the inverse of the Fréchet derivative of F, calculated at x_n. The following theorem provides conditions under which this Newton scheme converges to the solution of the problem (in the sense that $x_n \to x^*$, in the strong topology of X, where x^* satisfies $F(x^*) = 0$). Let $B(x_0, r)$ denote the open ball of centre x_0 and radius r.

THEOREM A.3.4 *Let $F : B(x_0, r) \to Y$ be continuously differentiable and such that*

(i) $F'(x_0)^{-1} \in \mathcal{L}(Y, X)$, $||F'(x_0)^{-1}F(x_0)||_X = \alpha$, $||F'(x_0)^{-1}||_X = \beta$,

(ii) $||F'(u) - F'(v)||_Y \leq k||u - v||_X$, $u, v \in B(x_0, r)$,

(iii) $2k\alpha\beta < 1$, $2\alpha < r$

hold. Then F has a unique zero $x^* \in \overline{B(x_0, 2\alpha)}$ and the Newton iterates (A.1) *converge to x^*.*

For the Fréchet derivative and its properties, see, e.g., [3], [60], and the appendix in [98].

A.3.2 The Bochner integral

The Bochner integral extends the theory of Lebesgue integration for functions that take values in a Banach space. Let $(X, \mathscr{F}, \mu)$ be a measure space and consider a function $F : X \to Y$, where X, Y are Banach spaces. A particular case of interest is when $(X, \mathscr{F}, \mu) = (\mathbb{R}, \mathscr{B}(\mathbb{R}), \mu_{\text{L}})$, where $\mathscr{B}$ is the Borel σ-algebra on $\mathbb{R}$ and μ_{L} is the Lebesgue measure. In the general case, however, where X is a Banach space, we need to be careful about the definition of the measure space since in general, one may define various measurability concepts. We will always consider the case where X is separable, where according to the Pettis theorem all these notions coincide.

DEFINITION A.3.5 *Consider the measurable spaces $(X, \mathscr{F})$ and $(Y, \mathscr{G})$. A function $f : X \to Y$ is called $\mathscr{F}$-simple if it can be expressed as*

$$f(x) = \sum_{k=1}^{n} f_k \, \chi_{A_k}(x), \quad x \in X, \; f_k \in Y, \; A_k \in \mathscr{F}.$$

The Bochner integral for an $\mathscr{F}$-simple function is defined as

$$\int_X f \, d\mu = \sum_{k=1}^{n} f_k \mu(A_k).$$

The Bochner integral takes values in Y.

It can be proved that any measurable function can be approximated by a sequence of simple functions. This motivates the following definition for the Bochner integral of any measurable function.

DEFINITION A.3.6 *Let $f : X \to Y$ be a measurable function and f_n be a sequence of $\mathscr{F}$-simple functions converging μ-a.e. to f in the strong (norm) topology of Y. If the sequence of Bochner integrals $\int_X f_n \, d\mu$ converges in the strong topology of Y, then the function f is called Bochner integrable and the limit is the Bochner integral of f.*

The space of Bochner integrable functions $f : X \to Y$ is denoted by $L^1(X, \mathscr{F}, \mu; Y)$. The following theorem characterises the Bochner integrable functions.

THEOREM A.3.7 *A function $f : X \to Y$ is Bochner integrable if and only if it is $\mathscr{F}$-measurable and $\int_X ||f||_X \, d\mu < \infty$. For a Bochner integrable function, we have*

$$\left|\left| \int_X f \, d\mu \right|\right|_Y \leq \int_X ||f||_Y \, d\mu.$$

The theory of the Bochner integral allows the generalisation of many of the powerful results of the Lebesgue integral to vector-valued functions. For instance, the following vector-valued version of the Lebesgue dominated convergence theorem is often useful:

THEOREM A.3.8 *Consider a sequence $f_n \in L^1(X, \mathscr{F}, \mu)$ converging μ-a.e. to f, so that $||f_n||_Y \leq g$, μ-a.e. for every $n \in \mathbb{N}$ and a real-valued function $g : X \to \mathbb{R}$, $g \in L^1(X, \mathscr{F}, \mu; \mathbb{R})$. Then $f \in L^1(X, \mathscr{F}, \mu; Y)$, and*

$$\lim_{n \to \infty} \int_X f_n \, d\mu = \int_X f \, d\mu.$$

As an application of the above, we have the following.

PROPOSITION A.3.9 *If a series $\sum_{n=1}^{\infty} f_n$ with $f_n \in L^1(X, \mathscr{F}, \mu; Y)$ for every $n \in \mathbb{N}$ is majorised by a convergent series of real numbers, then*

$$\sum_{n=1}^{\infty} \int_X f_n \, d\mu = \int_X \left(\sum_{n=1}^{\infty} f_n \right) d\mu.$$

An important question, is how does the Bochner integral behave with respect to the action of bounded operators? The following theorem provides the answer to this question.

THEOREM A.3.10 *Let $f : X \to Y$, $f \in L^1(X, \mathscr{F}, \mu; Y)$ and A be a bounded operator from Y to Z. Then $Af : X \to Z$ is Bochner integrable ($Af \in L^1(X, \mathscr{F}, \mu; Z)$) and*

$$A \int_X f \, d\mu = \int_X Af \, d\mu.$$

REMARK A.3.11 If $(X, \mathscr{F}, \mu) = (\Omega, \mathscr{F}, P)$ is a probability space, then a measurable function $f : \Omega \to Y$ is called a Y-valued random variable and the Bochner integral $\int_\Omega f \, dP$ is called the expectation of the random variable f, and is denoted by $\mathbb{E}_P[f]$.

A standard reference for the Bochner integral and its properties is [361].

A.4 BASIC ELEMENTS OF SPECTRAL THEORY

This section closely follows [368]. Some of the proofs of the results in the sequel can be found in [368]. For a comprehensive account of spectral theory one can consult, e.g., [189], [229], [364]; see also [367].

A.4.1 Resolvent, Spectrum

Let X be a complex normed linear space and let $A : X \supseteq D(A) \to X$.

DEFINITION A.4.1 (RESOLVENT OPERATOR, RESOLVENT SET, SPECTRUM)

(*i*) *For* $\lambda \in \mathbb{C}$, *the* RESOLVENT OPERATOR *of* A *is defined as* $R_\lambda(A) := (A - \lambda I)^{-1}$, *where* I *is the identity operator on* X.

(*ii*) *A* REGULAR VALUE *of* A *is a* $\lambda \in \mathbb{C}$ *such that* $R_\lambda(A)$ *exists, is bounded and is defined on a dense subset of* X.

(*iii*) *The set of all regular values of* A, *denoted by* $\rho(A)$, *is called the* RESOLVENT SET *of* A.

(*iv*) *The* SPECTRUM *of* A, $\sigma(A) := \mathbb{C} \setminus \rho(A)$. *The spectrum of* A *is the union of the following three disjoint sets:*

  (*a*) *The* POINT SPECTRUM, $\sigma_p(A) := \{\lambda \in \mathbb{C} : R_\lambda(A)$ *does not exist*$\}$.

  (*b*) *The* CONTINUOUS SPECTRUM, $\sigma_c(A) := \{\lambda \in \mathbb{C} : R_\lambda(A)$, *exists as an unbounded operator and is defined on a dense subset of* $X\}$.

  (*c*) *The* RESIDUAL SPECTRUM, $\sigma_r(A) := \{\lambda \in \mathbb{C} : R_\lambda(A)$, *exists as either a bounded or an unbounded operator, but in either case it is not defined on a dense subset of* $X\}$.

REMARK A.4.2 Let A be a linear operator on a Banach space X. If X is *finite dimensional*, then $\sigma_c(A) = \sigma_r(A) = \emptyset$.

REMARK A.4.3 Let A be a bounded linear operator on a Banach space X. We have the following results:

(*i*) $\rho(A)$ is open.

(*ii*) For all $\nu \in \rho(A)$ we have $R_\lambda(A) = \sum_{k=0}^{\infty} (\lambda - \nu)^k R_\nu^{k+1}(A)$. This series is absolutely convergent for every λ in the open disc $|\lambda - \nu| < ||R_\nu(A)||^{-1}$, which is a subset of $\rho(A)$ in $\mathbb{C}$.

(*iii*) $\sigma(A)$ is compact and lies in the disc $|\lambda| \leq ||A||$.

REMARK A.4.4 Let H be a complex separable Hilbert space and A a linear operator on H. Let λ be an eigenvalue of A. The set M_λ consisting of the zero element in H and all eigenvectors of A corresponding to λ is the eigenspace of A corresponding to λ. M_λ is a subspace of A. The eigenvalues are countable.

REMARK A.4.5 Let H be a complex separable Hilbert space and A a linear self-adjoint operator on H. Then $\sigma_r(A) = \emptyset$.

A.4.2 Spectral Decompositions of Infinite-Dimensional Spaces

Let H be a complex separable Hilbert space. Assume that there exists a nondecreasing family (i.e., $M_\lambda \subseteq M_\mu$ for $\mu < \lambda$) of subspaces $\{M_\lambda\}, \lambda \in \mathbb{R}$, such that $\cap_{\lambda \in \mathbb{R}} M_\lambda = 0$ and $\cup_{\lambda \in \mathbb{R}} M_\lambda$ is a dense subset of H.

DEFINITION A.4.6 *A family of projection operators $\{E_\lambda\}$ is called a spectral family if*

(*i*) $E_\lambda E_\mu = E_\mu E_\lambda = E_{min\{\lambda,\mu\}} = E_\mu$.

(*ii*) *If $\epsilon > 0$, then for any $\vartheta \in H$ and $\lambda \in \mathbb{R}$: $E_{\lambda+\epsilon}\vartheta \to E_\lambda \vartheta$, as $\epsilon \to 0$.*

(*iii*) *For any $\vartheta \in H$, $E_\lambda \vartheta \to 0$ as $\lambda \to -\infty$ and $E_\lambda \vartheta \to \vartheta$ as $\lambda \to \infty$.*

THEOREM A.4.7 (SPECTRAL THEOREM) *Let H be a complex separable Hilbert space.*

(*i*) *For each bounded self-adjoint operator A on H, there exists a unique spectral family $\{E_\lambda\}$ such that*

$$(A\psi, \phi) = \int_{-\infty}^{\infty} \lambda \, d(E_\lambda \psi, \phi) \, , \forall \phi, \psi \in H \, .$$

Equivalently, we write

$$A = \int_{-\infty}^{\infty} \lambda dE_\lambda \, ,$$

and call it the spectral decomposition of A.

(*ii*) *For each unitary operator U on H, there exists a unique spectral family $\{F_\lambda\}$ with $F_\lambda = \mathbb{O}$ for $\lambda \leq 0$ (where $\mathbb{O}$ is the zero operator on H) and $F_\lambda = I$ for $\lambda \geq 2\pi$, such that*

$$(U\psi, \phi) = \int_0^{2\pi} e^{i\lambda} \, d(F_\lambda \psi, \phi) \, , \forall \phi, \psi \in D(U) \subset H \, .$$

Equivalently, we write

$$U = \int_0^{2\pi} e^{i\lambda} dF_\lambda \, ,$$

and call it the spectral decomposition of U.

So we see that a spectral family $\{E_\lambda\}$ determines a self-adjoint operator A. In many applications it is essential to obtain the spectral family associated with a given self-adjoint operator. This can be achieved in view of the celebrated following theorem.

THEOREM A.4.8 (STONE'S FORMULA) *Let H be a complex separable Hilbert space and $A : H \to H$ be a self-adjoint operator. For all $f, g \in H$ and $a, b \in \mathbb{R}$, we have*

$$\Big(\big(E_b - E_a\big)f, g\Big) = \lim_{\delta \to 0} \lim_{\epsilon \to 0} \frac{1}{2\pi \mathrm{i}} \int_{a+\delta}^{b+\delta} \bigg(\Big(\big(A - (t + \mathrm{i}\epsilon)I\big)^{-1} - \big(A - (t - \mathrm{i}\epsilon)I\big)^{-1}\Big)f, g\bigg)\, dt\,.$$

In the following result we list some properties of spectral families.

THEOREM A.4.9 *Let H be a complex separable Hilbert space and $A : H \to H$ be a bounded, linear self-adjoint operator with an associated spectral family $\{E_\lambda\}$ and spectral decomposition $A = \int_{-\infty}^{\infty} \lambda dE_\lambda$.*

(i) *Then E_λ has a discontinuity at $\lambda = \mu$ if and only if μ is an eigenvalue of A.*

(ii) *Let μ be an eigenvalue of A and M_μ be the subspace spanned by the eigenvectors of A associated with μ. Further, let $P_\mu : H \to M_\mu$ denote the projection operator onto M_μ and h denote the Heaviside step function. Then*

   (a) $E_\lambda P_\mu = P_\mu \mathsf{h}(\lambda - \mu)$.

   (b) *For $\epsilon > 0$, we have $E_\mu \psi - E_{\mu-\epsilon}\psi \to P_\mu \psi$, $\forall \psi \in H$, as $\epsilon \to 0$.*

(iii) *$(\mathbb{R} \ni)\mu \in \rho(A)$ if and only if there exists a constant $c > 0$ such that $\{E_\lambda\}$ is constant on the interval $[\mu - c, \mu + c]$.*

(iv) *$(\mathbb{R} \ni)\mu \in \sigma_c(A)$ if and only $\{E_\lambda\}$ is continuous at μ and is not constant in any neighbourhood of μ.*

In view of the analogies between the two parts of the spectral theorem, property (*ii*)above, has an obvious version for linear unitary operators.

A.4.3 On functions of an operator

The spectral theorem allows us to form a large class of functions of a self-adjoint operator. Let H be a complex separable Hilbert space and $A : H \to H$ be a bounded, linear self-adjoint operator with an associated spectral family $\{E_\lambda\}$ and spectral decomposition $A = \int_{-\infty}^{\infty} \lambda dE_\lambda$. Let $\varphi : \mathbb{R} \to \mathbb{C}$ be a continuous function. We can define an operator $\varphi(A)$ by

$$D(\varphi(A)) = \{f \in H : \int_{-\infty}^{\infty} |\varphi(\lambda)|^2 d(E_\lambda f, f) < \infty\}\,,$$

$$\big(\varphi(A)f, g\big) = \int_{-\infty}^{\infty} \varphi(\lambda) d(E_\lambda f, g)\,,\ f \in D(\varphi(A))\,, g \in H\,.$$

Therefore, at least formally, we have

$$\varphi(A) = \int_{-\infty}^{\infty} \varphi(\lambda)\, dE_\lambda\,.$$

Let H be a complex separable Hilbert space and $A : H \to H$ be a bounded, linear self-adjoint operator with an associated spectral family $\{E_\lambda\}$ and spectral decomposition $A = \int_{-\infty}^{\infty} \lambda dE_\lambda$. Let $\varphi, \varphi_1, \varphi_2 : \mathbb{R} \to \mathbb{C}$ be continuous functions defined on the support of $\{E_\lambda\}$. Below we list some basic properties of the operator $\varphi(A)$.

1. $\varphi(\lambda) = 1 \Rightarrow \varphi(A) = I$ and $\varphi(\lambda) = \lambda \Rightarrow \varphi(A) = A$.
2. $\varphi(A)^* = \overline{\varphi}(A)$, where $\overline{\varphi}(\lambda) = \overline{\varphi(\lambda)}$.
3. $\varphi(\lambda) = \varphi_1(\lambda)\varphi_2(\lambda) \Rightarrow \varphi(A) = \varphi_1(A)\varphi_2(A)$.
4. $\varphi(\lambda) = c_1\varphi_1(\lambda) + c_2\varphi_2(\lambda) \Rightarrow \varphi(A) = c_1\varphi_1(A) + c_2\varphi_2(A)$.
5. $(\varphi_1\varphi_2)(\lambda) = \varphi_1(\lambda)\varphi_2(\lambda) \Rightarrow (\varphi_1\varphi_2)(A) = \varphi_1(A)\varphi_2(A)$.
6. $\varphi(A)^*\varphi(A) = \varphi(A)\varphi(A)^*$.
7. $\varphi(A)$ commutes with all bounded operators that commute with A.
8. Let $\varphi_z(\lambda) = (\lambda - z)^{-1}$, where $z \in \mathbb{C}$ with $\mathcal{I}\mathrm{m}\, z \neq 0$. Then $\varphi_z(A) = \int_{-\infty}^{\infty} (\lambda - z)^{-1} dE_\lambda = R_z(A) = (A - zI)^{-1}$. Further, $||\varphi_z(A)|| \leq 1/\mathcal{I}\mathrm{m}\, z$.

A.5 COMPACTNESS CRITERIA

This standard result can be found in any of the references cited at the beginning of this appendix.

THEOREM A.5.1 (ARZELÀ-ASCOLI THEOREM) *Let X be a Banach space. A subset F of $C([0,T];X)$ is relatively compact if and only if*

(i) *$F(t) := \{f(t) : f \in F\}$ is relatively compact in X for all $t \geq 0$.*

(ii) *F is uniformly equicontinuous, i.e., for all $\epsilon > 0$, there exists $\delta > 0$ such that*
$$||f(t_1) - f(t_2)||_X \leq \epsilon \;\forall t_1, t_2 \in [0,T], \text{ such that } |t_1 - t_2| \leq \delta.$$

For our purposes the following version of the above theorem is useful.

THEOREM A.5.2 *Let $Y \subset C([0,T];(L^2(\mathcal{O}))^3)$ have the following properties:*

(i) *$Y(t)$ is relatively compact in $(L^2(\mathcal{O}))^3$, $\forall t \in [0,T]$.*

(ii) *$\forall \varepsilon > 0 \; \exists \delta = \delta(\varepsilon) > 0 : \|y(t_1) - y(t_2)\|_Y \leq \varepsilon$, $\forall t_1, t_2 \in [0,T]$ such that $|t_1 - t_2| \leq \delta$ (equicontinuity of Y).*

(iii) *$\sup_{y \in Y, t \in [0,T]} \|y(t)\|_Y < \infty$ (uniform boundedness of Y).*

Then $\exists \{y_k\} \in Y$ and $y_0 \in C([0,T];(L^2(\mathcal{O}))^3)$ such that
$$\lim_{k \to \infty} \|y_k(\cdot) - y_0(\cdot)\|_{C([0,T];L^2(\mathcal{O}))} = 0.$$

REMARK A.5.3 A characterisation of compact subsets of $L^p([0,T];X)$ has been established by J. Simon [385].

A.6 COMPACT OPERATORS

This subsection is based mainly on [103] and [361]. The main impetus for the study of compact operators arose from the use of integral equations in attempting to solve the classical BVPs of mathematical physics.

DEFINITION A.6.1 *Let X and Y be Banach spaces, and $\mathcal{L}(X, Y)$ be the space of bounded linear operators $X \to Y$, endowed with the operator norm. An operator $A \in \mathcal{L}(X, Y)$ is called compact if A maps bounded sets in X into precompact sets (i.e., having compact closure $\overline{X}$) in Y. Equivalently, A is compact if and only if for every bounded sequence $\{x_n\} \subset X$, $\{Ax_n\}$ has a convergent subsequence in Y.*

Important properties of compact operators are given in the following:

THEOREM A.6.2 *(i) A compact operator maps weakly convergent sequences into norm-convergent sequences.*

(ii) If X is reflexive, then the converse of the above is true.

THEOREM A.6.3 *Let X, Y, Z be Banach spaces, $A_n \in \mathcal{L}(X, Y)$, $A \in \mathcal{L}(X, Y)$ and $S \in \mathcal{L}(Y, Z)$. Then*

(i) If A is compact then it is bounded (the converse is false [3]).

(ii) If A_n are compact and $c_n \in \mathbb{C}$ $n = 1, \ldots, N$, then the linear combination $\sum_{n=1}^{N} c_n A_n$ is compact.

(iii) If one of A or S is compact, then SA is compact.

(iv) A is compact if and only if its adjoint A^* is compact.

(v) If $\{A_n\}$ are compact and $A_n \to A$ in the norm topology, then A is compact.

(vi) If the range $A(X)$ is finite dimensional, then A is compact.

PROPOSITION A.6.4 (THE RIESZ LEMMA) *Let U be a proper closed subspace of a normed space X, and $\alpha \in (0, 1)$. Then there exists an element $\psi \in X$ with $||\psi|| = 1$ such that $||\psi - \phi|| \geq \alpha$ for all $\phi \in U$.*

A consequence of this result is the following theorem.

THEOREM A.6.5 *The identity operator $I : X \to X$ is compact if and only if X is finite dimensional.*

We are primarily interested in the case where $A : H \to H$ is a compact operator, where H is a separable Hilbert space. We use the symbol $\mathcal{L}(H)$ for $\mathcal{L}(H, H)$.

---

[3] This follows from Theorem A.6.5.

THEOREM A.6.6 *Every compact operator on H is the norm limit of a sequence of operators of finite rank.*

REMARK A.6.7 The compact operators on a separable Hilbert space H form a Banach space. The dual and double dual spaces of this Banach space are illustrative of the difference between the weak Banach space topology on $\mathcal{L}(H)$ and the weak operator topology.

The basic principle that makes compact operators important is the *Fredholm alternative*: if A is compact, then either $A\psi = \psi$ has a solution, or $(I - A)^{-1}$ exists. This property is not true for all bounded linear operators: e.g., if A is the operator $(A\phi)(x) = x\phi(x)$ on $L^2[0, 2]$, then $A\phi = \phi$ has no solutions, but $(I - A)^{-1}$ does not exist (as a bounded operator). As far as "solving equations" is concerned, the Fredholm alternative is a powerful tool: *compactness and uniqueness imply existence*!

Since the Fredholm alternative holds for finite-dimensional matrices, it is plausible to expect that it will be true for compact operators (in the Hilbert space case) in view of the fact that any compact operator A can be written in the form $A = F + S$, where F has finite rank and S has small norm.

Compactness combines very nicely with analyticity, as shown in the next theorem.

THEOREM A.6.8 (ANALYTIC FREDHOLM THEOREM) *Let* D *be an open connected subset of* $\mathbb{C}$. *Let* $f : \mathsf{D} \to \mathcal{L}(H)$ *be an analytic operator-valued function such that* $f(z)$ *is compact for each* $z \in \mathsf{D}$. *Then either*

(*i*) $(I - f(z))^{-1}$ *does not exist for any* $z \in \mathsf{D}$,

or

(*ii*) $(I - f(z))^{-1}$ *exists for all* $z \in \mathsf{D} \setminus \Sigma$, *where* Σ *is a discrete subset of* D *(i.e., a set that has no limit points in* D*). In this case,* $(I - f(z))^{-1}$ *is meromorphic in* D, *analytic in* $\mathsf{D} \setminus \Sigma$, *the residues at the poles are finite rank operators, and if* $z \in \Sigma$ *then* $f(z)\,\psi = \psi$ *has a nonzero solution in* H.

This theorem has four important consequences:

THEOREM A.6.9 (THE FREDHOLM ALTERNATIVE) *If A is a compact operator on H, then* either $(I - A)^{-1}$ *exists* or $A\psi = \psi$ *has a solution.*

THEOREM A.6.10 (THE RIESZ-SCHAUDER THEOREM) *Let A be a compact operator on H. Then its spectrum $\sigma(A)$ is a discrete set having no limit points, with the possible exception of $\lambda = 0$. In addition, any $\lambda \in \sigma(A) \setminus \{0\}$ is an eigenvalue of finite multiplicity.*

THEOREM A.6.11 (THE HILBERT-SCHMIDT THEOREM) *Let A be a self-adjoint compact operator on H. Then there is a complete orthonormal basis $\{\phi_n\}$ of H such that $A\phi_n = \lambda_n\phi_n$ and $\lim_{n\to\infty} \lambda_n = 0$.*

Theorem A.6.12 (Canonical form for compact operators) *Consider a compact operator A on H. Then there exist (not necessarily complete) orthonormal sets $\{\psi_n\}_{n=1}^N$ and $\{\phi_n\}_{n=1}^N$, and positive real numbers $\{\lambda_n\}_{n=1}^N$ with $\lim_{n\to\infty}\lambda_n = 0$ such that*

$$A = \sum_{n=1}^{N} \lambda_n (\psi_n, \cdot)\, \phi_n\,.$$

The sum, which may be finite or infinite, converges in norm. The numbers $\{\lambda_n\}$ are called the singular values *of A. In addition, the singular values of A are precisely the eigenvalues of $|A|$.*

The "standard" (as far as solving equations is concerned) statement of the above theory assumes usually the following form.

Proposition A.6.13 *Let $A : H \to H$ be a compact linear operator.*

(*i*) *If the* homogeneous *equation*

$$\phi - A\phi = 0$$

has only the trivial solution $\phi = 0$, then for all $f \in H$ the inhomogeneous *equation*

$$\phi - A\phi = f$$

has a unique solution $\phi \in H$ that depends continuously on f.

(*ii*) *If the* homogeneous *equation*

$$\phi - A\phi = 0$$

has the nontrivial linearly independent solutions $\phi_1, \phi_2, \ldots, \phi_m$, then the inhomogeneous *equation*

$$\phi - A\phi = f$$

is either unsolvable or its general solution is of the form

$$\phi = \widetilde{\phi} + \sum_{k=1}^{m} \alpha_k \phi_k\,,$$

where $\widetilde{\phi}$ is a particular solution of the inhomogeneous equation, and $\alpha_1, \alpha_2, \ldots, \alpha_m$ are arbitrary complex numbers.

(*iii*) *(First Fredholm theorem) The null spaces of $I - A$ and $I - A^*$ have the same finite dimension.*

(*iv*) *(Second Fredholm theorem) A necessary and sufficient condition for the inhomogeneous equation $\phi - A\phi = f$ to be solvable is*

$$(f, \psi_k) = 0\,, k = 1, \ldots, m\,,$$

for all solutions ψ_k of the homogeneous adjoint *equation*

$$\psi - A^*\psi = 0\,.$$

This statement also holds with the rôles of A and A^* interchanged.

Remark A.6.14 The above results remain valid when the operator $I - A$ is replaced by an operator of the form $L - A$, where L is a bounded linear operator having a bounded inverse L^{-1}.

A.6.1 Nuclear, trace class, and Hilbert-Schmidt operators

Consider two separable Hilbert spaces U and V and denote by $\mathcal{L}(U, V)$ the space of bounded linear operators $A : U \to V$. The adjoint operator A^* is an element of $\mathcal{L}(V, U)$ such that

$$(Ax, y) = (x, A^* y), \quad \forall x \in U, \ y \in V.$$

Two important classes of compact operators are given in the following.

DEFINITION A.6.15 (NUCLEAR AND TRACE CLASS OPERATORS)

(*i*) *An operator $Q \in \mathcal{L}(U, V)$ is called a nuclear operator if there exists a sequence $\{v_n\} \in V$ and a sequence $\{u_n\} \in U$ such that*

$$Qx = \sum_{n=1}^{\infty} v_n \, (u_n, x)_U \quad \forall\, x \in U, \quad \textit{and} \quad \sum_{n=1}^{\infty} ||v_n||_V \, ||u_n||_U < \infty.$$

(*ii*) *Let $U = V$. A nuclear operator Q that is non-negative (i.e., $(Lu, u) \geq 0$ for all $u \in U$) and symmetric (i.e., $(Lu, v) = (u, Lv)$ for all $u, v \in U$) is called a trace class operator.*

The following is a very useful property of nuclear operators.

PROPOSITION A.6.16 *Let $Q : U \to U$ be a nuclear operator and let $\{e_n\}$ be an orthonormal basis of U. Define the trace of the operator Q as the infinite series $Tr(Q) := \sum_{n=1}^{\infty} (Q\, e_n, e_n)$. Then $Tr(Q)$ is a well-defined finite quantity and independent of the choice of the orthonormal basis $\{e_n\}$.*

Trace class operators are interesting from the point of view of infinite-dimensional stochastic analysis since they can be considered the generalisation of the covariance matrix in infinite dimensions.

The solution of the eigenvalue problem for trace class operators provides us with an orthonormal basis for the Hilbert space U.

An interesting subclass of nuclear operators consists of the Hilbert-Schmidt operators.

DEFINITION A.6.17 *A bounded linear operator $Q : U \to V$ is called a Hilbert-Schmidt operator if $\sum_{n=1}^{\infty} ||Q\, e_n||^2 < \infty$, where $\{e_n\}$ is an orthonormal basis of U. We will denote the space of all Hilbert-Schmidt operators from U to V by $L_2(U, V)$.*

The space of Hilbert-Schmidt operators can be turned into a separable Hilbert space by defining the inner product

$$(Q_1, Q_2)_{L_2(U,V)} = \sum_{n=1}^{\infty} (Q_1 e_n, Q_2 e_n).$$

The following proposition helps us to define the "square root" of a trace class operator.

PROPOSITION A.6.18 *If* $: U \to U$ *is a trace class operator, then there exists a unique Hilbert-Schmidt operator* R *such that* $R \circ R = Q$. *We will use the notation* $R = Q^{1/2}$. *Furthermore,* $||Q||^2_{L_2(U)} = Tr(Q)$.

The operator $Q^{1/2}$ has the useful property that $L \circ Q^{1/2} \in L_2(U, V)$ for any $L \in \mathcal{L}(U, V)$.

DEFINITION A.6.19 *Let* $U_0 := Q^{1/2}(U)$. *When equipped with the inner product*

$$(u_0, v_0)_0 := (Q^{-1/2}u_0, Q^{-1/2}v_0)_U,$$

where $Q^{-1/2}$ *is the inverse of* $Q^{1/2}$ *(or the pseudo-inverse, in case* Q *is not one-to-one), this space is a subspace of* U *and a separable Hilbert space.*

A.7 THE BANACH-STEINHAUS THEOREM

THEOREM A.7.1 *Let* X, Y *be two Banach spaces and* $X_d \subset X$ *be a dense subspace of* X. *Assume further that* $\{A_n\}$ *is a sequence of linear operators in* $\mathcal{L}(X, Y)$ *such that*

(*i*) $\|A_n x\|_Y < C$ *for all* $n \in \mathbb{N}$ *(*C *independent of* n*),*

(*ii*) *the limit* $\lim_{n\to\infty} A_n x$ *exists for all* $x \in X_d$.

Then there exists a continuous linear operator $A : X \to Y$ *such that* $A_n x \to Ax$ *in* Y, *for every* $x \in X$.

The *principle of uniform boundedness* is often useful in the following.

THEOREM A.7.2 *Assume that* $\{A_n\} \subset \mathcal{L}(X, Y)$ *and that* $\{A_n x\}$ *is uniformly bounded in* n *for every* $x \in X$. *Then the set* $\{\|A_n\|_{\mathcal{L}(\mathcal{X},\mathcal{Y})}\}$ *is also uniformly bounded.*

A.8 SEMIGROUPS AND THE CAUCHY PROBLEM

This section is based on [56], [141], [346] and [363].

A.8.1 Semigroups of linear operators

DEFINITION A.8.1 *Consider a one-parameter family* $\{T(t)\}$, $t \in \mathbb{R}^+$ *of linear and bounded operators from a Banach space* X *onto itself. The family is called a strongly continuous, or*[4] C_0, *semigroup if the following properties hold:*

(*i*) $T(0) = I$, *where* I *is the identity operator in* X.

[4] C_0 is the abbreviation for "Cesàro summable of order 0".

(*ii*) *$T(t) \circ T(s) = T(t+s)$ for every $t, s \in \mathbb{R}^+$, where by $\circ$ we denote the composition of the operators.*

(*iii*) *The map $t \mapsto T(t)$ is strongly continuous, i.e., for every $x \in X$,*

$$\lim_{t\to 0^+} T(t)x \to x,$$

where the limit is taken in the strong topology of X.

Strongly continuous semigroups have important properties. One important such property is that they are bounded.

PROPOSITION A.8.2 *If $\{T(t)\}$ is a C_0 semigroup on X, then there exist $M \in \mathbb{R}^+$, $\theta \in \mathbb{R}$, such that*

$$||T(t)||_{\mathcal{L}(X)} \leq M\, e^{\theta\, t}, \;\; 0 \leq t < \infty. \tag{A.2}$$

DEFINITION A.8.3 *If $M = 1$ and $\theta = 0$ in (A.2), i.e., if*

$$||T(t)||_{\mathcal{L}(X)} \leq 1\,,$$

then $\{T(t)\}$ is called a contraction semigroup.

The characterisation of a C_0 semigroup can be obtained through an operator A that is the infinitesimal generator of the semigroup.

DEFINITION A.8.4 *Let $\{T(t)\}$ be a C_0 semigroup on X. Define the operator $A : D(A) \to X$ by*

$$A\,x = \lim_{t\to 0} \frac{T(t)\,x - x}{t}, \quad x \in D(A),$$

where

$$D(A) = \left\{ x \in X \;:\; \lim_{t\to 0^+} \frac{T(t)\,x - x}{t} \;\; exists \right\}.$$

The limit is taken in the strong topology of X. The operator A is a linear and possibly unbounded operator.

What types of operators generate C_0 semigroups? An answer to this question is provided by the following generalisation of the Hille-Yosida theorem.

THEOREM A.8.5 *Let $A : D(A) \to X$, where $D(A) \subseteq X$, be a linear operator on a Banach space. Then the following are equivalent:*

(*i*) *A is the generator of a C_0-semigroup $\{T(t)\}$ on X, satisfying the bound $||T(t)|| \leq Me^{\theta\, t}$, $t \geq 0$, for some $\theta \in \mathbb{R}$, $M \geq 1$.*

(*ii*) *A is a closed operator, densely defined, and for every $\lambda \in \mathbb{C}$ with $\mathcal{R}e\,\lambda > \theta$ we have that $\lambda \in \rho(A)$ and that*

$$||R_\lambda(A)^n|| \leq \frac{M}{(\mathcal{R}e\,\lambda - \theta)^n}, \;\; for\ all\ n \in \mathbb{N}\,.$$

REMARK A.8.6 If $M = 1$ it is enough to prove the above resolvent estimate for $n = 1$.

A.8.2 Groups of operators

For the study of the Maxwell equations, the concept of groups of operators is very useful. We summarise here the basic properties and a generation theorem. This section is based on [141].

DEFINITION A.8.7 *Consider a one-parameter family $\{T(t)\}$, $t \in \mathbb{R}$, of linear and bounded operators from a Banach space X onto itself. The family is called a strongly continuous (C_0) group if the following properties hold:*

(i)) $T(0) = I$, where I is the identity operator in X.

(ii) $T(t) \circ T(s) = T(t+s)$ for every $t, s \in \mathbb{R}$, where by $\circ$ we denote the composition of the operators.

(iii) The map $t \mapsto T(t)$ is strongly continuous, i.e., for every $x \in X$,
$$\lim_{t \to 0} T(t)x = x,$$
where the limit is taken in the strong topology of X.

Obviously, if $\{T(t)\}$ is a group of operators, it is also a semigroup of operators; therefore, all the properties of semigroups hold for groups of operators as well.

DEFINITION A.8.8 *Let $\{T(t)\}$ be a C_0 group on X. Define the operator $A : D(A) \to X$ by*
$$A\,x = \lim_{t \to 0} \frac{T(t)\,x - x}{t}, \quad x \in D(A),$$
where
$$D(A) = \left\{ x \in X \,:\, \lim_{t \to 0^+} \frac{T(t)\,x - x}{t} \text{ exists} \right\}.$$
The limit is taken in the strong topology of X. The operator A is a linear and possibly unbounded operator.

The following theorem is very useful in the characterisation of groups of operators.

THEOREM A.8.9 *Let $A : D(A) \to X$, where $D(A) \subseteq X$, be a linear operator on a Banach space. Then the following are equivalent:*

(i) A is the generator of a C_0 group $\{T(t)\}$ on X, satisfying the bound $||T(t)|| \le M\, e^{\theta\,|t|}$, for some $\theta \in \mathbb{R}$ and $M \ge 1$.

(ii) A is a closed operator, densely defined, and for every $\lambda \in \mathbb{C}$ with $|\mathcal{R}e\,\lambda| > \theta$ we have that $\lambda \in \rho(A)$ and
$$||R_\lambda(A)^n|| \le \frac{M}{\big(|\mathcal{R}e\,\lambda| - \theta\big)^n}.$$

The following theorem is often useful.

THEOREM A.8.10 (STONE THEOREM) *Let H be a Hilbert space and A be a skew adjoint ($A^* = -A$) linear operator that is densely defined in H. Then A is the generator of a strongly continuous group of unitary operators $e^{t\,A}$.*

A.8.3 Semigroups and the Cauchy problem

The theory of semigroups is intimately linked with the existence and regularity properties for the abstract Cauchy problem. This connection renders semigroup theory an indispensable tool in the theory of differential equations. This section is based in [141] (see also [128]).

Let X be a Banach space and $A : D(A) \to X$, $D(A) \subset X$ be a closed operator. The problem

$$u' = A\,u\,, \;\; u(0) = u_0\,, \tag{A.3}$$

is called the abstract (i.e., Banach space-valued) Cauchy problem associated with $A, D(A)$ and u_0.

DEFINITION A.8.11 (CLASSICAL AND MILD SOLUTIONS OF (A.3))

(*i*) *A function $u : \mathbb{R}^+ \to X$ is called a classical solution of the abstract Cauchy problem (A.3) if u is continuously differentiable, $u \in D(A)$ and (A.3) holds for $t \geq 0$.*

(*ii*) *A function $u : \mathbb{R}^+ \to X$ is called a mild solution of the abstract Cauchy problem (A.3), if $\int_0^t u(s)\,ds \in D(A)$ for all $t \in \mathbb{R}^+$ and*

$$u(t) = u_0 + A\int_0^t u(s)\,ds\,.$$

THEOREM A.8.12 *Assume that A is the generator of a C_0 semigroup, $\{T(t)\}$, on X. Then*

(*i*) *For every $u_0 \in X$ there exists a unique mild solution of (A.3), which can be represented as*

$$u(t) = T(t)\,u_0.$$

(*ii*) *If, furthermore, $u_0 \in D(A)$, then there exists a unique classical solution of (A.3), which can be represented as*

$$u(t) = T(t)\,u_0.$$

The problem

$$u' = A\,u + F\,, \;\; u(0) = u_0\,, \tag{A.4}$$

is called the abstract nonhomogeneous Cauchy problem.

DEFINITION A.8.13 (CLASSICAL AND MILD SOLUTIONS OF (A.4))

(*i*) *A function $u : \mathbb{R}^+ \to X$ is called a classical solution of the abstract Cauchy problem (A.4) if u is continuously differentiable, $u \in D(A)$, and (A.4) holds for $t \geq 0$.*

(*ii*) *A function* $u : \mathbb{R}^+ \to X$ *is called a mild solution of the abstract Cauchy problem (A.4) if* $\int_0^t u(s)\, ds \in D(A)$ *for all* $t \in \mathbb{R}^+$ *and*

$$u(t) = u_0 + A \int_0^t u(s)\, ds + \int_0^t F(s)\, ds\,.$$

The following regularity results are often needed (see, e.g., [363], Theorem 12.16).

THEOREM A.8.14 *Assume that* A *is the generator of a* C_0 *semigroup* $\{T(t)\}$ *on a Banach space* X*. Here we think of* $D(A)$ *as a Banach space equipped with the graph norm. Then.*

(*i*) *For every* $u_0 \in X$ *and* $F \in L^1([0,T], X)$*, there exists a unique mild solution of (A.3), that can be represented as*

$$u(t) = T(t)\, u_0 + \int_0^t T(t-s)\, F(s)\, ds.$$

(*ii*) *If* $u_0 \in D(A)$ *and either* $F \in C([0,T]; X) \cap L^1([0,T]; D(A))$*, or* $F \in C([0,T]; H) \cap W^{1,1}([0,T]; X)$*, then there exists a unique classical solution of (A.3) that can be represented as*

$$u(t) = T(t)\, u_0 + \int_0^t T(t-s)\, F(s)\, ds.$$

A.9 SOME FIXED POINT THEOREMS

THEOREM A.9.1 (BANACH CONTRACTION THEOREM) *Let* X *be a Banach space. Consider a map* $f : X \to X$ *which is a contraction map, i.e., it satisfies the property*

$$||f(x) - f(y)||_X \le \alpha\, ||x - y||_X,$$

for all $x, y \in X$*, with* $\alpha \in (0,1)$*. Then* f *has a unique fixed point,* $x_* \in X$*.*

REMARK A.9.2 This result guarantees the unique solvability of the operator equation $f(x) = x$.

The following theorem is an important generalisation of the Banach contraction theorem.

THEOREM A.9.3 (SCHAUDER FIXED POINT THEOREM) *Consider a map* $f : X \to X$ *with the following properties:*

(*i*) f *is continuous,*

(*ii*) $f(X)$ *has compact closure in* X*.*

Then f *has a (possibly nonunique) fixed point.*

The above theorem finds many uses in the theory of both deterministic and stochastic integral and differential equations.

Finally, the following nonlinear alternative theorem is often very useful[5].

THEOREM A.9.4 (LERAY-SCHAUDER ALTERNATIVE)
Let H be a Hilbert space, $D \subseteq H$ a convex set, and U an open subset of D, such that $0 \in U$. Then each continuous compact mapping $f : \overline{U} \to D$ has at least one of the following properties:

(i) f has a fixed point.

(ii) There is $(x_*, \lambda_*) \in \partial U \times (0, 1)$ such that $x_* = \lambda_* f(x_*)$.

A.10 THE LAX-MILGRAM LEMMA

The Lax-Milgram lemma is a very useful result for the treatment of variational problems (see, e.g., [125]).

LEMMA A.10.1 (LAX-MILGRAM LEMMA) *Let H be a Hilbert space and $a : H \times H \to \mathbb{C}$ be a bounded and coercive sesquilinear form. Then the variational problem $a(u, v) = (f, v)$ has a solution in H for all $f \in H'$.*

The following generalisation of the Lax-Milgram lemma is often useful.

THEOREM A.10.2 *Let H_1 and H_2 be two Hilbert spaces and $a : H_1 \times H_2 \to \mathbb{C}$ be a bounded sesquilinear form such that*

(i) There is a constant $C > 0$ such that

$$\inf_{u \in A_1} \sup_{v \in A_2} |a(u, v)| \geq C\,,$$

where $A_1 := \{u \in H_1 \,:\, ||u||_{H_1} = 1\}$ and $A_2 := \{v \in H_2 \,:\, ||v||_{H_2} \leq 1\}$.

(ii) For every $v \in H_2$, $v \neq 0$,

$$\sup_{u \in H_1} |a(u, v)| > 0\,.$$

Then for every $f \in H_2'$ there exists a unique $u \in H_1$ such that

$$a(u, \phi) = (f, \phi), \quad \forall \phi \in H_2.$$

The solution to this problem satisfies the bound

$$||u||_{H_1} \leq \frac{C_0}{C}\, ||f||_{H_2'}\,.$$

[5]It is interesting to view this classical (1934) result in the recent framework of *complementarity problems*. Complementarity theory has deep relations with several chapters of fundamental mathematics, e.g., with fixed-point theory, theory of variational inequalities, topological degree, functional analysis, and theory of topological ordered vector spaces. Each complementarity problem is a mathematical model for several kinds of practical problems from economics, optimisation, game theory, engineering and mechanics. See, e.g., [207].

REMARK A.10.3 The first condition on a is called an *inf-sup*, or *Babuška-Brezzi condition* (see, e.g., [324]).

Consider two Hilbert spaces H_1, H_2 and two bounded sesquilinear forms

$$a : H_1 \times H_1 \to \mathbb{C}, \ \ b : H_2 \times H_2 \to \mathbb{C}.$$

Define the subset $\mathcal{H}$ of H_1 by

$$\mathcal{H} = \{u \in H_1 \, : \, b(u, w) = 0, \ \forall w \in H_2\}.$$

We define a weaker coercivity condition as follows.

DEFINITION A.10.4 *We say that a is $\mathcal{H}$-coercive if there exists a constant C_1 such that*

$$|a(u, u)| \geq C_1 \, ||u||_{H_1} \quad \forall \, u \in \mathcal{H}.$$

We now consider the mixed variational problem:
Given $f \in H_1'$ and $g \in H_2'$, find $u \in H_1$ and $p \in H_2$ such that

$$\begin{aligned} a(u, \phi) + b(\phi, p) &= \langle f, \phi \rangle, \quad \forall \, \phi \in H_1, \\ b(u, \psi) &= \langle g, \psi \rangle, \quad \forall \, \psi \in H_2. \end{aligned} \tag{A.5}$$

THEOREM A.10.5 *Suppose that the sesquilinear form b satisfies the Babuška-Brezzi condition and that there exists $C_2 > 0$ such that*

$$\sup_{w \in H_1} \frac{|b(w, p)|}{||w||_{H_1}} \geq C_2 \, ||p||_{H_2} \, , \ \forall \, p \in H_2 \, .$$

Then there exists a unique solution (u, p) to problem (A.5) satisfying the bound

$$||u||_{H_1} + ||p||_{H_2} \leq C \, (||f||_{H_1'} + ||g||_{H_2'}).$$

A.11 GRONWALL'S INEQUALITY

For the proofs of the following results we refer the reader to [115], where more integral inequalities of this type can also be found.

THEOREM A.11.1 *Let $k, h, y \in C([t_0, T))$, $T \leq \infty$ and $k(t) \geq 0$, $t \in [t_0, T)$. If*

$$y(t) \leq h(t) + \int_{t_0}^{t} k(s) \, y(s) ds \, , \ t \in [t_0, T) \, ,$$

then

$$y(t) \leq h(t) + \int_{t_0}^{t} k(s) \, h(s) \exp\Big(\int_{s}^{t} k(\tau) d\tau\Big) ds \, , \, t \in [t_0, T).$$

There is a very useful special case.

COROLLARY A.11.2 *Let k and y be as above, and*

$$y(t) \le m + \int_{t_0}^{t} k(s)\, y(s) ds\,,\ t \in [t_0, T)\,,$$

where m is a constant. Then

$$y(t) \le m \exp\Big(\int_{t_0}^{t} k(s) ds \Big)\,,\ t \in [t_0, T).$$

Let us also mention the so-called "generalised Gronwall's inequality", below.

THEOREM A.11.3 *Let $y \in C([t_0, T])$ and $y(t) \ge 0\,,\ t \in [t_0, T]$. Let $\sigma(t)$ be a nondecreasing function on $[t_0, T]$ such that $\sigma(t) = \sigma(t + 0)$. Let, also, m be a positive constant such that*

$$y(t) \le m + \int_{t_0}^{t} y(s) d\sigma(s)\,.$$

Then

$$y(t) \le m \exp(\sigma(t) - \sigma(t_0))\,.$$

A.12 NONLINEAR OPERATORS

In this section we collect some necessary concepts regarding nonlinear operators.

Let X be a real reflexive Banach space.

DEFINITION A.12.1 *A (possibly nonlinear) operator $\mathsf{N} : X \to X'$ is called*

(*i*) MONOTONE, *if $\langle \mathsf{N}(u) - \mathsf{N}(v), u - v \rangle \ge 0\,,\ \forall u, v \in X$,*

(*ii*) STRICTLY MONOTONE, *if $\langle \mathsf{N}(u) - \mathsf{N}(v), u - v \rangle > 0\,,\ \forall u, v \in X : u \ne v$,*

(*iii*) COERCIVE, *if $\lim_{||u|| \to \infty} \frac{\langle \mathsf{N}(u), u \rangle}{||u||} = +\infty$,*

(*iv*) WEAKLY COERCIVE, *if $\lim_{||u|| \to \infty} ||\mathsf{N}(u)|| = +\infty$,*

(*v*) HEMICONTINUOUS, *if the map $\tau \mapsto \langle \mathsf{N}(u + \tau v), w \rangle$ is continuous on $[0, 1]\,, \forall u, v, w \in X$,*

(*vi*) OF M-TYPE, *if $u_n \rightharpoonup u$, $\mathsf{N}(u_n) \rightharpoonup f$ and $\limsup_n \langle \mathsf{N}(u_n), u_n \rangle \le \langle f, u \rangle$ implies that $\mathsf{N}(u) = f$.*

THEOREM A.12.2 ([451]) *Let $\mathsf{N} : X \to X'$ be monotone, hemicontinuous and weakly coercive. Then for each $f \in X'$, the equation $\mathsf{N}(u) = f$ has a solution. In addition, if N is strictly monotone the solution is unique.*

THEOREM A.12.3 ([380]) *An operator N that is monotone and hemicontinuous is of M-type.*

Appendix B

Some Facts from Stochastic Analysis

For a thorough introduction to stochastic analysis and stochastic differential equations in finite dimensions we refer to the excellent book of Karatzas and Shreve [224]. The infinite-dimensional theory, needed for the study of stochastic partial differential equations and stochastic integrodifferential equations, can be found in the monograph of Da Prato and Zabczyk [120]; see also [159] and [352] for more recent developments. For a general introduction to probability theory see, [300].

This appendix focuses on the results from stochastic analysis that are necessary for the study of stochastic PDEs and is based on [120] and [352].

B.1 PROBABILITY IN HILBERT SPACES

Let $(\Omega, \mathscr{F}, P)$ be a probability space and H be a separable Hilbert space.

DEFINITION B.1.1 *An H-valued random variable is a map $X : \Omega \to H$ that is measurable with respect to $\mathscr{F}$.*

REMARK B.1.2 We assume separability of H to simplify the subtle issues concerning different types of measurability in infinite-dimensional spaces (according to the Pettis theorem).

DEFINITION B.1.3 *The expectation of X is defined as*

$$\mathbb{E}_P[X] = \int_\Omega X \, dP \,,$$

where the integral is interpreted in the Bochner sense.

For any $h_1, h_2 \in H$, define the operator $(h_1 \otimes h_2)$ by $(h_1 \otimes h_2)h := h_1(h_2, h)$, $h \in H$. With the aid of this operator we may define the correlation operator between two Hilbert space random variables as follows.

DEFINITION B.1.4 *Let $X_1, X_2 : \Omega \to H$ be two H-valued random variables. The covariance operator of X_1 and X_2 is defined as*

$$cov(X_1, X_2) = \mathbb{E}[(X_1 - \mathbb{E}[X_1]) \otimes (X_2 - \mathbb{E}[X_2])] \,.$$

If $X_1 = X_2 = X$, the above definition gives the variance operator $cov(X)$ for the random variable X. The covariance operator is a positive and symmetric operator, i.e., $cov(X_1, X_2)^* = cov(X_2, X_1)$ and $cov(X) \geq 0$. Furthermore, $Tr(cov(X_1, X_2)) = \mathbb{E}[(X_1, X_2)] - (\mathbb{E}[X_1], \mathbb{E}[X_2])$ and if $\Phi \in \mathcal{L}(H, U)$

where U is another Hilbert space, $Tr(cov(\Phi X)) = Tr((cov(X)\Phi^*\Phi) = Tr(\Phi^*\Phi(cov(X))$. All the above hold for random variables in $L^2(\Omega, P, H)$.

For sequences of random variables we will encounter the following concepts of convergence:

1. $X_n \to X$, P-a.s. if $P(\{\omega : X_n(\omega) \not\to X(\omega)\}) = 0$.

2. $X_n \to X$ in $L^2(\Omega, \mathscr{F}, P; H)$ if $\mathbb{E}[||X_n - X||_H^2] \to 0$.

The generalisation of these concepts for weak convergence in H is obvious.

We now define the concept of conditional expectation.

DEFINITION B.1.5 *Let $X \in L^1(\Omega, \mathscr{F}, P; H)$ and let $\mathscr{G}$ be a σ-subalgebra of $\mathscr{F}$. The conditional expectation of X with respect to the σ-algebra $\mathscr{F}$ is the unique $\mathscr{G}$-measurable random variable $\mathbb{E}[X \mid \mathscr{G}]$ such that*

$$\int_G \mathbb{E}[X \mid \mathscr{G}]\, dP = \int_G X\, dP, \ \forall\, G \in \mathscr{G}.$$

In the case where $\mathscr{G} = \{\emptyset, \Omega\} =: \mathscr{O}$, the trivial σ-algebra, the conditional expectation coincides with the expectation. The conditional expectation has the meaning of the best predictor (in the L^2 sense) for the random variable X, given the information contained in the σ-algebra $\mathscr{G}$. For random variables $X \in L^2(\Omega, \mathscr{F}, P; H)$, the conditional expectation $\mathbb{E}[X \mid \mathscr{G}]$ can be interpreted as the projection of X in the subspace of square integrable $\mathscr{G}$-measurable random variables.

Another fundamental concept is the concept of *independence*.

DEFINITION B.1.6 *Consider the family of events $\{F_\beta\}_{\beta \in \mathcal{I}}$, $F_\beta \in \mathscr{F}$ where $\mathcal{I}$ is an index set. The events are said to be independent if*

$$P(\bigcap_{i=1}^n F_{\beta_i}) = \prod_{i=1}^n P(F_{\beta_i})$$

for any finite collection $\{F_{\beta_i}\}_{i=1}^n$ of them.

The concept of independence can be extended from events to σ-algebras and from that to random variables, through the concept of the σ-algebra generated by a random variable. For instance, the σ-subalgebras $\mathscr{F}_\beta$ of $\mathscr{F}$, $\beta \in \mathcal{I}$, are said to be independent if the events $\{F_\beta\}_{\beta \in \mathcal{I}}$ are independent for all $F_\beta \in \mathscr{F}_\beta$. A family of random variables is called independent if the σ-algebras generated by them[1] are independent.

By the definition of independence, we see that

1. If X_1, X_2 are independent, then $\mathbb{E}[(X_1, X_2)] = (\mathbb{E}[X_1], \mathbb{E}[X_2])$, or equivalently $cov(X_1, X_2) = 0$.

2. If X is independent of the σ-algebra $\mathscr{G}$, then $\mathbb{E}[X \mid \mathscr{G}] = \mathbb{E}[X]$.

---

[1]That is, the smaller σ-algebras that make these random variables measurable.

B.2 STOCHASTIC PROCESSES AND RANDOM FIELDS

In what follows we assume that H is a separable Hilbert space. An H-valued *stochastic process* is a family of H-valued random variables $\{X(t)\}$ parametrised by a parameter $t \in \mathcal{I}$ where $\mathcal{I}$ is a family of indices. A common choice is $\mathcal{I} = \mathbb{R}^+$. In many cases the index set is multidimensional. Then the stochastic process is usually called a *random field.* A family of σ-algebras $\{\mathscr{F}_t\}_{t\in\mathcal{I}}$ is called a *filtration* if it has the property $\mathscr{F}_s \subseteq \mathscr{F}_t$ for $s \le t$. A filtration is called *right-continuous* if $\mathscr{F}_{t+} := \cap_{s\ge t}\mathscr{F}_s = \mathscr{F}_t$. A stochastic process is said to be *cádlág*[2] if it is right continuous with limits from the left. A filtration is called *complete* if for $G \in \mathscr{F}$, $P(G) = 0$ implies $G \in \mathscr{F}_0$. A complete and right-continuous filtration is called a *normal* filtration.

DEFINITION B.2.1 *A stochastic process $\{X(t)\}$ with the property that $X(s)$ is measurable with respect to $\mathscr{F}_s$, for all $s \in \mathcal{I}$, is called adapted.*

DEFINITION B.2.2 *An H-valued stochastic process $X : [0,T] \times \Omega \to H$ is called predictable if it is measurable with respect to the σ-algebra:*

$$\mathcal{P}_T := \sigma\big((s,t] \times F_s : 0 \le s < t,\ F_s \in \mathscr{F}_s\} \cup \{\{0\} \times F_0 : F_0 \in \mathscr{F}_0\}\big)\,.$$

This σ-algebra is the one that makes $X : [0,T] \times \Omega \to H$ left-continuous and adapted to $\mathscr{F}_t$, $t \in [0,T]$.

DEFINITION B.2.3 *A process X is called progressive if for all $t \ge 0$ the restriction of the map $(t,\omega) \mapsto X(t,\omega)$ to $[0,t]\times\Omega$ is $\mathscr{B}([0,t])\otimes\mathscr{F}_t$-measurable.*

The following connection between the above types of stochastic processes holds.

PREDICTABLE $\longrightarrow$ ADAPTED AND cádlág $\longrightarrow$ PROGRESSIVE VERSION

Furthermore (see, e.g., [120]),

CONTINUOUS AND ADAPTED $\longrightarrow$ PREDICTABLE VERSION

An important class of stochastic processes, with many uses in stochastic analysis, is that of martingales.

DEFINITION B.2.4 *Let $(\Omega, \mathscr{F}, P)$ be a probability space and $\{\mathscr{F}_t\}_{t\in\mathcal{I}}$ a filtration on $\mathscr{F}$. A stochastic process $\{X(t)\}_{t\in\mathcal{I}}$, adapted with respect to this filtration, is called a martingale if it is integrable with respect to P for all $t \in \mathcal{I}$ and $\mathbb{E}[X(t) \mid \mathscr{F}_s] = X(s)$, for all $s \le t$.*

A martingale is a process for which the best prediction of future values, given the information available by time s, is the value at s. A martingale is the model for the winnings of a fair game. The following quantities are of interest for a Hilbert space-valued martingale.

[2] Continue á droite limites á gauche.

DEFINITION B.2.5 *Let $\{M(t)\}_{t \in \mathbb{R}^+}$ be a square integrable martingale, i.e., a martingale such that $\mathbb{E}[\,||M(t)||_H^2] < \infty$ for all $t \in \mathbb{R}^+$. Then*

(*i*) *The (real-valued) process $\langle M \rangle_t$ is the unique continuous adapted increasing process such that $\langle M \rangle_0 = 0$ and $||M(t)||_H - \langle M \rangle_t$ is a continuous martingale.*

(*ii*) *The quadratic variation process $\langle\langle M \rangle\rangle_t$ is adapted and continuous taking values in the space of non-negative trace class operators on H such that $\langle\langle M \rangle\rangle_0 = 0$ and for all $u_1, u_2 \in H$,*

$$\langle M(t), u_1 \rangle_H \langle M(t), u_2 \rangle_H - (\langle\langle M \rangle\rangle_t(u_1), u_2)_H$$

is a martingale.

It is possible to show (see, e.g., [159]) that $\langle M \rangle_t = Tr(\langle\langle M \rangle\rangle_t)$. Furthermore, if $\{t_i^n\}$ is a sequence of partitions of the interval $[0, t]$ such that $\Delta_n = \max\{|t_{i+1}^n - t_i^n|\, 1 \leq i \leq n-1\} \to 0$, then $\sum_{i=1}^n ||M(t_{i+1}^n) - M(t_i^n)||_H^2 \to \langle M \rangle_t$ where the convergence is P-a.s.

B.3 GAUSSIAN MEASURES

Let $Q : U \to U$ be a trace class operator (see Section A.6.1, Appendix A).

DEFINITION B.3.1 *Let U be a Hilbert space and $\mathscr{B}(U)$ be its Borel σ-algebra. A probability measure μ on $(U, \mathscr{B}(U))$ is called a Gaussian measure if for all $v \in U$, the mapping $v' : U \to \mathbb{R}$ defined by $v'(u) = (u, v)$, $u \in U$, has a Gaussian law, i.e., there exist $\mu \in \mathbb{R}$, $\sigma > 0$, such that*

$$(\mu \circ (v')^{-1})(A) = \mu(v' \in A) = \frac{1}{\sqrt{2\pi\sigma^2}} \int_A e^{-\frac{(x-m)^2}{2\sigma^2}}\, dx, \quad \forall A \in \mathscr{B}(\mathbb{R})\,.$$

If X is a U-valued random variable, we say that it follows a Gaussian law if for every $v \in U$, the real-valued random variable (X, v) follows a Gaussian law.

A well-known theorem provides the representation of a Gaussian measure on a Hilbert space in terms of a trace class operator.

THEOREM B.3.2 *A measure μ on $(U, \mathscr{B}(U))$ is Gaussian if and only if its Fourier transform (characteristic functional)*

$$\widetilde{\mu}(u) := \int_U e^{i(u,v)}\, \mu(dv) = e^{i(m,u) - \frac{1}{2}(Qu,u)}, \quad \forall\, u \in U$$

for some $m \in U$ and some trace class operator $Q : U \to U$. In this case the measure μ will be denoted by $N(\mu, Q)$.

PROPOSITION B.3.3 *The U-valued random variable X follows a Gaussian law $N(\mu, Q)$, i.e., $X \sim N(\mu, Q)$, if and only if*

$$X = \sum_{n=1}^{\infty} \sqrt{\lambda_n}\, z_n\, e_n + m,$$

where the equality holds in $L^2(\Omega, \mathscr{F}, P; U)$, λ_n and e_n are the eigenvalues and eigenfunctions, respectively, of the operator Q,

$$Q\, e_n = \lambda_n\, e_n, \;\; \lambda_n \geq 0, \;\; n \in \mathbb{N}$$

and z_n real-valued independent and identically distributed (i.i.d.) random variables $z_n \sim N(0, 1)$.

B.4 THE Q- AND THE CYLINDRICAL WIENER PROCESS

Let U be a Hilbert space, $(\Omega, \mathscr{F}, P)$ be a probability space, and $Q : U \to U$ be a linear symmetric non-negative nuclear operator (see Section A.6.1).

DEFINITION B.4.1 *A U-valued stochastic process $W(t)$, $t \in [0, T]$ on the probability space $(\Omega, \mathscr{F}, P)$ is called a Q-Wiener process if*

(*i*) $W(0) = 0$.

(*ii*) *W has P-a.s continuous trajectories.*

(*iii*) *The increments of W are independent.*

(*iv*) *The increments follow the Gaussian law $W(t) - W(s) \sim N(0, (t-s)\, Q)$.*

The nuclear operator Q serves as the covariance operator of the Wiener process.

EXAMPLE B.4.2 *In the particular case where $U = \mathbb{R}$ and Q degenerates to the identity operator $Q = I : \mathbb{R} \to \mathbb{R}$, the Q-Wiener process becomes the standard real-valued Wiener process.*

EXAMPLE B.4.3 *In the particular case where $U = \mathbb{R}^d$ and $Q = I_d$, the identity $d \times d$ matrix, the Q-Wiener process is the standard d-dimensional Wiener process.*

The Q-Wiener process has a useful series representation, as shown in the next proposition.

PROPOSITION B.4.4 *A U-valued stochastic process $W(t)$ is a Q-Wiener process if and only if it has the series representation $W(t) = \sum_n \sqrt{\lambda_n}\, w_n(t)\, e_n$, where λ_n, e_n are the eigenvalues and eigenfunctions of the operator Q, respectively, and $w_n(t)$ are independent standard Wiener processes in $\mathbb{R}$.*

The convergence of the above series can be shown to extend to the space $L^2(\Omega, \mathscr{F}, P; C([0,T]; U))$, thus guaranteeing the existence of continuous representatives of paths for the Q-Wiener process. Furthermore, the Q-Wiener process is a continuous square integrable martingale, with $\langle W \rangle_t = t\,Tr(Q)$ and $\langle\langle W \rangle\rangle_t = t\,Q$.

In a number of applications, generalisations of the Q-Wiener process are needed. These generalisations deal with the case where Q has the properties expected by a covariance operator but is not necessarily a trace class operator, e.g., we may need to consider cases where $\sum_n \sqrt{\lambda_n} = \infty$ (cylindrical Wiener process) or where $U = L^2(\mathbb{R}^d)$ and $\lambda_k = 1$ (space-time white noise process). More precisely:

Definition B.4.5 *A cylindrical Wiener process on a Hilbert space V is a family of mappings $\breve{W}(t,\omega) : V \to L^2(\Omega, \mathscr{F}, P; \mathbb{R})$ such that for every $v \in V$, the real-valued random variable $\breve{W}(t,\omega)(v) := \langle \breve{W}(t,\omega), v\rangle$ follows the centered normal distribution $N(0,t)$ and $\mathbb{E}[\breve{W}(t,\omega)(v_1)\breve{W}(t,\omega)(v_2)] = t\,(v_1, v_2)_V$.*

Let us note (see [339], Remark 1, p. 4) that $\breve{W}$ is linear. The cylindrical Wiener process has a formal series representation in terms of the basis of V, $\langle \breve{W}(t,\omega), v\rangle = \sum_{n=1}^{\infty} (v, e_n)\,\langle \breve{W}(t,\omega), e_n\rangle$, but *cannot* be realised[3] as a random variable on V, as is the case for the Q-Wiener process. The Q-Wiener process and the cylindrical Wiener process are related in the sense that if $\breve{W}(t,\omega)$ is a cylindrical Wiener process, then $\breve{W}(t,\omega) \circ Q^{1/2} = \sum_{n=1}^{\infty} \lambda_n^{1/2} \langle \breve{W}(t,\omega), e_n\rangle e_n$, where $Q^{1/2}$ is an operator such that $Q^{1/2} \circ Q^{1/2} = Q$ (see Proposition A.6.18 in Appendix A), is a Q-Wiener process. On the other hand if $W(t,\omega)$is a Q-Wiener process with a representation as in Proposition B.4.4, then $\breve{W}(t,\omega) = \sum_n^{\infty} (v, e_n) w_n(t)$ is a cylindrical Wiener process.

B.5 THE ITŌ INTEGRAL

Let U be a separable Hilbert space, $Q : U \to U$ be a trace class operator, and $W(t)$ be a Q-Wiener process on the filtered probability space $(\Omega, \mathscr{F}, P)$. In this section we consider the expression $\int_0^T \Phi(t)\,dW(t)$, where $\Phi(t)$ is a stochastic process. Our intention is to understand this quantity as the limit (in a sense to be properly defined shortly) of sums of the form $\sum_{j=0}^{n} \Phi(t_j)\,(W(t_{j+1}) - W(t_j))$, on a partition $\{t_j\}$ of $[0,T]$. These sums will represent random variables, which will have to be U-valued, so $\Phi(t)$ will have to be an operator-valued stochastic process taking values on an appropriate subset of $\mathcal{L}(U,U)$.

To develop a theory for the Itō integral in Hilbert spaces, we need to recall (see Proposition A.6.18, Appendix A) that for a trace class oper-

[3] In the sense that there may not exist a V-valued random variable X such that $\langle \breve{W}(t,\omega), v\rangle = (X, v)$.

ator Q there exists an operator $Q^{1/2}$ such that the decomposition $Q = Q^{1/2} \circ Q^{1/2}$, and define $U_0 = Q^{1/2}(U)$, which is a separable Hilbert space; when equipped with the inner product $(u_0, v_0)_0 := (Q^{-1/2}u, Q^{-1/2}v)_U = \sum_{n=1}^{\infty} \lambda_n^{-1}(u, e_n)_U(v, e_n)_U$.

We further need the following definition.

DEFINITION B.5.1 *Let $L_2^0 := L_2(U_0, V)$, the (separable) Hilbert space of Hilbert-Schmidt operators $\mathsf{F} : U_0 \to V$, equipped with the inner product*

$$(\mathsf{F}_1, \mathsf{F}_2)_{L_2^0} = Tr((\mathsf{F}_1 Q^{1/2})(\mathsf{F}_2 Q^{1/2})^*).$$

This Hilbert space is the appropriate space for the operator-valued stochastic processes $\Phi(t)$ over which the Itō integral is defined. Any bounded operator $\mathsf{F} \in \mathcal{L}(U, V)$ can be considered a Hilbert-Schmidt operator $\mathsf{F} \in L_2^0$, so that $\mathcal{L}(U, V) \subset L_2^0$, and therefore we have the alternative representation $(\mathsf{F}_1, \mathsf{F}_2)_{L_2^0} = Tr(\mathsf{F}_1 Q \mathsf{F}_2)$ (see, e.g., [159]). The elements of L_2^0 considered as operators from U to V are unbounded operators.

We are now ready to construct the Hilbert space-valued Itō integral ([159], [352]). This is done in the following steps:

STEP 1 Approximate the L_2^0, valued operator process $\Phi(t)$ by the sequence $\Phi_n(t)$ of *elementary processes*

$$\Phi_n = \sum_{k=1}^{M_n} \Phi_k^{(n)} \mathbf{1}_{A_k^{(n)}} \text{ with } \Phi_k^{(n)} \in L_2^0,\ A_k^{(n)} \in \mathcal{P}_T.$$

This is always feasible when convergence is taken in the sense of the norm $||\Phi||_T := \left(\mathbb{E}\left[\int_0^T ||\Phi(s)||_{L_2^0}^2 \, ds\right]\right)^{1/2}$.

STEP 2 We define the stochastic integral for each member of the sequence $\Phi_n(t)$ as a random sum. For instance, if[4] $\Phi_n(t) = \sum_{m=0}^{k-1} \Phi_m^{(n)} \mathbf{1}_{(t_m, t_{m+1}]}(t)$, then we define

$$\Psi_n := \int_0^T \Phi_n(t)\, dW(t) := \sum_{m=0}^{k-1} \Phi_m^{(n)} \left(W(t_{m+1}) - W(t_m)\right)$$

This is a sequence of V-valued random variables.

STEP 3 For each step of the approximation procedure we have the *Itō isometry* $\mathbb{E}[\,||\Psi_n||_V^2] = ||\Phi_n||_T^2$, which by definition of the norm $||\cdot||_T$ involves the covariance operator Q of the Wiener process.

STEP 4 We now go to the limit as $n \to \infty$ and define the Itō integral $\int_0^T \Phi(t)\, dW(t)$ as the limit in $L^2(\Omega, \mathscr{F}, P; V)$ of the sequence Ψ_n. This can be shown to exist.

By this construction the Itō integral inherits some useful properties, the most interesting of which is the extension of the Itō isometry. Furthermore, when considered as a stochastic process, $M(t) := \int_0^t \Phi(s)\, dW(s) :=$

[4] We omit the n dependence on the intervals $(t_m, t_{m+1}]$ out of which we may construct $A_k^{(n)} \in \mathcal{P}_T$ from the right-hand side to ease notation.

$\int_0^T \phi(s)\mathbf{1}_{[0,\,t]}(s)\,dW(s)$ is a V-valued martingale. The important properties are summarised in the following.

THEOREM B.5.2 *The Itō integral satisfies the following properties:*

(*i*) $\mathbb{E}[\int_0^T \Phi(s)dW(s)] = 0.$

(*ii*) $\mathbb{E}[\| \int_0^T \Phi(s)dW(s)\|_V^2\, ds] = \int_0^T \mathbb{E}[Tr(\Phi(s)Q\Phi^*(s))]ds.$

(*iii*) *The process* $M(t) := \int_0^t \Phi(s)\,dW(s)$ *is a square integrable martingale and*

$$\langle M\rangle_t = \int_0^t \|\Phi(s)\|^2_{L_2^0} ds, \quad \textit{and } \langle\langle M\rangle\rangle_t = \int_0^t (\Phi(s)Q^{1/2})(\Phi(s)Q^{1/2})^* ds.$$

Furthermore, the following estimates (*Burkholder-Davis-Gundy inequalities*) are often useful,

$$\mathbb{E}[\sup_{t\in[0,T]} \| \int_0^t \Phi(s)dW(s)\|_V^2\, ds] \le 4\,\mathbb{E}[\| \int_0^T \Phi(s)dW(s)\|_V^2\, ds],$$

$$\mathbb{E}[\sup_{t\in[0,T]} \| \int_0^t \Phi(s)dW(s)\|_V\, ds] \le 3\left(\int_0^T \mathbb{E}[Tr(\Phi(s)Q\Phi^*(s))ds]\right)^{1/2}.$$

An alternative representation for the stochastic integral is in terms of the series expansion

$$\int_0^T \Phi(s)\,dW(s) = \sum_{n=1}^{\infty} \int_0^T \Phi(s)e_n\, d(W(s), e_n),$$

where $\{e_n\}_{n\,\in\,\mathbb{N}}$ is an orthonormal basis of U consisting of eigenvectors of Q.

REMARK B.5.3 The construction of the stochastic integral performed above holds for predictable integrands. It may be generalised for simply adapted and not necessarily predictable processes, and further generalised for progressively measurable integrands. Furthermore, the condition

$$\mathbb{E}[\int_0^T \|\Phi(s)\|^2_{L_0^2} ds] < \infty$$

can be extended to

$$P(\int_0^T \|\Phi(s)\|^2_{L_0^2} ds < \infty) = 1\,.$$

However, in this case the stochastic integral may no longer be a square integrable martingale but a *local martingale* instead (see [159]).

The Itō stochastic integral can be generalised for integrals of the form $\int_0^T \Phi(s_d\check{W}(s)$, where $\check{W}(t)$ is the cylindrical Wiener process. This corresponds to the case where $Q = I$, and therefore the integrands $\Phi(s)$ are now Hilbert-Schmidt operators in $L_2(U,V)$ (rather than in $L_2^0 := L_2(Q^{1/2}(U),V)$.

The construction follows more or less the same steps as above, using elementary processes and the Itō isometry

$$\mathbb{E}[\,\|\int_0^T \Phi(s)\,\check{W}(s)\,ds\|_V^2] = \int_0^T \mathbb{E}[\,\|\Phi(s)\|^2_{L_2(U,\,V)}]ds\,.$$

In this case the Itō integral has similar properties as the ones presented in Theorem B.5.2, replacing the L_2^0 norm by the $L_2(U,V)$ norm and setting $Q = I$ (see, [159]). Alternatively, the stochastic integral with respect to the cylindrical Wiener process may be defined by a series representation as

$$\int_0^T \Phi(s)\,d\check{W}(s) = \sum_{n=1}^{\infty} \int_0^T \Phi(s)e_n\,d\langle \check{W}(s,\omega), e_n\rangle,$$

where $\{e_n\}_{n\,\in\,\mathbb{N}}$ is an orthonormal basis of the Hilbert space U and $\Phi(s)e_n$ is the action of the operator $\Phi(s)$ on e_n.

B.6 ITŌ FORMULA

Let $\Phi(\cdot)$ be an L_2^0-valued Itō integrable[5] process and $\mu(\cdot)$ be a Bochner integrable V-valued process. The stochastic process

$$X(t) = X(0) + \int_0^t \mu(s)\,ds + \int_0^t \Phi(s)\,dW(s)$$

is called an *Itō process.*

Consider now a real-valued function $F : [0,T] \times V \to \mathbb{R}$. If we define the stochastic process $Y(t) := F(t, X(t))$, is that an Itō process, and if so what does it look like?

This question is answered by the famous Itō's lemma, which provides the new calculus rules required for the stochastic integral.

THEOREM B.6.1 *Assume that F is continuous and further that the Fréchet*[6] *partial derivatives $\partial_t F, \partial_x F, \partial_{xx} F$ of F are continuous and bounded on bounded subsets of $[0,T] \times V$. Then the following equality holds P-a.s. for all $t \in [0,T]$:*

$$\begin{aligned} F(t, X(t)) = F(0, X(0)) &+ \int_0^t (\partial_x(s, X(s)), \Phi(s)\,dW(s))\,ds \\ &+ \int_0^t \{\partial_t F(s, X(s)) + (\partial_x F(s, X(s)), \mu(s))\}\,ds \\ &+ \int_0^t \frac{1}{2}\{Tr[\partial_x^2 F(s, X(s)))\,(\Phi(s)\,Q^{1/2})\,(\Phi(s)\,Q^{1/2})^*]\}\,ds, \end{aligned}$$

where the last term is commonly referred to as the Itō drift.

There exists a version of the Itō formula for Itō processes driven by the cylindrical Wiener process, with the necessary modifications ($Q = I$; see, e.g., Theorem 2.10 in [159]).

---

[5]That is, either $\mathbb{E}[\int_0^T \|\Phi(s)\|^2_{L_2^0} ds] < \infty$ or $P(\int_0^T \|\Phi(s)\|^2_{L_2^0} ds < \infty) = 1$.

[6]See Section A.3.1, Appendix A.

B.7 STOCHASTIC CONVOLUTION

Let W be a Q-Wiener process, $B(t) : U \to H$ be a family of bounded linear operator, and $S_A(t)$ be a C_0 semigroup on H, with generator A.

DEFINITION B.7.1 *The stochastic process*

$$W_A(t) = \int_0^t S_A(t-s)\, B(s)\, dW(s) = \int_0^t e^{(t-s)\,A}\, B(s)\, dW(s)$$

is called a stochastic convolution.

We make the following assumption.

ASSUMPTION B.7.2

$$\int_0^T ||S_A(r)B(r)||^2_{L_2^0}\, dr = \int_0^T Tr[S_A(r)\, B(r)\, Q\, B^*(r)\, S_A^*(r)]\, dr \ < \infty\, .$$

We now provide an important result for the properties of the stochastic convolution as a stochastic process.

THEOREM B.7.3 *If Assumption B.7.2 holds, then*

(*i*) *The process $W_A(\cdot)$ is Gaussian, continuous in mean square and has a predictable version.*

(*ii*) *The covariance of $W_A(\cdot)$ is given by*

$$Cov(W_A(t)) = \int_0^t S_A(r)\, B(r)\, Q\, B^*(r)\, S_A^*(r)\, dr, \ \ t \in [0,T]\, .$$

Furthermore, if $B(t) \in D(A)$, then $A \int_0^T B(t) dW(t) = \int_0^T A\, B(t)\, dW(t)$, P-a.s.

B.8 SDES IN HILBERT SPACES

Consider the linear differential equation

$$dX(t) = [A\, X(t) + f(t)]\, dt + B\, dW(t)\, . \tag{B.1}$$

DEFINITION B.8.1 *An H-valued predictable process $X(\cdot)$ is said to be a weak solution of (B.1) if the trajectories of $X(\cdot)$ are P-a.s. Bochner integrable and if for all $\zeta \in D(A^*)$ and all $t \in [0,T]$ we have*

$$(X(t), \zeta) = (x, \zeta) + \int_0^t \{(X(s), A^*\zeta) + (f(s), \zeta)\}\, ds\, . + (BW(t), \zeta).$$

The weak solutions of additive SDEs in Hilbert space may be expressed in terms of the stochastic convolution.

THEOREM B.8.2 *Under the assumptions of the previous section, equation (B.1) has a unique weak solution given by*

$$X(t) = S(t)\,x + \int_0^t S(t-s)\,f(s)\,ds + W_A(t),\ t \in [0,T]\,.$$

It can be shown that, under certain conditions, this solution has a continuous version.

The above construction may be generalised for SDEs of the general form

$$dX(t) = (A\,X(t) + F(t,X)))\,dt + B(t,X)\,dW(t)\,,$$

where F and B are in general nonlinear mappings

$$F : \Omega \times [0,T] \times C([0,T],H) \to H,$$
$$B : \Omega \times [0,T] \times C([0,T],H) \to L_2(Q^{1/2}(U),H),$$

satisfying appropriate regularity conditions (mainly Lipschitz continuity-type conditions). Then a solution can be constructed using the formal representation

$$X(t) = S_A(t)\,x + \int_0^t S_A(t-s)\,f(s,X(s))\,ds + \int_0^t S_A(t-s)B(s,X(s))dW(s),$$

in terms of the stochastic convolution and using appropriate fixed point schemes. We do not enter into details but rather refer the reader to [120] or [159].

B.9 MARTINGALE REPRESENTATION THEOREM

As we have seen in Theorem B.5.2, the Itō integral can be considered an "operation" that maps a square integrable martingale (the Wiener process) into another square integrable martingale. The following theorem, the celebrated *Martingale representation theorem*, is in some sense the converse statement.

THEOREM B.9.1 ([159]) *Let U, V be two real separable Hilbert spaces, $Q : U \to V$ a trace class operator, and W a Q-Wiener process on U. For every square integrable martingale $\{M(t)\}$ with values in V there exists a square integrable predictable process $\{\sigma(s)\}$ with values in $L_2^0 := L_2(Q^{1/2}(U),V)$ and with the property $\mathbb{E}[\int_0^T \|\sigma(s)\|^2_{L_2^0} ds] < \infty$ such that*

$$M(t) = M(0) + \int_0^t \sigma(s)\,dW(s)\,.$$

The martingale representation theorem holds in the case where $Q = I$, (i.e., when W is replaced by the cylindrical Wiener process $\check{W}$) with the necessary modifications (replacing L_2^0 by $L_2(U,V)$); see [88]).

Appendix C

Some Facts from Elliptic Homogenisation Theory

C.1 SPACES OF PERIODIC FUNCTIONS

In this section we collect a few key results on spaces of periodic functions that will be used for homogenisation problems. Our presentation closely follows [97].

Let Y denote the interval in $\mathbb{R}^N$ defined by

$$Y = (0, \ell_1) \times (0, \ell_2) \times \ldots (0, \ell_N), \tag{C.1}$$

where $\ell_1, \ell_2, \ldots, \ell_N$ are given positive numbers. Y is referred to as the reference period. The following definition introduces the notion of periodicity for functions that are defined almost everywhere.

DEFINITION C.1.1 *Let Y be defined by (C.1) and let f be a function defined almost everywhere on $\mathbb{R}^N$. The function f is called Y-periodic iff*

$$f(x + m\ell_j e_j) = f(x), \text{ a.e. on } \mathbb{R}^N, \ \forall m \in \mathbb{Z}\ , \ \forall j \in \{1, 2, \ldots, N\},$$

where $\{e_1, e_2, \ldots, e_N\}$ is the canonical basis of $\mathbb{R}^N$.

The mean value of a periodic function is essential when studying periodic oscillating functions.

DEFINITION C.1.2 *Let $\mathcal{O}$ be a bounded open set in $\mathbb{R}^N$ and let f be a function in $L^1(\mathcal{O})$. The mean value of f over $\mathcal{O}$ is the real number $\langle f \rangle$ given by*

$$\langle f \rangle = \frac{1}{|\mathcal{O}|} \int_{\mathcal{O}} f(y) dy.$$

The mean value of a periodic function can be computed on any translated set of the reference period, as detailed in the following proposition.

PROPOSITION C.1.3 *Let f be a Y-periodic function in $L^1(Y)$. Let y_0 be a fixed point in $\mathbb{R}^N$, and let Y_0 be the translated set of Y, defined by $Y_0 = y_0 + Y$. Define $f_\epsilon(x) := f\left(\frac{x}{\epsilon}\right)$ a.e. on $\mathbb{R}^N$. Then*

(i) $\int_{Y_0} f(y)dy = \int_Y f(y)dy$,

(ii) $\int_{\epsilon Y_0} f_\epsilon(x)dx = \int_{\epsilon Y} f_\epsilon(x)dx = \epsilon^N \int_Y f(y)dy$.

The following theorem is a key result.

THEOREM C.1.4 *Let $1 \leq p \leq +\infty$ and let f be a Y-periodic function in $L^p(Y)$. Then*

(i) *If $p < +\infty$, $f_\epsilon \rightharpoonup \langle f \rangle$ in $L^p(\mathcal{O})$, as $\epsilon \to 0$, where $\mathcal{O}$ is any bounded open subset of $\mathbb{R}^N$.*

(ii) *If $p = +\infty$, $f_\epsilon \overset{*}{\rightharpoonup} \langle f \rangle$ in $L^\infty(\mathbb{R}^N)$, as $\epsilon \to 0$.*

The weak convergences given in Theorem C.1.4 cannot be strong unless f is a constant and $|Y| = 1$. Indeed, strong convergence implies that $\langle f^p \rangle = \langle f \rangle^p$. Nevertheless, it is easy to check that, for $p > 1$, one has $\langle f^p \rangle \neq \langle f \rangle^p$. An interesting per se by-product of the proof of the above theorem is the following result: let p and f be as in Theorem C.1.4 and $\mathcal{I}$ be any open interval of $\mathbb{R}^N$ containing at least a translated set of Y. Then there exists a constant C depending only on the spatial dimension n such that

$$\|f_\epsilon\|^p_{L^p(\mathcal{I})} \leq C \frac{|\mathcal{I}|}{|Y|} \|f\|^p_{L^p(Y)},$$

for ϵ small enough.

We can now introduce a notion of periodicity for functions in the Sobolev space H^1.

DEFINITION C.1.5 *Let $C^\infty_{per}(Y)$ be the subset of $C^\infty(\mathbb{R}^N)$ of Y-periodic functions. We denote by $H^1_{per}(Y)$ the closure of $C^\infty_{per}(Y)$ in the H^1-norm.*

DEFINITION C.1.6 *Let g be a function defined a.e. on Y. We denote by $g^\#$ its extension by periodicity to the whole of $\mathbb{R}^N$, defined by*

$$g^\#(x + m\ell_j e_j) = g(x), \text{ a.e. on } Y, \ \forall m \in \mathbb{Z}\ , \forall j \in \{1, 2, \dots, N\}, \qquad \text{(C.2)}$$

where $\{e_1, e_2, \dots, e_N\}$ is the canonical basis of $\mathbb{R}^N$.

We have the following properties.

PROPOSITION C.1.7 *Let $u \in H^1_{per}(Y)$.*

(i) *u has the same trace on opposite faces of Y.*

(ii) *For any bounded open subset $\mathcal{O}$ of $\mathbb{R}^N$, $u^\#$ is in $H^1(\mathcal{O})$.*

DEFINITION C.1.8 *The quotient space $\mathcal{W}_{per}(Y) = H^1_{per}(Y)/\mathbb{R}$ is defined as the space of equivalence classes with respect to the relation $u \simeq v \iff u - v$ is a constant, $\forall u, v \in H^1_{per}(Y)$. We denote by $\dot{u}$ the equivalence class represented by u.*

PROPOSITION C.1.9 *The following properties hold:*

(i) *The quantity*

$$\|\dot{u}\|_{\mathcal{W}_{per}(Y)} = \|\text{grad}\, u\|_{L^2(Y)}\ , \ \forall u \in \dot{u}\ , \ \dot{u} \in \mathcal{W}_{per}(Y),$$

defines a norm on $\mathcal{W}_{per}(Y)$.

(ii) *The dual space $(\mathcal{W}_{per}(Y))'$ can be identified with the set*

$$\{F \in (H^1_{per}(Y))' : F(c) = 0, \forall c \in \mathbb{R}\},$$

with

$$\langle F, \dot{u} \rangle_{(\mathcal{W}_{per}(Y))', \mathcal{W}_{per}(Y)} = \langle F, u \rangle_{(H^1_{per}(Y))', H^1_{per}(Y)}\ , \ \forall u \in \dot{u}\ , \forall \dot{u} \in \mathcal{W}_{per}(Y).$$

C.2 COMPENSATED COMPACTNESS

The following *compensated compactness* result allows us to obtain information concerning the behaviour of products of weakly convergent sequences.

THEOREM C.2.1 *Let $\mathcal{O} \subset \mathbb{R}^3$ be bounded and let $\{u_n\}_{n\in\mathbb{N}}$ and $\{v_n\}_{n\in\mathbb{N}}$ be two sequences of vector fields in $H(\mathrm{div}, \mathcal{O})$ and $H(\mathrm{curl}, \mathcal{O})$ respectively. Suppose that $u_n \rightharpoonup u$ in $H(\mathrm{div}, \mathcal{O})$, and $v_n \rightharpoonup v$ in $H(\mathrm{curl}, \mathcal{O})$. Then $v_n \cdot v_n \to u \cdot v$ in $\mathcal{D}'(\mathcal{O})$.*

This is a kind of restatement of the celebrated *div-curl lemma* that is the cornerstone of the theory of compensated compactness. It was developed by François Murat and Luc Tartar in the late 1970s and is still a very active area of research; see [80]. If $v = (v^1, \ldots, v^N)^{tr}$ is a vector field with values in $\mathbb{R}^N$, let $\mathrm{div} v = \sum_{i=1}^{N} \partial_{x_i} v^i$ and $(\mathrm{curl} v)_{ij} = \partial_{x_j} v^i - \partial_{x_i} v^i$ be the generalisation of the usual curl. In its classical form (see, e.g., [329], [330], [403]), this lemma reads as follows.

THEOREM C.2.2 (THE DIV-CURL LEMMA) *If $\{u_n\}_{n\in\mathbb{N}}$ and $\{v_n\}_{n\in\mathbb{N}}$ are sequences in $L^2(\mathcal{O}, \mathbb{R}^N)$ that converge weakly in $L^2(\mathcal{O}, \mathbb{R}^N) \simeq (L^2(\mathcal{O}))^N$ to u and v, respectively, and if $\{\mathrm{div}\, u_n\}_{n\in\mathbb{N}}$ is compact in $H^{-1}(\mathcal{O})$ and $\{\mathrm{curl}\, v_n\}_{n\in\mathbb{N}}$ is compact in $H^{-1}(\mathcal{O}, \mathbb{M}^{N\times N}) \simeq (H^{-1}(\mathcal{O}))^{N^2}$, then $u_n \cdot v_n \rightharpoonup u \cdot v$ in $\mathcal{D}'(\mathcal{O})$.*

REMARK C.2.3 A natural generalisation concerns sequences that are bounded in $L^p(\mathcal{O}, \mathbb{R}^N)$ and $L^{p'}(\mathcal{O}, \mathbb{R}^N)$, respectively (where p, p' are conjugate exponents), with $\{\mathrm{div}\, u_n\}$ compact in $W^{-1,p}(\mathcal{O}, \mathbb{M}^{N\times N})$ and $\{\mathrm{curl}\, v_n\}$ compact in $W^{-1,q}(\mathcal{O}, \mathbb{M}^{N\times N})$ (see [330]).

REMARK C.2.4 In another interesting recent result (see [111]) it is shown that $u_n \cdot v_n \rightharpoonup u \cdot v$, if $u_n \rightharpoonup u$ in L^p and $v_n \rightharpoonup v$ in L^q, under the additional assumptions that the sequences $\{\mathrm{div}\, u_n\}$ and $\{\mathrm{curl}\, u_n\}$ are compact in the dual space of $W_0^{1,\infty}$ and that $\{u_n \cdot v_n\}$ is equi-integrable.

C.3 HOMOGENISATION OF ELLIPTIC EQUATIONS

Since they are essential to the following discussion, we present here the very basic principles of the theory of homogenisation of elliptic equations. The notation and the involved notions (e.g., Y-periodicity) are considered as in Section C.1. Let $\mathcal{O}$ be an open set in $\mathbb{R}^N$.

DEFINITION C.3.1 *For $0 < \alpha_1 < \alpha_2$, we denote by $\mathbb{M}_N(\alpha_1, \alpha_2, \mathcal{O})$ the set of $N \times N$ matrices $A = (a_{ij})_{1\le i,j\le N} \in (L^\infty(\mathcal{O}))^{N\times N}$ such that*

(*i*) $(A(x)\lambda, \lambda) \ge \alpha_1 |\lambda|^2$,

(*ii*) $|A(x)\lambda| \le \alpha_2 |\lambda|$,

for any $\lambda \in \mathbb{R}^N$ and a.e. on $\mathcal{O}$.

Consider a Y-periodic matrix $A = (a_{ij}) \in \mathbb{M}_{\mathrm{N}}(\alpha_1, \alpha_2, Y)$ and define $A^\epsilon(x) := A\left(\frac{x}{\epsilon}\right)$. Further, assume that $f \in H^{-1}(\mathcal{O})$. We wish to describe the asymptotic behaviour as $\epsilon \to 0$ of the following problems:

$$-\mathrm{div}(A^\epsilon \mathrm{grad}\, u^\epsilon) = f \text{ in } \mathcal{O}\,,$$
$$u^\epsilon = 0 \text{ on } \partial\mathcal{O}\,.$$

Let $\mathbb{P}^o_{\mathrm{per}} := \{\eta : \eta \text{ is } Y\text{-periodic and } \langle\eta\rangle = 0\}$. Now define the functions $\widehat{\theta}_\lambda$ and θ_λ as the solutions in $\mathbb{P}^o_{\mathrm{per}}$, of the following problems

$$-\mathrm{div}_y(A(y)\,\mathrm{grad}_y\,\widehat{\theta}_\lambda) = -\mathrm{div}_y(A(y)\,\lambda) \text{ in } Y\,,$$

and

$$-\mathrm{div}_y(A^{tr}(y)\,\mathrm{grad}_y\,\theta_\lambda) = -\mathrm{div}_y(A^{tr}(y)\,\lambda) \text{ in } Y\,.$$

By the Lax-Milgram lemma, both θ_λ and $\widehat{\theta}_\lambda$ are unique.

Further, define the functions $\widehat{w}_\lambda$ and w_λ as the (unique) solutions in $\mathbb{P}^o_{\mathrm{per}}$, respectively, of the following problems:

$$-\mathrm{div}_y(A(y)\,\mathrm{grad}_y\,\widehat{w}_\lambda) = \mathrm{div}_y(A(y)\lambda) \text{ in } Y\,,$$

and

$$-\mathrm{div}_y(A^{tr}(y)\,\mathrm{grad}_y\,w_\lambda) = \mathrm{grad}_y(A^{tr}(y)\lambda) \text{ in } Y\,.$$

REMARK C.3.2 It can be shown that $\widehat{\theta}_\lambda, \theta_\lambda, \widehat{w}_\lambda, w_\lambda$ satisfy the corresponding equations of the above problems not only in Y but (by periodicity) in $\mathcal{D}'(\mathbb{R}^N)$ as well.

These functions play an essential rôle in the homogenisation of the above problem since the homogenised matrix A^h is expressed in terms of $\widehat{\theta}_\lambda, \theta_\lambda, \widehat{w}_\lambda, w_\lambda$.

The main convergence result is as follows.

THEOREM C.3.3 *Let $f \in H^{-1}(\mathcal{O})$, A^ϵ and u^ϵ be defined as above. Then*

(i) $u^\epsilon \rightharpoonup u^*$, *in* $H^1_0(\mathcal{O})$,

(ii) $A^\epsilon \mathrm{grad}\, u^\epsilon \rightharpoonup A^h \mathrm{grad}\, u^*$, *in* $(L^2(\mathcal{O}))^N$,

where u^* is the unique solution in $H^1_0(\mathcal{O})$ of the constant coefficients (homogenised) problem

$$-\mathrm{div}(A^h\,\mathrm{grad}\, u^*) = f\,,\ \textit{in } \mathcal{O}\,,$$
$$u^* = 0\,,\ \textit{on } \partial\mathcal{O}\,.$$

The constant matrix $A^h = (a^h_{ij})_{1\le i,j\le N}$ is coercive and given by

$$A^h\,\lambda = \langle (A\lambda + A\,\mathrm{grad}\,\widehat{w}_\lambda)\rangle\,,\ \forall\lambda \in \mathbb{R}^N\,,$$

or, equivalently, by

$$(A^h)^{tr}\,\lambda = \langle (A^{tr}\lambda + A^{tr}\,\mathrm{grad}\,w_\lambda)\rangle\,,\ \forall\lambda \in \mathbb{R}^N\,,$$

where $\widehat{w}_\lambda$ and w_λ are the auxiliary functions defined above.

REMARK C.3.4 As can be seen in the proof of the above theorem (see, e.g., [97]), convergence (*ii*) is deduced from convergence (*i*). This fact is a particularity of the periodic case because of the explicit computation of the homogenised coefficients. In the general nonperiodic case convergence (*ii*) is not a consequence of convergence (*i*), and has to be proved separately. We state this convergence in the theorem since it is one of the important homogenisation results.

REMARK C.3.5 The well-known result stated in the above theorem can be proved by a variety of methods. One of them is the variational method of oscillating test functions due to Tartar, and another is the two-scale method (which takes into account the two scales x and $\frac{x}{\epsilon}$ of the problem) of Nguetseng and Allaire. The convergence in this sense implies the weak convergence. There is also the formal method of asymptotic expansions (known as the multiple-scale method). Tartar's method is a general one and is based on the construction of a suitable set of oscillating test functions that allows us to pass eventually to the limit. This method is related to the notion of compensated compactness. In particular, for the case of periodic coefficients, the test functions are periodic and are explicitly constructed in terms of $\widehat{w}_\lambda$. By passing to the limit one obtains the homogenised matrix A^h. There are further convergence properties based on this method, e.g., the convergence of energies and a corrector result.

In terms of the macroscopic scale x and the microscopic scale $\frac{x}{\epsilon}$ (describing the micro-oscillations), we look for an expansion of u^ϵ in the form

$$u^\epsilon(x) = \sum_{j=0}^{\infty} \epsilon^j\, u_j\left(x, \frac{x}{\epsilon}\right),$$

where $u_j = u_j(x,y)$ are Y-periodic in the second variable y. One first obtains that u_* depends on x only, and then one shows that this u_* is actually the solution of the homogenised problem with A^h defined as above. The interest in this is that in general, it permits us to "guess" formally the homogenised problem. Some natural questions arise at this point: How "far" is u^ϵ from u_0, i.e., what is the error (in a suitable norm) when replacing u^ϵ by u_*? What is the estimate when replacing u^ϵ by a finite sum $\sum_{j=0}^{M} \epsilon^j\, u_j\left(x, \frac{x}{\epsilon}\right)$? We give here an error estimate for the case $M = 2$ under some additional regularity assumptions on the data and on the boundary of $\partial\mathcal{O}$.

THEOREM C.3.6 *Let $f \in H^{-1}(\mathcal{O})$, A^ϵ and u^ϵ be defined as above. Then u^ϵ admits the following asymptotic expansion:*

$$u^\epsilon = u^* - \epsilon \sum_{k=1}^{N} \widehat{\theta}_k\left(\frac{x}{\epsilon}\right)\frac{\partial u^*}{\partial x_k} + \epsilon^2 \sum_{k,m=1}^{N} \widehat{\varrho}^{km}\left(\frac{x}{\epsilon}\right)\frac{\partial^2 u^*}{\partial x_k \partial x_m} + \mathcal{R}_{N,\epsilon},$$

where u^* is the solution of the problem of Theorem C.3.3, $\widehat{\theta}_\lambda \in \mathcal{W}_{per}(Y)$ is defined as above, and $\widehat{\varrho}^{km}$ is the solution in $\mathbb{P}^o_{per}$ of

$$-\mathrm{div}(A(y)\,\mathrm{grad}\,\widehat{\varrho}^{km}) = -a^h_{km} - \sum_{i,j=1}^{N} \frac{\partial(a_{ij}\delta_{ki}\widehat{\theta}_m)}{\partial y_i} - \sum_{j=1}^{N} a_{kj}\frac{\partial(\widehat{\theta}_m - y_m)}{\partial y_j} \quad in\ Y.$$

Moreover, if $f \in C^\infty(\overline{\mathcal{O}})$, $\partial\mathcal{O}$ *is of class* $\in C^\infty$, *and, additionally,*

$$\widehat{\theta}_k, \widehat{\varrho}^{km} \in W^{1,\infty}(Y), \forall k, m = 1, \ldots, N,$$

then there exists a constant C *independent of* ϵ *such that*

$$||\mathcal{R}_{N,\epsilon}||^2_{H^1(\mathcal{O})} \leq C\epsilon.$$

REMARK C.3.7 To prove the above error estimate, more regularity on the functions $\widehat{\theta}_k$ and $\widehat{\varrho}^{km}$ has to be assumed, namely, that their first derivatives are bounded functions. It is plausible then to ask under what hypotheses we can establish this property. Actually, it can be deduced from classical elliptic regularity (see, e.g., [113], [161]), under strong regularity assumptions on the matrix A, namely, that it is at least assumed to be continuous. Nevertheless, this is not true in general for composite materials (see, e.g., Chapter 5 of [97]).

C.4 RANDOM ELLIPTIC HOMOGENISATION THEORY

The homogenisation theory for elliptic problems may be extended for random media when the condition for periodicity is replaced by suitable ergodicity assumptions. The ergodic theorem can then be used to provide results for the limiting behaviour of the solutions when $\epsilon \to 0$.

DEFINITION C.4.1 *Let* $F \in L^1_{loc}(\mathbb{R}^N)$. *A number* F_{M} *is called the mean value of* F *if*

$$\lim_{\epsilon \to 0} \frac{1}{|K|} \int_K F\left(\frac{x}{\epsilon}\right) dx = F_{\mathsf{M}}$$

for any Lebesgue-measurable bounded set $K \subset \mathbb{R}^N$ *and* $|K|$ *stands for the Lebesgue measure of* K.

THEOREM C.4.2 (BIRKHOFF ERGODIC THEOREM) *Let* $(\Omega, \mathscr{F}, P)$ *be a probability space and let* $\tau = \{\tau_x\}_{x \in \mathbb{R}}$ *be a measure-preserving group of transformations on* Ω. *If* $F \in L^p(\Omega, \mathscr{F}, P)$, $p \geq 1$, *then the realisation* $\check{F}(x, \omega) := F(\tau_x \omega)$ *for almost all* $\omega \in \Omega$ *possesses a mean value*[1] $F_{\mathsf{M}}(\omega)$ *in the sense that*

$$F\left(\frac{x}{\epsilon}, \omega\right) \rightharpoonup F_{\mathsf{M}}(\omega), \quad in\ L^p_{loc}(\mathbb{R}^N), \quad P-a.s.$$

The mean value $F_{\mathsf{M}}(\omega)$ *is invariant under* τ *in the sense that* $F_{\mathsf{M}}(\tau_x\omega) = F_{\mathsf{M}}(\omega)$ *for all* $x \in \mathbb{R}$, P*-a.s. and we also have that* $\mathbb{E}_P[F] = \mathbb{E}_P[F_{\mathsf{M}}]$. *If, furthermore, the transformation* τ_x *is ergodic, then* F_{M} *is no longer a random variable*[2], *and* $F_{\mathsf{M}} = \mathbb{E}_P[F]$.

[1]The mean value is in general a random variable.
[2]That is, it is independent of $\omega \in \Omega$.

The ergodic theorem in some sense replaces Proposition C.1.4 in the case of a random coefficients. The above theorem implies that (see, e.g., [67])

$$F\left(\frac{x}{\epsilon}, \omega\right) \stackrel{*}{\rightharpoonup} \mathbb{E}_P[F(x,\omega)], \text{ as } \epsilon \to 0, \text{ in } L^\infty(\mathbb{R}^N), \ P - a.s.$$

The periodic elliptic homogenisation theory has a counterpart for random media, with ergodic coefficients. The basic theoretical tools of homogenisation theory, e.g., the div-curl lemma, have generalisations in the case where the fields involved are random variables. For these generalisations one may consult, e.g., [216], which allows the identification of the homogenised coefficients.

Then the elliptic homogenisation theory can be generalised as follows.

THEOREM C.4.3 *Let $A = A(\omega)$ be a matrix defined on a probability space $(\Omega, \mathscr{F}, P)$, $A(\omega) = \{a_{ij}(\omega)\}$, $a_{ij} \in L^\infty(\Omega)$, satisfying the ellipticity condition a.s. Then, for almost all ω, the matrix $A(x) = A(\tau_x\omega)$ admits homogenisation and the homogenised matrix A^h is independent of ω.*

The explicit form of the homogenised matrix A^h can be given in terms of averages of the coefficients with respect to the solutions of the cell problems. The cell problems are now random problems of similar form as their periodic counterparts,

$$-\text{div}(A(x)\, \text{grad}\, w_\lambda) = \text{div}(A(x)\lambda), \quad \lambda \in \mathbb{R}^N,$$

but now the periodicity boundary condition is replaced by the conditions that $\text{grad} w_\lambda(\tau_x\omega)$ is invariant and $\langle w_\lambda \rangle = \mathbb{E}_P[w] = 0$, and the average diffusivity is given formally by the same expression as in Theorem C.3.3, where now the averaging operator is to be understood in the above sense.

Appendix D

Some Facts from Dyadic Analysis

(*by George Dassios*)

The dyadic representation is very useful in the following situations:

1. when we deal with more than one basis,
2. when we use curvilinear coordinates, where differential operators need to act on the components as well as on the basis elements,
3. in representing solutions of equations in a way that separates the variables from the parameters of the problem,
4. in performing successive algebraic operations in a systematic way,
5. in presenting complicated expressions in a compact and meaningful form.

The theory of dyadics was introduced and developed by J. W. Gibbs and first appeared in the celebrated book on vector calculus, first published in 1901, by E. B. Wilson [434]. Nevertheless, the most complete introduction to polyadics is contained in the monumental book *Vector and Tensor Analysis* by L. Brand [78].

If $\widehat{\mathbf{x}}_1$, $\widehat{\mathbf{x}}_2$, $\widehat{\mathbf{x}}_3$ are the basis vectors[1] of a Cartesian system in $\mathbb{R}^3$, then any vector $\mathbf{x} \in \mathbb{R}^3$ is represented, with respect to this basis, as the linear combination

$$\mathbf{x} = x_1\widehat{\mathbf{x}}_1 + x_2\widehat{\mathbf{x}}_2 + x_3\widehat{\mathbf{x}}_3 \,, \tag{D.1}$$

where for $i = 1, 2, 3$,

$$x_i = \mathbf{x} \cdot \widehat{\mathbf{x}}_i \tag{D.2}$$

are the projections of $\mathbf{x}$ on the basis vectors.

Substituting (D.2) into (D.1) and using the commutative property of the inner product, we obtain

$$\mathbf{x} = \widehat{\mathbf{x}}_1(\widehat{\mathbf{x}}_1 \cdot \mathbf{x}) + \widehat{\mathbf{x}}_2(\widehat{\mathbf{x}}_2 \cdot \mathbf{x}) + \widehat{\mathbf{x}}_3(\widehat{\mathbf{x}}_3 \cdot \mathbf{x}) \,. \tag{D.3}$$

We can interpret the right-hand side of (D.3) as the inner product from the right of an entity consisting of the sum of three pairs of *ordered* basis vectors.

[1] In this appendix, contrary to the notation adopted in the rest of the book, the (traditional) boldface notation is used for vectors.

That is, if we denote the order relation by the tensor product symbol "$\otimes$", then (D.3) is written as

$$\mathbf{x} = (\widehat{\mathbf{x}}_1 \otimes \widehat{\mathbf{x}}_1 + \widehat{\mathbf{x}}_2 \otimes \widehat{\mathbf{x}}_2 + \widehat{\mathbf{x}}_3 \otimes \widehat{\mathbf{x}}_3) \cdot \mathbf{x} . \tag{D.4}$$

More generally, if $\mathbf{a}_i$ and $\mathbf{b}_i$, $i = 1, 2, \ldots, n$, are any vectors in $\mathbb{R}^3$, then we can form the sum of ordered vectors

$$\widetilde{\widetilde{\mathbf{A}}} = \sum_{i=1}^{n} \mathbf{a}_i \otimes \mathbf{b}_i \, , \tag{D.5}$$

which defines[2] a *dyadic*. Every term $\mathbf{a}_i \otimes \mathbf{b}_i$ of the dyadic is called a *dyad*. The vectors $\mathbf{a}_i$ are called *antecedents* and the vectors $\mathbf{b}_i$ are called *consequents*.

The dyadic

$$\widetilde{\widetilde{\mathbf{I}}} = \sum_{i=1}^{3} \widehat{\mathbf{x}}_i \otimes \widehat{\mathbf{x}}_i, \tag{D.6}$$

where $\widehat{\mathbf{x}}_i$, $i = 1, 2, 3$, are orthonormal basis vectors, represents the *identity dyadic* in $\mathbb{R}^3$, having the property

$$\widetilde{\widetilde{\mathbf{I}}} \cdot \mathbf{x} = \mathbf{x} \cdot \widetilde{\widetilde{\mathbf{I}}} = \mathbf{x}, \;\; \mathbf{x} \in \mathbb{R}^3 . \tag{D.7}$$

In analogy with a vector, which is an ordered set of scalars, a dyadic is an ordered set of vectors. It is actually a linear map $\mathbf{f}$ that acts on vectors through the inner product from the right, i.e.,

$$\mathbf{f}(\mathbf{x}) = \widetilde{\widetilde{\mathbf{A}}} \cdot \mathbf{x} = \left(\sum_{i=1}^{n} \mathbf{a}_i \otimes \mathbf{b}_i \right) \cdot \mathbf{x} = \sum_{i=1}^{n} \mathbf{a}_i (\mathbf{b}_i \cdot \mathbf{x}) . \tag{D.8}$$

The dyadic

$$(\widetilde{\widetilde{\mathbf{A}}})^{tr} = \textstyle\sum_{i=1}^{n} \mathbf{b}_i \otimes \mathbf{a}_i \tag{D.9}$$

is called the *transpose* of $\widetilde{\widetilde{\mathbf{A}}}$. The dyadic $\widetilde{\widetilde{\mathbf{A}}}$ is called *symmetric* if it coincides with its transpose, i.e., $\widetilde{\widetilde{\mathbf{A}}} = (\widetilde{\widetilde{\mathbf{A}}})^{tr}$, and it is called *antisymmetric* if it coincides with the opposite of its transpose, i.e., $\widetilde{\widetilde{\mathbf{A}}} = -(\widetilde{\widetilde{\mathbf{A}}})^{tr}$. Every dyadic can be decomposed into the sum of a symmetric and an antisymmetric dyadic according to the standard decomposition

$$\widetilde{\widetilde{\mathbf{A}}} = (\widetilde{\widetilde{\mathbf{A}}})_{\text{sym}} + (\widetilde{\widetilde{\mathbf{A}}})_{\text{ant}} = \frac{\widetilde{\widetilde{\mathbf{A}}} + (\widetilde{\widetilde{\mathbf{A}}})^{tr}}{2} + \frac{\widetilde{\widetilde{\mathbf{A}}} - (\widetilde{\widetilde{\mathbf{A}}})^{tr}}{2} . \tag{D.10}$$

Obviously, the two vectors that define any dyad do not commute, but the action of $(\widetilde{\widetilde{\mathbf{A}}})^{tr}$ on the vector $\mathbf{x}$ is equivalent to the action of $\widetilde{\widetilde{\mathbf{A}}}$ on the vector $\mathbf{x}$ from the left, i.e.,

$$(\widetilde{\widetilde{\mathbf{A}}})^{tr} \cdot \mathbf{x} = \mathbf{x} \cdot \widetilde{\widetilde{\mathbf{A}}} , \;\; \mathbf{x} \in \mathbb{R}^3. \tag{D.11}$$

---

[2]The notation $\widetilde{\widetilde{\mathbf{A}}}$, and not the standard $\widetilde{\mathbf{A}}$, for a dyadic is adopted in this book, since the symbol tilde is reserved to denote the Fourier transform.

Two dyadics $\widetilde{\widetilde{\mathbf{A}}}$ and $\widetilde{\widetilde{\mathbf{B}}}$ are *equal* if they act the same way on any vector, i.e., when

$$\widetilde{\widetilde{\mathbf{A}}} \cdot \mathbf{x} = \widetilde{\widetilde{\mathbf{B}}} \cdot \mathbf{x}\,, \;\; \mathbf{x} \in \mathbb{R}^3\,, \tag{D.12}$$

or equivalently, when

$$\mathbf{x} \cdot \widetilde{\widetilde{\mathbf{A}}} = \mathbf{x} \cdot \widetilde{\widetilde{\mathbf{B}}}\,, \;\; \mathbf{x} \in \mathbb{R}^3\,. \tag{D.13}$$

The zero dyadic $\widetilde{\widetilde{\mathbf{0}}}$ is defined by the relations

$$\widetilde{\widetilde{\mathbf{0}}} \cdot \mathbf{x} = \mathbf{x} \cdot \widetilde{\widetilde{\mathbf{0}}} = \mathbf{0}\,, \;\; \mathbf{x} \in \mathbb{R}^3\,. \tag{D.14}$$

The vector space operations of addition and scalar multiplication, as well as their properties, are transfered to the dyadics in a straightforward way. Indeed, we define the addition of dyadics as

$$(\widetilde{\widetilde{\mathbf{A}}} + \widetilde{\widetilde{\mathbf{B}}}) \cdot \mathbf{x} = \widetilde{\widetilde{\mathbf{A}}} \cdot \mathbf{x} + \widetilde{\widetilde{\mathbf{B}}} \cdot \mathbf{x}\,, \;\; \mathbf{x} \in \mathbb{R}^3 \tag{D.15}$$

and the scalar multiplication of dyadics as

$$(\lambda \widetilde{\widetilde{\mathbf{A}}}) \cdot \mathbf{x} = \widetilde{\widetilde{\mathbf{A}}} \cdot (\lambda \mathbf{x}), \lambda \in \mathbb{R}\,, \;\; \mathbf{x} \in \mathbb{R}^3\,, \tag{D.16}$$

and it is easily shown that these two operators satisfy all the "standard" properties. Therefore, the set of all dyadics in $\mathbb{R}^3$, equipped with the operation (D.15) and (D.16), defines a vector space $\mathcal{D}$ called the *dyadic space*.

Consider the vectors

$$\begin{aligned} \mathbf{a}_i &= a_i^1 \widehat{\mathbf{x}}_1 + a_i^2 \widehat{\mathbf{x}}_2 + a_i^3 \widehat{\mathbf{x}}_3 \;, i = 1, 2, \ldots, n \;, \\ \mathbf{b}_i &= b_i^1 \widehat{\mathbf{x}}_1 + b_i^2 \widehat{\mathbf{x}}_2 + b_i^3 \widehat{\mathbf{x}}_3 \;, i = 1, 2, \ldots, n. \end{aligned}$$

Then the dyadic $\widetilde{\widetilde{\mathbf{A}}}$ is written as

$$\begin{aligned} \widetilde{\widetilde{\mathbf{A}}} &= \sum_{i=1}^{n} \mathbf{a}_i \otimes \mathbf{b}_i = \left(\sum_{i=1}^{n} \mathbf{a}_i b_i^1 \right) \otimes \widehat{\mathbf{x}}_1 + \left(\sum_{i=1}^{n} \mathbf{a}_i b_i^2 \right) \otimes \widehat{\mathbf{x}}_2 + \left(\sum_{i=1}^{n} \mathbf{a}_i b_i^3 \right) \otimes \widehat{\mathbf{x}}_3 \\ &= \widehat{\mathbf{x}}_1 \otimes \left(\sum_{i=1}^{n} a_i^1 \mathbf{b}_i \right) + \widehat{\mathbf{x}}_2 \otimes \left(\sum_{i=1}^{n} a_i^2 \mathbf{b}_i \right) + \widehat{\mathbf{x}}_3 \otimes \left(\sum_{i=1}^{n} a_i^3 \mathbf{b}_i \right), \end{aligned} \tag{D.17}$$

which shows that every dyadic in $\mathbb{R}^3$ can be written as a sum of three dyads. Similarly, using the distributive law, we can write any dyadic in the form

$$\widetilde{\widetilde{\mathbf{A}}} = \sum_{i,j=1}^{3} A_{ij} \widehat{\mathbf{x}}_i \otimes \widehat{\mathbf{x}}_j\,, \tag{D.18}$$

which implies immediately that the set

$$\mathcal{B} = \{ \widehat{\mathbf{x}}_i \otimes \widehat{\mathbf{x}}_j : i = 1, 2, 3 \;\text{ and }\; j = 1, 2, 3 \} \tag{D.19}$$

provides a basis for the dyadic space $\mathcal{D}$.

Note that when there is no ambiguity about the reference system, we specify a vector as ordered scalars, i.e., $\mathbf{x} = (x_1, x_2, x_3)$, where x_1, x_2, x_3 are the components of $\mathbf{x}$ with respect to the suppressed basis vectors $\widehat{\mathbf{x}}_1$, $\widehat{\mathbf{x}}_2$,

$\widehat{\mathbf{x}}_3$. If we try to do the same with the dyadic $\widetilde{\widetilde{\mathbf{A}}}$ given in (D.18), we see that because of the double index in the components A_{ij}, we need to order these components in two dimensions, and this ordering yields the matrix

$$\mathbb{A} = [A_{ij}], \tag{D.20}$$

which is called the *nonion form* of the dyadic $\widetilde{\widetilde{\mathbf{A}}}$ with respect to the basis $\widehat{\mathbf{x}}_1, \widehat{\mathbf{x}}_2, \widehat{\mathbf{x}}_3$. Therefore, a dyadic is nothing more than the representation of a linear map in such a way that not only the components A_{ij} but also the basis dyads $\widehat{\mathbf{x}}_i \otimes \widehat{\mathbf{x}}_j$ are "visible" in the representation. In the same way, the representation (D.1) of a vector shows both the components and the basis vectors, while in the representation $\mathbf{x} = (x_1, x_2, x_3)$, the basis vectors are suppressed. In this respect, a dyadic is formed by ordering vectors.

A dyadic is called *complete* if its range is three-dimensional, i.e., if the dyadic maps $\mathbb{R}^3$ to $\mathbb{R}^3$; otherwise it is called *singular*. Then, a dyadic is *invertible* if and only if it is complete. The inverse of a dyadic is defined by the property

$$\widetilde{\widetilde{\mathbf{A}}} \cdot (\widetilde{\widetilde{\mathbf{A}}})^{-1} = (\widetilde{\widetilde{\mathbf{A}}})^{-1} \cdot \widetilde{\widetilde{\mathbf{A}}} = \widetilde{\widetilde{\mathbf{I}}}. \tag{D.21}$$

As a general rule, the properties and the conditions that one needs to know in order to perform calculations at the dyadic level are identical with the corresponding ones at the matrix level. For example, a dyadic is invertible if and only if its nonion form is invertible. Similarly, a dyadic is singular if and only if there is a nonzero vector that the dyadic maps to zero. Furthermore, one can show that the dyadic

$$\widetilde{\widetilde{\mathbf{A}}} = \mathbf{a}_1 \otimes \mathbf{b}_1 + \mathbf{a}_2 \otimes \mathbf{b}_2 + \mathbf{a}_3 \otimes \mathbf{b}_3 \tag{D.22}$$

is complete if and only if $\mathbf{a}_1, \mathbf{a}_2, \mathbf{a}_3$ and $\mathbf{b}_1, \mathbf{b}_2, \mathbf{b}_3$ form two sets of linearly independent vectors. It is straightforward to show that if only two of the three vectors $\mathbf{a}_1, \mathbf{a}_2, \mathbf{a}_3$ are linearly independent, then the dyadic can be written as the sum of two dyads. Such a dyadic is called *planar* because its range is two-dimensional. If only one of the vectors $\mathbf{a}_1, \mathbf{a}_2, \mathbf{a}_3$ is linearly independent, then the dyadic can be represented by a single dyad. In this case, the dyadic is called *linear*, since its range is restricted to a line.

The following products, which are defined for dyads, can be extended by linearity arguments to any general dyadic

$$(\mathbf{a} \otimes \mathbf{b}) \cdot (\mathbf{c} \otimes \mathbf{d}) = (\mathbf{b} \cdot \mathbf{c})\mathbf{a} \otimes \mathbf{d}. \tag{D.23}$$

$$(\mathbf{a} \otimes \mathbf{b}) : (\mathbf{c} \otimes \mathbf{d}) = (\mathbf{b} \cdot \mathbf{c})(\mathbf{a} \cdot \mathbf{d}). \tag{D.24}$$

$$(\mathbf{a} \otimes \mathbf{b}) \times \mathbf{c} = \mathbf{a} \otimes (\mathbf{b} \times \mathbf{c}). \tag{D.25}$$

$$\mathbf{c} \times (\mathbf{a} \otimes \mathbf{b}) = (\mathbf{c} \times \mathbf{a}) \otimes \mathbf{b}. \tag{D.26}$$

$$(\mathbf{a} \otimes \mathbf{b}) \overset{\cdot}{\times} (\mathbf{c} \otimes \mathbf{d}) = (\mathbf{b} \cdot \mathbf{c}) \otimes (\mathbf{a} \times \mathbf{d}). \tag{D.27}$$

$$(\mathbf{a} \otimes \mathbf{b}) \overset{\times}{\times} (\mathbf{c} \otimes \mathbf{d}) = (\mathbf{b} \times \mathbf{c}) \otimes (\mathbf{a} \times \mathbf{d}). \tag{D.28}$$

The most important characteristics of a dyadic are its *invariants*. These are particular expressions of the antecedents and the consequents of the dyadic that stay invariant under any change of the basis vectors. Therefore, they carry the structure of the dyadic and are independent of the choice of basis. Every dyadic has six invariants - three scalar, two vector and one dyadic. The dyadic

$$\tilde{\tilde{\mathbf{A}}} = \mathbf{a} \otimes \boldsymbol{\ell} + \mathbf{b} \otimes \mathbf{m} + \mathbf{c} \otimes \mathbf{n} \tag{D.29}$$

has the *first*, *second* and *third scalar invariants*

$$\varphi_1 = \mathbf{a} \cdot \boldsymbol{\ell} + \mathbf{b} \cdot \mathbf{m} + \mathbf{c} \cdot \mathbf{n}, \tag{D.30}$$

$$\varphi_2 = (\mathbf{a} \times \mathbf{b}) \cdot (\boldsymbol{\ell} \times \mathbf{m}) + (\mathbf{b} \times \mathbf{c}) \cdot (\mathbf{m} \times \mathbf{n}) + (\mathbf{c} \times \mathbf{a}) \cdot (\mathbf{n} \times \boldsymbol{\ell}), \tag{D.31}$$

$$\varphi_3 = (\mathbf{a} \times \mathbf{b} \cdot \mathbf{c})(\boldsymbol{\ell} \times \mathbf{m} \cdot \mathbf{n}), \tag{D.32}$$

the *first* and *second vector invariants*

$$\boldsymbol{\phi}_1 = \mathbf{a} \times \boldsymbol{\ell} + \mathbf{b} \times \mathbf{m} + \mathbf{c} \times \mathbf{n}, \tag{D.33}$$

$$\boldsymbol{\phi}_2 = (\mathbf{a} \times \mathbf{b}) \times (\boldsymbol{\ell} \times \mathbf{m}) + (\mathbf{b} \times \mathbf{c}) \times (\mathbf{m} \times \mathbf{n}) + (\mathbf{c} \times \mathbf{a}) \times (\mathbf{n} \times \boldsymbol{\ell}), \tag{D.34}$$

and the *dyadic invariant*

$$\tilde{\tilde{\boldsymbol{\Phi}}} = (\mathbf{a} \times \mathbf{b}) \otimes (\boldsymbol{\ell} \times \mathbf{m}) + (\mathbf{b} \times \mathbf{c}) \otimes (\mathbf{m} \times \mathbf{n}) + (\mathbf{c} \times \mathbf{a}) \otimes (\mathbf{n} \times \boldsymbol{\ell}) \,. \tag{D.35}$$

Note that φ_2 is the first scalar invariant and $\boldsymbol{\phi}_2$ is the first vector invariant of the dyadic invariant $\tilde{\tilde{\boldsymbol{\Phi}}}$. The scalar invariants φ_1 and φ_3 are the trace and the determinant, respectively, of any nonion form of $\tilde{\tilde{\mathbf{A}}}$.

We can extend the theory of dyadics to include linear combinations of three ordered vectors $\mathbf{a} \otimes \mathbf{b} \otimes \mathbf{c}$, which defines a *triadic*; four ordered vectors $\mathbf{a} \otimes \mathbf{b} \otimes \mathbf{c} \otimes \mathbf{d}$, which defines a *tetradic*; five ordered vectors $\mathbf{a} \otimes \mathbf{b} \otimes \mathbf{c} \otimes \mathbf{d} \otimes \mathbf{e}$, which defines a *pentadic*; and in general to include n-ordered vectors $\mathbf{a}_1 \otimes \mathbf{a}_2 \otimes \cdots \otimes \mathbf{a}_n$, which defines a *polyadic of order n*. The terms of a polyadic acts on vectors as

$$\mathbf{a}_1 \otimes \mathbf{a}_2 \otimes \cdots \otimes \mathbf{a}_n \cdot \mathbf{x} = (\mathbf{a}_n \cdot \mathbf{x})\, \mathbf{a}_1 \otimes \mathbf{a}_2 \otimes \cdots \otimes \mathbf{a}_{n=1} \,. \tag{D.36}$$

The nonion form of a polyadic of order n is a tensor of order n. In every case, the relative inner and cross products are acting between the neighbouring vectors, as the following examples show;

$$(\mathbf{a} \otimes \mathbf{b} \otimes \mathbf{c} \otimes \mathbf{d}) \overset{\times}{\underset{\times}{}} (\mathbf{e} \otimes \mathbf{f} \otimes \mathbf{g}) = \mathbf{a} \otimes \mathbf{b} \otimes (\mathbf{d} \times \mathbf{e}) \otimes (\mathbf{c} \otimes \mathbf{f}) \otimes \mathbf{g}. \tag{D.37}$$

$$(\mathbf{a} \otimes \mathbf{b} \otimes \mathbf{c}) \vdots (\mathbf{e} \otimes \mathbf{f} \otimes \mathbf{g}) = (\mathbf{c} \cdot \mathbf{e})(\mathbf{b} \cdot \mathbf{f})(\mathbf{a} \cdot \mathbf{g}). \tag{D.38}$$

$$(\mathbf{a} \otimes \mathbf{b} \otimes \mathbf{c}) \overset{\times}{\underset{\times}{\cdot}} (\mathbf{e} \otimes \mathbf{f} \otimes \mathbf{g} \otimes \mathbf{h}) = (\mathbf{b} \cdot \mathbf{f})(\mathbf{c} \otimes \mathbf{e}) \otimes (\mathbf{a} \times \mathbf{g}) \otimes \mathbf{h} \,. \tag{D.39}$$

The associative and distributive laws can extend these definitions to any combination of terms.

Observe that the top cross product comes first in the order of the vectors, then the second one, and so on. It is obvious that the inner product annihilates two orders and the cross product annihilates one order from the total polyadic formed by the factor polyadics.

Just as a dyadic is a linear operator that maps vectors to vectors, a triadic is a linear operator that maps vectors to dyadics, a tetradic is a linear operator that maps vectors to triadics and so on. A polyadic of order n is an operator that maps vectors to polyadics of order $n-1$. By using double, triple, or higher-order inner products we can arrange to map polyadics of any order to polyadics of any other order. For example, relation (D.38) defines an operator that maps triadics to scalars.

For any symmetric dyadic $\tilde{\tilde{\mathbf{A}}}$ there exist an orthonormal set of vectors $\widehat{\mathbf{e}}_1$,$\widehat{\mathbf{e}}_2$, $\widehat{\mathbf{e}}_3$ and three real numbers λ_1, λ_2, λ_3, in terms of which the dyadic has the representation

$$\tilde{\tilde{\mathbf{A}}} = \lambda_1 \widehat{\mathbf{e}}_1 \otimes \widehat{\mathbf{e}}_1 + \lambda_2 \widehat{\mathbf{e}}_2 \otimes \widehat{\mathbf{e}}_2 + \lambda_3 \widehat{\mathbf{e}}_3 \otimes \widehat{\mathbf{e}}_3 \,. \tag{D.40}$$

If $\mathbb{A}$ represents a nonion form of $\tilde{\tilde{\mathbf{A}}}$, then λ_1, λ_2, λ_3 are the eigenvalues and $\widehat{\mathbf{e}}_1$,$\widehat{\mathbf{e}}_2$, $\widehat{\mathbf{e}}_3$ are the eigenvectors of the symmetric matrix $\mathbb{A}$.

If the vectors that enter the expression of a polyadic are functions of some particular variables, then the linear operators of differentiation and integration act on any polyadic component-wise according to the known properties of the differentiation and integration.

Polyadics appear naturally in analysis as follows. Suppose we want to calculate the directional derivative of the function $f = f(\mathbf{r})$ in the direction $\widehat{\mathbf{e}} = (\cos\alpha, \cos\beta, \cos\gamma)$ at the point $\mathbf{r}_0$, where α, β, γ are the angles that the direction $\widehat{\mathbf{e}}$ makes with the three Cartesian axes. The ray that emanates from $\mathbf{r}_0$ in the direction $\widehat{\mathbf{e}}$ is given by the equation $\mathbf{r} = \mathbf{r}_0 + s\widehat{\mathbf{e}}$, and by the chain rule we obtain

$$\begin{aligned}
\frac{df(\mathbf{r}_0)}{ds} &= \frac{\partial f(\mathbf{r}_0)}{\partial x_1}\frac{dx_1}{ds} + \frac{\partial f(\mathbf{r}_0)}{\partial x_2}\frac{dx_2}{ds} + \frac{\partial f(\mathbf{r}_0)}{\partial x_3}\frac{dx_3}{ds} \\
&= \frac{\partial f(\mathbf{r}_0)}{\partial x_1}\cos\alpha + \frac{\partial f(\mathbf{r}_0)}{\partial x_2}\cos\beta + \frac{\partial f(\mathbf{r}_0)}{\partial x_3}\cos\gamma \\
&= \frac{\partial f(\mathbf{r}_0)}{\partial x_1}(\widehat{\mathbf{x}}_1 \cdot \widehat{\mathbf{e}}) + \frac{\partial f(\mathbf{r}_0)}{\partial x_2}(\widehat{\mathbf{x}}_2 \cdot \widehat{\mathbf{e}}) + \frac{\partial f(\mathbf{r}_0)}{\partial x_3}(\widehat{\mathbf{x}}_3 \cdot \widehat{\mathbf{e}}) \\
&= \widehat{\mathbf{e}} \cdot \left(\widehat{\mathbf{x}}_1 \frac{\partial f(\mathbf{r}_0)}{\partial x_1} + \widehat{\mathbf{x}}_2 \frac{\partial f(\mathbf{r}_0)}{\partial x_2} + \widehat{\mathbf{x}}_3 \frac{\partial f(\mathbf{r}_0)}{\partial x_3} \right).
\end{aligned} \tag{D.41}$$

Consequently, we arrive at the well-known formula[3]

$$D_{\widehat{\mathbf{e}}} f(\mathbf{r}_0) = \widehat{\mathbf{e}} \cdot \nabla f(\mathbf{r}_0) \,, \tag{D.42}$$

where $D_{\widehat{\mathbf{e}}}$ denotes the directional derivative in the direction $\widehat{\mathbf{e}}$. Formula (D.42) shows that the infinite set of directional derivatives of the *scalar* function f is replaced by the *vector* function ∇f, and every directional derivative is synthesised via this formula. In a similar way, for the directional derivative of the vector function $\mathbf{f}$ we obtain

$$D_{\widehat{\mathbf{e}}} \mathbf{f}(\mathbf{r_0}) = \widehat{\mathbf{e}} \cdot \big(\nabla \otimes \mathbf{f}(\mathbf{r_0}) \big) \,, \tag{D.43}$$

[3] In this appendix, contrary to the notation adopted in the rest of the book, the notation ∇ is used instead of grad for the gradient operator, and similarly for the other differential operators of vector and dyadic analysis.

which shows that the infinite set of directional derivatives of the *vector* function $\mathbf{f}$ is replaced by the *dyadic* function $\nabla \otimes \mathbf{f}$. In the general case, the infinite set of directional derivatives of the *polyadic* $\tilde{\tilde{\mathbf{P}}}$ *of order* n *is replaced by the* polyadic $\nabla \otimes \tilde{\tilde{\mathbf{P}}}$ of order $n+1$.

The significance of the basic differential operators of divergence $\nabla \cdot \mathbf{f}$ and rotation $\nabla \times \mathbf{f}$ is that these two forms are the first scalar and the first vector invariants of the dyadic $\nabla \otimes \mathbf{f}$, which incorporates all directional derivatives of the vector field $\mathbf{f}$. If $\mathbf{f} = \nabla \phi$, then the two basic invariants of the dyadic field $\nabla \otimes \nabla \phi$ provide the formulae

$$\nabla \cdot \nabla \phi = \nabla^2 \phi = \Delta \phi = \frac{\partial^2 \phi}{\partial x_1^2} + \frac{\partial^2 \phi}{\partial x_2^2} + \frac{\partial^2 \phi}{\partial x_3^2} \tag{D.44}$$

and

$$\nabla \otimes \nabla \phi = \mathbf{0}. \tag{D.45}$$

Similarly, if $\mathbf{f} = \nabla \times \mathbf{g}$, then

$$\nabla \cdot (\nabla \times \mathbf{g}) = 0 \tag{D.46}$$

and

$$\nabla \times (\nabla \times \mathbf{g}) = \nabla (\nabla \cdot \mathbf{g}) - \boldsymbol{\Delta} \mathbf{g}. \tag{D.47}$$

The general Gauss theorem for the continuously differentiable polyadic $\tilde{\tilde{\mathbf{P}}}$ of any order in a smooth domain V with boundary S and outward unit normal $\widehat{\mathbf{n}}$ assumes the form

$$\int_V \nabla \otimes \tilde{\tilde{\mathbf{P}}} dv = \oint_S \widehat{\mathbf{n}} \otimes \tilde{\tilde{\mathbf{P}}} ds. \tag{D.48}$$

Similarly, the general Stokes theorem for the continuously differentiable polyadic $\tilde{\tilde{\mathbf{P}}}$ of any order on a smooth surface S with unit normal $\widehat{\mathbf{n}}$, having boundary C with tangential unit vector $\widehat{\mathbf{T}}$, assumes the form

$$\int_S \widehat{\mathbf{n}} \times \nabla \otimes \tilde{\tilde{\mathbf{P}}} ds = \oint_C \widehat{\mathbf{T}} \otimes \tilde{\tilde{\mathbf{P}}} dl, \tag{D.49}$$

where the integration over the line integral is taken in the positive direction with respect to the orientation of the surface S that has $\widehat{\mathbf{n}}$ as its unit normal.

Obviously, by taking invariants of the general theorems of Gauss and Stokes, we obtain other forms of integral theorems. For example, if $\tilde{\tilde{\mathbf{P}}}$ is the vector field $\mathbf{f}$, then the first scalar invariants of the dyadic equations (D.48) and (D.49) recover the classical theorems of Gauss and Stokes

$$\int_V \nabla \cdot \mathbf{f} dv = \oint_S \widehat{\mathbf{n}} \cdot \mathbf{f} ds \tag{D.50}$$

and

$$\int_S \widehat{\mathbf{n}} \times \nabla \cdot \mathbf{f} ds = \int_S \widehat{\mathbf{n}} \cdot \nabla \times \mathbf{f} ds = \oint_C \widehat{\mathbf{T}} \cdot \mathbf{f} dl, \tag{D.51}$$

respectively.

Appendix E

Notation and abbreviations

Following is a list of notations and abbreviations used in this book for quick reference for the convenience of the reader.

- ▷ $\mathcal{O} \subset \mathbb{R}^3$ is an open bounded domain.
- ▷ $\Gamma = \partial\mathcal{O}$ the boundary of $\mathcal{O}$.
- ▷ $\mathcal{O}_e$ is the complement of $\mathcal{O}$ in $\mathbb{R}^3$, an exterior domain.
- ▷ $L^p(\mathcal{O})$ is the Lebesgue space of p-integrable functions $u : \mathcal{O} \to \mathbb{R}$.
- ▷ $W^{m,p}(\mathcal{O})$ is the Sobolev space of functions $u : \mathcal{O} \to \mathbb{R}$, with generalised derivatives up to m order in $L^p(\mathcal{O})$.
- ▷ X is a Banach space, $||\cdot||_X$ its norm, X' its dual, $\langle\cdot,\cdot\rangle_{X',X}$ the duality pairing between them.
- ▷ $\mathbb{H}$ is used as a symbol for a "generic" Hilbert space, and $(\cdot,\cdot)$ is used for the inner product in this space.
- ▷ $L^p([0,T],\mathbb{H})$ is the space of p-integrable functions $u : [0,T] \to \mathbb{H}$.
- ▷ $C^k([0,T],\mathbb{H})$, $k \in \mathbb{N}_0$ is the space of k-times continuously differentiable functions $u : [0,T] \to \mathbb{H}$.
- ▷ $W^{m,p}([0,T],\mathbb{H})$ is the Sobolev-Bochner space of m-times weakly differentiable in $L^p([0,T],\mathbb{H})$ functions $u : [0,T] \to \mathbb{H}$.
- ▷ $\to$ denotes strong convergence.
- ▷ $\rightharpoonup$ denotes weak convergence.
- ▷ $\overset{*}{\rightharpoonup}$ denotes weak-$\star$ convergence.
- ▷ $\star$ denotes convolution.
- ▷ $\overset{c}{\hookrightarrow}$ is the symbol used for a compact embedding.
- ▷ $\widehat{u}$ denotes the Laplace transform of a vector-valued function u.
- ▷ $\widetilde{u}$ denotes the Fourier transform of a vector-valued function u.
- ▷ $(\Omega,\mathscr{F},P)$ is used for a probability space, and ω denotes a realisation of the random variable $X : \Omega \to \mathbb{H}$.

- ▷ $\mathbb{E}_P[X] = \int_\Omega X\, dP$ is the expectation of the random variable X.
- ▷ $\mathbb{E}_P[X \mid \mathscr{G}]$ is the conditional expectation of X with respect to the σ-subalgebra $\mathscr{G}$ of $\mathscr{F}$.
- ▷ $L^p(\Omega, \mathscr{F}, P; \mathbb{H})$ is the space of $\mathbb{H}$-valued, $\mathscr{F}$-measurable random variables, $X : \Omega \to \mathbb{H}$, such that $\mathbb{E}_P[\,||X||^p_{\mathbb{H}}\,] < \infty$.
- ▷ $W(t)$ is the Q-Wiener process and $\check{W}(t)$ is the cylindrical Wiener process.
- ▷ $\int_0^t \mathsf{Q}_{\mathrm{A}}\, dW(s)$ is the Itō integral of the operator valued process Q_{A}.
- ▷ $H(\mathrm{div}, \mathcal{O}) = \{u \in (L^2(\mathcal{O}))^3 \ : \ \mathrm{div}\, u \in L^2(\mathcal{O})\}$.
- ▷ $H_0(\mathrm{div}, \mathcal{O}) = \{u \in H(\mathrm{div}, \mathcal{O}) \ : \ u \cdot n|_\Gamma = 0\}$.
- ▷ $H(\mathrm{div}0, \mathcal{O}) = \left\{u \in (L^2(\mathcal{O}))^3 : \ \mathrm{div}\, u = 0\right\}$.
- ▷ $H_0(\mathrm{div}0, \mathcal{O}) := \{u \in H(\mathrm{div}0, \mathcal{O}) \ : \ u \cdot n = 0 \text{ on } \Gamma\}$.
- ▷ $H(\mathrm{curl}, \mathcal{O}) = \{u \in (L^2(\mathcal{O}))^3 \ : \ \mathrm{curl}\, u \in (L^2(\mathcal{O}))^3\}$.
- ▷ $H_0(\mathrm{curl}, \mathcal{O}) = \{u \in H(\mathrm{curl}, \mathcal{O}) \ : \ n \times u|_\Gamma = 0\}$.
- ▷ $\mathsf{X} := (L^2(\mathcal{O}))^3$.
- ▷ $\mathbb{X} := \mathsf{X} \times \mathsf{X}$.
- ▷ $\mathbb{X}_{\mathsf{M}} := H_0(\mathrm{curl}, \mathcal{O}) \times H(\mathrm{curl}, \mathcal{O})$.
- ▷ $\mathfrak{X}_1 := H_0(\mathrm{curl}, \mathcal{O}) \cap H(\mathrm{div}0, \mathcal{O})$.
- ▷ $\mathfrak{X}_2 := H(\mathrm{curl}, \mathcal{O}) \cap H_0(\mathrm{div}0, \mathcal{O})$.
- ▷ $\mathfrak{X}_{\mathsf{M}} := \mathfrak{X}_1 \times \mathfrak{X}_2$.
- ▷ M is the Maxwell operator, $\mathsf{M} := \begin{pmatrix} 0 & \mathrm{curl} \\ -\mathrm{curl} & 0 \end{pmatrix}$.
- ▷ $\{\mathsf{T}_{\mathsf{M}}(t)\}_{t \in \mathbb{R}^+}$ is the semigroup generated by M.
- ▷ $\mathsf{A}_{\mathrm{or}}(x) = \begin{pmatrix} \varepsilon(x) & \xi(x) \\ \zeta(x) & \mu(x) \end{pmatrix}$ is the 6×6 optical response matrix, and M_{A} is the modified Maxwell operator, $\mathsf{M}_{\mathrm{A}} = \mathsf{A}_{\mathrm{or}}^{-1}\mathsf{M}$.
- ▷ $\{\mathsf{T}_{\mathsf{M}_{\mathrm{A}}}(t)\}_{t \in \mathbb{R}^+}$ is the semigroup generated by M_{A}.

Bibliography

[1] R. A. Adams, *Sobolev Spaces*, Academic Press, New York, 1975.

[2] G. Allaire, Homogenization and two-scale convergence, *SIAM J. Math. Anal.* 23, 1992, 1482–1518.

[3] A. Ambrosetti, G. Prodi, *A Primer of Nonlinear Analysis*, Cambridge University Press, Cambridge, 1993.

[4] H. Ammari, G. Bao, Analysis of the diffraction from periodic chiral structures, *C. R. Acad. Sci. Paris, Sér. I, Math.* 326, 1998, 1371–1376.

[5] H. Ammari, G. Bao, Maxwell's equations in periodic chiral structures, *Math. Nachr.* 251, 2003, 3–18.

[6] H. Ammari, G. Bao, Coupling of finite element and boundary element methods for the scattering by periodic chiral structures, *J. Comput. Math.* 26, 2008, 261–283.

[7] H. Ammari, K. Hamdache, Global existence and regularity of solutions to a system of nonlinear Maxwell equations, *J. Math. Anal. Appl.* 286, 2003, 51–63.

[8] H. Ammari, K. Hamdache, J.-C. Nédélec, Chirality in the Maxwell equations by the dipole approximation method, *SIAM J. Appl. Math.* 59, 1999, 2045–2059.

[9] H. Ammari, H. Kang, H. Lee, *Layer Potential Techniques in Spectral Analysis*, Mathematical Surveys and Monographs no. 153, American Mathematical Society, Providence, RI, 2009.

[10] H. Ammari, M. Laouadi, J.-C. Nédélec, Low frequency behavior of solutions to electromagnetic scattering problems in chiral media, *SIAM J. Appl. Math.* 58, 1998, 1022–1042.

[11] H. Ammari, J.-C. Nédélec, Time-harmonic electromagnetic fields in chiral media, in *Modern Mathematical Methods in Diffraction Theory and Its Applications in Engineering* (ed. E. Meister), Methoden und Verfahren der Mathematischen Physik 42, Peter Lang, Frankfurt am Main, 1997, 174–202.

[12] H. Ammari, J.-C. Nédélec, Small chirality behavior of solutions to electromagnetic scattering problems in chiral media, *Math. Methods Appl. Sci.* 21, 1998, 327–359.

[13] H. Ammari, J.-C. Nédélec, Time-harmonic electromagnetic fields in thin chiral curved layers, *SIAM J. Appl. Math.* 29, 1998, 395–423.

[14] H. Ammari, J.-C. Nédélec, Analysis of the diffraction from chiral gratings, in *Mathematical Modelling in Optical Science*, Frontiers in Applied Mathematics no. 22 (eds. G. Bao, L. Cowsar, W. Masters), SIAM, Philadelphia, 2001, 179–206.

[15] N. Antonić, *On Some Questions Related to Memory Effects in Homogenisation and Propagation of Singularities*, PhD Thesis, Carnegie Mellon University, 1992.

[16] F. Arago, Mémoire sur une modification remarquable qu'éprouvent les rayons lumineux dans leur passage à travers certais corps diaphanes, et sur quelques autres nouveaux phénomènes d'optique, *Mém. Sci. Math. Phys. Inst.*, 1811, 93–134.

[17] T. Arbogast, J. Douglas, U. Hornung, Derivation of the double porosity model of a single phase flow via homogenization theory, *SIAM J. Math. Anal.* 21, 1990, 823–836.

[18] M. Artola, M. Cessenat, Un problème raide avec homogénéization en electromagnétisme, *C. R. Acad. Sci. Paris* 310, 1990, 9–14.

[19] C. Athanasiadis, On the far field patterns for electromagnetic scattering by a chiral obstacle in a chiral environment, *J. Math. Anal. Appl.* 309, 2005, 517–533.

[20] C. Athanasiadis, N. Berketis, Scattering relations for point-source excitation in chiral media, *Math. Methods Appl. Sci.* 29, 2006, 27–48.

[21] C. Athanasiadis, G. Costakis, I. G. Stratis, Electromagnetic scattering by a homogeneous chiral obstacle in a chiral environment, *IMA J. Appl. Math.* 64, 2000, 245–258.

[22] C. Athanasiadis, G. Costakis, I. G. Stratis, On some properties of Beltrami fields in chiral media, *Rep. Math. Phys.* 45, 2000, 257–271.

[23] C. Athanasiadis, G. Costakis, I. G. Stratis, Low frequency electromagnetic scattering from a nonchiral object in a chiral environment, in *Scattering Theory and Biomedical Engineering* (eds. G. Dassios, D. I. Fotiadis, C. V. Massalas, K. Kiriaki), World Scientific, Hackensack, NJ, 2000, 38–49.

[24] C. Athanasiadis, G. Costakis, I. G. Stratis, Electromagnetic scattering by a perfectly conducting obstacle in a homogeneous chiral environment: solvability and low frequency theory, *Math. Methods Appl. Sci.* 25, 2002, 927–944.

[25] C. Athanasiadis, G. Costakis, I. G. Stratis, Transmission problems in contrasting chiral media, *Rep. Math. Phys.* 53, 2004, 143–156.

[26] C. Athanasiadis, S. Giotopoulos, The Atkinson-Wilcox expansion theorem for electromagnetic chiral waves, *Appl. Math. Lett.* 16, 2003, 675–681.

[27] C. Athanasiadis, S. Giotopoulos, Scattering theorems for dyadic chiral fields, *Rep. Math. Phys.* 53, 2004, 339–350.

[28] C. Athanasiadis, E. Kardasi, Beltrami Herglotz functions for electromagnetic scattering theory in chiral media, *Appl. Anal.* 84, 2005, 145–163.

[29] C. Athanasiadis, E. Kardasi, Electromagnetic Herglotz dyadics in chiral media, *Rep. Math. Phys.* 55, 2005, 385–395.

[30] C. Athanasiadis, E. Kardasi, On the far field operator for electromagnetic scattering in chiral media, *Appl. Anal.* 85, 2006, 623–639.

[31] C. Athanasiadis, E. Kardasi, Inverse electromagnetic scattering by a perfect conductor in a chiral environment, *J. Inv. Ill-Posed Problems* 16, 2008, 1–18.

[32] C. Athanasiadis, K. Kiriaki, I. G. Stratis, A spectral-theoretic approach for the solution of Maxwell's equations in stratified media, *Math. Methods Appl. Sci.* 21, 1998, 685–700.

[33] C. Athanasiadis, P. A. Martin, A. Spyropoulos, I. G. Stratis, Scattering relations for point sources: acoustic and electromagnetic waves, *J. Math. Phys.* 43, 2002, 5683–5697.

[34] C. Athanasiadis, P. A. Martin, I. G. Stratis, Electromagnetic scattering by a homogeneous chiral obstacle: boundary integral equations and low-chirality approximations, *SIAM J. Appl. Math.* 59, 1999, 1745–1762.

[35] C. Athanasiadis, P. A. Martin, I. G. Stratis, Electromagnetic scattering by a homogeneous chiral obstacle: scattering relations and the far-field operator, *Math. Methods Appl. Sci.* 22, 1999, 1175–1188.

[36] C. Athanasiadis, P. A. Martin, I. G. Stratis, On the scattering of point generated electromagnetic waves by a perfectly conducting sphere and related near-field inverse problems, *Z. Angew. Math. Mech.* 83, 2003, 129–136.

[37] C. Athanasiadis, G. F. Roach, Time dependent potential scattering in chiral media, *J. Math. Anal. Appl.* 310, 2005, 1–15.

[38] C. Athanasiadis, G. F. Roach, I. G. Stratis, A time-domain analysis of wave motions in chiral materials, *Math. Nachr.* 250, 2003, 3–16.

[39] C. Athanasiadis, I. G. Stratis, Electromagnetic scattering by a chiral obstacle, *IMA J. Appl. Math.* 58, 1997, 83–91.

[40] C. E. Athanasiadis, V. I. Sevroglou, K. I. Skourogiannis, The direct electromagnetic scattering problem by a mixed impedance screen in chiral media, *Appl. Anal.*, in press.

[41] C. Athanasiadis, I. G. Stratis, The conductive transmission problem for a chiral scatterer, *Electromagn. Waves Electron. Syst.* 2, 1997, 71–79.

[42] C. Athanasiadis, I. G. Stratis, Uniqueness of the inverse scattering problem by a chiral obstacle, *Int. J. Appl. Electromagn. Mech.* 9, 1998, 123–133.

[43] C. Athanasiadis, I. G. Stratis, Low frequency electromagnetic scattering theory for a multi-layered chiral scatterer, *Meth. Appl. Anal.* 6, 1999, 437–450.

[44] C. Athanasiadis, I. G. Stratis, A transmission problem for bi-isotropic media, *Appl. Anal.* 77, 2001, 195–209.

[45] C. Athanasiadis, N. L. Tsitsas, Radiation relations for electromagnetic excitation of a layered chiral medium by an interior dipole, *J. Math. Phys.* 49, 2008, 1–14.

[46] K. Atkinson, W. Han, *Theoretical Numerical Analysis: A Functional Analysis Framework*, 2nd ed., Springer, New York, 2005.

[47] J. P. Aubin, *Applied Functional Analysis*, Wiley, New York, 2000.

[48] A. Babin, A. Figotin, Nonlinear Maxwell equations in inhomogeneous media, *Commun. Math. Phys.* 241, 2003, 519–581.

[49] G. Bal, *Waves in Random Media*, Lecture Notes, Columbia University, 2006.
http://www.columbia.edu/~gb2030/COURSES/E6901-Waves/Lectures-Waves.pdf

[50] A. Barbagallo, S. Pia, Weighted variational inequalities in non-pivot Hilbert spaces, *Comput. Optim. Appl.*, in press.

[51] G. Barbatis, I. G. Stratis, Homogenization of Maxwell's equations in dissipative bianisotropic media, *Math. Methods Appl. Sci.*, 26, 2003, 1241–1253.

[52] G. Barbatis, I. G. Stratis, A. N. Yannacopoulos, Homogenization for random complex media in electromagnetics, in preparation.

[53] V. Barbu, S. I. Grossman, Asymptotic behavior of linear integro-differential systems, *Trans. Amer. Math. Soc.* 173, 1972, 277–288.

[54] V. Barbu, A. Rascanu, G. Tesitore, Carleman estimates and controllability of linear stochastic heat equations, *Appl. Math. Optim.* 47, 2003, 97–120.

[55] F. G. Bass, G. Ya. Slepyan, S. A. Maksimenko, A. Lakhtakia, Time-domain Beltrami-Maxwell solitons in certain nonlinear chiral media, *Microw. Opt. Technol. Lett.* 9, 1995, 218–221.

[56] A. Belleni-Morante, *Applied Semigroups and Evolution Equations*, Oxford University Press, Oxford, 1979.

[57] A. Bensoussan, On the general theory of exact controllability for skew symmetric operators, *Acta Appl. Math.* 20, 1990, 197–229.

[58] A. Bensoussan, G. Da Prato, M. C. Delfour, S. K. Mitter, *Representation and Control of Infinite Dimensional Systems*, 2nd ed., Birkhäuser, Boston, 2007.

[59] A. Bensoussan, J.-L. Lions, G. Papanicolaou, *Asymptotic Analysis for Periodic Structures*, North Holland, New York, 1978.

[60] M. S. Berger, *Nonlinearity and Functional Analysis*, Academic Press, New York, 1977.

[61] E. Bermani, S. Caorsi, M. Raffetto, Uniqueness to the solution of electromagnetic boundary value problems in the presence of partly lossy chiral media, *J. Electromagn. Waves Appl.* 15, 2001, 797–811.

[62] A. Bernland, A. Luger, M. Gustafsson, Sum rules and constraints on passive systems, Technical Report LUTEDX/(TEAT-7193)/1-28, Lund University, 2010. http://www.eit.lth.se

[63] J.-B. Biot, Mémoire sur un nouveau genre d'oscillation que les molécules de la lumière éprouvent en traversant certains cristaux, *Mém. Sci. Math. Phys. Inst.*, 1812, 1–372.

[64] M. Sh. Birman, M. Z. Solomyak, L_2-Theory of the Maxwell operator in arbitrary domains, *Uspekhi Mat. Nauk* 42, 1987, 61–76 (English trans.: *Russian Math. Surv.* 42, 1987, 75–96).

[65] M. Sh. Birman, M. Z. Solomyak, The selfadjoint Maxwell operator in arbitrary domains, *Algebra Analiz* 1, 1989, 96–110 (English trans.: *Leningrad Math. J.* 1, 1990, 99–115).

[66] X. Blanc, C. Le Bris, P.-L. Lions, Une variante de la théorie de l'homogénéisation stochastique des opérateurs elliptiques, *C. R. Acad. Sci. Paris, Sér. I*, 343, 2006, 717–724.

[67] X. Blanc, C. Le Bris, P.-L. Lions, Stochastic homogenization and random lattices, *J. Math. Pures Appl.* 88, 2007, 34–63.

[68] A. Bossavit, *Computational Electromagnetism: Variational Formulations, Complementarity, Edge Elements*, Academic Press, San Diego, 1998.

[69] A. Bossavit, *Électromagnétisme, en vue de la modélisation*, Springer-Verlag, Paris, 2004.

[70] A. Bossavit, G. Griso, B. Miara, Modelling of periodic electromagnetic structures: bianisotropic materials with memory effects, *J. Math. Pures Appl.* 84, 2005, 819–850.

[71] A. Boström, G. Kristensson, S. Ström, Transformation properties of plane, spherical and cylindrical scalar and vector wave functions, in *Field Representations and Introduction to Scattering* (eds. V. V. Varadan, A. Lakhtakia, V. K. Varadan), Elsevier, Amsterdam, 1991, 165–210.

[72] T. Z. Boulmezaoud, T. Amari, On the existence of non-linear force-free fields in three-dimensional domains, *Z. Angew. Math. Phys.* 51, 2000, 942–967.

[73] A. Boutet de Monvel, D. Shepelsky, Direct and inverse problem for a stratified non–reciprocal chiral medium, *Inv. Prob.* 13, 1997, 239–251.

[74] A. Boutet de Monvel, D. Shepelsky, Inverse scattering approach for stratified chiral media, in *Inverse Problems of Wave Propagation*, Lecture Notes in Physics no. 486, Springer-Verlag, Berlin, 1997, 47–57.

[75] A. Boutet de Monvel, D. Shepelsky, Inverse scattering problem for stratified bi-isotropic medium at oblique incidence, *Inv. Prob.* 14, 1998, 29–40.

[76] A. Boutet de Monvel, D. Shepelsky, A frequency-domain inverse problem for a dispersive stratified chiral medium, *J. Math. Phys.* 41, 2000, 6116–6129.

[77] S. Brahim-Otsmane, G. Francfort, F. Murat, Correctors for the homogenization of the wave and heat equations, *J. Math. Pures Appl.* 71, 1992, 197–231.

[78] L. Brand, *Vector and Tensor Analysis*, Wiley, New York, 1947.

[79] H. Brezis, *Functional Analysis, Sobolev Spaces and Partial Differential Equations*, Springer-Verlag, New York, 2010.

[80] M. Briane, J. Casado-Diaz, F. Murat, The div-curl lemma "trente ans après": an extension and an application to the G-convergence of unbounded monotone operators, *J. Math. Pures Appl.* 91, 2009, 476–494.

[81] T. Buchukuri, R. Duduchava, D. Kapanadze, D. Natroshvili, On the uniqueness of a solution to anisotropic Maxwell's equations, *Oper. Theory Adv. Appl.* 203, 2009, 137–164.

[82] T. A. Burton, *Stability and Periodic Solutions of Ordinary and Functional Differential Equations*, Dover, New York, 2005.

[83] C. Byrne, *A Brief History of Electromagnetism*, 2010. http://faculty.uml.edu/cbyrne/EMHIST.pdf

[84] F. Cakoni, D. Colton, P. Monk, *The Linear Sampling Method in Inverse Electromagnetic Scattering*, SIAM, Philadelphia, 2011.

[85] R. H. Cameron, W. T. Martin, The orthogonal development of non-linear functionals in series of Fourier-Hermite functionals, *Ann. Math.* 48, 1947, 385–392.

[86] L. Cao, Y. Zhang, W. Allegretto, Y. Lin, Multiscale asymptotic method for Maxwell's equations in composite materials, *SIAM J. Numer. Anal.* 47, 2010, 4257–4289.

[87] G. Carbou, B. Hanouzet, Relaxation approximation for the Kerr model for the three dimensional initial boundary value problem, *J. Hyperbol. Differ. Eq.* 6, 2009, 577–614.

[88] R. Carmona, M. Tehranchi, *Interest-Rate Models: An Infinite-Dimensional Stochastic Analysis Perspective*, Springer-Verlag, Berlin, 2006.

[89] R. W. Carroll, R. E. Showalter, *Singular and Degenerate Cauchy Problems*, Academic Press, New York, 1976.

[90] A. L. Cauchy, Mémoire sur l'application de l'analyse mathématique à la recherche des lois générales observées par les physiciens, et, en particulier, sur les lois de polarisation circulaire, *C. R. Acad. Sci. Paris* 15, 1842, 910–918.

[91] M. Cessenat, *Mathematical Methods in Electromagnetism*, World Scientific, Singapore, 1996.

[92] P. Ciarlet Jr., G. Legendre, Well-posedness of the Drude-Born-Fedorov model for chiral media, *Math. Models Methods Appl. Sci.* 17, 2007, 461–484. Erratum: *Math. Models Methods Appl. Sci.* 19, 2009, 173–174.

[93] P. Ciarlet Jr., G. Legendre, S. Nicaise, Time-domain study of the Drude-Born-Fedorov model for a class of heterogeneous chiral materials, preprint, 2009.
http://hal.archives-ouvertes.fr/hal-00430196/fr

[94] P. Ciarlet Jr., F. Lefévre, S. Lohrengel, S. Nicaise, Weighted regularization for composite materials in electromagnetism, *ESAIM: M2AN* 44, 2010, 75–108.

[95] D. Cioranescu, A. Damlamian, G. Griso, Periodic unfolding and homogenization, *C. R. Math. Acad. Sci. Paris* 335, 2002, 99–104.

[96] D. Cioranescu, A. Damlamian, G. Griso, The periodic unfolding in homogenization, *SIAM J. Math. Anal.* 40, 2008, 1585–1620.

[97] D. Cioranescu, P. Donato, *An Introduction to Homogenization*, Oxford University Press, Oxford, 1999.

[98] S. N. Chandler-Wilde, R. Potthast, The domain derivative in rough surface scattering and rigorous estimates for first order perturbation theory, *Proc. R. Soc. Lond. A* 458, 2002, 2967–3001.

[99] J. J. Chao, Y. J. Liu, F. J. Rizzo, P. A. Martin, L. Udpa, Regularized integral equations and curvilinear boundary elements for electromagnetic wave scattering in three dimensions, *IEEE Trans. Antennas Propag.* AP-43, 1995, 1416–1422.

[100] D. Cheng, Solitons in Kerr-type nonlinear materials with uniaxial chirality, *Microw. Opt. Technol. Lett.* 14, 1997, 351–354.

[101] F. Collino, B. Despres, Integral equations via saddle point problems for time-harmonic Maxwell's equations, *J. Comp. Appl. Math.* 150, 2003, 157–192.

[102] D. Colton, A. Kirsch, Dense sets and far field patterns in acoustic wave propagation, *SIAM J. Math. Anal.* 15, 1984, 996–1006.

[103] D. Colton, R. Kress, *Integral Equation Methods in Scattering Theory*, Wiley, New York, 1983.

[104] D. Colton, R. Kress, Dense sets and far field patterns in electromagnetic wave propagation, *SIAM J. Math. Anal.* 16, 1985, 1049–1060.

[105] D. Colton, R. Kress, Eigenvalues of the far field operator and inverse scattering theory, *SIAM J. Math. Anal.* 26, 1995, 601–615.

[106] D. Colton, R. Kress, *Inverse Acoustic and Electromagnetic Scattering Theory*, 2nd ed., Springer-Verlag, Berlin, 1998.

[107] D. Colton, R. Kress, On the denseness of Herglotz wave functions and electromagnetic Herglotz pairs in Sobolev spaces, *Math. Methods Appl. Sci.* 22, 2001, 1289–1303.

[108] D. Colton, P. Monk, A novel method for solving the inverse scattering problem for time-harmonic acoustic waves in the resonance region, *SIAM J. Appl. Math.* 45, 1985, 1039–1053.

[109] D. Colton, P. Monk, A novel method for solving the inverse scattering problem for time-harmonic acoustic waves in the resonance region II, *SIAM J. Appl. Math.* 46, 1986, 506–523.

[110] C. Conca, M. Vanninathan, Fourier approach to homogenization problems, *ESAIM: Control, Optimisation and Calculus of Variations* 8, 2002, 489–511.

[111] S. Conti, G. Dolzmann, S. Müller, The div-curl lemma for sequences whose divergence and curl are compact in $W^{-1,1}$, *arXiv: 0907.0397.v1 [math.AP]*, 2009.

[112] J.-M. Coron, *Control and Nonlinearity*, Mathematical Surveys and Monographs no. 136, American Mathematical Society, Providence, RI, 2007.

[113] M. Costabel, M. Dauge, S. Nicaise, *GLC Project I: Corner Singularities and Analytic Regularity for Linear Elliptic Systems, Part I: Smooth Domains*, preprint, 2010. http://perso.univ-rennes1.fr/monique.dauge/publis/
BeDaMa08a.html

[114] M. Costabel, F. Le Louër, Shape derivatives of boundary integral operators in electromagnetic scattering, *arXiv:1002.1541v2 [math.NA]*, 2010.

[115] C. Corduneanu, *Principles of Differential and Integral Equations*, 2nd ed., Chelsea, New York, 1977.

[116] P. Courilleau, T. Horsin Molinaro, On the controllability for Maxwell's equations in specific media, *C. R. Acad. Sci. Paris, Sér. I*, 341, 2005, 665–668.

[117] P. Courilleau, T. Horsin Molinaro, I. G. Stratis, On the controllability of time-harmonic electromagnetic fields in chiral media, *Adv. Math. Sci. Appl.* 16, 2006, 491–502.

[118] P. Courilleau, T. Horsin Molinaro, I. G. Stratis, On the controllability of Maxwell's equations in a class of complex media, *Bull. Greek Math. Soc.* 54, 2007, 97–113.

[119] G. Dal Maso, *An Introduction to* Γ*-convergence*, Birkhäuser, Basel, 1993.

[120] G. Da Prato, J. Zabczyk, *Stochastic Equations in Infinite Dimensions*, Cambridge University Press, Cambridge, 1992.

[121] G. Dassios, R. Kleinman, *Low Frequency Scattering*, Clarendon Press, Oxford, 2000.

[122] G. Dassios, Z. Rigou, Elastic Herglotz functions, *SIAM J. Appl. Math.* 55, 1995, 1345–1361.

[123] G. Dassios, Z. Rigou, On the reconstruction of a rigid body in the theory of elasticity, *Z. Angew. Math. Mech.* 77, 1997, 911–923.

[124] R. Dautray, J.-L. Lions, *Mathematical Analysis and Numerical Methods for Science and Technology, Volume 1: Physical Origins and Potential Theory*, Springer-Verlag, Berlin, 1990.

[125] R. Dautray, J.-L. Lions, *Mathematical Analysis and Numerical Methods for Science and Technology, Volume 2: Functional and Variational Methods*, Springer-Verlag, Berlin, 1988.

[126] R. Dautray, J.-L. Lions, *Mathematical Analysis and Numerical Methods for Science and Technology, Volume 3: Spectral Theory and Applications*, Springer-Verlag, Berlin, 1990.

[127] R. Dautray, J.-L. Lions, *Mathematical Analysis and Numerical Methods for Science and Technology, Volume 4: Integral Equations and Numerical Methods*, Springer-Verlag, Berlin, 1990.

[128] R. Dautray, J.-L. Lions, *Mathematical Analysis and Numerical Methods for Science and Technology, Volume 5: Evolution Problems I*, Springer-Verlag, Berlin, 1992.

[129] E. De Giorgi, Sulla converegenza di alcune successioni di integrali di tipo dell'area, *Rend. Mat.* 8, 1975, 277–294.

[130] E. De Giorgi, T. Franzoni, Su un tipo di convergenza variazionale, *Atti Accad. Naz. Linzei Rend. Cl. Sci. Mat.* 58, 1975, 842–850.

[131] A. T. de Hoop, *Handbook of Radiation and Scattering of Waves*, Academic Press, London, 1995.

[132] A. de La Bourdonnaye, Décomposition de $TH_{div}^{-1/2}(\partial\Omega_2)$ et nature de l'operateur de Steklov-Poincaré du problème extérieur de l'électromagnétisme, *C. R. Acad. Sci. Paris, Sér. I* 316, 1993, 369–372.

[133] W. Desch, R. Grimmer, W. Schappacher, Some considerations for linear integrodifferential equations, *J. Math. Anal. Appl.* 104, 1984, 219–234.

[134] E. DiBenedetto, *Degenerate Parabolic Equations*, Springer-Verlag, New York, 1993.

[135] G. Duvaut, J.-L. Lions, *Inequalities in Mechanics and Physics*, Springer-Verlag, Berlin, 1976.

[136] D. M. Eidus, The principle of limiting absorption, *Math. Sb.* 57 (99), 1962 (English trans.: *Amer. Math. Soc. Transl.* 47, 1965, 157–191).

[137] D. M. Eidus, The principle of limiting amplitude, *Uspekhi Nauk* 24, 1969, 91–56 (English trans.: *Russian Math. Surv.* 24, 1969, 97–167).

[138] M. M. Eller, Continuous observability for the anisotropic Maxwell system, *Appl. Math. Optim.* 55, 2007, 185–201.

[139] M. M. Eller, J. E. Lagnese, S. Nicaise, Stabilization of heterogeneous Maxwell's equations by linear or nonlinear boundary feedbacks, *Electron. J. Diff. Eq.* 2002, 1–26.

[140] M. M. Eller, J. E. Masters, Exact boundary controllability of electromagnetic fields in a general region, *Appl. Math. Optim.* 45, 2002, 99–123.

[141] K. J. Engel, R. Nagel, *One-Parameter Semigroups for Linear Evolution Equations*, Springer-Verlag, New York, 2000.

[142] N. Engheta, D. L. Jaggard, Electromagnetic chirality and its applications, *IEEE Antennas Propag. Soc. Newslett.* 30, 1988, 6–12.

[143] N. Engheta, P. Pelet, Mode orthogonality in chirowaveguides, *IEEE Trans. Microw. Theory Tech.* 38, 1990, 1631–1634.

[144] L. C. Evans, *Partial Differential Equations*, Graduate Studies in Mathematics no. 19, American Mathematical Society, Providence, RI, 1998.

[145] M. Fabrizio, A. Morro, *Electromagnetism of Continuous Media (Mathematical Modelling and Applications)*, Oxford University Press, New York, 2003.

[146] H. O. Fattorini, *Infinite Dimensional Linear Control Systems: The Time Optimal and Norm Optimal Problems*, Elsevier, Amsterdam, 2005.

[147] T. R. Field, *Electromagnetic Scattering from Random Media*, Oxford University Press, Oxford, 2009.

[148] A. S. Fokas, *A Unified Approach to Boundary Value Problems*, SIAM, Philadelphia, 2008.

[149] J.-P. Fouque, J. Garnier, G. Papanicolaou, K. Sølna, *Wave Propagation and Time Reversal in Randomly Layered Media*, Springer-Verlag, New York, 2007.

[150] M. Fowler, *Historical Beginnings of Theories of Electricity and Magnetism*, 1997. http://galileo.phys.virginia.edu/classes/109N/more_stuff/E&M_Hist.pdf

[151] D. J. Frantzeskakis, A. Ioannidis, G. F. Roach, I. G. Stratis, A. N. Yannacopoulos, On the error of the optical response approximation in chiral media, *Appl. Anal.* 82, 2003, 839–856.

[152] D. J. Frantzeskakis, I. G. Stratis, A. N. Yannacopoulos, Bright-dark vector solitons in chiral media, *Phys. Scripta* 66, 2002, 280–283.

[153] D. J. Frantzeskakis, I. G. Stratis, A. N. Yannacopoulos, On equilibria of the two-fluid model in magnetohydrodynamics, *Math. Phys. Anal. Geom.* 7, 2004, 97–117.

[154] A. Fresnel, Extrait d'un mémoire sur la double réfraction particulière que presente le cristal de roche dans la direction de son axe, *Bull. Soc. Philomat.* 1822, 191–198.

[155] J. Fridén, G. Kristensson, Transient external 3D excitation of a dispersive and anisotropic slab, *Inv. Prob.* 13, 1997, 691–709.

[156] J. Fridén, G. Kristensson, A. Sihvola, Effect of dissipation on the constitutive relations of bi-anisotropic media: the optical response, *Electromagnetics* 17, 1997, 251–267.

[157] T. L. Gao, F. Ma, D. Zhang, Analysis of the linear sampling method for boundary reconstruction of chiral obstacle in electromagnetic scattering problems, *Northeast. Math. J.* 22, 2006, 491–502.

[158] M. Gardner, *The New Ambidextrous Universe*, 3rd ed., W. H. Freeman, New York, 1990.

[159] L. Gawarecki, V. Mandrekar, *Stochastic Differential Equations in Infinite Dimensions with Applications to Stochastic Partial Differential Equations*, Springer-Verlag, Berlin, 2011.

[160] T. Gerlach, The two-dimensional electromagnetic inverse scattering problem for chiral media, *Inv. Prob.* 15, 1999, 1663–1675.

[161] D. Gilbarg, N. S. Trudinger, *Elliptic Partial Differential Equations of Second Order*, Springer-Verlag, Berlin, 1998.

[162] M. L. Goldberger, K. M. Watson, *Collision Theory*, Wiley, New York, 1967.

[163] A. Gómez, A. Lakhtakia, J. Margineda, G. J. Molina-Cuberos, M. J. Núñez, J. S. Ipiña, A. Vegas, Full-wave hybrid technique for 3D isotropic-chiral-material discontinuities in rectangular waveguides: theory and experiment, *IEEE Trans. Microw. Theory Tech.* 56, 2008, 2815–2825.

[164] A. Gómez, A. Lakhtakia, A. Vegas, M. A. Solano, Hybrid technique for analysing metallic waveguides containing isotropic chiral materials, *IET Microw. Antennas Propag.* 4, 2010, 305–315.

[165] A. Gómez, A. Vegas, M. A. Solano, A brief discussion on the different formulations of the coupled mode method in chiral media: application to the parallel-plate chirowaveguide, *Microw. Opt. Technol. Lett.* 42, 2004, 181–185.

[166] D. Goreak, Approximate controllability for linear stochastic differential equations in infinite dimensions, *Appl. Math. Optim.* 60, 2009, 105–132.

[167] T. E. Govindan, Autonomous semilinear stochastic Volterra integrodifferential equations in Hilbert spaces, *Dynam. Syst. Applic.* 3, 1994, 51–74.

[168] W. Grecksch, C. Tudor, *Stochastic Evolution Equations: A Hilbert Space Approach*, Akademie-Verlag, Berlin, 1995.

[169] R. Grimmer, Resolvent operators for integral equations in a Banach space, *Trans. Amer. Math. Soc.* 273, 1982, 333–349.

[170] R. Grimmer, J. Prüss, On linear Volterra equations in Banach spaces, *Comput. Math. Appl.* 11, 1985, 189–205.

[171] G. Grubb, *Distributions and Operators*, Springer-Verlag, New York, 2009.

[172] S. M. Grudsky, K. V. Khmelnytskaya, V. V. Kravchenko, On a quaternionic Maxwell equation for the time-dependent electromagnetic field in a chiral medium, *J. Phys. A* 37, 2004, 4641–4647.

[173] B. Z. Guo, G. Q. Xu, Expansion of solution in terms of generalised eigenfunctions for a hyperbolic system with static boundary condition, *J. Funct. Anal.* 231, 2006, 245–268.

[174] K. Gürlebeck, W. Sprössig, *Quaternionic Analysis and Elliptic Boundary Value Problems*, Akademie-Verlag, Berlin, 1989.

[175] K. Gürlebeck, W. Sprössig, *Quaternionic and Clifford Calculus for Physicists and Engineers*, Wiley, Chichester, 1997.

[176] M. Gustafsson, *Wave Splitting in Direct and Inverse Scattering Problems*, PhD Thesis, Lund University, 2000.

[177] J. Hadamard, Sur les problémes aux dérivées partielles et leur signification physique, *Princeton University Bulletin* 13, 1902, 49–52.

[178] P. Hähner, An exterior boundary value problem for the Maxwell's equations with boundary data in a Sobolev space, *Proc. R. Soc. Edinb. A* 123, 1988, 271–296.

[179] S. Halkos, A second order differential equation approach in the time domain for electromagnetic fields in a bounded chiral medium, *Bull. Greek Math. Soc.*, in press.

[180] G. W. Hanson, A. B. Yakovlev, *Operator Theory for Electromagnetics: An Introduction*, Springer-Verlag, New York, 2002.

[181] A. Haraux, *Semi-Linear Hyperbolic Problems in Bounded Domains*, Harwood Academic Publishers, Chur, 1987.

[182] A. Haraux, On a completion problem in the theory of distributed control of wave equations, in *Nonlinear Partial Differential Equations and Their Applications*, Collège de France Seminar (1987 - 88), Pitman Research Notes in Mathematics no. 220, Addison Wesley Longman, London, 1991, 241–271.

[183] R. F. Harrington, Boundary integral formulations for homogeneous material bodies, *J. Electrom. Waves Appl.* 3, 1989, 1–15.

[184] P. Hartman, On the solutions of $\Delta V + V = 0$ in an exterior domain, *Math. Z.* 71, 1959, 251–257.

[185] P. Hartman, C. Wilcox, On solutions of the Helmholtz equation in exterior domains, *Math. Z.* 75, 1961, 228–255.

[186] K. Hayata, M. Koshiba, Chirosolitons: unique spatial solitons in chiral media, *IEEE Trans. Microw. Theory Tech.* 43, 1995, 1814–1818.

[187] E. Hecht, *Optics*, 4th ed., Addison Wesley, San Francisco, 2002.

[188] R. A. Hegstrom, D. K. Kondepundi, The handedness of the universe, *Scientific American* 262, January 1990, 108–115.

[189] G. Helmberg, *Introduction to Spectral Theory in Hilbert Space*, Elsevier, New York, 1969.

[190] F. Hettlich, *The Domain Derivative in Inverse Obstacle Problems*, Habilitation Thesis, Erlangen, 1999.

[191] K. Hizanidis, D. J. Frantzeskakis, Reductive perturbation analysis of short pulse propagation in a nonlinear dielectric slab: the role of material dispersion in bright-to-dark soliton transitions, *IEEE J. Quantum Electron.* QE-29, 1993, 286–295.

[192] S. Heumann, *The Factorization Method for Chiral Media*, PhD in progress, Department of Mathematics, Karlsruhe Institute of Technology.

[193] A. Holmbom, J. Silfver, N. Svanstedt, N. Wellander, On two-scale convergence and related sequential compactness topics, *Appl. Math.* 51, 2006, 247–262.

[194] K. Hori, S. Katz, A. Klemm, R. Pandharipande, R. Thomas, C. Vafa, R. Vakil, E. Zaslow, *Mirror Symmetry*, Clay Mathematical Monographs no. 1, American Mathematical Society, Providence, RI, 2003.

[195] T. Horsin, *Quelques problèmes de contrôlabilité*, Thése de Habilitation, Université de Versailles Saint-Quentin en Yvelines, 2010.

[196] T. Horsin, I. G. Stratis, A. N. Yannacopoulos, On the approximate controllability of the stochastic Maxwell equations, *IMA J. Math. Control Inform.* 27, 2010, 103–118.

[197] G. C. Hsiao, W. L. Wendland, *Boundary Integral Equations*, Springer-Verlag, Berlin, 2008.

[198] Y. Hu, S. Peng, Adapted solution of a backward semilinear stochastic evolution equation, *Stoch. Anal. Appl.* 9, 1991, 445–449

[199] A. Ichikawa, Stability of semilinear stochastic evolution equations, *J. Math. Anal. Appl.* 90, 1982, 12–44.

[200] E. Infeld, G. Rowlands, *Nonlinear Waves, Solitons and Chaos*, Cambridge University Press, Cambridge, 1992.

[201] A. Ioannidis, *Μαθηματικά* Προβλήματα *Διάδοσης και Σκέδασης Ηλεκτρομαγνητικών Κυμάτων σε Σύνθετα Μέσα (Mathematical Problems in Propagation and Scattering of Electromagnetic Waves in Complex Media)*, PhD Thesis, National and Kapodistrian University of Athens, 2006 (in Greek).

[202] A. Ioannidis, On the cavity problem for the general linear medium in electromagnetic theory, *Bull. Greek Math. Soc.* 57, 2010, 199–209.

[203] A. Ioannidis, G. Kristensson, D. Sjöberg, On the dispersion equation for a homogeneous bi-isotropic waveguide of arbitrary cross-section, *Microw. Opt. Techn. Lett.* 51, 2009, 2701–2706.

[204] A. Ioannidis, G. Kristensson, D. Sjöberg, The propagation problem in a bi-isotropic waveguide, *Progr. Electrom. Res. B* 19, 2010, 21–40.

[205] A. Ioannidis, G. Kristensson, I. G. Stratis, On the well-posedness of the Maxwell system for linear bianisotropic media, *SIAM J. Math. Anal.*, in press.

[206] A. Ioannidis, I. G. Stratis, A. N. Yannacopoulos, Electromagnetic wave propagation in dispersive bianisotropic media, in *Advances in Scattering and Biomedical Engineering* (eds. D. I. Fotiadis, C. V. Massalas), World Scientific, Hackensack, NJ, 2004, 295–304.

[207] G. Isac, Leray-Schauder type alternatives and the solvability of complementarity problems, *Topological Methods in Nonlinear Analysis (Journal of the Juliusz Schauder Center)* 18, 2001, 191–204.

[208] V. Isakov, *Inverse Problems for Partial Differential Equations*, Springer-Verlag, Berlin, 1998.

[209] A. Ishimaru, *Wave Propagation and Scattering in Random Media, volume 1: Single Scattering and Transport Theory*, Academic Press, New York, 1978.

[210] A. Ishimaru, *Wave Propagation and Scattering in Random Media, volume 2: Multiple Scattering, Turbulence, Rough Scattering and Remote Sensing*, Academic Press, New York, 1978.

[211] J. D. Jackson, *Classical Electrodynamics*, 3rd ed., John Wiley, New York, 1999.

[212] J. Jacques, *The Molecule and Its Double*, McGraw-Hill, New York, 1993.

[213] D. L. Jaggard, N. Engheta, Chirality in electrodynamics: modeling and applications, in *Directions in Electromagnetic Wave Modeling* (eds. H. L. Bertoni, L. B. Felsen), Plenum, New York, 1991, 485–493.

[214] D. L. Jaggard, J. C. Liu, The matrix Riccati equation for scattering from stratified chiral spheres, *IEEE Trans. Antennas and Propagation* 47, 1999, 1201–1207.

[215] J.-S. Jiang, C.-K. Lin, C.-H. Liu, Homogenization of the Maxwell's system for conducting media, *Discrete Contin. Dyn. Syst. Ser. B* 10, 2008, 91–107.

[216] V. V. Jikov, S. M. Kozlov, O. A. Oleinik, *Homogenization of Differential Operators and Integral Functionals*, Springer-Verlag, New York, 1994.

[217] W. Jing, *Corrector Theory in Random Homogenization of Partial Differential Equations*, PhD Thesis, Columbia University, 2011.

[218] D. S. Jones, *Acoustic and Electromagnetic Scattering*, Clarendon Press, Oxford, 1986.

[219] J. Kaipio, E. Somersalo, *Statistical and Computational Inverse Problems*, Springer-Verlag, Berlin, 2004.

[220] E. Kalpinelli, N. E. Frangos, A. N. Yannacopoulos, Wiener chaos solutions for hyperbolic SPDEs, *Stoch. Anal. Appl.* 29, 2011, 237–258.

[221] D. Kannan, A. T. Bharucha-Reid, On a stochastic integrodifferential evolution equation of Volterra type, *J. Integr. Eq.* 10, 1985, 351–379.

[222] I. Kant, *Prolegomena to Any Future Metaphysics* (ed. G. Hartfield), rev. ed., Cambridge University Press, Cambridge, 2004.

[223] B. V. Kapitonov, Stabilization and exact boundary controllability for Maxwell's equations, *SIAM J. Control Optim.* 32, 1994, 308–320.

[224] I. Karatzas, S. Shreve, *Brownian Motion and Stochastic Calculus*, Springer-Verlag, New York, 2004.

[225] A. Karczewska, *Convolution Type Stochastic Volterra Equations*, Lecture Notes in Nonlinear Analysis 10, J. Schauder Center for Nonlinear Studies, Nicolaus Copernicus University, Toruń, 2007.

[226] A. Karlsson, G. Kristensson, Constitutive relations, dissipation and reciprocity for the Maxwell equations in the time domain, *J. Electrom. Waves Appl.* 6, 1992, 537–551.

[227] A. Karlsson, G. Kristensson, Constitutive relations, dissipation and reciprocity for the Maxwell equations in the time domain, Technical Report LUTEDX/(TEAT-7005)/1-36, Lund University, 1999. ([227] is an expanded version of [226]). http://www.eit.lth.se

[228] G. J. Karonis, D. I. Kaklamani, N. K. Uzunoglu, Scattering of electromagnetic waves by a chiral object of ellipsoidal shape, *J. Electromagn. Waves Appl.* 13, 1999, 817–845.

[229] T. Kato, *Perturbation Theory for Linear Operators*, Springer-Verlag, Berlin, 1995.

[230] N. Kavallaris, I. G. Stratis, A. N. Yannacopoulos, On the well posedness of a class of nonlinear complex media in electromagnetics, in preparation.

[231] D. N. Keck, M. A. McKibben, Functional integro-differential stochastic evolution equations in Hilbert space, *J. Appl. Math. Stoch. Anal.* 16, 2003, 127–147.

[232] D. N. Keck, M. A. McKibben: On a McKean-Vlasov stochastic integro-differential evolution equation of Sobolev-type, *Stoch. Anal. Appl.* 21, 2003, 1115–1139.

[233] D. N. Keck, M. A. McKibben: Abstract stochastic delay integrodifferential equations, *J. Appl. Math. Stoch. Anal.* 3, 2005, 275–305.

[234] Kelvin, *Baltimore Lectures on Molecular Dynamics and the Wave Theory of Light*, C . J. Clay and Sons, London, 1904.

[235] B. N. Khatir, M. Al-Kanhal, A. Sebak, Electromagnetic wave scattering by elliptic chiral cylinder, *J. Electromagn. Waves Appl.* 20, 2006, 1377–1390.

[236] K. V. Khmelnytskaya, V. V. Kravchenko, Biquaternions for analytic and numerical solution of equations of electromagnetic theory, in *Some Topics on Value Distribution and Differentiability in Complex and p-adic Analysis* (eds. A. Escassut, W. Tutschke, C. C. Yang), Science Press, Beijing, 2008, 301–322.

[237] K. V. Khmelnytskaya, V. V. Kravchenko, H. Oviedo, Quaternionic integral representations for electromagnetic fields in chiral media, *Telecommun. Radio Eng.* 56, 2001, 53–61.

[238] K. V. Khmelnytskaya, V. V. Kravchenko, V. S. Rabinovich, Quaternionic fundamental solutions for electromagnetic scattering problems and application, *Z. Anal. Anw.* 22, 2003, 147–166.

[239] H. Kiili, *Transmission Problem for the Electromagnetic Scattering by a Chiral Obstacle*, PhD Thesis, University of Oulu, 2011.

[240] J. U. Kim, Approximate controllability of a stochastic wave equation, *Appl. Math. Optim.* 49, 2004, 81–98.

[241] A. Kirsch, Surface gradients and continuity properties for some integral operators in classical scattering theory, *Math. Methods Appl. Sci.* 11, 1989, 789–804.

[242] A. Kirsch, *An Introduction to the Mathematical Theory of Inverse Problems*, Springer-Verlag, Berlin, 1996.

[243] Yu. S. Kivshar, B. Luther-Davies, Dark optical solitons: Physics and applications, *Phys. Rep.* 298, 1998, 81–197.

[244] V. Komornik, *Exact Controllability and Stabilization: The Multiplier Method*, Masson, Paris, 1994.

[245] J. A. Kong, *Theory of Electromagnetic Waves*, Wiley, New York, 1975.

[246] V. V. Kravchenko, On the relation between holomorphic biquaternionic functions and time-harmonic electromagnetic fields, Deposited in UkrINTEI, 29.12.1992, #2073-Uk-92 (in Russian).

[247] V. V. Kravchenko, Quaternion-valued integral representations for time-harmonic electromagnetic and spinor fields, *Dokl. Akad. Nauk* 341, 1995, 603–605 (Engl. trans. *Russ. Acad. Sci., Dokl. Math.* 51, 1995, 287–289).

[248] V. V. Kravchenko, On the relation between the Maxwell system and the Dirac equation, *WSEAS Trans. Syst.* 1, 2002, 115–118.

[249] V. V. Kravchenko, On Beltrami fields with nonconstant proportionality factor, *J. Phys. A* 36, 2003, 1515–1522.

[250] V. V. Kravchenko, Quaternionic equation for electromagnetic fields in inhomogeneous media, in *Progress in Analysis, volume 1* (eds. H. Begehr, R. Gilbert, M. Wah Wong), World Scientific, Singapore, 2003, 361–366.

[251] V. V. Kravchenko, *Applied Quaternionic Analysis*, Research and Exposition in Mathematics Series vol. 28, Heldermann-Verlag, Leipzig, 2003.

[252] V. V. Kravchenko, *Applied Pseudoanalytic Function Theory*, Birkhäuser, Basel, 2009.

[253] V. V. Kravchenko, H. Oviedo, On a quaternionic reformulation of Maxwell's equations for chiral media and its applications, *Z. Anal. Anw.* 22, 2003, 569–589.

[254] V. V. Kravchenko, H. Oviedo, On Beltrami fields with nonconstant proportionality factor on the plane, *Rep. Math. Phys.* 61, 2008, 29–38.

[255] V. V. Kravchenko, H. Oviedo, Time-dependent electromagnetic fields in chiral media, *J. Phys. A Math. Theor.* 43, 2010, 455213.

[256] V. V. Kravchenko, M. V. Shapiro, *Integral Representations for Spatial Models of Mathematical Physics*, Pitman Research Notes in Mathematics vol. 351, Addison Wesley Longman, London, 1996.

[257] R. Kress, On the boundary operator in electromagnetic scattering, *Proc. R. Soc. Edinb. A* 103, 1986, 91–98.

[258] R. Kress, Wave scattering, in *Scattering* (eds. R. Pike, P. Sabatier), Academic Press, New York, 2001, 175–210.

[259] G. Kristensson, *Second Order Differential Equations; Special Functions and Their Classification*, Springer-Verlag, New York, 2010.

[260] G. Kristensson, *Scattering of Electromagnetic Waves*, book in preparation.

[261] G. Kristensson, N. Wellander, Personal communication to one of the authors (IGS), August 2010.

[262] A. Kufner, O. John, S. Fučik, *Function Spaces*, Noordhoff, Leyden, 1977.

[263] O. A. Ladyzenskaja, V. A. Solonikov, N. N. Ural'ceva, *Linear and Quasilinear Equations of Parabolic Type*, American Mathematical Society Translations no. 23, American Mathematical Society, Providence, RI, 1968.

[264] J. E. Lagnese, Exact boundary controllability of Maxwell's equations in a general region, *SIAM J. Control Optim.* 27, 1989, 374–388.

[265] J. E. Lagnese, The Hilbert Uniqueness Method: a retrospective, in *Optimal Control of Partial Differential Equations*, Lecture Notes in Control and Information Science no. 149, Springer-Verlag, Berlin, 1991, 158–181.

[266] A. Lakhtakia (ed.), *Selected Papers on Natural Optical Activity*, SPIE Optical Engineering Press, Bellingham, WA, 1990.

[267] A. Lakhtakia, On the Huyghens's principles and the Ewald-Oseen extinction theorems for, and the scattering of, Beltrami fields, *Optik* 91, 1992, 35–40.

[268] A. Lakhtakia, *Beltrami Fields in Chiral Media*, World Scientific, Singapore, 1994.

[269] A. Lakhtakia, Time-dependent Beltrami fields in material continua: the Beltrami-Maxwell postulates, *Int. J. Infrared Millim. Waves* 15, 1994, 369–394.

[270] A. Lakhtakia: Viktor Trkal, Beltrami fields and Trkalian flows, *Czech. J. Phys.* 44, 1994, 89–96.

[271] A. Lakhtakia (ed.), *Selected Papers on Linear Optical Composite Materials*, SPIE Optical Engineering Press, Bellingham, WA, 1996.

[272] A. Lakhtakia, A mini-review on isotropic chiral mediums, in *Electromagnetic Fields in Unconventional Materials and Structures* (eds. O. N. Singh, A. Lakhtakia), Wiley, New York, 2000, 125–149.

[273] A. Lakhtakia, V. K. Varadan, V. V. Varadan, *Time-Harmonic Electromagnetic Fields in Chiral Media*, Lecture Notes in Physics no. 335, Springer-Verlag, Berlin, 1989.

[274] A. Lakhtakia, W. S. Weiglhofer, Maxwell Garnett formalism for weakly nonlinear, bianisotropic, dilute, particulate composite media, *Int. J. Electron.* 87, 2000, 1401–1408.

[275] I. Lasiecka, R. Triggiani, *Differential and Algebraic Riccati Equations with Application to Boundary/ Point Control Problems: Continuous Theory and Approximation Theory*, Springer-Verlag, Berlin, 1991.

[276] P. D. Lax, *Functional Analysis*, Wiley, New York, 2002.

[277] P. D. Lax, R. S. Phillips, *Scattering Theory*, Academic Press, New York, 1967.

[278] R. Leis, *Initial Boundary Value Problems in Mathematical Physics*, Teubner & Wiley, Stuttgart, 1986.

[279] A. Lejay, Homogenization of divergence-form operators with lower order terms in random media, *Probab. Theory Related Fields* 120, 2001, 255–276.

[280] J. Le Rousseau, G. Lebeau, On Carleman estimates for elliptic and parabolic operators. Applications to unique continuation and control of parabolic equations, preprint, 2011.

[281] S. Li, An inverse problem for Maxwell's equations in bi-isotropic media, *SIAM J. Math. Anal.* 37, 2005, 1027–1043.

[282] K. B. Liaskos, I. G. Stratis, A. N. Yannacopoulos, Well posedness of the stochastic Drude-Born-Fedorov model in electromagnetics, *Bull. Greek Math. Soc.* 54, 2007, 207–220.

[283] K. B. Liaskos, I. G. Stratis, A. N. Yannacopoulos, Pseudoparabolic equations with additive noise and applications, *Math. Methods Appl. Sci.* 32, 2009, 963–985.

[284] K. B. Liaskos, I. G. Stratis, A. N. Yannacopoulos, A priori estimates for a singular limit approximation of the constitutive laws in the time domain, *J. Math. Anal. Appl.* 355, 2009, 288–302.

[285] K. B. Liaskos, I. G. Stratis, A. N. Yannacopoulos, Stochastic integrodifferential equations in Hilbert spaces with applications in electromagnetics, *J. Integr. Eq. Appl.* 22, 2010, 559–590.

[286] K. B. Liaskos, I. G. Stratis, A. N. Yannacopoulos, Stochastic PDE models for chiral media: well posedness, singular limits and a priori estimates for their validity, *Bull. Greek Math. Soc.* 57, 2010, 265–279.

[287] I. V. Lindell, *Methods for Electromagnetic Field Analysis*, Clarendon Press, Oxford, 1992.

[288] I. V. Lindell, A. H. Sihvola, J. Kurkijarvi, Karl F. Lindman: The last Hertzian, and a harbinger of electromagnetic chirality, *IEEE Antennas Propag. Mag.* 34, 1992, 24–30.

[289] I. V. Lindell, A. H. Sihvola, S. A. Tretyakov, A. J. Viitanen, *Electromagnetic Waves in Chiral and Bi-isotropic Media*, Artech House, Boston, 1994.

[290] J.-L. Lions, Exact controllability, stabilization and perturbations for distributed systems, *SIAM Rev.* 30, 1988, 1–68.

[291] J.-L. Lions, *Contrôlabilité exacte, perturbations et stabilisation de systémes distribués, Tomes 1 et 2*, Masson, Paris, 1988.

[292] S. Lototsky, B. Rozovskii, Wiener chaos solutions of linear stochastic evolution equations, *Ann. Prob.* 34, 2006, 638–662.

[293] S. Lototsky, B. Rozovskii, Stochastic differential equations: a Wiener chaos approach, in *From Stochastic Calculus to Mathematical Finance: The Shirayev Festschrift* (eds. Y. Kabanov, R. Lipster, J. Stoyanov), Springer-Verlag, Berlin, 2006, 433–507.

[294] S. Lototsky, B. Rozovskii, Stochastic partial differential equations driven by purely spatial noise, *SIAM J. Math. Anal.* 41, 2009, 1295–1322.

[295] R. J. Loy, Continuity of linear operators commuting with shifts, *J. Funct. Anal.* 16, 1974, 48–60.

[296] T. G. Mackay, A. Lakhtakia, W. S. Weiglhofer, Strong-property fluctuation theory for homogenization of bianisotropic composites: formulation, *Phys. Rev. E* 62, 2000, 6052–6064.

[297] T. G. Mackay, A. Lakhtakia, W. S. Weiglhofer, Homogenisation of isotropic, cubically nonlinear, composite mediums by the strong-property-fluctuation theory: Third-order considerations, *Opt. Commun.* 204, 2002, 219–228.

[298] T. G. Mackay, A. Lakhtakia, W. S. Weiglhofer, The strong-property-fluctuation theory for cubically nonlinear isotropic chiral composite mediums, *Electromagnetics* 23, 2003, 455–479.

[299] T. G. Mackay, A. Lakhtakia, *Electromagnetic Anisotropy and Bi-Anisotropy: A Field Guide*, World Scientific, Singapore, 2010.

[300] P. Malliavin, *Integration and Probability*, Springer-Verlag, New York, 1995.

[301] X. Mao, *Stochastic Differential Equations and Applications*, 2nd ed., Woodhead, Abington, 2007.

[302] P. A. Markowich, F. Poupaud, The Maxwell equation in a periodic medium: homogenization of the energy density, *Ann. Sc. Norm. Sup. Pisa Cl. Sci.* 23, 1996, 301–324.

[303] P. A. Martin, *Multiple Scattering: Interaction of Time-Harmonic Waves with N Obstacles*, Cambridge University Press, Cambridge, 2006.

[304] P. A. Martin, P. Ola, Boundary integral equations for the scattering of electromagnetic waves by a homogeneous dielectric obstacle, *Proc. R. Soc. Edinb. A* 123A, 1993, 185–208.

[305] P. A. Martin, F. J. Rizzo, On boundary integral equations for crack problems, *Proc. R. Soc. Lond. A* 421, 1989, 341–355.

[306] J. C. Maxwell, *A Treatise on Electricity and Magnetism, volumes 1 and 2*, Oxford University Press, Oxford, 1998.

[307] S. R. Mc Dowall, Boundary determination of material parameters from electromagnetic boundary information, *Inv. Prob.* 13, 1997, 153–163.

[308] S. R. Mc Dowall, An inverse problem in chiral media, *Trans. Amer. Math. Soc.* 352, 2000, 2993–3013.

[309] W. McLean, *Strongly Elliptic Systems and Boundary Integral Equations*, Cambridge University Press, Cambridge, 2000.

[310] V. K. Melnikov, Wave emission and absorption in a nonlinear integrable system, *Phys. Lett. A* 118, 1986, 22–24.

[311] V. K. Melnikov, Exact solutions of the Korteweg - de Vries equation with a self-consistent source, *Phys. Lett. A* 128, 1988, 488–492.

[312] V. K. Melnikov, Integration method of the Korteweg - de Vries equation with a self-consistent source, *Phys. Lett. A* 133, 1988, 493–496.

[313] H. Mesnil, F. Hache, Experimental evidence of third-order nonlinear dichroism in a liquid of chiral molecules, *Phys. Rev. Lett.* 85, 2000, 4257–4260.

[314] H. Mesnil, M. C. Schanne-Klein, M. Alexandre, G. Lemercier, C. Andraud, Wavelength dependence of nonlinear circular dichroism in a chiral ruthenium-tris (bipyridyl) solution, *Phys. Rev. A* 66, 2002, 013802 (1–9).

[315] G. Métivier, *The Mathematics of Nonlinear Optics*, 2008. http://www.math.u-bordeaux.fr/ metivier/NLO2008v2.pdf

[316] B. Michel, A. Lakhtakia, Strong property fluctuation theory for homogenizing chiral particular composites, *Phys. Rev. E* 51, 1995, 5701–5707.

[317] B. Michel, Recent developments in the homogenization of linear bianisotropic composite materials, in *Electromagnetic Fields in Unconventional Materials and Structures* (eds. O. N. Singh, A. Lakhtakia), Wiley, New York, 2000, 39–81.

[318] S. G. Mikhlin, *Mathematical Physics: an Advanced Course*, North Holland, Amsterdam, 1970.

[319] S. G. Mikhlin, S. Prössdorf, *Singular Integral Operators*, Springer-Verlag, Berlin, 1986.

[320] A. Milani, R. Picard, Evolution equations with nonlinear constitutive laws and memory effects, *Diff. Integr. Equ.* 15, 2002, 327–344.

[321] R. K. Miller, Volterra integral equations in a Banach space, *Funkcial. Ekvac.* 18, 1975, 163–194.

[322] G. W. Milton, *The Theory of Composites*, Cambridge Monographs on Applied and Computational Mathematics no. 6, Cambridge University Press, Cambridge, 2004.

[323] M. Mitrea, The method of layer potentials for electromagnetic waves in chiral media, *Forum Math.* 13, 2001, 423–446.

[324] P. Monk, *Finite Element Methods for Maxwell's Equations*, Clarendon Press, Oxford, 2003.

[325] P. M. Morse, H. Feshbach, *Methods of Theoretical Physics, parts I and II*, McGraw-Hill, New York, 1953.

[326] H. E. Moses, Eigenfunctions of the curl operator, rotationally invariant Helmholtz theorem and applications to electromagnetic theory and fluid mechanics *SIAM J. Appl. Math.* 21, 1971, 114–144.

[327] C. Müller, Über die ganzen lösungen der wellengleichung, *Math. Ann.* 124, 1952, 235–264.

[328] C. Müller, *Foundations of the Mathematical Theory of Electromagnetic Waves*, Springer-Verlag, Berlin, 1969.

[329] F. Murat, Compacité par compensation, *Ann. Sc. Norm. Sup. Pisa Cl. Sci., IV. Ser.*, 5, 1978, 489–507.

[330] F. Murat, Compacité par compensation: condition necessaire et suffisante de continuite faible sous une hypothèse de rang constant, *Ann. Sc. Norm. Sup. Pisa, Cl. Sci., IV. Ser.*, 8, 1981, 69–102.

[331] F. Murat, H-*Convergence*, mimeographed notes, Séminaire d'Analyse Fonctionnelle et Numérique de l'Université d'Alger, 1978.

[332] F. Murat, A survey on compensated compactness, in *Contributions to the Modern Calculus of Variations* (ed. L. Cesari), Pitman Research Notes in Mathematics 148, Longman, Harlow, 1987, 145–183.

[333] F. Murat, L. Tartar, H-Convergence, in *Topics in the Mathematical Modelling of Composite Materials* (eds. A. Cherkaev, R. Kohn) Birkhäuser, Basel, 1997, 21–43.

[334] J.-C. Nédélec, *Acoustic and Electromagnetic Equations*, Springer-Verlag, Berlin, 2001.

[335] J.-C. Nédélec, Mixed finite elements in $\mathbb{R}^3$, *Numer. Math.* 35, 1980, 315–341.

[336] S. Nicaise, Exact boundary controllability of Maxwell's equations in heterogeneous media and an application to an inverse source problem, *SIAM J. Control Optim.* 38, 2000, 1145–1170.

[337] G. Nguetseng, A general convergence result for a functional related to the theory of homogenization, *SIAM J. Math. Anal.* 20, 1989, 608–623.

[338] G. Nguetseng, Homogenization structures and applications I, *Z. Anal. Anw.* 22, 2003, 73–107.

[339] D. Nualart, *The Malliavin Calculus and Related Topics*, 2nd ed., Springer-Verlag, Berlin, 2006.

[340] B. Øksendal, Optimal control of stochastic partial differential equations, *Stoch. Anal. Appl.*, 23, 2005, 165–179.

[341] P. Ola, Boundary integral equations for the scattering of electromagnetic waves by a homogeneous chiral obstacle, *J. Math. Phys.* 35, 1994, 3969–3980.

[342] P. Ola, L. Päivärinta, E. Somersalo, An inverse boundary value problem in electrodynamics, *Duke Math. J.* 70, 1993, 617–653.

[343] J. M. Ottino, *The Kinematics of Mixing: Stretching, Chaos, and Transport*, Cambridge University Press, Cambridge, 1989.

[344] G. C. Papanicolaou, Diffusion in random media, in *Surveys in Applied Mathematics* (eds. J. B. Keller, D. McLaughlin, G. C. Papanicolaou), Plenum, New York, 1995, 205–255.

[345] L. Pasteur, Sur les relations qui peuvent exister entre la forme cristalline, la composante chimique et le sens de la polarisation rotatoire, *Ann. Chim. Phys.* 24, 1848, 442–459.

[346] A. Pazy, *Semigroups of Linear Operators and Applications to Partial Differential Equations*, Springer-Verlag, New York, 1983.

[347] S. Peszat, J. Zabczyk, *Stochastic Partial Differential Equations with Lévy Noise*, Cambridge University Press, Cambridge, 2007.

[348] R. E. Peters II, *A Brief Outline of the History of Electromagnetism*, 2000.
http://www.vuse.vanderbilt.edu/~rap2/papers/em_history.pdf

[349] R. Picard, On a selfadjoint realization of curl in exterior domains, *Math. Z.* 229, 1998, 319–338.

[350] C. Pignotti, Observability and controllability of Maxwell's equations, *Rend. Mat.* 19, 2000, 523–546.

[351] V. Prelog, Chirality in hemistry, Nobel lecture, 1975.
http://nobelprize.org/nobel_prizes/chemistry/laureates/1975/prelog-lecture.pdf

[352] C. Prévôt, M. Röckner, *A Concise Course on Stochastic Partial Differential Equations*, Springer-Verlag, Berlin, 2007.

[353] E. Priola, J. Zabczyk, On linear evolution equations with cylindrical Lévy noise, preprint, 2009.

[354] R. Potthast, Domain derivatives in electromagnetic scattering, *Math. Methods Appl. Sci.* 19, 1996, 1157–1175.

[355] R. Potthast, *Point Sources and Multipoles in Inverse Scattering Theory*, Chapman and Hall, London, 2001.

[356] R. Potthast, I. G. Stratis, On the domain derivative for scattering by impenetrable obstacles in chiral media, *IMA J. Appl. Math.* 68, 2003, 621–635.

[357] J. Prüss, *Evolutionary Integral Equations and Applications*, Birkhäuser, Basel, 1993.

[358] A. G. Ramm, *Scattering by Obstacles*, Reidel, Dordrecht, 1986.

[359] A. G. Ramm, *Inverse Problems: Mathematical and Analytical Techniques with Applications to Engineering*, Springer-Verlag, New York, 2005.

[360] A. G. Ramm, *Wave Scattering by Small Bodies of Arbitrary Shapes*, World Scientific, Singapore, 2005.

[361] M. Reed, B. Simon, *Methods of Modern Mathematical Physics, volume 1: Functional Analysis*, rev. ed., Academic Press, New York, 1980.

[362] M. Reed, B. Simon, *Methods of Modern Mathematical Physics, volume 3: Scattering Theory*, Academic Press, New York, 1979.

[363] M. Renardy, R. C. Rogers, *An Introduction to Partial Differential Equations*, 2nd ed., Springer, New York, 2004.

[364] F. Riesz, B. Sz.-Nagy, *Functional Analysis*, Ungar, New York, 1981.

[365] S. Rikte, G. Kristensson, M. Andersson, Propagation in bianisotropic media: reflection and transmission, Technical Report LUTEDX/(TEAT-7067)/1-32, Lund University, 1998. http://www.eit.lth.se

[366] S. Rikte, B. Sauviac, G. Kristensson, F. Mariotte, Reconstruction of bi-isotropic material parameters from experimental data, Technical Report LUTEDX/(TEAT-7072)/1-22, Lund University, 1998. http://www.eit.lth.se

[367] G. F. Roach, *An Introduction to Linear and Nonlinear Scattering Theory*, Pitman Monographs and Surveys in Pure and Applied Mathematics vol. 78, Longman, Essex, 1995.

[368] G. F. Roach, *Wave Scattering by Time Dependent Perturbations*, Princeton Series in Applied Mathematics, Princeton University Press, Princeton, NJ, 2007.

[369] G. F. Roach, *An Introduction to Echo Analysis: Scattering Theory and Wave Propagation*, Springer Monographs in Mathematics, Springer-Verlag, London, 2008.

[370] T. M. Roberts, Causality theorems, in *Invariant Imbeddings and Inverse Problems* (eds. J. P. Corones, G. Kristensson, P. Nelson, D. L. Seth), SIAM, Philadelphia, 1992, 114–128.

[371] F. N. H. Robinson, *Macroscopic Electromagnetism*, Pergamon Press, Oxford, 1973.

[372] R. G. Rojas, Integral equations for the scattering of a three–dimensional inhomogeneous chiral body, *J. Electrom. Waves Appl.* 6, 1992, 733–750.

[373] T. Roubíček, *Nonlinear Partial Differential Equations with Applications*, Birkhäuser, Basel, 2005.

[374] J. Sanchez-Hubert, Étude de certaines équations intégrodifférentielles issues de la théorie de l'homogénéisation, *Boll. U.M.I.* 16, 1979, 857–875.

[375] J. Sanchez-Hubert, E. Sanchez-Palencia, *Introduction aux méthodes asymptotiques et à l' homogénéisation*, Masson, Paris, 1992.

[376] E. J. P. G. Schmidt, Spectral and scattering theory for Maxwell's equations in an exterior domain, *Arch. Rat. Mech. Anal.*, 28, 1968, 284–322.

[377] E. J. P. G. Schmidt, On scattering by time-dependent perturbations, *Indiana Univ. Math. J.*, 24, 1975, 925–935.

[378] A. Serdyukov, I. Semchenko, S. Tretyakov, A. Sihvola, *Electromagnetics of Bi-Anisotropic Materials: Theory and Applications*, Gordon and Breach, Amsterdam, 2001.

[379] D. Sheen, D. Shepelsky, Uniqueness in a frequency-domain inverse problem of a stratified uniaxial bianisotropic medium, *Wave Motion* 31, 2000, 371–385.

[380] R. E. Showalter, *Monotone Operators in Banach Spaces and Nonlinear Partial Differential Equations*, Mathematical Surveys and Monographs vol. 49, American Mathematical Society, Providence, RI, 1997.

[381] A. H. Sihvola, *Electromagnetic Mixing Formulas and Applications*, Electromagnetic Waves Series vol. 47, Institute of Electrical Engineers, London, 1999.

[382] A. H. Sihvola, O. P. M. Pekonen, Effective medium formulae for bi-anisotropic mixtures, *J. Phys. D Appl. Phys.* 29, 1996, 514–521.

[383] A. H. Sihvola, Electromagnetic emergence in metamaterials, in *Advances in Electromagnetics of Complex Media and Metamaterials* (eds. S. Zouhdi, A. H. Sihvola, M. Arsalane), NATO Science Series II: Mathematics, Physics and Chemistry, Kluwer, Dordrecht, 2002, 3–17.

[384] A. H. Sihvola, S. Zouhdi, Handedness in plasmonics: electrical engineer's perspective, in *Metamaterials and Plasmonics: Fundamentals, Modelling, Applications* (eds. S. Zouhdi, A. H. Sihvola, A. P. Vinogradov), NATO Science for Peace and Security Series B: Physics and Biophysics, Springer, Dordrecht, 2009, 3–20.

[385] J. Simon, Compact sets in the space $L^p(0,T;B)$, *Ann. Mat. Pura Appl.* 146, 1987, 65–96.

[386] D. Sjöberg, Shock structure for electromagnetic waves in bianisotropic, nonlinear materials, *Radio Sci.* 38, 2003, 8015–8015.

[387] D. Sjöberg, Homogenization of dispersive material parameters for Maxwell's equations using a singular value decomposition, *Multiscale Model. Simul.* 4, 2005, 760–789.

[388] D. Sjöberg, On uniqueness and continuity for the quasi-linear, bianisotropic Maxwell equations, using an entropy condition, *Prog. Electromagn. Res.* 71, 2007, 317–339.

[389] D. Sjöberg, A modified Drude-Born-Fedorov model for isotropic chiral media, obtained by finite scale homogenization, *J. Phys. D Appl. Phys.* 41, 2008, 155412.

[390] D. Sjöberg, C. Engstrom, G. Kristensson, D. J. N. Wall, N. Wellander, A Floquet-Bloch decomposition of Maxwell's equations applied to homogenization, *Multiscale Model. Simul.* 4, 2005, 149–171.

[391] S. Spagnolo, Sul limite delle soluzioni di problemi di Cauchy relativi all'equazione di calore, *Ann. Sc. Norm. Sup. Pisa* 21, 1967, 657–699.

[392] J. Stöhr, H. C. Siegmann, *Magnetism: From Fundamentals to Nanoscale Dynamics*, Springer-Verlag, Berlin, 2006.

[393] I. G. Stratis, Electromagnetic scattering problems in chiral media: a review, *Electromagnetics* 19, 1999, 547–562.

[394] I. G. Stratis, A. N. Yannacopoulos, Electromagnetic fields in linear and nonlinear chiral media: a time-domain analysis, *Abstr. Appl. Anal.* 2004, 2004, 471–486.

[395] I. G. Stratis, A. N. Yannacopoulos, Homogenisation theory for deterministic and random bianisotropic media, *Compos. Part B-Eng*, in press.

[396] J. A. Stratton, *Electromagnetic Theory*, McGraw-Hill, New York, 1972.

[397] S. Ström, The scattered field, in *Field Representations and Introduction to Scattering* (eds. V. V. Varadan, A. Lakhtakia, V. K. Varadan), North Holland, Amsterdam, 1991, 143–164.

[398] M. A. Sumbatyan, A. Scalia, *Equations of Mathematical Diffraction Theory*, Chapman and Hall / CRC, Boca Raton, FL, 2005.

[399] G. Sviridyuk, V. E. Fedorov, *Linear Sobolev Type Equations and Degenerate Semigroups of Operators*, VSP, Utrecht, 2003.

[400] C. T. Tai, *Dyadic Green Functions in Electromagnetic Theory*, IEEE Press, Piscataway, NJ, 1994.

[401] S. Tang, X. Zhang, Null controllability for forward and backward stochastic parabolic equations, *SIAM J. Control Optim.* 48, 2009, 2191–2216.

[402] A. Tarantola, *Inverse Problem Theory and Model Parameter Estimation*, SIAM, Philadelphia, 2005.

[403] L. Tartar, Compensated compactness and applications to partial differential equations, in *Nonlinear Mechanics and Analysis: Heriot-Watt Symposium, volume IV* (ed. R. J. Knops), Research Notes in Mathematics no. 39, Pitman, London, 1979, 136–212.

[404] L. Tartar, Cours Peccot au Collège de France, 1997 (unpublished), partially appearing in [331].

[405] L. Tartar, An introduction to the homogenization method in optimal design, in *Optimal Shape Design*, Lecture Notes in Mathematics no. 1740, Springer-Verlag, Berlin, 2000, 47–156.

[406] L. Tartar, *An Introduction to Sobolev Spaces and Interpolation Spaces*, Springer-Verlag, Berlin, 2007.

[407] L. Tartar, *The General Theory of Homogenization, A Personalized Introduction*, Springer-Verlag, Berlin, 2009.

[408] J. J. Telega, Stochastic homogenization: Convexity and non convexity, in *Nonlinear Homogenization and Its Applications to Composites, Polycrystals and Smart Materials* (eds. P. Ponte Castaneda, J. J. Telega, B. Gambin), Kluwer, Dordrecht, 2004, 305–349.

[409] G. Tessitore, Existence, uniqueness and space regularity of the adapted solutions of a backward SPDE, *Stoch. Anal. Appl.*, 14, 1996, 461–486.

[410] L. Tsang, J. A. Kong, K.-H. Ding, *Scattering of Electromagnetic Waves: Theories and Applications*, Wiley, New York, 2000.

[411] L. Tsang, J. A. Kong, K.-H. Ding, C. O. Ao, *Scattering of Electromagnetic Waves: Numerical Simulations*, Wiley, New York, 2001.

[412] L. Tsang, J. A. Kong, *Scattering of Electromagnetic Waves: Advanced Topics*, Wiley, New York, 2001.

[413] N. L. Tsitsas, A. Lakhtakia, D. J. Frantzeskakis, Vector solitons in nonlinear isotropic chiral metamaterials, *J. Phys. A: Math. Theor.*, 44, 2011, 435203.

[414] M. Tucsnak, G. Weiss, *Observation and Control for Operator Semigroups*, Birkhäuser, Basel, 2009.

[415] V. Twersky, Scattering of waves by two objects, in *Electromagnetic Waves* (ed. R. E. Langer), University of Wisconsin Press, Madison, 1962, 361–369.

[416] V. Twersky, Multiple scattering by arbitrary configurations in three dimensions, *J. Math. Phys.* 3, 1962, 83–91.

[417] V. Twersky, Multiple scattering of electromagnetic waves by arbitrary configurations, *J. Math. Phys.* 3, 1967, 589–598.

[418] S. Vänskä, *Direct and Inverse Scattering for Beltrami Fields*, PhD Thesis, University of Helsinki, 2006.

[419] M. Veneroni, Stochastic homogenization of subdifferential inclusions via scale integration, *Int. J. Struct. Changes Solids*, in press.

[420] V. Veselago, The electrodynamics of substances with simultaneously negative values of ε and μ, *Sov. Phys. Uspekhi* 10, 1968, 509–514 (the original paper, in Russian, appeared in *Uspekhi Fiz. Nauk* 92, 1964, 517–526).

[421] A. Visintin, Electromagnetic processes in doubly-nonlinear composites, *Comm. PDE* 33, 2008, 808–841.

[422] N. Weck, Exact boundary controllability of a Maxwell problem, *SIAM J. Control Optim.* 38, 2000, 736–750.

[423] R. Weder, *Spectral and Scattering Theory for Wave Propagation in Perturbed Stratified Media*, Springer-Verlag, New York, 1991.

[424] W. Wei, H. M. Yin, Global solvability for a singular nonlinear Maxwell's equations, *Comm. Pure Appl. Anal.* 4, 2005, 431–444.

[425] N. Wellander, Homogenization of the Maxwell equations: Case I. Linear theory, *Appl. Math.* 46, 2001, 29–51.

[426] N. Wellander, Homogenization of the Maxwell equations: Case II. Nonlinear conductivity, *Appl. Math.* 47, 2002, 255–283.

[427] N. Wellander, The two-scale Fourier transform approach to homogenization; periodic homogenization in Fourier space, *Asymptot. Anal.* 62, 2009, 1–40.

[428] N. Wellander, G. Kristensson, Homogenization of the Maxwell equations at fixed frequency, *SIAM J. Appl. Math.* 64, 2003, 170–195.

[429] D. V. Widder, *The Laplace Transform*, Princeton University Press, Princeton, NJ, 1946.

[430] C. H. Wilcox, A generalization of theorems of Rellich and Atkinson, *Proc. Amer. Math. Soc.* 7, 1956, 271–276.

[431] C. H. Wilcox, An expansion theorem for electromagnetic fields, *Comm. Pure Appl. Math.* 9, 1956, 115–134.

[432] C. H. Wilcox, *Wave Scatering Theory for the d'Alembert Equation in Exterior Domains*, Lecture Notes in Mathematics no. 442, Springer-Verlag, Berlin, 1975.

[433] P. Wilde, A. Willers, Dense sets and far field patterns for the vector Helmholtz equation under transmission boundary conditions, *SIAM J. Math. Anal.* 19, 1988, 348–357.

[434] E. B. Wilson, *Vector Analysis: A Text-Book for the Use of Students of Mathematics and Physics, Founded upon the Lectures of J. Willard Gibbs*, Dover, New York, 1960.

[435] T. X. Wu, D. L. Jaggard, A comprehensive study of discontinuities in chirowaveguides, *IEEE Trans. Microw. Theory Tech.* 50, 2002, 2320–2330.

[436] T. X. Wu, D. L. Jaggard, Scattering of chiral periodic structure, *IEEE Trans. Antennas Propag.* 52, 2004, 1859–1870.

[437] J. Xin, Front propagation in heterogeneous media, *SIAM Rev.* 42, 2000, 161–230.

[438] X. Yang, T. X. Wu, D. L. Jaggard, Physical properties of wave scattering by a chiral grating, *J. Opt. Soc. Amer. A* 21, 2004, 2109–2116.

[439] A. N. Yannacopoulos, N. E. Frangos, I. Karatzas, Wiener chaos solutions for linear backward stochastic evolution equations, *SIAM J. Math. Anal.* 43, 2011, 68–113.

[440] H. M. Yin, On a singular limit problem for nonlinear Maxwell's equations, *J. Diff. Eq.* 156, 1999, 355–375.

[441] Z. Yoshida, Applications of Beltrami functions in plasma physics, *Nonlinear Anal. Theory Methods Appl.* 30, 1997, 3617–3627.

[442] Z. Yoshida, Y. Giga, Remarks on spectra of operator rot, *Math. Z.* 204, 1990, 235–240.

[443] Z. Yoshida, S. Mahajan, Simultaneous Beltrami conditions in coupled vortex dynamics, *J. Math. Phys.* 40, 1999, 5080–5091.

[444] D. Zhang, F. Ma, An integral equation method for electromagnetic scattering by a periodic chiral structure, *Northeast. Math. J.* 21, 2005, 165–174.

[445] D. Zhang, F. Ma, Electromagnetic scattering by a chiral grating in a homogeneous chiral environment and its finite element method with perfectly matched absorbing layers, *Numer. Math. J. Chin. Univ. (Engl. Ser.)* 15, 2006, 164–179.

[446] D. Zhang, F. Ma, Electromagnetic scattering by a perfectly conducting grating in a homogeneous chiral environment, *Appl. Math. J. Chinese Univ., Ser. B*, 21, 2006, 127–134.

[447] X. Zhang, Carleman and observability estimates for stochastic wave equations, *SIAM J. Math. Anal.* 40, 2008, 851–868.

[448] X. Zhang, A unified controllability/observability theory for some stochastic and deterministic partial differential equations, *arXiv:1003.5819v1 [math.OC]*, 2010.

[449] Y. Zhang, L.-Q. Cao, Y.-S. Wong, Multiscale computations for 3D time-dependent Maxwell's equations in composite materials, *SIAM J. Sci. Comput.* 32, 2010, 2560–2583.

[450] J. Zhao, Analysis of finite element approximation for time-dependent Maxwell problems, *Math. Comput.* 73, 2003, 1089–1105.

[451] E. Zeidler, *Nonlinear Functional Analysis and Its Applications, volume II A: Linear Monotone Operators*, *volume II B: Nonlinear Monotone Operators*, Springer-Verlag, New York, 1990.

[452] A. H. Zemanian, *Distribution Theory and Transform Analysis*, Dover, New York, 1987.

[453] E. Zuazua, Controllability and observability of partial differential equations: Some results and open problems, in *Handbook of Differential Equations* volume 3 (eds. C. M. Dafermos, E. Feireisl), North-Holland Mathematical Library, Elsevier, 2006, 527–621.

Index

Absorption cross section, 116, 122
Ampère's law, 10
Atkinson-Wilcox expansion theorem, 114, 124

Babuška-Brezzi lemma
  application, 68
Backward stochastic partial differential equation, 269
Beltrami field, 86, 87
  left-handed, 87
  right-handed, 87
  spherical representation, 96
Beltrami operator, 125
Beltrami-Herglotz functions, 131
Bilinear form, 64, 75
  time-harmonic interior problem, 63
  time-harmonic problem, 65
Bohren decomposition, 87, 99
Boundary conditions
  Leontovich, 24
  perfect conductor, 23
  Silver-Müller, 27
  transmission conditions, 26
Boundary integral equations, 101
  generalised Müller, 106
  perfect conductor problem, 101
  transmission problem, 103
Boundary integral operators, 100

Calderón operator
  chiral exterior, 76
  exterior, 75
Cameron-Martin basis, 258
Carleman estimates, 272
Chiral Herglotz pairs, 132
Chiral Müller equations, 107
Chirality measure, 20
Classical solutions, 154
  existence, 155
Compact embeddings, 53
Constitutive relations, 11, 13–23
  convolution form, 14
  frequency domain, 18
    Drude-Born-Fedorov, 19
    Tellegen, 19
  Kerr nonlinearity, 22
  nonlinear media, 21
    compact form, 22
  operator form, 12
  periodic media, 30
  plausible physical hypotheses, 13
    causality, 13
    determinism, 13
    linearity, 13
    locality in space, 13
    time-translation invariance, 13
  polynomial nonlinearity, 22
  random media, 20
    complex, 36
  saturable nonlinearity, 22
  time-harmonic fields, 19
Contrasting chiral media, 109
Controllability, 9
  deterministic
    abstract approach, 176
    achiral media, 165–174
    auxiliary forward problem, 165
    backward adjoint problem, 165
    backward problem, 168
    boundary, 178
    complex media, 174
    control-to-state operator, 176
    controllability operator, 177
    fixed point scheme, 163
    forward problem, 171
    Hilbert uniqueness method, 172
    observability inequality, 177
    operator Λ, 166, 172
    optimal control, 178
    time-harmonic, 179
  stochastic, 35, 263
    additive noise, 263
    approximate, 264
    auxiliary uncontrolled problem, 265
    backward adjoint problem, 265
    backward stochastic partial differential equations, 269
    boundary, 272
    fixed point scheme, 265
    multiplicative noise, 272
    operator Λ, 265
    optimal control, 273

random partial differential equations, 266
subtleties, 264
Current j, 12

De Rham sequences, 54
Decomposition
Bohren, 87, 99
Helmholtz, 52
Hodge, 52
normal, tangential, 53
Dielectric susceptibility kernel, $\chi^{e}(t)$, 16
Differential scattering cross section, 116
Dissipative media, 16, 17
Drude-Born-Fedorov constitutive relations, 19
Dyadic Green function, 124

Eigenvalue problem, 69
weak formulation, 69
Electric charge density, ρ, 10
Electric current density, J, 10
Electric field, E, 9
Electric flux density, D, 9
Electromagnetic field, u, 12
Electromagnetic flux density, d, 12
Electromagnetic materials
dissipative, 17
passive, 17
Electromagnetic media classification, 15
anisotropic, 15
bianisotropic, 15
biisotropic, 15
isotropic, 15
Equation
chiral Müller, 107
Helmholtz, 24, 97, 98, 113
Ergodic action, 276
Expansion
Atkinson-Wilcox, 114, 124
Exterior domain problems, 25
Extinction cross section, 117, 122

Far-field behaviour
low-frequency, 128
Far-field operator, 120, 135
electric, 136
Fréchet differentiability, 137
Newton scheme, 139
T-matrix, 120
Far-field pattern
reciprocity principle, 119, 134
Far-field pattern generator
spherical, 121
Far-field patterns
left-handed, 136
right-handed, 136

Faraday's law, 10
Finite elements, 79
Fréchet differentiability of scattering operators, 138
Fredholm alternative
time-harmonic problem, 66
Fredholm integral equation
first kind, 94
second kind, 94
Fredholm theory
application in electromagnetics, 101
electromagnetics, 66
Functions
Beltrami-Herglotz, 131
Herglotz
scalar wave, 130
vector wave, 130
normalised real scalar spherical harmonic
even, 96
odd, 96
normalised real vector spherical harmonic, 97
normalised spherical vector wave
outgoing, 97
spherical Bessel, first kind, 96
spherical Bessel, second kind, 96
spherical Hankel, 120
first kind, 96
spherical wave, 96

Gauss divergence theorem, 18
Gauss's laws, 10
Gelfand triple, 44, 177, 294
Green function
dyadic, 124
Green's formulae, 46, 55

Hölder seminorms, 41
Hankel function, spherical, 120
Harmonic time dependence, 18
Helmholtz equation, 24, 97, 98
Herglotz kernel, 130
electric, 132
vector, 130
Herglotz norm, 129
Herglotz pairs, chiral, 132
Herglotz scalar wave function, 130
Herglotz vector wave function, 130
Herglotz wave functions, 129
Hilbert uniqueness method (HUM), 167
Homogenisation, 9
deterministic
Bloch wave method, 211
formulation, 181
general bianisotropic media, 199
optical response region, 188
periodic unfolding, 203

two-scale convergence, 207
two-scale expansion, 184–187
periodic media, 30
random
two-scale expansion, 282–284
random media, 36
stochastic
ergodic media, 276
ergodicity, 275, 276
formulation, 279
solvability, 284
stationarity, 276

Image
curl, 52
grad, 52
Integral identities, 46
Integral representation
Beltrami fields, 98
exterior, 99
interior, 98
scattered field, 124
scattering problems, 98
Stratton-Chu, 99
Interior domain problems, 28
Interior problem
saddle point formulation, 73
Interior problems, time-harmonic, 23
Inverse problems, 144
Itō drift, 324
Itō process, 324
Itō's lemma, 324
Itō integral, 34

Jump relations, 47, 55

Kernel
curl, 52, 55
div, 52, 55
Kernels
dielectric susceptibility, $\chi^{\mathrm{e}}(t)$, 16
Herglotz, 130
electric, 132
vector, 130
magnetic susceptibility, $\chi^{\mathrm{m}}(t)$, 16
magnetoelectric, $\chi^{\mathrm{em}}(t)$, $\chi^{\mathrm{me}}(t)$, 16

Lax-Milgram lemma, application in electromagnetics, 65
Layer potentials, 100
Local model
stochastic
finite-energy divergence-free solutions, 255
finite-energy solutions, 255
Low chirality approximation, 108
Low chirality behaviour, 70
Low-frequency theory, 125

far-field behaviour, 128
scattering problems, 125

Magnetic field, H, 9
Magnetic flux density, B, 9
Magnetic susceptibility kernel, $\chi^{\mathrm{m}}(t)$, 16
Magnetoelectric kernels, $\chi^{\mathrm{em}}(t)$, $\chi^{\mathrm{me}}(t)$, 16
Maxwell equations, 9, 23
abstract form, 12
control, 29
discretised versions, 81
exterior problems, 25
initial conditions, 10
interior problems, 28
nonlinear media, 31
perfect conductor boundary condition, 10
periodic media, 30
radiation condition, 106
random media, 36
scattering problems, 26
stochastic, 32, 34
time-harmonic
weak formulation, 63
transmission problems, 25
Maxwell operator
(M), 12
modified (M_{A}), 58
group generation, 58
Maxwell operator (M), 56
group generation, 57
properties, 57
Maxwell system, 10
Media
dissipative, 16
ergodic, 276, 278
Mild solutions, 153
stochastic, 251

Noise
additive, 248
multiplicative, 248
Nonlinear
a priori estimates, 233
dispersion effect, 240
Faedo-Galerkin method, 233
finite-dimensional approximation, 236
global existence, 241
soliton-type solutions, 243
solvability, 238
steep kernels assumption, 240
stochastic problems, 262
Nonlinear deterministic complex media
constitutive relations, 231
Nonlinearity
convex, 233
monotone, 232

Normalised outgoing spherical vector wave functions, 97
Normalised spherical harmonic functions
  real scalar
    even, 96
    odd, 96
  real vector, 97

Ohm's law, 11
Operator
  M-type, 315
  Curl, 47, 55
  Div, 48, 55
  Grad, 47, 55
  curl, 45, 54
  div, 45, 54
  grad, 45, 54
  $\overrightarrow{\text{Curl}}$, 55
  Λ, 166, 172
  Beltrami, 125
  boundary integral, 100
  coercive, 315
  diagrams, 46, 54
  far-field, 120
  hemicontinuous, 315
  Hilbert-Schmidt, 307
  Maxwell (M), 12, 56
    group generation, 57
    properties, 57
  Maxwell modified (M_{A}), 58
    group generation, 58
  Moisil-Teodorescu, 140
  monotone, 233, 315
  nonlinear, 315
  normal trace, 46
  nuclear, 307
  strictly monotone, 315
  tangential components trace, 46
  tangential trace, 46
  trace, 46
    normal, 51
    tangential, 51
    tangential components, 51
  trace class, 307
    square root, 307
  weakly coercive, 315
Operators of vector analysis
  range, 54
Optical response, A_{or}, 14
Optimal control, 29

Passive medium, 17
Perfect conductor problem
  existence uniqueness, 99
Periodic structures
  scattering problems, 142
Polarised wave
  circularly
    left(LCP), 85
    right(RCP), 85
  elliptically, 84
  linearly, 85
Poynting theorem, 16
Poynting vector, 16
Pseudoparabolic equations, 162

Quaternia
  biquaternia, 140
  chiral media, 140
  Moisil-Teodorescu operator, 140

Radiation condition, 25
  Silver-Müller, 25
  Sommerfeld, 25, 113, 222
Reciprocity principle
  chiral media, 133
  far-field pattern, 119
Reciprocity relation
  spherical waves, 122
Resonating cavity (interior) problem, 24

Saddle point formulation
  interior problem, 73
Scattering, 9
  evolution problems
    asymptotically equal solutions, 216
    basic strategies, 214
    free problem, 213
    generalised eigenfunction expansion, 218
    limiting absorption arinciple, 222
    perturbed problem, 213
    wave operator, 216
  limiting absorption principle, 106
  limiting amplitude principle, 105
  time-harmonic problems
    Herglotz functions, 129
    Atkinson-Wilcox expansion, 114
    Atkinson-Wilcox expansion theorem, 124
    Beltrami fields, 86
    Beltrami-Herglotz functions, 130
    boundary integral equations, 101, 106
    domain derivative, 136
    far-field operator, 118
    far-field operators, 135
    far-field pattern, 115, 118
    Herglotz pairs, 132
    Herglotz wave functions, 129
    incident fields, 88
    integral representations, 98
    low chirality approximation, 108
    low-frequency theory, 125
    optical theorem, 117, 118

perfect conductor problem, 89
periodic structures, 142
quaternionic approach, 140
scatterer boundary dependence, 138
scattering problem formulation, 88
solvability, 99
spherical waves, 120
statement, 26
transmission problem, 90
Scattering cross section, 116, 122
Scattering relations
spherical waves, 120
Scattering theorem, 119
spherical waves, 122
Silver-Müller radiation conditions, 25, 27
Six-vector notation, 12
current, j, 12
electromagnetic field, u, 12
electromagnetic flux density, d, 12
Small chirality behaviour, 74
limit as $\beta \to 0$, 72, 73
Sobolev embeddings, 53
Solvability
deterministic
evolution families approach, 158
Faedo-Galerkin approach, 159
nonlinear, 232–241
semigroup approach, 153
stochastic, 250
evolution families approach, 255
Faedo-Galerkin approach, 256
semigroup approach, 251
Wiener chaos approach, 257
time-dependent problems
deterministic, 153
time-harmonic
boundary integral operators, 100
exterior problem, 74, 77
interior problem, 62, 68
layer potentials, 100
Spaces, 38
$H^1_{\mathrm{Div},\,loc}(\Gamma)$, 51
$H^1_{\mathrm{Div},\,loc}(\mathcal{O}_e)$, 51
$H^1_{\mathrm{Div}}(\mathcal{O})$, 51
$H^{\mathsf{s}}_0(\mathcal{O})$, 41
$W^{m,p}_0(\mathcal{O})$, 40
$\mathbb{T}^2_{\mathrm{L}}(S^2)$, 130
$C^{0,1}(\mathcal{O})$, 39
$C^\infty(\mathcal{O})$, 39
$C^{k,\alpha}(\mathcal{O})$, 39
$C^k(\mathcal{O})$, 38
$C^k_0(\mathcal{O})$, 38
$C^\infty_0(\mathcal{O})$, 39
$C_t(\Gamma)$, 50
$C_t(\partial\mathcal{O})$, 39
$H(\mathrm{curl}0,\mathcal{O})$, 49
$H(\mathrm{curl},\mathcal{O})$, 49
$H(\mathrm{curl},\mathrm{div}0,\mathcal{O})$, 49
$H(\mathrm{curl},\mathrm{div},\mathcal{O})$, 49
$H(\mathrm{div}0,\mathcal{O})$, 48
$H(\mathrm{div},\mathcal{O})$, 48
$H^m(\mathcal{O})$, 40
$H^{-1/2}(\mathrm{curl},\Gamma)$, 50
$H^{-1/2}(\mathrm{curl},\Gamma)$, 50
$H^{-1/2}(\mathrm{div},\Gamma)$, 50
$H^{-\mathsf{s}}(\Gamma)$, 42
$H^{-\mathsf{s}}_t(\Gamma)$, 50
$H^{-s}(\mathcal{O})$, 41
$H^{1/2}(\mathrm{curl},\Gamma)$, 50
$H^{1/2}(\mathrm{div},\Gamma)$, 50
$H^1(\mathrm{curl},\mathcal{O})$, 50
$H^1(\mathrm{div},\mathcal{O})$, 50
$H^{\mathsf{s}}(\Gamma)$, 42
$H^{\mathsf{s}}(\mathcal{O})$, 41
$H^{\mathsf{s}}_0(\Gamma)$, 42
$H^{\mathsf{s}}_0(\mathcal{O})$, 41
$H^{\mathsf{s}}_t(\Gamma)$, 50
$H_0(\mathrm{curl},\mathrm{div},\mathcal{O})$, 49
$H_0(\mathrm{curl}0,\mathcal{O})$, 49
$H_0(\mathrm{curl},\mathcal{O})$, 49
$H_0(\mathrm{div}0,\mathcal{O})$, 48
$H_0(\mathrm{div},\mathcal{O})$, 48
$H^m_0(\mathcal{O})$, 40
$H^1_n(\mathcal{O})$, 50
$H^1_t(\mathcal{O})$, 50
$L^2(\Omega,\mathscr{F},P;\mathbb{H})$, 45
$L^2_n(\Gamma)$, 50
$L^2_t(\Gamma)$, 50
$L^p(I,V)$, 43
$L^p(\mathcal{O})$, 39
$T^2_{\mathrm{LR}}(S^2)$, 132
$W^{0,p}(\mathcal{O})$, 40
$W^{1,p,q}(I,V_1,V_2)$, 44
$W^{\mathsf{s},p}(\mathcal{O})$, 41
$W^{m,p}(\mathcal{O})$, 40
$\mathfrak{X}_{\mathsf{M}}$, 57
$\mathfrak{X}_1$, 56
$\mathfrak{X}_2$, 56
$\mathbb{X}$, 56
$\mathbb{X}_{\mathsf{M}}$, 56
X, 56
$\mathbb{T}^2_{\mathrm{R}}(S^2)$, 130
$\mathcal{C}([0,T];\mathbb{H})$, 45
$\mathcal{D}'(\mathcal{O})$, 39
Lebesgue-Bochner, 43
compact embeddings, 53
decomposition
$(L^2(\mathcal{O}))^3$, 52
normal, tangential, 53
decompositions, 52–53
for electromagnetics, 48–51
Lebesgue, 39
random fields, 44
Sobolev, 40

Sobolev-Bochner, 44
Spherical Bessel functions, first kind, 96
Spherical Bessel functions, second kind, 96
Spherical Hankel function, first kind, 96
Spherical representation
Beltrami fields, 96
Spherical waves, 120
State space, 12
Stationary random variable, 277
Stochastic partial differential equations (SPDEs), 34
Stratton-Chu integral representation, 99, 106
Strong solutions, 154
existence, 155
stochastic, 252
existence, 254
Superposition, incident electric fields, 133
Susceptibility kernel, G_{d}, 14

T-matrix
far-field operator, 120
Tangential differential operators
definitions, 47
properties, 55
Tellegen constitutive relations, 19
Theorem
analytic Fredholm, 305
Arzelà-Ascoli, 303
Atkinson-Wilcox, 114, 124
Banach contraction, 312
Banach-Steinhaus, 308
Birkhoff ergodic, 333
Burkholder-Davis-Gundy inequalities, 323
Céa's lemma, 81
chain rule in Banach spaces, 297
compensated compactness, 329
div-curl lemma, 329
divergence, 46
Eberlein-Šmulian, 296
Fredholm alternative, 305
Friedrichs' inequality, 40
Gauss divergence, 18
general scattering, 117, 119, 121
generalised Gronwall's inequality, 315
Green's formulae, 46, 55
Gronwall's inequality, 314
Hahn-Banach, 293
Hilbert-Schmidt, 305
Hille-Yosida, 309
Itō integral, 321
Itō's lemma, 324
Lax-Milgram, 313
Lebesgue dominated convergence
vector-valued, 299
Leray-Schauder alternative, 313
martingale reperesentation, 326
optical, 117
Poincaré lemma, 55
Poynting, 16
principle of uniform boundedness, 308
reciprocity principle, 119
Rellich-Kondrachov, 53
Riesz representation, 294
Riesz-Schauder, 305
Schauder fixed point, 312
Sobolev embedding, 41
fractional order, 42
spectral, 301
Stone, 310
Stone's formula, 301
Young inequality for Lebesgue-Bochner spaces, 43
Total cross section, 116
Trace operators, 46
normal, 46
tangential, 46
tangential components, 46
Trace spaces, 50
Traces, 51–52
properties, 52
Transmission conditions, 26
Transmission problem
achiral scatterer, achiral environment, 91
achiral scatterer, chiral environment, 91
chiral scatterer, achiral environment, 91
boundary integral equations, 106
existence uniqueness, 99
chiral scatterer, chiral environment, 91
existence uniqueness, 99
Transmission problems, 25
Trkalian flow, 86
Twersky notation, 119

Waveguides, 140
Weak formulation
eigenvalue problem, 69
exterior time-harmonic problem, 75
interior time-harmonic problem, 63
Weak solutions, 153
existence, 154
stochastic, 251
existence, 252
Wiener process, 33

Princeton Series in Applied Mathematics

Chaotic Transitions in Deterministic and Stochastic Dynamical Systems: Applications of Melnikov Processes in Engineering, Physics, and Neuroscience, Emil Simiu

Selfsimilar Processes, Paul Embrechts and Makoto Maejima

Self-Regularity: A New Paradigm for Primal-Dual Interior-Point Algorithms, Jiming Peng, Cornelis Roos, and Tamás Terlaky

Analytic Theory of Global Bifurcation: An Introduction, Boris Buffoni and John Toland

Entropy, Andreas Greven, Gerhard Keller, and Gerald Warnecke, editors

Auxiliary Signal Design for Failure Detection, Stephen L. Campbell and Ramine Nikoukhah

Thermodynamics: A Dynamical Systems Approach, Wassim M. Haddad, Vijay Sekhar Chellaboina, and Sergey G. Nersesov

Optimization: Insights and Applications, Jan Brinkhuis and Vladimir Tikhomirov

Max Plus at Work, Modeling and Analysis of Synchronized Systems: A Course on Max-Plus Algebra and Its Applications, Bernd Heidergott, Geert Jan Olsder, and Jacob van der Woude

Impulsive and Hybrid Dynamical Systems: Stability, Dissipativity, and Control, Wassim M. Haddad, VijaySekhar Chellaboina, and Sergey G. Nersesov

The Traveling Salesman Problem: A Computational Study, David L. Applegate, Robert E. Bixby, Vasek Chvátal, and William J. Cook

Positive Definite Matrices, Rajendra Bhatia

Genomic Signal Processing, Ilya Shmulevich and Edward R. Dougherty

Wave Scattering by Time-Dependent Perturbations: An Introduction, G. F. Roach

Algebraic Curves over a Finite Field, J.W.P. Hirschfeld, G. Korchmáros, and F. Torres

Distributed Control of Robotic Networks: A Mathematical Approach to Motion Coordination Algorithms, Francesco Bullo, Jorge Cortés, and Sonia Martínez

Robust Optimization, Aharon Ben-Tal, Laurent El Ghaoui, and Arkadi Nemirovski

Control Theoretic Splines: Optimal Control, Statistics, and Path Planning, Magnus Egerstedt and Clyde Martin

Matrices, Moments, and Quadrature with Applications, Gene H. Golub and Gérard Meurant

Totally Nonnegative Matrices, Shaun M. Fallat and Charles R. Johnson

Matrix Completions, Moments, and Sums of Hermitian Squares, Mihály Bakonyi and Hugo J. Woerdeman

Modern Anti-windup Synthesis: Control Augmentation for Actuator Saturation, Luca Zaccarian and Andrew W. Teel

Graph Theoretic Methods in Multiagent Networks, Mehran Mesbahi and Magnus Egerstedt

Stability and Control of Large-Scale Dynamical Systems: A Vector Dissipative Systems Approach, Wassim M. Haddad and Sergey G. Nersesov

Mathematical Analysis of Deterministic and Stochastic Problems in Complex Media Electromagnetics, G. F. Roach, I. G. Stratis, and A. N. Yannacopoulos